Rame
B. 74
Gt Falls VA 22066
MW00843693

Dietrich G. Altenpohl

ALUMINUM:

TECHNOLOGY, APPLICATIONS, AND ENVIRONMENT

A Profile of a Modern Metal

Aluminum from Within—the Sixth Edition

Dietrich G. Altenpohl

ALUMINUM:

TECHNOLOGY, APPLICATIONS, AND ENVIRONMENT

A Profile of a Modern Metal

Aluminum from Within—the Sixth Edition

Technical Editor—J. Gilbert Kaufman
TMS Editor—Subodh K. Das

Translators
Robert Bridges
Andrew Bushnell
Robert & Susan Dean
Malcolm Hill

The Aluminum Association, Inc.
900 19th Street, Suite 300
Washington, D.C.
and
The Minerals, Metals & Materials Society (TMS)
420 Commonwealth Drive
Warrendale, Pennsylvania

©1998 The Aluminum Association, Inc.

No part of this publication may be reproduced or transmitted in any form or by any means, whether electronic or mechanical, including by photocopying, recording or any information storage and retrieval system, without written permission from the copyright owner.

Printed in the United States of America
Library of Congress Catalog Number 97-75874
ISBN Number 0-87339-406-2

Published by The Aluminum Association and
The Minerals, Metals & Materials Society.

TMS
Minerals • Metals • Materials

The Aluminum Association, Inc.
900 19th Street, Suite 300
Washington, D.C.
(202) 862-5100

The Minerals, Metals & Materials Society (TMS)
420 Commonwealth Drive
Warrendale, Pennsylvania
(412) 776-9000

Contents

FOREWORD
by
Dr. Rodney E. Hanneman
Chairman, Technical Advisory Committee
The Aluminum Association

Aluminum: Technology, Applications, and Environment is an impressive book that has evolved remarkably through six editions and multilingual translations over the last four decades, with global recognition as the definitive educational text and reference book for aluminum industry participants, a broad range of aluminum fabricators and users, students, and the scientific, engineering, and academic community. This extraordinary book incorporates significant inputs from outstanding aluminum industry and academic participants throughout North America and Europe.

The text builds from a brief history of aluminum through its various production and processing steps with a clear and refreshing description of relationships between processing steps, structure, and properties of aluminum alloys.

Expert attention is given to various casting processes and the role of metal quality and casting parameters and methods. The text includes clear descriptions of key mechanical test methods and property relationships, along with valuable descriptions of major industrial forming processes and their underlying thermomechanical principles. The fundamental principles of alloying aluminum with various elements and the use of heat treating methods to achieve specific properties are also included, along with an excellent treatment of corrosion principles and a broad range of methods used to enhance corrosion protection. An effective description of modern joining technologies and principles for the manufacture of various aluminum structures is included for the practitioner.

Various examples are given regarding the utilization of composition controls, microstructure, and manufacturing process controls to achieve the desired combinations of properties for various applications, including the case of can making. The significance of computer-aided materials design, computer-aided engineering of components, and computer-aided manufacturing methods are recognized.

The author also treats the current relative competitive properties and trade-offs regarding aluminum versus magnesium, titanium, plastics, composite materials, and steel.

One of the most significant additions to the 6th edition of this book is a highly informative description of a wide array of emerging applications for aluminum, ranging from aerospace, buildings, bridges, infrastructure, and automotive, to marine, rail, packaging, and durable goods.

In the penultimate chapter, a rich tapestry is woven of key ecological and energy issues and challenges faced by the aluminum industry, along with relevant life cycle considerations.

Finally, the author concludes with new paradigms for the aluminum industry that range from advanced intelligent processing concepts and conservation to more emphasis on systems orientation, flexibility and speed to market, and sustainable development.

In summary, *Aluminum: Technology, Applications, and Environment* transcends the previous five editions of *Aluminum From Within* and will serve as the benchmark and definitive text and reference book on aluminum for years to come.

Technical Editor's Introduction for The Aluminum Association, Inc.

The aluminum industry has had a great need for a comprehensive educational text on the attributes and advantages of the world's most plentiful, versatile, and broadly useful material. Thirty years ago, ASM International (then better known as the American Society of Metals) and the Aluminum Company of America produced a three-volume treatise on *Aluminum*. It served for a generation as a fine reference text for the industry, but has long been out of print and out of date in many respects.

In the meantime, the needs of an Alusuisse metallurgist, Dr. rer. nat. Dietrich Altenpohl, in training his own staff and enlightening those he encountered in prodigious travels around the world has led to the gradual evolvement of the excellent and extensive text contained in this volume, *Aluminum: Technology, Applications, and Environment*. This clearly meets the needs for a new educational text. Its value comes not only from its breadth in the technology, applications, and ecological advantages enumerated therein, but also from the author's care in including the "why" of material behavior as well as the "what" and "how," as incorporated in the original title of the book *Aluminium von Innen* or *Aluminum from Within*.

The fundamental thesis underlying the way the first part of the book is laid out is that in order to understand the properties, performance, and advantages of our favorite metal, you must understand its metallurgical structure and, in turn, the role of each step in its production in creating that microstructure. And so a key element of the book focuses on looking "inside" the material, at its microstructure, and understanding both what processes create that microstructure, and how that microstructure impacts finished product performance. For this reason, the subtitle was retained in this volume *Aluminum from Within;* it reflects not some "insider's" perspective, but the foundation of the "edifice of knowledge" that the author builds in these pages.

The Aluminum Association is pleased to be able to bring this learning tool to the industry and to the university educational system that supports the industry. It is of value not only to students considering the aluminum industry as a career, but also to those who, as architects, designers, engineers, manufacturers, and even consumers, will take advantage of aluminum. Aluminum is all around us, in packaging, in our transportation system (automobiles, buses, trains, planes, ships, et al.), in buildings and highway structures, in electrical systems, and in countless other components of everyday life. We should understand this as well as we can, and I believe this volume will be helpful in enabling us all to do that.

J. Gilbert Kaufman
The Aluminum Association
Washington, D.C.

Author's Preface

I came as a young metallurgist to a German plant of the Alusuisse group. The Singen plant produced sheet, foil, extrusions, and forgings from a wide variety of aluminum alloys and composites.

In a plant with up to 5000 people, a sizable development program, and growing end-uses in the marketplace, enormous differences in background and understanding existed whenever a dialogue took place in house or with customers about specific properties of aluminum alloys, for example, on how to increase the yield strength or corrosion resistance, or how to define the elasticity of a specially designed component.

Accordingly, in 1956 I wrote for the company journal a series of articles titled "Aluminium from Within," an easily understandable introduction into aluminum processing and the related physical metallurgy.

When this endeavor came to the attention of the Aluminium Zentrale in Dusseldorf, they found the content to be exactly what was needed in the marketplace to help fabricators and users of aluminum products better understand the possibilities and limitations of aluminum and its alloys, and they decided to print the compilation as a book with the same title (about 200 pages).

Four editions of this small book were printed from 1957 through 1992 in five languages, and no other book on aluminum was in a wider distribution, providing an indication that the purpose and structure of our text served a need.

In comparison to the first four editions, the 5th edition, printed in 1994 in German, was enlarged to twice its size by describing aluminum in relation to its principal production processes and end products as well. The whole life cycle of the aluminum industry was brought in focus and described in relation to many ongoing innovations.

It came as somewhat of a surprise when this structure could accommodate the experiences from 25 contributors with outstanding know-how in the enlarged 5th edition (contributors mainly from German speaking countries) and now again in a 6th edition, with additional contributions from North American authors. Thus, the reader will see that the purpose of the book has been extended to also look at "Aluminum from the Outside," and to describe technology relevant for the aluminum industry at large. This is reflected by the title of our 6th edition. The underlying basis of describing the related principles of physical metallurgy of aluminum was maintained and extended into understanding the ongoing conversion from products to systems.

The development of this book truly reflects the new pattern of behavior of the whole aluminum industry away from its previous "ton pushing" mode to one of development and delivery of tailor-made products for highly demanding markets.

As in previous editions, we made certain that:

- Each chapter can be understood on its own;
- The whole book is easily understandable and can serve a wide range of readership with quite different backgrounds and interests in aluminum; and
- An explanation of all technical terms is provided chapter-by-chapter to be of help to a newcomer to aluminum or to its processing or end-uses.

Our book again can serve as a bridge between the vast "knowledge building" about aluminum and its pragmatic use in so many different applications.

Dietrich G. Altenpohl, Feldmeilen, November 1997

Acknowledgements and Contributors

For the generous advancement of this book, the author thanks the Board of Directors and management of The Aluminum Association, Inc. in Washington, D.C., and all of the contributors cited below who have provided about one quarter of the text based upon their long experience in the industry.

This book could not have been realized without the advice and encouragement of top personalities from major aluminum companies and colleagues from academia and industry. I will just limit myself to mentioning a few names:

Alcan—Dr. D. Moore
Alcoa—Dr. P. Bridenbaugh
Alusuisse—Dr. P. Furrer
Hydro Aluminium—H. Knoblauch, Dr. R. Marstrander
Reynolds—Dr. R. Hanneman
VAW—Dr. G. Scharf
Dr. J. Nitsche, Editor of the 5th Edition
Prof. M. Speidel, ETH Zurich

Last but not least, my special and very personal thanks go to J.G. Kaufman, the editor of this book, for his untiring engagement for restructuring our manuscript and making this book into a venture with credibility, and also to our translators: R. Bridges, A. Bushnell, R. Dean and S. Dean, and M. Hill for their many contributions for upgrading our text.

Contributors to the 5th Edition, 1994 and 6th Edition, 1997:

Aluisse-Lonza
Dr. Kurt Buxmann (Chippis)
Dr. Peter Furrer (Neuhausen)
Dr. Harald Severus (Neuhausen)
Dipl.-Ing. Reinhold Gitter (Singen)
The late Dipl.-Ing. Guido Angehrn (Zurich)
Dipl.-Ing. Jurg Zehnder (Zurich)

AMAG, Ranshofen
Dr. Frank Ellermann
Dr. Martin Hoyas
Dr. Wolfgang Kuehlein
Dipl.-Ing. Alfred Mundl

Berndorf Technologie GmbH, Berndorf
Ing. Gerd Klaus Gregor

Geilinger AG, Winterthur
Dipl.-Ing. Erik Hartmann

Salzburger Aluminum AG (SAG)
Dr. Karin Woehrer

VAW Aluminium, Bonn
Dipl.-Ing. Gerd Bulian
Dr. Axel Blecher
Dipl.-Ing. Wolf-Dieter Finkelnburg
Dr. Werner Huppatz
Prof. Dr. Edgar Lossack
Dr. Gunther Scharf
Dr. Wolfgang Schneider
Dr. Gerhard Tempus

Fachhochschule Giessen-Friedberg
Prof. Dr. Friedhelm Kahn

Universitat Stuttgart, Institut fur Umformtechnik
Prof. Dr. Klaus Siegert

University of Virginia, Dept. of Materials Science
Charlottesville VA
Prof. Dr. Heiner Wilsdorf

Individual Contributors (retired from Alusuisse)
Dr. Jean Schrade, Zurich
Dipl.-Ing. Rudolf Vogtlin, Urdorf / Zurich

Contributors to the 6th Edition 1997:

Alcan International, Inc.
Dr. David Moore

Alusuisse
Robert Dean
Werner Stelzer
Klaus D. Waldeck

BMW
Michael Weitzer (consultant)

Reynolds Metals Company
Dr. Rodney Hanneman
Rajeev G. Kamat

Hydro Aluminium
Dr. Peter Erz

The Aluminum Association, Inc.
Dr. John A.S. Green
J. Gilbert Kaufman

For the Aluminum Association
Elwin Rooy

For The Minerals, Metals & Materials Society (TMS)
Dr. Subodh K. Das, ARCO Aluminum Company
Randal K. Broyles, ARCO Aluminum Company
Dr. James G. Morris, University of Kentucky

Chapter 1. Introduction

1.1 Aluminum: A Young Material

1.1.1 History of aluminum

The development of civilization is closely linked with that of metals, and in our age of industrialization no metal has proved so versatile as aluminum. Among the common metals, aluminum now cedes first place only to steel. If measured by volume rather than by weight, it now exceeds in quantity all other non-ferrous metals combined, including copper and its alloys, lead, tin, and zinc.

There are many reasons for the rapid and sustained growth in the use of aluminum. Because of its unique properties, from the beginning aluminum has competed with and replaced much older, established materials, such as wood, copper, and steel. Aluminum has won this position even though its industrial production began only in the late 19th century; therefore, it is a latecomer among common metals.

The reason aluminum was not used earlier was the difficulty of extracting it from its ore. It combines strongly with oxygen in a compound that, unlike iron, cannot be reduced in a reaction with carbon. Sometime between 1808–1812, Englishman Sir Humphrey Davy was the first to concentrate what he suspected to be a new metal mixed with iron from its naturally occurring ores. Davy named the new element "aluminum," derived from alum, its bisulfate salt, which was known already to the ancient Egyptians for its use in dyeing. It is still called aluminum in the USA, but in Britain and many other countries, its spelling ends in "ium."

Hans Christian Ørsted first succeeded in making aluminum on a laboratory scale in Denmark in 1825; Friedrich Wöhler did the same in Germany a little later. Small scale production started in 1855, when Henri Sainte-Claire Deville made aluminum by a chemical process. The first ingot of aluminum produced by this process was 96–97% pure and was presented at the World Exhibition in Paris in 1855, where it aroused public interest in the new metal. The Emperor Napoleon III, hoping to equip his armies with aluminum helmets and armor, supported further development by Deville. Being expensive, however, it remained a "precious metal" that Napoleon III used instead of silver for his tableware.

Two almost identical electrolytic processes succeeded in 1886, at first on a laboratory scale. Simultaneously but independently on opposite sides of the Atlantic, the American Charles Martin Hall and the Frenchman Paul T. Héroult both electrolyzed aluminum oxide (alumina) dissolved in molten cryolite.[1] Providing the large amount of electric power needed was no longer an obstacle by then, since General Electric and Siemens had simultaneously developed the dynamo. Economic production of aluminum still awaited a further step: the large scale extraction of refined alumina. In 1892, the Austrian Karl Joseph Bayer developed an efficient process that used caustic soda to extract the alumina from bauxite, a natural ore containing 35–50% of alumina.

[1] Cryolite is a naturally occurring sodium aluminum fluoride, which dissolves alumina at above 900°C. Originally mined in Greenland, it is today synthesized for use in aluminum electrolysis.

Fig.1.1: Pioneers of aluminum: Top row (left to right): H. Davy, H.C. Ørsted, F. Wöhler, and H. St. Claire Deville. Bottom row (left to right): P.T. Héroult, C.M. Hall, W.V. Siemens, and K.J. Bayer (source: Aluminum-Verlag)

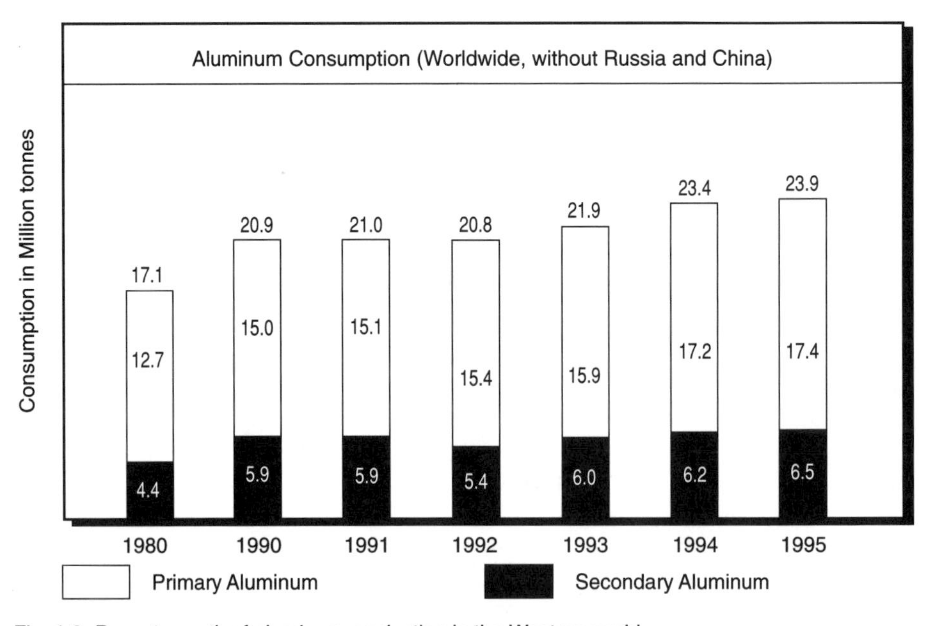

Fig. 1.2: Recent growth of aluminum production in the Western world.

All aluminum smelters in the world still use essentially the same process principles invented by Bayer, Hall, and Héroult between 1886 and 1892. Fig. 1.1 shows the pioneers of the technology of today's aluminum industry.

1.1.2 Growth of aluminum production and uses

Industrial production of aluminum began in 1888 at about the same time in America and in Europe: in the USA, at Pittsburgh, Pennsylvania, using Hall's process and in Switzerland at Neuhausen using Héroult's process. Since then, aluminum production by the Hall-Héroult process has multiplied amazingly; in 1918, it had already reached the 180,000 tonne level. It has maintained steady long-term growth since, although this was interrupted by recessions in 1920, 1930, and at the end of the World War II. From that time through the mid-1970s, the production and consumption of aluminum grew, on average, at more than 8% per year, with lesser rates thereafter. Total aluminum consumption in the Western world reached 2 million tonnes in 1952 and 20 million tonnes in 1989.

Fig. 1.2 shows recent growth in the Western world separated into primary (potroom) and secondary (recycled) aluminum. As may be seen from Fig. 1.3, annual consumption of primary aluminum has more than doubled in the past 25 years.

Growth rate varies from one application to another. Between 1970 and 1995 the relative importance changed among the main segments of use (Fig. 1.4). Of course, these percentage figures do not reveal the absolute growth, since total consumption more than doubled in this period. This makes the increases in transport and packaging all the more significant. Both are likely to remain strong growth areas.

The decreasing share (but not tonnage) in electrical applications can be explained. In the first few years following World War II, the industrialized countries of the Western world rapidly built up their electric power grids. These networks use large quantities of aluminum conductor material. This expansion phase then leveled off, resulting in lower relative growth. Chapter 15 deals with the main market segments in greater detail.

1.2 Properties of Aluminum and Resulting End Uses

The properties of aluminum that contribute to its widespread use are:

- Aluminum is light; its density is only one third that of steel.
- Aluminum is resistant to weather, common atmospheric gases, and a wide range of liquids.
- Aluminum can be used in contact with a wide range of foodstuffs.
- Aluminum has a high reflectivity and, therefore, finds many decorative uses.
- Aluminum alloys can equal or even exceed the strength of normal construction steel.
- Aluminum has high elasticity, which is an advantage in structures under shock loads.
- Aluminum keeps its toughness down to very low temperatures, without becoming brittle like carbon steels.
- Aluminum is easily worked and formed; it can be rolled to very thin foil.
- Aluminum conducts electricity and heat nearly as well as copper.

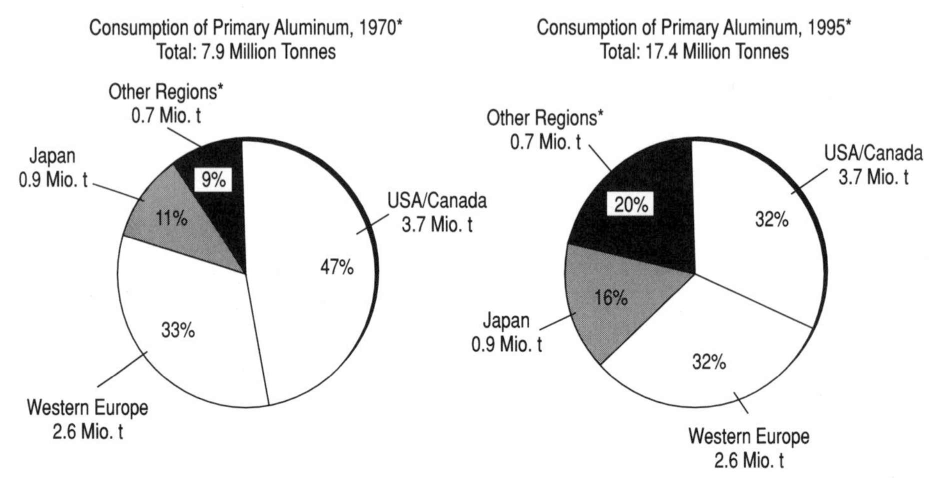

Fig. 1.3: Primary aluminum production by region (data from EAA, MCG).

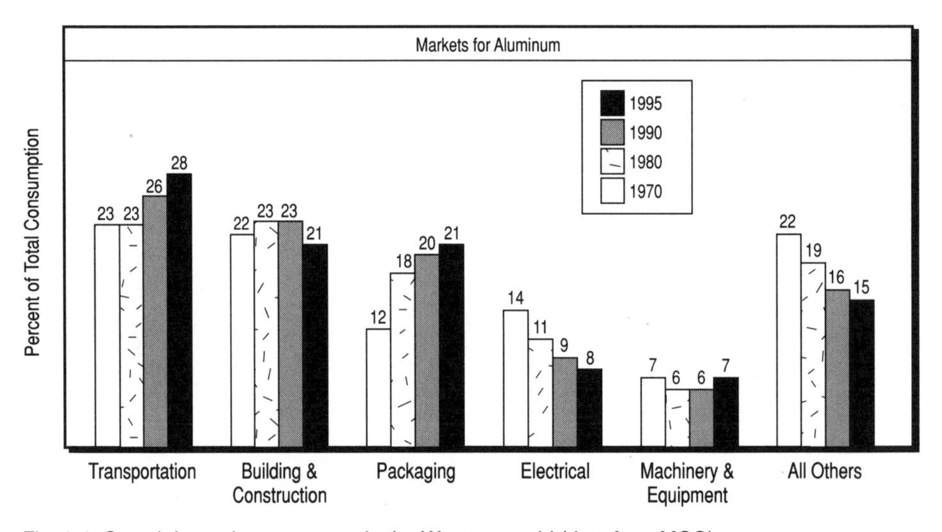

Fig. 1.4: Growth by end use category in the Western world (data from MCG).

These characteristics account for aluminum's main applications. Table 1.1 shows that besides its physical properties and decorative appearance, its formability and the many ways by which it can be processed help to ensure a wide range of applications.

The above account presents aluminum's chief advantages for various uses. A limitation to its use is its loss of strength at elevated temperatures. This will be discussed in detail later in the chapters that describe how the properties of alloys can be influenced by altering their microstructure.

Table 1.1: Characteristics of aluminum and their importance for different end uses

	Characteristics				Type of semi-fabricated products					
Field of use	**Lightness**	**Good heat and electrical conductivity**	**Resistance to corrosion**	**Decorative aspects (with or without surface treatment)**	**Castings or forgings**	**Formed sheet**	**Impact extrusions**	**Extruded sections**	**Wire and cable**	**Foil**
Transport	X		o	o	o	o		o		
Building	o		o	X		o		o		
Packaging	+	+	X	X			o			o
Electrical	+	X	o				o	o	o	o
Household	o	X	X	o		o				o
Machines, appliances	X	o	o	o	o	o		o		
Chemicals & food	o	o	X	o	+	o		o		o

+ desirable
o important
X very important

1.3 Purpose and Resulting Structure of This Book

Like the first four editions of *Aluminium from Within,* this volume focuses on the "why" of aluminum technology as well as the "what" and "how." The intention is to enable readers to understand the characteristics and advantages of aluminum by understanding the processes that influence its internal structure.

In the fifth edition and in this sixth edition, the scope has been broadened to provide a description of aluminum in relation to its end products and the resulting systems as well. The whole life cycle of the aluminum industry has been brought in focus and is described in relation to many ongoing innovations. This extended purpose was matched by integrating a number of new chapters to outline the "Edifice of Knowledge about Aluminium" and its impact on alloy, process, and systems development. The underlying basis of describing the related principles of physical metallurgy of aluminum was extended into understanding the ongoing conversion from products to systems.

To enhance the understanding of cause and effect during aluminum processing, the following structuring of the text was applied consequentially and runs like a "silver thread" through the first thirteen chapters:

- Measures actively taken during casting or fabrication of intermediate or final products;
- Resulting changes in the structure of the metal; and
- Resulting properties and characteristics of products and systems.

The intent is to "look at the whole battlefield" of the aluminum industry by surveying all main product segments about ongoing innovation. This is accomplished by reviewing first, in Chapters 2 through 12, the principal production technologies and their influence on both metallurgical structure and properties. Then, in Chapter 13, the whole "Edifice of Knowledge" is described, with examples of its application in alloy, process, and system development.

Chapters 14 and 15 illustrate the position of aluminum in the world of materials competition as well as the increasing number of successful applications of aluminum and its broadening use, even today, in new products, applications, and systems.

The last two chapters look into the total environment of the aluminum industry within the context of its positioning within the "Issue Triangle" of resources, energy consumption, and ecology. The intent is to outline how technology changes and innovation result in a sustainable development of this industry at large through an understanding of the growth phases of the industry.

"Aluminum in Change" is the message of the concluding Chapter 17, outlining the importance of intelligent technologies for the transformations to provide systems instead of products as important structural elements in the emerging "new paradigm" of the aluminum industry.

Chapter 2. Production and Processing of Aluminum

2.1 *Extraction of Aluminum*

The extraction of aluminum from its ore and subsequent processing into finished products takes place in a series of successive operations, each largely independent of the other. Generally the various processes are carried out at different plant sites. A summary of production steps from the bauxite mine through casting is given in Fig. 2.1.

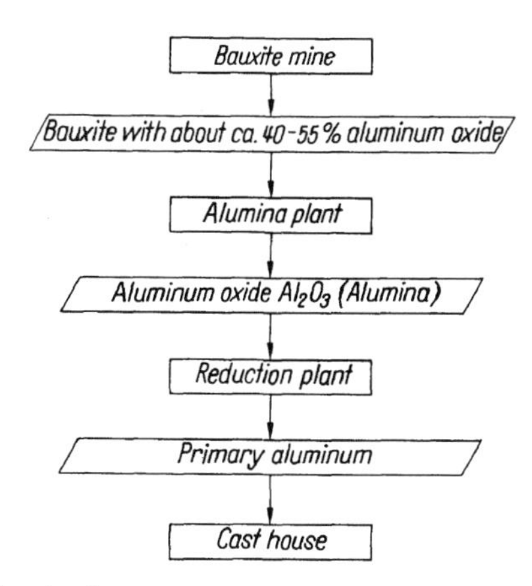

Fig. 2.1: Production steps for aluminum.

2.1.1 Bauxite mining

Aluminum comprises approximately 8% of the earth's crust, making it second only to silicon (27.7%). Iron is third at about 5%. Metallic aluminum is not found in nature; it occurs in the form of hydrated oxides or silicates (clays). The principal ore from which aluminum is extracted is called bauxite after the town of Les Baux in southern France where the ore was originally discovered. Bauxite occurs mainly in the tropics and in some Mediterranean countries. Today, the main mining locations are in Latin America, Australia, India, and Africa.

Bauxite is a weathered rock containing two forms of hydrated aluminum oxide, either mostly a monohydrate AlO(OH) in caustic bauxite, or mostly a trihydrate $Al(OH)_3$ in lateric bauxite. Besides these compounds, bauxite contains iron oxide, which usually gives it a reddish-brown colour, as well as silicates (clay, quartz) and titanium oxide. The crystal structure also contains 12–20% by weight of water. Tropical monohydrate bauxite grades yielding 35–55% Al_2O_3 will no doubt continue to be the most favored aluminum ores for many decades.

Laterite rocks similar to bauxite, but with lower alumina content, are available in large quantities. Clays became a source of alumina to a limited extent in Germany during the

Second World War. In addition, many other types of rock contain considerable amounts of alumina, such as kaolin, nepheline, andalusite, leucite, labradorite, and alunite. The former Soviet Union exploited such ores to maintain partial autonomy, but these ores play no significant role in today's aluminum production. Chapter 16 will show that the world's bauxite supplies are guaranteed into the distant future.

2.1.2 The alumina plant

The starting material for electrolytic smelting of aluminum is pure, anhydrous aluminum oxide (Al_2O_3) called alumina. In the Western World, the Bayer[1] process, invented in the 19th century, is by far the most important process used in the production of aluminum oxide from bauxite. The process has been refined and improved since its inception. Fig. 2.2 shows that the production of alumina is a complex chemical process. The alumina content of bauxite ores varies from one deposit to another, and methods of treatment differ accordingly (see 16.1.2). This means that each alumina plant is almost tailor-made to suit a particular bauxite. The processes are nevertheless basically similar, and a general description is given in the following. The bauxite from the mine is crushed and ground. It is then mixed with a solution of caustic soda and pumped into large autoclaves. There, under pressure and at a temperature of 110–270°C, the alumina contained in the ore is dissolved to form sodium aluminate. The silica in the bauxite reacts and precipitates from solution as sodium-aluminum-silicate. Iron and titanium oxide and other impurities are not affected chemically, and being solid, settle out of solution. This waste material, known as red mud, is separated from the sodium aluminate solution, washed to recover the caustic soda, and then pumped to disposal areas.

The disposal of red mud can present an environmental problem simply because there is so much of it. From a few alumina plants, red mud is deposited on the sea bed under strictly controlled conditions. One very common method of disposal is to contain the mud in an area surrounded by dikes. After an interval of some years, these ponded areas can be recultivated to eliminate "visual pollution." Although a great deal of effort has been expended on finding and developing various uses for red mud, no bulk application of commercial value has yet been found.

Adding the weak soda washed out of the red mud to the sodium aluminate solution dilutes it and cools it to about 100°C. With stirring and cooling to 60°C, aluminum hydroxide $Al(OH)_3$ (hydrargillite) precipitates. Seeding the liquor with crystals from a previous cycle helps to control precipitation. Vacuum filters separate the hydroxide precipitate, which is then washed with pure water. Calcination in rotary kilns or in fluidized beds at 1100°C to 1300°C finally converts the hydroxide to a dry, white powder. This powder is technical purity alumina, containing as impurities at most 0.01–0.02% SiO_2, 0.01–0.03% Fe_2O_3, and 0.3–0.6% NaO_2.

The grade of the alumina (particle size, α- and γ-Al_2O_3 content) can be influenced by precipitation and calcining conditions, and it is usual to differentiate between two main grades, i.e. "floury" alumina, which is highly calcined and contains mostly α-Al_2O_3, and "sandy" alumina, which calcined to a lesser degree with mainly γ-Al_2O_3 in the hydrated

[1] Named after the Austrian K.J Bayer.

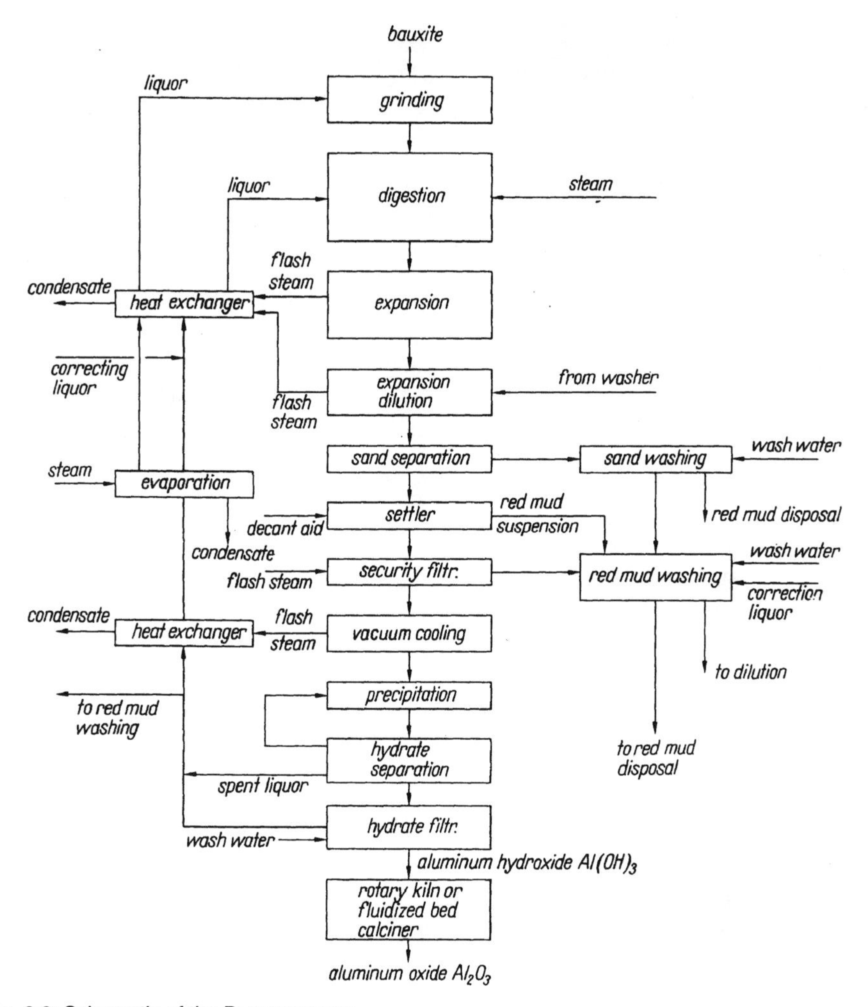

Fig. 2.2: Schematic of the Bayer process.

form. The sandy alumina has a large, active surface area, which makes it suitable for use in dry scrubber systems for fluoride abatement at aluminum reduction plants. There is a clear trend toward the production of increased quantities of sandy alumina.

In Russia, due to a lack of bauxite, a process using nepheline as feed-stock has been used to produce alumina. Essentially the technique consists of sintering a nepheline ore, or concentrate, with limestone. The resultant sinter-cake consists of sodium and potassium aluminates and dicalcium silicate. This material is crushed, ground, and leached. After leaching, the aluminate liquor is desilicated and decomposed by carbonation. Alumina hydrate is separated from the liquor and calcined to obtain alumina. After evaporation and crystallization, the carbonate liquor yields soda and potash. These are centrifuged, dried, and packed for shipment. Limestone is added to the slime from sinter-leaching to produce Portland cement in a second calcination step. Processing about 4–4.5 tonnes of nepheline ore yields 1 tonne of alumina, 9–11 tonnes of cement, 0.6–0.8 tonnes of soda ash, and 0.2–0.3 tonnes of potash.

According to the grade of the bauxite ore, 2–3 tonnes of ore yield one tonne of alumina and about one tonne of red mud (dry weight). When designing an alumina plant, factors other than the type of bauxite ore to be used as feed material and the form of alumina to be produced have to be taken into consideration. A high silica content of the bauxite is undesirable because insoluble sodium-aluminum-silicate will form, causing losses of caustic soda and alumina which increases input material costs. Energy consumption is another consideration. The economical operation of the Bayer process requires the rational use of energy for steam generation and calcining. Inexpensive fuel is desirable because the process needs a large amount of thermal energy. The end product of the alumina plant is a dry white powder that is the feedstock for aluminum smelting.

2.2 Primary Aluminum Production

Throughout the world, primary aluminum is still produced by the electrolysis of alumina in molten fluoride salt. This is, in essence, the process that Hall and Héroult invented, and which is named after them, but its efficiency has been significantly improved over the years. The electrolysis plant—the aluminum smelter—needs large amounts of electrical energy. Therefore, besides good bulk transport facilities, abundant inexpensive electric power is essential. Because hydroelectric power is a relatively inexpensive and clean source of energy, aluminum smelters are mostly built in countries with readily available hydroelectric power, such as Canada, Norway, Venezuela, and Brazil, or in countries with abundant deposits of low-grade coal such as Australia or the Republic of South Africa (RSA). Furthermore, the same regions lack other local industry to use this energy, and it is impracticable to transport electric power over very long distances to the industrialized regions that could use it. Considering their large energy consumption, aluminum smelters are major customers that guarantee a stable base load and, thereby, help to reduce their power suppliers' unit costs. At present, hydroelectric sources produce the power for about two-thirds of world aluminum production, although placing a smelter next to a nuclear power plant can also be economically attractive, as at Dunkerque in France. Fig. 2.3 shows an aluminum smelter with its long potrooms and tall alumina silos.

2.2.1 The electrolysis process[2]

In the Hall-Héroult process, the electrolyte is molten cryolite (Na_3AlF_6) in which 2–8% of alumina (Al_2O_3) is dissolved. To lower the melting point, industrial cryolite-alumina mixtures also contain various amounts of other salts, such as aluminum fluoride (AlF_3) and calcium fluoride (CaF_2); sometimes lithium carbonate (Li_2CO_3) is present and, less frequently, magnesium fluoride (MgF_2) is introduced. These additions also improve current efficiency and reduce evaporation losses. For each tonne of aluminum produced, the smelting process consumes, in addition to electrical energy, about 1.95 tonnes of alumina, 0.5 tonnes of anode coke, and small amounts of fluoride salts.

[2] By G.W. Bulian.

Fig. 2.3: View of a modern aluminum smelter in Canada, showing the harbor and alumina silos in the foreground and the potrooms behind.

The electrolysis cell, or "pot," shown schematically in Fig. 2.4, is shaped like a shallow rectangular basin. It consists of a steel shell with a lining of fireclay brick for heat insulation, which is, in turn, lined with carbon bricks to hold the fused salt electrolyte. Steel bars carry the electric current through the insulating bricks into the carbon cathode floor of the cell. Carbon anode blocks are suspended on steel rods, and dip into the electrolyte. As the electric current flows through the electrolyte, it breaks down the dissolved alumina into its component elements as metallic aluminum and oxygen gas. The oxygen reacts with the carbon anodes, forming bubbles of CO and CO_2 gas. Liquid aluminum settles on the bottom of the cell since it is denser (specific gravity 2.3 at 960°C) than the electrolyte (specific gravity 2.1). Periodically, this aluminum is siphoned off by vacuum into crucibles. To replace the alumina consumed in the reaction, more alumina must be added. Today, computer-controlled devices called point feeders automatically inject the alumina powder through the top surface crust of solidified electrolyte. Pots may each have two or more point feeders, depending on their size.

At 4–4.5 volts per cell, the operating voltage is considerably higher than the theoretical decomposition voltage of aluminum oxide. The difference is due to various voltage losses, which are unavoidable under industrial conditions. The resulting excess power generates heat, which maintains electrolyte temperature. More heat comes from the slow burning of the carbon anodes.

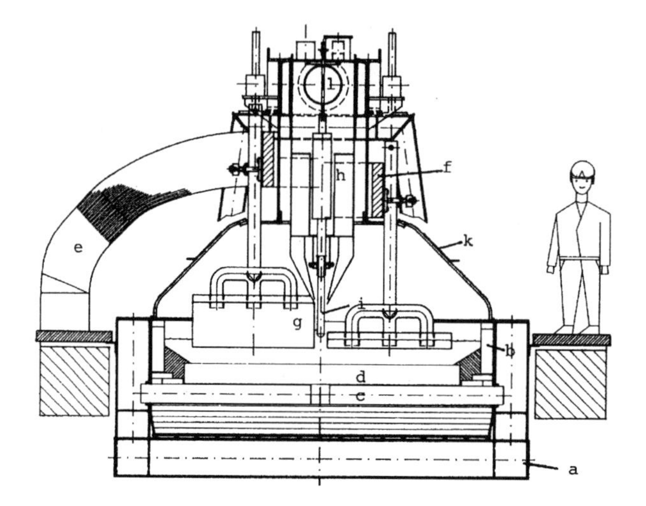

Fig. 2.4: View of the cross-section of an electrolysis cell with point feeding: (a) steel cathode shell; (b) insulation; (c) steel cathode collector conductor bar; (d) cathode; (e) riser conductor; (f) bridge, height-adjustable traverse bar supporting anodes; (g) anode; (h) alumina feed funnel; (i) point feeder; (k) pot covering hood; (l) pot gas and dust extraction. (VAW)

The cell is controlled mainly by regulating the anode/cathode distance and the direct current, which can be up to 300,000 A in modern cells. In modern smelters, process-control computers connected to remote sensors ensure optimal operation, this being one of the main reasons for today's high energy efficiency. The individual cells are connected in series, bringing the supply voltage to over 1000 V, which is the optimum operating voltage of thyristor power supplies. Thus, a modern potline consists typically of 264 cells in series, supplied at 1150 V. Aluminum busbars carry the current from one cell to the next.

As explained above, anode material is consumed in the classical Hall-Héroult reaction. Most smelters use prebaked carbon anode blocks. These are manufactured by first compacting blocks from a paste of calcined petroleum coke and tar pitch. Formerly made by pressing, the blocks were of uneven density, and tended to fail in service. Today, vibrating the paste under load, and often under vacuum, produces blocks of more uniform density. After this compacting process, the anodes are prebaked in a calcining furnace.

Instead of prebake anodes, some older smelters use Söderberg anodes. These are continuously formed in-situ by feeding "green" (raw) paste into the top of a sheet steel shell over the cell. Heat from the process bakes the paste to solid anode coke. There are two variants of the Söderberg process. These are referred to as either vertical- or horizontal-stud processes, depending on the positioning of the steel pins that deliver electric current into the anode. Disadvantages of this process are: low current efficiency of about 86–90% (compared with 95% in modern prebake smelters) and more fume emissions than with prebake anodes. Such fumes create problems with workers' health. No more Söderberg cells are being built, and those existing are progressively being shut down, converted, or replaced. What follows will only deal with modern plant using prebake anodes.

Fumes leaving the cells contain mainly CO_2, CO, and SO_2 if the anode coke contains sulfur, together with smaller amounts of fluorine compounds and dust. Most of the gaseous fluorine exists as hydrogen fluoride HF, and the dust is mostly fluorine compounds such as cryolite and aluminum fluoride, as well as some alumina.

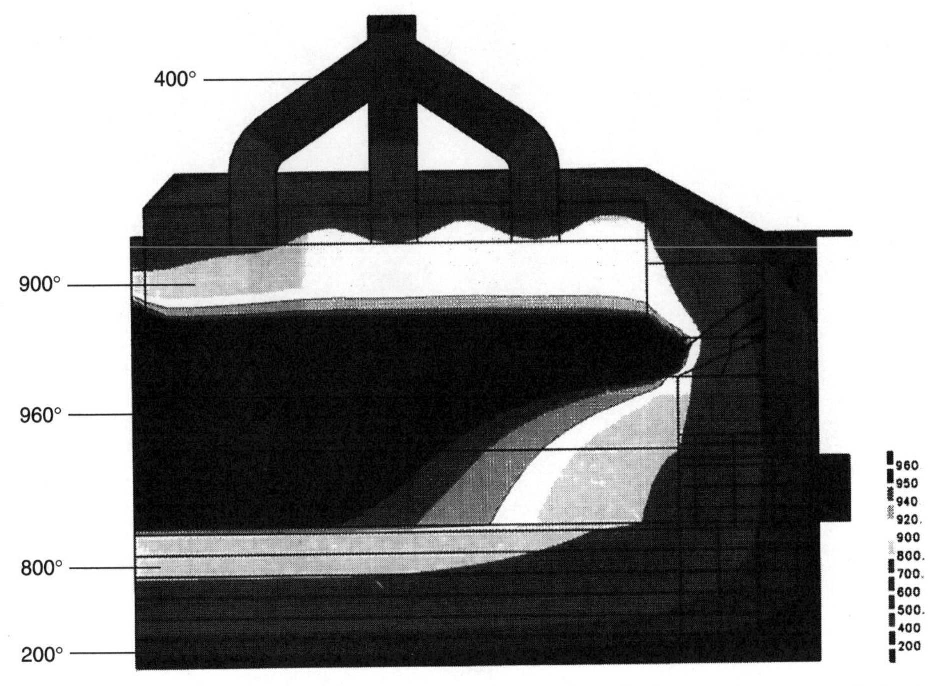

Fig. 2.5: Temperature distribution in a 180 kA electrolysis cell, drawn using a computerized mathematical model. (VAW) A color version of this figure appears on page 425.

Under unfavorable conditions, the pots may produce small quantities of the fluorocarbon compounds CF_4 and C_2F_6, which are known to take part in the "greenhouse effect" of the upper atmosphere. However, this emission happens only during the so-called "anode effect," which occurs when the alumina concentration drops below a critical threshold. During the anode effect, the cell voltage climbs from the normal 4.5 volts to over 40 volts. In modern electrolysis pots, which are fitted with pneumatic alumina transport and feeding systems, the alumina concentration can be held at an almost constant level. With the aid of modern, microprocessor-controlled potroom control, the frequency of anode effects, and hence the emission of fluorocarbons, can be much reduced.[3] Formerly, about one per day and per cell, anode effects can now be reduced to one every two months by automated alumina feeding.

Since emissions of fluorine compounds, either as gases or dust, can harm the environment, these fumes are collected by hoods over modern cells. In a process called dry scrubbing, the fluorides bind to sandy alumina during intensive mixing. This fluoride-loaded alumina then goes to dust separators and to electrostatic dust filters, and finally back to the potline as feed material. The fluoride-free remainder is exhausted to atmosphere. Although technically difficult and capital intensive, dry scrubbing meets the low emission limits required for modern smelters. In recovering lost fluorine compounds and recycling them to the cell, dry scrubbing shows that environmental protection and economy

[3] Such a control system is the "ELIAS" (Electrolysis Automation System) of the VAW Company. Other leading primary aluminum producers use similar or equivalent process-control sysems. See also Chapter 16.

Fig. 2.6: Inside view of a potroom with cells arranged crosswise; in the foreground is a 240 kA pot with electrical connection through risers at the side. (VAW)

are not necessarily opposed, but can aim in the same direction (see Chapter 16).

The aluminum industry has made considerable technical advances since the Hall-Héroult process was introduced and is working on further improvements. Specific energy consumption has fallen in the last 35 years from 21 to almost 13 kilowatt-hours per kilogram of aluminum produced. This was made possible by advanced computer control of all the relevant parameters in the electrolysis cell. Development of new cells today involves computer modeling to optimize chemical, electrical, magnetohydrodynamic, and thermal conditions. Fig. 2.5 shows a calculated temperature distribution as an example of the computer-aided development of electrolysis cells.

Formerly, manual operation of the cells required wide access routes for vehicles between the potlines, but in modern smelters (Fig. 2.6) manipulator cranes perform most tasks. In contrast to the longitudinal alignment of anode beams in earlier potlines, the newer designs have anode beams arranged transverse to the line of cells. This arrangement has several advantages: it better compensates magnetic fields in the cell, requires less floor space in the potroom, and reduces the length of conductors, thereby reducing resistance losses. Computer modeling has shown that to reduce magnetic fields, electrical busbar connections should be through the transverse "risers" rather than as previously through the ends of the anode beams.

2.2.2 Other processes for producing aluminum

Other processes for extracting aluminum have been developed. Here, it is worth mentioning two electrolysis processes (using aluminum chloride or, alternatively, aluminum sulfide electrolytes) and two metallothermic ones (the Toth process, in which manganese reduces aluminum chloride, and the carbothermic reduction process, in which carbon reduces alumina). Except for the aluminum chloride electrolysis, technical or economic reasons have prevented these processes from developing beyond the laboratory or pilot scale.

For a few years from 1976, Alcoa operated an aluminum chloride electrolysis plant with a capacity of 15,000 tonnes per year. However, this was later shut down, the reason given being the excessive cost of producing anhydrous aluminum chloride feedstock by chlorinating alumina. Aluminum chloride feedstock is dissolved in an electrolyte consisting mainly of sodium chloride (NaCl) and potassium chloride (KCl) or lithium chloride (LiCl). Electrolysis releases aluminum metal and chlorine gas. The latter is recycled by chlorinating alumina. The main advantage of the chloride process over the Hall-Héroult process is a saving of 30% in electric power consumption. In addition, the process avoids fluoride emissions and makes more optimum use of anode coke, since it can work with multipolar graphite electrodes. At 700°C, the temperature is much lower than that in the Hall-Héroult cells, so that it needs less heat energy. In spite of these advantages, its chances of success remain uncertain because, as mentioned above, the economics of producing aluminum chloride feedstock remain unclear. Some observers have noted two further disadvantages: the tower-like multistory building, and corrosion problems within the system for circulating the electrolyte.

2.2.3 Primary aluminum

Smelters produce primary aluminum (as opposed to secondary, or recycled, aluminum) with a purity of 99.7–99.9%. The main impurities are iron and silicon, together with smaller amounts of zinc, magnesium, manganese, and titanium. Typical analyses also show traces of copper, chromium, gallium, sodium, lithium, calcium, vanadium, and boron. Passing chlorine gas through the molten aluminum can remove traces of sodium, lithium, calcium, and, if necessary, magnesium. Filtering can remove suspended particles, such as oxides and carbides. Hydrogen, the only gas soluble to any extent in aluminum, can be removed by degassing with chlorine, nitrogen, or, better still, argon (see Chapter 5).

Aluminum for electrical use must not exceed fairly low maximum levels of titanium, vanadium, manganese, and chromium, because these elements greatly reduce conductivity. Conductor-grade aluminum is generally produced by selecting the purer metal available from the best cells. If the level of these elements is still too high, adding boron can precipitate them as insoluble borides, which have little effect on conductivity.

International standards distinguish two types of unalloyed aluminum: "pure aluminum" of 99.0–99.9% and "high-purity" aluminum of at least 99.97%, which is produced by further refinement. Table 2.1 shows the different grades of pure and high-purity aluminum, as classified by Aluminum Association standards.

Table 2.1: Composition of unalloyed aluminum ingots*

(Composition in weight percent; maximum unless a range is shown)†

Al Assoc. designation	Si	Fe	Cu	Mn	Mg	Cr	Ni	Zn	Ti	Others Each	Others Total	Al. (min.)
1050	0.025	0.40	0.05	0.05	0.05	—	—	0.05	0.03	0.03		99.50
1060	0.025	0.35	0.05	0.03	0.03	—	—	0.05	0.03	0.03	—	99.60
1100	0.095 Si + Fe		0.05–0.20	0.05	—	—	—	0.10	—	0.05	0.15	99.00
1145	0.055 Si + Fe		0.05	0.05	0.05	—	—	0.05	0.03	0.03	—	99.45
1175	0.015 Si + Fe		0.10	0.02	0.02	—	—	0.04	0.02	0.02	—	99.75
1200	1.00 Si + Fe		0.05	0.05	—	—	—	0.10	0.05	0.05	0.15	99.00
1230	0.70 Si + Fe		0.10	0.05	0.05	—	—	0.10	0.03	0.03	—	99.30
1235	0.65 Si + Fe		0.05	0.05	0.05	—	—	0.10	0.06	0.03	—	99.35
1345	0.30	0.40	0.10	0.05	0.05	—	—	0.05	0.03	0.03	—	99.45
1350	0.10	0.40	0.05	0.01	—	0.01	—	0.05	—	0.03	0.10	99.50

* From Table 6.2 of *Aluminum Standards & Data* (the Aluminum Association, 1997).

† See original reference for additional limit restrictions

2.2.4 High-purity aluminum

For most applications,the purity of aluminum as it comes from the potroom (i.e., up to 99.9%) is adequate. High-purity aluminum of at least 99.97% aluminum content is necessary for certain special purposes (e.g. reflectors or electrolytic capacitors); for such applications, the potroom metal has to be further refined in an additional process. Less than one percent of the total volume of primary metal undergoes this second stage of refining.

High-purity aluminum is produced by so-called three-layer electrolysis (the Hoopes cell). This cell, in contrast to the two layers of the Hall-Héroult process, operates with three liquid layers. The lowest layer, called anode metal, receives the input of normal primary aluminum to which has been added about 30% copper to increase the specific gravity to 3.4–3.7. The second layer is the molten electrolyte with a specific gravity of about 2.7–2.8, and the uppermost layer is the separated high-purity liquid aluminum with a specific gravity of 2.3. The cell has a siphon chamber for loading the anode metal.

Aluminum produced in this way is 99.99% pure. Higher purities of up to 99.9999% ("six-nines" aluminum) can be obtained using one or two additional zone-refining operations. Zone-refining traps impurities in a molten zone that moves gradually from one end to the other of a specially prepared ingot. Lesser purities in the range 99.97–99.98% are today produced in limited quantities by fractional crystallization. Here, any impurities that form an eutectic system with aluminum can be concentrated in the liquid melt, which can be separated from the primary crystals of aluminum, and the purer aluminum separated once it has crystallized out.

Another way of producing 99.97–99.98% purity is simply to mix higher-purity metal with that of a lower purity. The organic electrolysis which was formerly used for making the highest-purity metal is seldom if ever used because the highly inflammable electrolyte requires extreme safety measures and no economic advantage over zone-refining.

2.3 The Production of Secondary Aluminum

A used aluminum part, whether extrusion, sheet or plate, forging or casting, or a used finished product such as a can or a cast wheel, can be efficiently remelted and reconverted via the appropriate fabrication route—ingot casting, extruding, rolling or die-casting—into a new usable form. The resulting material loss by surface oxidation, called melt loss, varies from a few tenths of one percent in the case of clean, uncoated, massive castings or forgings to as much as 10% for light-gage coated packaging scrap. Melt loss depends very much on the type of feedstock: its shape and gage and the type and thickness of lacquer or other form of coating—all of these factors have a major influence on the amount of metal "lost." Melt loss also depends very much on the chosen method of melting.

The economics of recycling, together with improved techniques of scrap preparation and melting provided higher yields and thus led to the further development of the secondary aluminum industry. This field grew rapidly during the 1950s, with the blossoming of the nonmilitary uses of aluminum; today, it meets 35% of the total aluminum metal demand

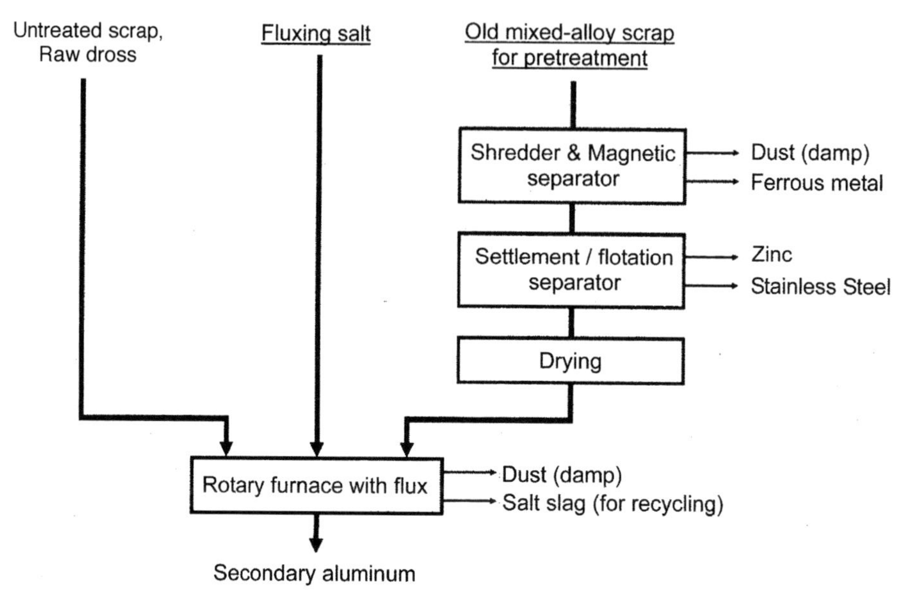

Fig. 2.7: Schematic showing the traditional processing of old scrap for making secondary aluminum.

in Europe. The supply of raw material to this secondary aluminum industry is a highly organized business involving collection and sorting networks and scrap trading.

2.3.1 Feedstock

Aluminum recycling forms the basis of the secondary aluminum industry. Today, secondary smelters serve chiefly to supply die-casting plants with casting alloys, using metal recovered from mixed scrap and dross. Much secondary metal goes into cast parts for the car industry, where demand both from independent foundries and the integrated in-plant casthouses of the car plants themselves has grown steadily over the years. Fig. 2.7 shows schematically the traditional method of scrap preparation. There has been some confusion in the terminology, so it is as well to be clear as to what is meant by "scrap" in the context of recycling. The principal feedstocks for the secondary aluminum industry are:

- Process scrap generated during the manufacture of finished aluminum products. This will be referred to as "process scrap" or "runaround scrap"
- Old scrap arising from products that have reached the end of their useful lives. This, often representing the greater part of the feedstock, will be referred to simply as "scrap."
- The residues from result of skimming and cleaning furnaces in aluminum casthouses, consisting mainly of a mixture of metal and oxides. This is usually called "dross," and it will be referred to as such.

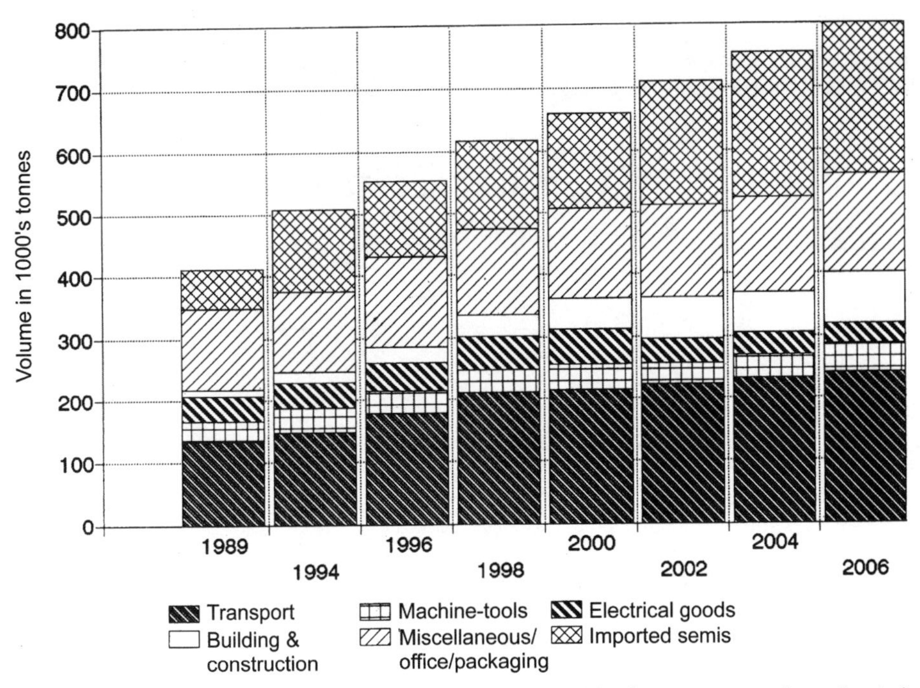

Fig. 2.8: Increase in the volume of scrap in western Europe arising from an assumed growth rate in aluminum consumption of two percent per year. Assumed service life: building—30 years; machine tools—20 years; electrical—15 years; transport—11 years.

2.3.2 Process technologies

Scrapped aluminum products are broken into small pieces and separated from dirt and foreign materials so as to yield feedstock suitable for remelting. This is done using breakers, shredders, magnetic, and settlement/flotation separators. Such scrap typically contains alloys of many types, all mixed together. A more sophisticated kind of recycling was developed in the seventies and eighties for process scrap and used beverage cans (UBC). By selectively collecting scrap in targeted alloy categories, the goal was to recycle the material back into products similar to those from which it originated. Thus, the casthouses of extrusion plants produce extrusion billets from process scrap and from recycled scrap extrusions. Similarly, the high rate of recovery of used beverage cans from the consumer, most notably in the USA, enables a large proportion of canstock coils to be made from UBC.

The demand for more selective groupings of scrap from the growing volume of material available for recycling will drive the scrap metal industry to develop refined and automated techniques of sorting scrap into the various alloy families. The steady growth in the consumption of aluminum in the past few decades will result in a higher rate of aluminum-containing goods reaching the end of their service lives and a correspondingly steady increase in the volume of scrap available for recycling as more and more aluminum-containing goods reach the end of their service lives. This trend will intensify

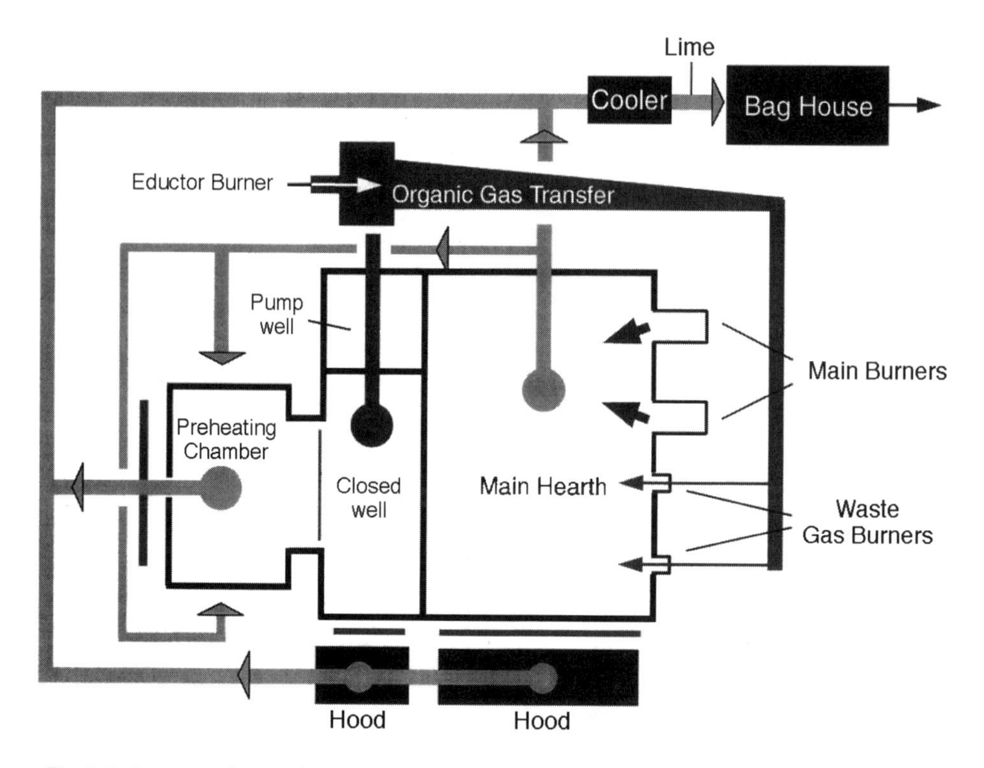

Fig. 2.9: A process for treating coated or lacquered scrap using the energy of gas generated by pyrolyzing the coatings (R-furnace). Capacity: 20,000 tonnes per year; feedstock: 98% scrap, 2% flux; metal recovery: 90%. (VAW)

in the near future because of the projected increase in the use of aluminum in cars (see Fig. 2.8). It will no longer be necessary to recycle these larger volumes of scrap only into casting alloys in the traditional way. High-volume products such as cars will be systematically dismantled and their different components collected separately to be grouped together in scrap grades of similar alloy type. Shredded scrap will be subjected to similar sorting processes that will deliver scrap by alloy grade. Future schemes will strive to recover the original alloy from the scrap material and will try to ensure a minimum of process emissions and by-products during the recycling process.

Strong differentiation of aluminum scrap by alloy type favors the development of specialized scrap preparation and melting technologies. This can already be seen in UBC recycling plants that use technology dedicated to that type of scrap. The secondary aluminum industry's standard remelt furnace, the rotary hearth with its molten flux layer, has been replaced in UBC recycling to a very large extent by the delacquering line with specially designed integrated melting furnace. The ultimate goal is flux-free remelting. Contamination from organic coatings tends to be less with castings than with extrusion and sheet scrap, such as coated or insulated building products and car body sheet. In this respect, packaging materials from household refuse present a particular problem in that a relatively small volume of aluminum is strongly bonded to plastic and paper materials. It is essential to pretreat such types of scrap before melting in order to separate the metal from the organic material. Another technique for handling scrap for recycling in bulk uses some form of rapid analysis method coupled with mechanical sorting. In this way, scrap segregated into the main alloy families can be subjected to special pretreatment and melting processes so as to remove organic coatings of all sorts while ensuring the recovery of a higher fraction of the aluminum content.

Fig. 2.9 illustrates a process known as the "R-Furnace." Coatings in the form of lacquer or laminated plastic are separated from the metal by pyrolyzation and the resulting gases, rich in energy, burn to supplement the primary fuel. The amount of salt needed to tie up the dross is reduced to a minimum. Comparison with the conventional process makes this clear: the rotary hearth furnace uses 400 kg of salt per tonne of scrap; the R-furnace uses only 25 kg per tonne.

This and similar new technologies for melting coated scrap are environmentally friendly and they do not generate excessive amounts of polluting by-products such as salt slag. They yield high-value aluminum alloys with a good recovery rate, thus enabling a more complete aluminum material cycle than was possible with the older techniques. The arrival of these processes changed aluminum recycling in a fundamental and far-reaching way (see, also, Chapter 16).

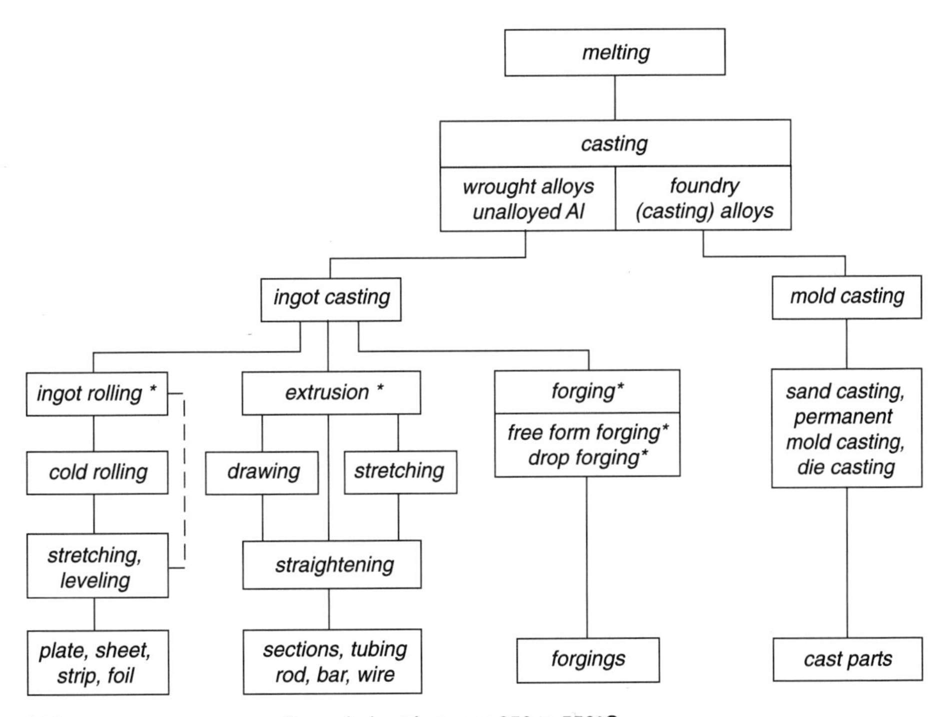

* These processes are usually carried out between 350 to 550°C

Figure 2.10: Operations and products in semifabricated products plants and foundries. Special casting processes like strip casting or wire/bar casting are not shown.

2.4 *Processing of Aluminum*

2.4.1 Cast houses and foundries

2.4.1.1 Ingot casting (DC casting)

Primary aluminum and also scrap are cast into rolling ingot (slab), extrusion ingot (billet) and wire bar ingot, and, to a lesser extent, forging stock in the casthouses of the reduction plants or semifabricated products plants. Appropriate alloying elements are added in the melting or holding furnaces, after which the metal is cleaned and cast. Reduction plant cast houses also produce pigs from a part of the primary metal.

2.4.1.2 Mold casting

In foundries, cast products are usually produced from prealloyed metal supplied by secondary smelters. In some cases, casting alloys are prepared from a primary metal base for products that must meet rigid requirements that can be achieved only with minor amounts of impurities (for example, iron).

There are three main aluminum casting processes: sand casting, permanent mold casting, and die casting, which usually produces a finished part in one step. Unlike semifabricated products plants, foundries may deliver a finished product which requires no further forming. For this reason, foundries are not usually classified as semifabricated products plants and are shown separately in Fig. 2.10.

2.4.1.3 Special casting processes

The continuous casting of strip, wire, and rod belong to this family. Lately, continuous strip casting processes are experiencing the most rapid growth.

2.4.2 Semifabricated products plants

An aluminum plant for the production of semifabricated products (sometimes called a "semis" plant) may receive ingots for fabrication directly from the reduction plant or from their own remelt shop. Typical fabrication steps in a semis plant are shown in Fig. 2.10. The first operation is the hot deformation of the cast ingot at temperatures between 350°C and 550°C. Depending on the process, the deformation may be executed by hot rolling, extrusion, or forging. Such hot-working is often followed by cold deformation such as the cold rolling of sheet or drawing of tube. Some semifinished products are supplied in the as-fabricated condition in the form of extruded shapes, forged parts, and hot-rolled sheet or coils. Prior to delivery, extruded shapes are usually stretched for straightening and stess relieval, which imparts a small amount of cold work to the material.

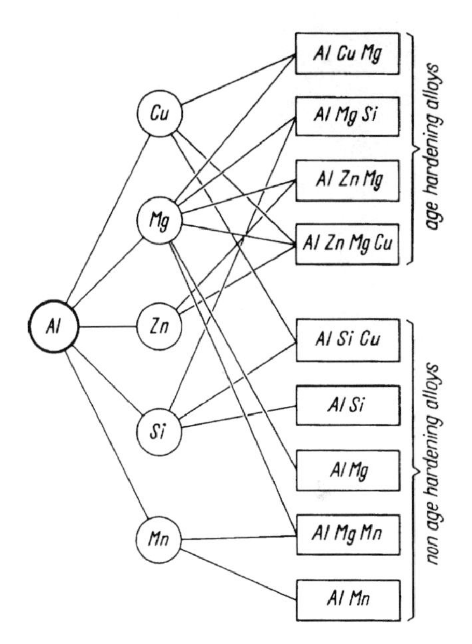

Figure 2.11: Synopsis of the principal aluminum alloys (commercial alloys on the basis of primary aluminum).

2.4.3 Subsequent fabrication of aluminum semifabricated products

Semifabricated products undergo further fabrication before they are sold to the consumer as a finished product. The fabricators use a great number of techniques in finishing aluminum. The primary objective of the fabricator is to impart the desired shape to the work piece. The techniques may produce chips (milling, turning, boring, etc.) or be chipless (deep drawing, stretch forming, impact extrusion, blanking, bending, spinning). Surface treatment of aluminum is of special importance and may include mechanical polishing, etching, brightening of the aluminum by electrolytic or chemical processes and the generation of thicker oxide layers through anodizing.

The picture would not be complete without mentioning joining techniques, which have been greatly improved in recent years, especially in the field of welding under protective gas, as well as in the use of adhesives.

2.5 *Types of Aluminum Alloys*

As mentioned earlier, the production of semifabricated products utilizes three different types of aluminum, namely super purity, commercial purity, and alloys. Alloys are used for producing castings or fabricating wrought products. The alloys used for castings contain a greater amount of alloying additions than those used for wrought products. The addition of alloying elements has the effect of strengthening the wrought alloys and improving the castability of the casting alloys.

Table 2.2: The Aluminum Association alloy designation system

Wrought alloys	Series	Cast alloys	Series
Al (99.00 % miniumum or greater)	1xxx	Al (99.00 % miniumum or greater)	1xx.x
Alloys grouped by major alloying elements			
Cu	2xxx	Cu	2xx.x
Mn	3xxx	Si + Cu or Mg	3xx.x
Si	4xxx	Si	4xx.x
Mg	5xxx	Mg	5xx.x
Mg and Si	6xxx	Zn	7xx.x
Zn	7xxx	Sn	8xx.x
Other element	8xxx	Other element	9xx.x
Unused series	9xxx	Unused series	6xx.x

Wrought alloy designations: Four digits are used to identify wrought aluminum and wrought aluminum alloys. The alloy group is identified by the first digit. Modifications of the original alloy and impurity limits are indicated by the second digit. In the case of the 1xxx group, the last two digits indicate the minimum aluminum percentage. For the 2xxx through 8xxx groups, the last two digits serve to further identify individual aluminum alloys. For experimental alloys, the prefix "X" is added, as in a designation (e.g., X2037). When the alloy ceases to be experimental the prefix is dropped.

Cast alloy designations: The designation system for cast aluminum alloys is similar to that for wrought products, in that the first digit indicates the major alloy group. Digits two and three indicate the aluminum purity or further identify the alloy. The digit to the right of the decimal place indicates the product form, either a casting or ingot. Modifications to original casting alloys are indicated by a serial letter before the numerical designation (e.g., A356.0 or B413.0).

Today, more than half of the semi-finished production is delivered in alloys, most of the remainder in commercial-purity aluminum with only a small quantity of super-purity aluminum being produced.

It is necessary to differentiate between wrought and casting alloys. Both of these alloy types are subdivided into those alloys that are solution heat treatable and those that are not. Wrought alloys are genarally used for further fabrication, for example, rolling, forging, extrusion, and drawing. Casting alloys are used for cast parts and have the flow characteristics favorable for this process. For example, an alloy with good castability must be able to fill the mold completely and have a low sensitivity toward cracking during casting.

The most important elements that are added to aluminum, in alphabetical order, are bismuth (Bi), boron (B), chromium (Cr), copper (Cu), iron (Fe), lead (Pb), magnesium (Mg), manganese (Mn), nickel (Ni), silicon (Si), titanium (Ti), zinc (Zn), and zirconium (Zr). Magnesium is the most frequent addition to aluminum. In some alloys, two or more elements are used in combination (e.g., magnesium together with silicon or manganese). There are also alloys of aluminum containing only manganese or only silicon. The alloying elements are added to bring about changes in the properties of aluminum, as described in Sections 2.5.1 and 2.5.2.

Fig. 2.11 shows a schematic survey of the most commonly used aluminum alloys. Table 2.2 illustrates the Aluminum Association and American National Standard Institute (ANSI) alloy designation system for wrought and cast alloys, described more fully in Appendix C and in full in the Aluminum Association's *Aluminum Standards and Data.* There is an international accord on wrought alloy designations, recognizing the Aluminum Association designations virtually worldwide.

There is not an equivalent international accord on cast alloy designations, so there are not exact equivalents for these alloy designations in different countries. For this reason, at various places in this book, both the US and international designations for alloys and alloy types are sometimes utilized to help the reader. Similarly, there is no international accord on temper designations (also described in Appendix C), so these too may vary from country to country.

2.5.1 Wrought alloys

Strength is increased by the addition of magnesium (up to a maximum of 7%) and also by additions of zinc, copper, and / or silicon in addition to magnesium. Good high-temperature strength is attained by the addition of copper (up to 4%) and / or nickel, manganese, or iron up to one percent each. Good chemical resistance is shown by alloys with additions of magnesium, manganese, or a combination of magnesium and silicon. Machinability is greatly improved by the addition of lead and bismuth up to 0.6% each. A fine-grained structure is obtained especially through the addition of titanium and boron (up to 0.1%). A fine grain upon recrystallization of semifabricated products is often obtained through additions of chromium or zirconium (up to approximately 0.1%). Other elements, for example iron and manganese, have the tendency to act as grain refiners as well.

2.5.2 Casting alloys

Castability is improved through the addition of silicon up to 13%. Dimensional stability upon heating (pistons) is assured by silicon contents up to 25%.

There are also various specialty groups of alloys that make up only a very small part of the total production of aluminum semifabricated products. These are not included in the schematic in Fig. 2.11. Piston alloys and "free-machining" alloys belong to this special group. The latter group includes alloys with lead and other additions that cause the turnings to break into small pieces during machining; this is desirable from a fabricating standpoint. Having covered the main factors in the production and processing of aluminum, we turn now to the main theme of this book.

Chapter 3. The Internal Structure of Aluminum

3.1 Examining the Structure

In order to understand the properties of aluminum, a certain knowledge is required of the smallest building blocks of which aluminum is composed. The way these components fit together is called the metal structure.

In order to examine the structure, a section is cut out of an aluminum sample. The section is polished as required and usually etched with suitable reagents. For a metallographic investigation of the structural composition, three different enlargement ranges may be used. These are shown schematically in Fig. 3.1 using, as an example, the structure of a cold-worked and then annealed sheet sample.

3.2 Macrostructure

After etching the sample surface, one can see the individual grains of the metal with the naked eye. This corresponds to the macrostructure in the left section of Fig. 3.1. A single grain in the plastically deformed state usually has a diameter between 0.01–1.0 mm and can be much larger in the cast structure. The grains are usually elongated in the cold-worked state. A new grain structure arises during annealing. This will be discussed in more detail later.

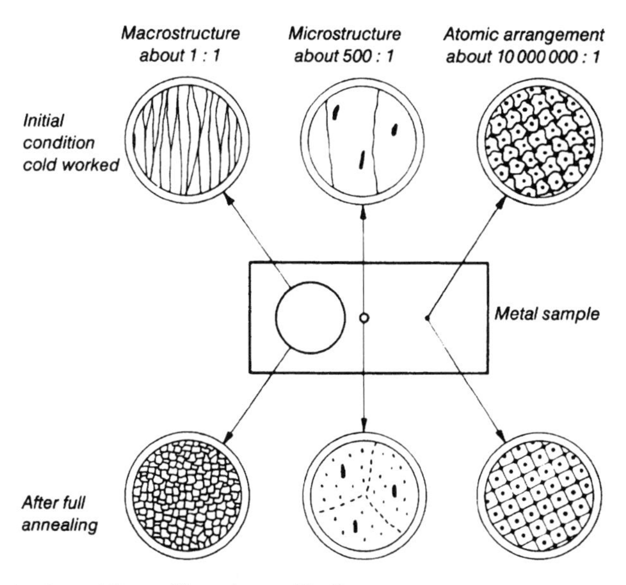

Fig. 3.1: Metal structure at three different magnifications.

3.3 Microstructure

With an optical microscope, magnifications of 50–1000× are normally used in the investigation of the structure. In this range, the inclusions in the aluminum structure (heterogeneities) that originate from the alloying elements or from naturally occurring impurities in the metal can be seen. Inclusions of this type may be changed by a thermal treatment illustrated in the 500× enlargement in Fig. 3.1. There may also be voids in the metal structure (porosity or large cavities) that can sometimes be seen with the naked eye.

For more detailed or scientific investigations, an electron microscope may be used in addition to an optical microscope. The electron microscope can magnify the structure more than 100,000 times. Even the finest heterogeneities, especially "precipitates" that form in the solid state as well as so-called "subgrains," become clearly visible at this high magnification. Subgrains arise during a partial anneal or during plastic deformation and have a slightly different orientation than neighboring portions of the same crystal. Generally, they are separated by low-angle boundaries.

3.4 Atomic Structure

The atoms in solid aluminum are arranged in a three-dimensional lattice type arrangement. The arrangement is typical for all crystalline materials, including all solid metals. There are also amorphous substances in which the atoms have no definite arrangement. Liquids are examples of amorphous bodies, but there may also be solid amorphous bodies (e.g., glass).

To investigate the crystallographic structure, it is preferable to carry out the fine structure analysis using x-rays. Fig. 3.2 depicts the face-centered lattice structure of aluminum.

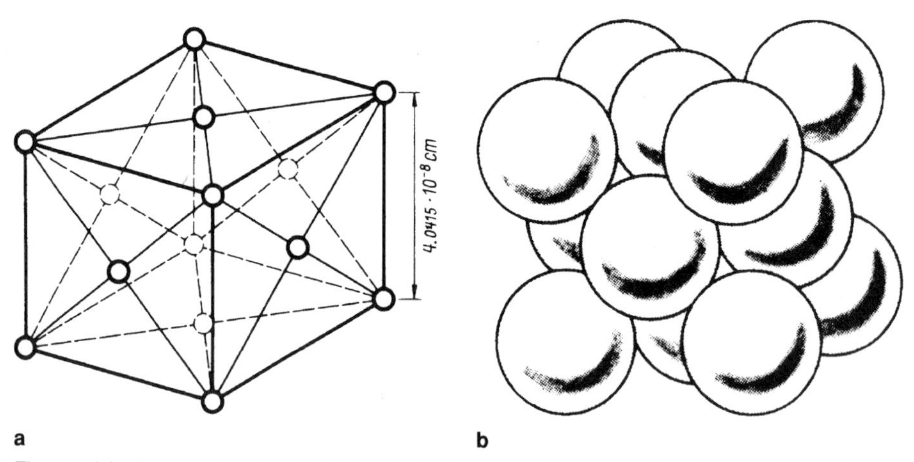

Fig. 3.2: (a) A face-centered cubic lattice of aluminum showing the position of atoms. Three faces of the cube are visible; dashed lines are used to indicate the position of the atoms in the other faces. Although 14 atoms are shown, the unit cell contains only four atoms. (b)The densest spherical packing of atoms in the aluminum lattice, where the atoms actually touch each other.

In the right-hand section of Fig. 3.1, it was shown schematically that cold work produces certain lattice distortions. A grain in the structure of pure and undeformed aluminum has an almost continuous lattice with rectilinear lattice planes and can be considered a uniform crystal. The direction of the lattice planes changes at the boundary with a neighboring crystal. Two neighboring crystals are thus said to have different "orientations." (This situation is shown schematically in Chapter 4, Fig. 4.4.) As can be seen in the figure, certain lattice irregularities occur at the grain boundaries. Lattice imperfections that form irrespective of grain boundaries, but strongly affect the properties of the metal, will be discussed later.

3.5 *Why View Aluminum from Within?*

A general understanding of aluminum metallurgy is of interest to the metal fabricating industry, especially for the following problem areas.

3.5.1 Changes in the metal structure during the fabrication of aluminum

In order to get a clear picture of what takes place during casting, forming, or annealing of aluminum, it is necessary to go beyond the normal standpoint (i.e., no longer view aluminum from the outside, but rather from within). In following up a special problem (e.g., in the production of semi-finished products), it is an advantage to make a distinction among the following three groups of factors and observations.

A. Production procedures, such as alloying, casting, and the variables in forming or heat treatment of the metal (i.e., measures actively taken).
B. Changes in the metallic structure resulting from one or more of the measures listed under A.
C. Properties of the intermediate or final products, such as formability, mechanical properties, and corrosion behavior.

In following up a plant quality problem, an examination of the end product may provide a key to the necessary corrective action (causal link from C-A). In a development project or in a concentrated plant effort to improve quality, the A-C links often come under consideration. In both cases, an investigation of the structure (B) is often insufficient or even nonexistent.

The main purpose of the first part of this book is to give the reader a certain understanding of the collective phenomena in group B. This makes it possible to call on the entire causal chain A→B→C or C→B→A when clarifying a particular problem in the fabrication of aluminum resulting from actively taken measures (A) and the resulting microstructure (B).

3.5.2 Specifying properties of aluminum products

Products fabricated from aluminum may have to meet a number of specifications with regard to strength, formability, or surface properties. This brings up certain questions between supplier and fabricator that must be clarified in order for the salesman, purchaser, plant-operating personnel, and laboratory personnel to carry on meaningful discussions. These specialists often discuss typical properties or structures in detail.

In many cases, it is not sufficient to be guided by standards or company brochures listing the characteristics of the material in question; rather, a certain knowledge of aluminum metallurgy is required as a basis for understanding the phenomena influencing the properties of the metal.

3.5.3 Structure and properties

Aluminum in the solid state has the following physical properties in common with other metals:

- Crystalline structure
- High electrical and thermal conductivity
- Formability
- Good surface reflectivity

Technically important properties and characteristics of aluminum can be divided into two groups: those independent of the microstructure and those dependent upon the microstructure.

3.5.4 Properties independent of the structure

Of prime importance in this category are the melting point and density (specific gravity). These properties are not affected by the type of structure present. Also, the thermal conductivity is independent of the structure and composition to a great extent. The modulus of elasticity of common wrought alloys, which is important for the practical application of aluminum, is not changed by differences in structure. For most aluminum alloys, the elastic modulus is given as 7.2×10^4 MPa.

However, the modulus of elasticity of aluminum alloys can be increased or decreased as a result of certain alloying additions (see Chapter 15.4.3, the influence of lithium additions on modulus). In the case of alloys produced by powder metallurgy techniques, where the amount of alloying addition can be greater than that obtainable by conventional means, the modulus can also be changed significantly.

3.5.5 Properties dependent on the structure

A number of technically important characteristics of aluminum can be changed by an order of magnitude or more by suitable means such as alloying additions, plastic deformation, or heat treatment. This is especially true with regard to strength, formability, and corrosion resistance.

Electrical conductivity also is very dependent on the structure and may vary by a ratio of 1:2 (99.99% aluminum compared to a 12% silicon alloy). The alloys used to conduct electricity may show changes of ±20% as a function of structure and alloying elements. This variation has a great deal of significance to the electrical industry and to the selection of production processes for conductor materials.

In the following chapters, the structure-dependent properties will be discussed in detail in order to clarify the causal chain between production procedures (A), structure (B), and the properties (C) using specific examples. The first example of this type will be the solidification of liquid aluminum.

Chapter 4. Formation of the Cast Structure

4.1 Atomic View of the Solidification of an Aluminum Melt

Aluminum loses its crystalline structure when heated above its melting point. The atoms are in complete disarray in the melt, but if the melt is cooled to the solidification temperature, the atoms again assume their position in the normal crystal lattice (Fig. 3.2 and Fig. 4.1). During solidification the atoms arrange themselves so that if one imagines a cube, an atom would be located at each of the corners and at the center of each face of the cube (Fig. 4.1). The formation of the crystalline structure of aluminum from the melt may be illustrated by the following example. Imagine an empty football stadium with its arrangement of seats around the playing field. If one stands atop the stadium before the gates are opened and watches the people milling around, this can be compared to the amorphous state of the liquid aluminum. Now, if the gates are opened and all the people begin to take their seats, this is the process of solidification, or the crystallization of aluminum during the cooling of the metal. Naturally, it takes a certain amount of time until everyone has found his seat. It also takes a certain amount of time for the atoms to find their places when aluminum solidifies.

At this point, the comparison begins to break down. The people seeking their places are able to think, but how do the atoms know in which order to arrange themselves? The explanation is that atoms attract each other, and they have a definite diameter that is different for each metal. The atoms are drawn together during solidification due to the attraction for each other and arrange themselves so that they lie as close to each other as possible (Fig. 3.2.b). This automatically yields the lattice structure of aluminum, which has already been described, in which a very large number of atoms arrange themselves in a face-centered cubic structure.

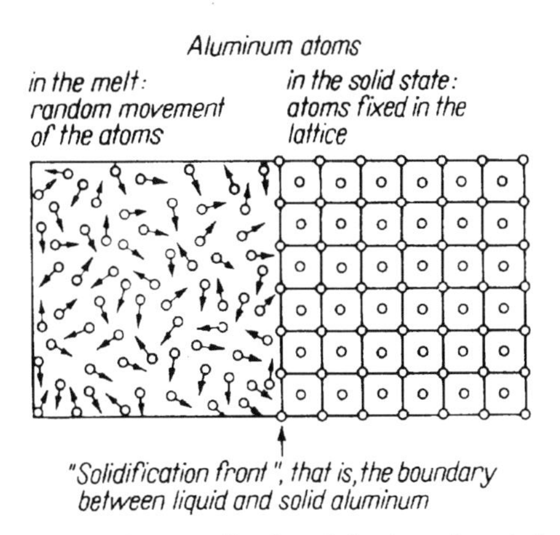

Fig. 4.1: Behavior of atoms during the crystallization of aluminum (i.e., during the solidification of an aluminum melt).

4.2 A Heat Balance

Each aluminum atom has a definite amount of energy associated with it, which varies depending on whether the metal is liquid or solid.

During the melting process, a definite amount of heat energy is required to bring the temperature of the aluminum (99.99% Al) to its melting point of 660°C. However, additional energy is required in order to spring the atoms loose from the lattice bonds and transfer them into the amorphous, or liquid state.

In order to heat one kilogram of aluminum from 20°C to 660°C (293 to 933 K), 670 kilojoules are required. This heat energy can be calculated from the specific heat of aluminum, which is defined as the number of joules required to raise the temperature of one gram of metal one degree Kelvin. In the range between room temperature and the melting point, aluminum requires approximately 1.05 joules/g · K. In order to convert the aluminum from solid to liquid without any increase in temperature at 660°C, an additional 396 joules per gram of aluminum are required (heat of fusion). Thus, one can see that the energy of the aluminum atoms increases very significantly in the transfer from solid to liquid. The higher energy of atoms in the liquid state can be recognized through their increased movement; that is to say that the atoms in the liquid state have a higher kinetic energy than those in the solid structure.

If we look at the opposite process, that is the solidification of the melt at 660°C, it is necessary that the 396 joules per gram of aluminum be removed by cooling, in order to get the atoms back in the low-energy level of the crystalline state. The "heat of solidification" is just exactly the same amount of energy as the "heat of fusion." The melting curve and the solidification curve of aluminum are mirror images of each other (Fig. 4.2). During either melting or solidification there is a delay in the temperature change. The temperature remains unchanged despite an introduction of heat or removal of heat until either the total heat of fusion is added or the heat of solidification is removed. One can see from these data that the largest part of the heat removed in the temperature range for the solidification of aluminum comes from the heat of solidification rather than the cooling of the solidified metal.

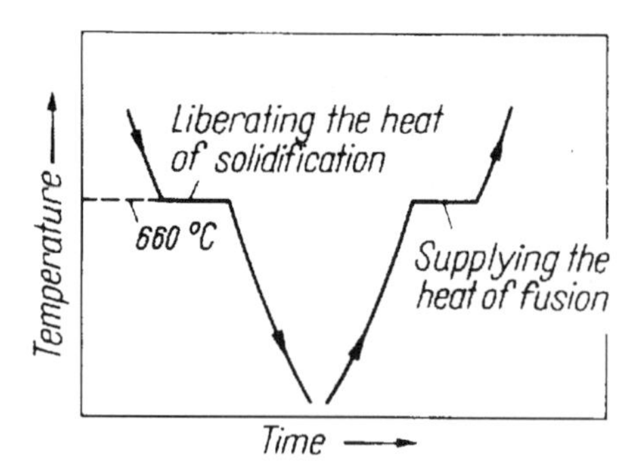

Fig. 4.2: Cooling (solidification) and heating (melting) curves for 99.99% aluminum.

4.3 How Do Cast Grains Form?

Going from the liquid to the solid state, the metal crystallizes, so that the atoms arrange themselves in the closest packing arrangement. The process of crystallization starts at the so-called "nuclei." As soon as the melt has been cooled to 660°C, crystallization nuclei form at individual points within the melt. These small nuclei increase in size very rapidly in that aluminum atoms continue to arrange themselves around each nucleus. The heat released in this process must be conducted away. In commercial casting operations, continuous heat removal is provided so that the crystallization nuclei grow very rapidly until they are stopped by neighboring crystals.

Fig. 4.3 schematically shows the nucleation and growth of neighboring grains during the course of solidification. The dark background corresponds to the liquid melt, a small white square to a "unit cell," such as already shown in Fig. 3.2. At the upper left, we see the beginning of the solidification. Seven crystallization nuclei have formed, of which six have already started to grow through the arrangement of further unit cells. The nucleus in the middle has just formed. The succeeding pictures show the growth of the crystals with time, until finally the total melt is consumed so that the resulting crystals (called cast grains) now join at grain boundaries. The lattice of the individual cast grains is arranged at differing angles in space.

Within one grain, the aluminum atoms are arranged in a uniform lattice except for certain local discontinuities. The most important lattice planes run parallel or are located at 90 degrees or 45 degrees to each other within the grain (Fig. 4.4.a). Another nucleus, for example, the one that caused the growth of a neighboring grain usually has a different orientation at the moment of origin in the melt. This explains why the uniform lattice is interrupted at the grain boundaries, and then starts again at a different angle as has already been described. There is a definite degree of disarray at the grain boundaries which

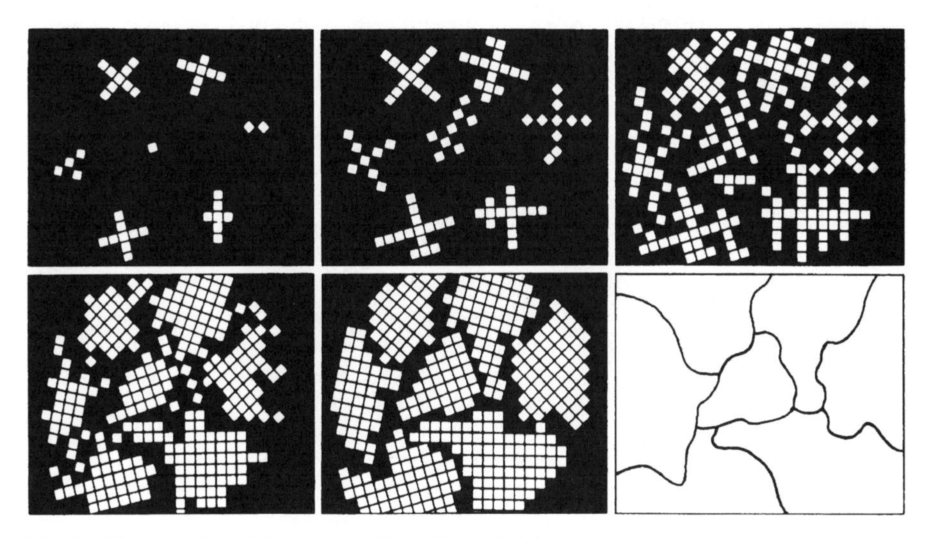

Fig. 4.3: The formation of the cast structure. (Rosenhain)

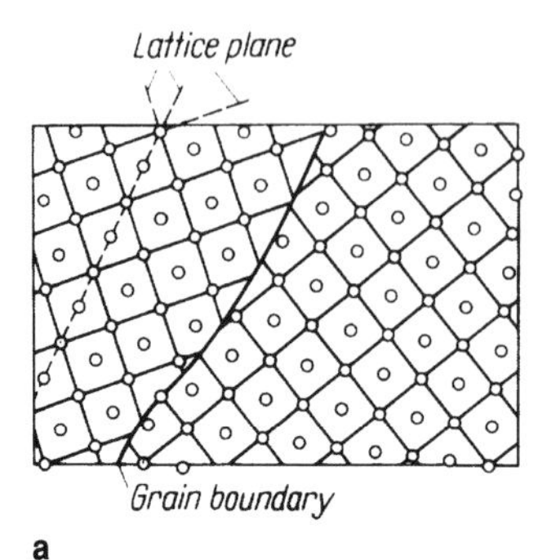

a

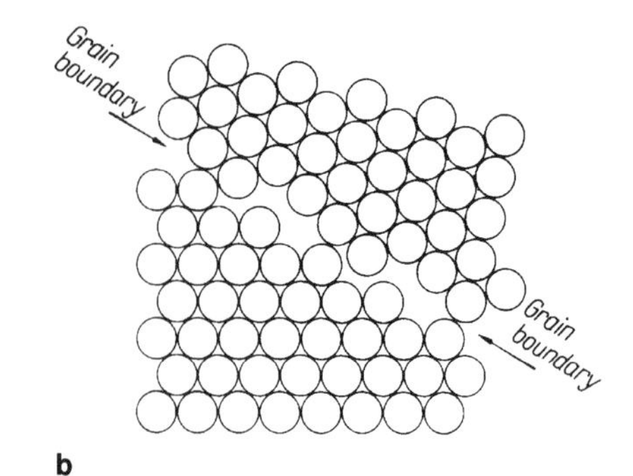

b

Fig. 4.4: (a) Schematic representation of the atomic arrangement in two neighboring crystals (grains). The lattice is interrupted at the grain boundaries. (b) Enlarged section from Fig. 4.4.a, illustrating the interruption of the atomic arrangement at a grain boundary. Inside the grain, the aluminum atoms achieve the "closest packing of spheres." The atomic disarray at the grain boundary favors migration (diffusion) of foreign atoms.

is aggravated by a concentration of foreign atoms during solidification of commercially utilized metals. For this reason, the grain boundaries often represent weak points in the structure with regard to chemical resistance.

The number of atoms within a grain is very large. On the average, a grain is approximately the size of a grain of rice but the size may vary from much smaller to much greater. Within an average grain are approximately 10^{21} atoms (10^{21} is one followed by 21 zeroes). During solidification, this large number of individual atoms can find their correct places within even less than one second.

In order to visualize the great speed with which the atoms arrange themselves during solidification, we need to remember that one must imagine an aluminum atom as a sphere with a diameter of 0.000286 angstrom (1 angstrom equals 1/1000 of a millimeter). If one were to throw a large number of such atoms into a basket and shake them for just an instant, the spheres would arrange themselves in the most dense arrangement possible. In the case of aluminum atoms, this leads automatically to the lattice structure of the densest spherical packing as described in Fig. 3.2.b in a very short time. The contents of our basket correspond to one grain. The shaking has the same effect as the innumerable rapid movements of the atoms. In this way it is easier to understand how such a large number of individual atoms can arrange themselves in a uniform lattice in such a short time. (A closer observation shows that the atoms in the melt already have a definite arrangement that approaches a "shaky" lattice. This speeds up the location of the atoms on their lattice points during solidification. It is only in the vapor state that the atoms achieve free movement. This may also be seen by the fact that the vaporization of aluminum requires 25 times as much heat input as melting.)

During solidification of the cast structure, there are a whole series of changes to be observed, which will only be highlighted in the following. Before the start of solidification, the melt reaches a definite undercooling below the solidification temperature (liquidus temperature). The greater the undercooling, the faster the heat removal during solidification and the fewer the nuclei-forming alloying constituents present in the melt. In commercial alloys the nucleus formation usually results from crystallization nuclei already present in the melt before true solidification sets in. These nuclei may be particles containing either titanium, boron, or iron, or consist of oxides or aluminum crystals that were not completely melted. The latter occurs when solid aluminum is added to the melt below approximately 700–720°C before the start of solidification, since the solution of the aluminum requires a definite time. Especially below 700°C, the lattice remnants are retained that serve as nuclei in subsequent solidification. The same lattice remnants are present if the melt does not exceed 700°C during melting. The process of solidification presented in Fig. 4.3 represents a certain simplification with regard to nucleus formation.

4.4 Revealing the Cast Structure by Etching

The cast structure consists of grains that are firmly joined at grain boundaries (Fig. 4.5.a). In order to reveal the grain structure, etching is carried out with a suitable reagent. After etching, the different grains reflect the light in different ways and, therefore, can be easily distinguished. The etching causes small etching steps (Fig. 4.6.a) as the etchant removes atom layers parallel to or in a definite angle to the cube surface (Fig. 4.6.b). Therefore, it is understandable that the individual grains reflect the light differently after etching. The resulting small etching steps measure at most a few thousandths of a millimeter in height. The crystallization nucleus in general grows opposite to the direction of the heat removal. For this reason, the cast grains often are not equiaxed but columnar.

Especially in the outer zones of the cast billet or shape, one finds a non-uniform structure (Fig. 4.5.a). The first crystallization nuclei form in the vicinity of the cooled wall of the mold. Near the mold surface, numerous crystallization nuclei form simultaneously, which leads to a fine-grained zone (Fig. 4.5.b). The subsequent structure zone often consists of columnar crystals. The crystals have been built up in layers from the melt and then serve to conduct the liberated heat of solidification to the mold wall. In the center zone of the ingot one finds mostly equiaxed grains.

Generally, the higher the purity, the larger is the grain size. This is because the impurities found in the melt act as crystallization nuclei so that the higher the purity, the fewer the nucleation sites.

Crystallization normally does not produce grains that are equiaxed or columnar in shape. The growth of so-called dendrites results in another crystal shape. Dendrites are crystals with a tree-like branching pattern. Their growth is caused by the fact that the conditions for further growth are more favorable on the corners and edges of a growing crystal, which was originally cubic, than on the center areas. During solidification, this leads to a growth on the edges and corners of the cubes and produces star-shaped or crystals with an irregular outline (Fig. 4.7). Alloys often solidify with dendritic structure. A cast grain can have many dendritic arms. If a cross-section is made of the cast structure, the grains are often divided into cells. These cells are sections through the dendrite arms (see Section 4.7.1).

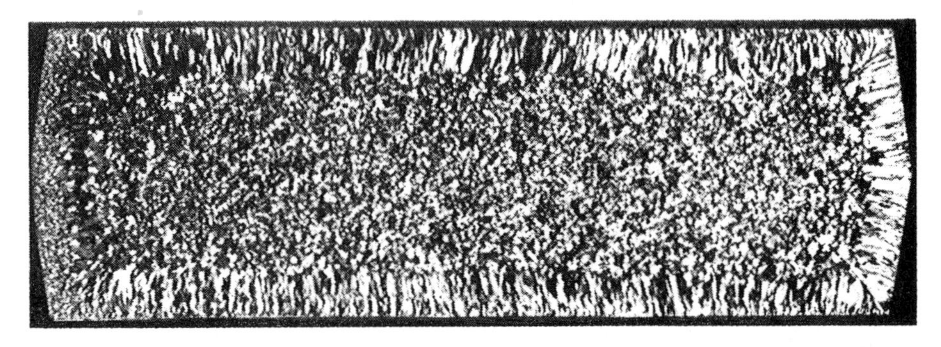

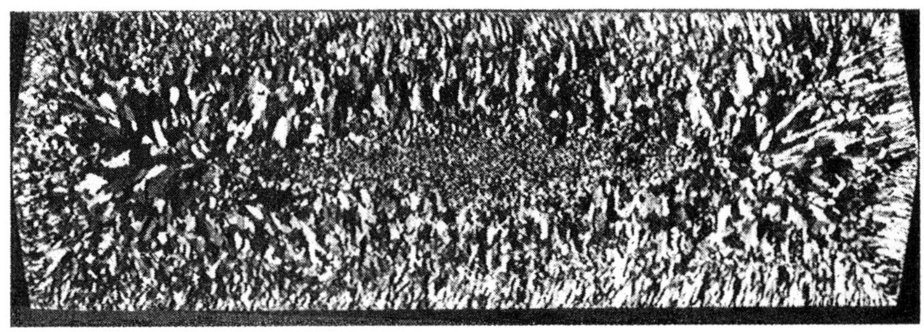

a

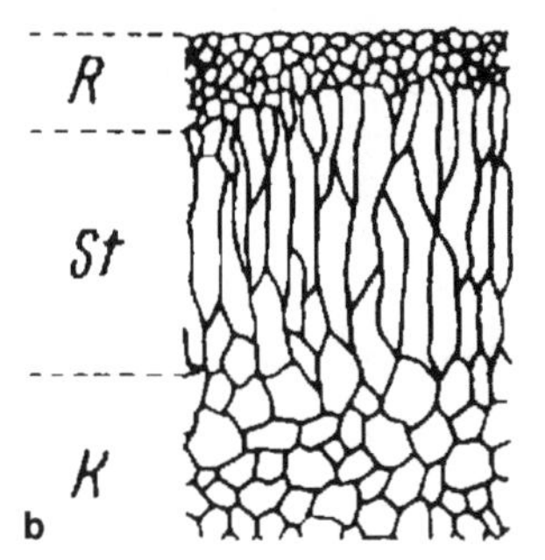

Fig. 4.5: (a) Cross-sections through cast, commercial purity aluminum ingots (DC cast in rolling ingot shape). The grain structure has been revealed through etching. Columnar crystals (grains) can be seen in the outer zones, especially in the upper ingot. The columnar grains grow in a direction opposite to the removal of heat. (b) Composition of the cast structure near the ingot surface. R = narrow exterior band of fine crystals, due to rapid cooling of the surface of the casting; St = zone of columnar grains, with their axes parallel to the heat flow; K = grain zone without directional cooling—"equiaxed grain structure."

4.5 Structure Formation in the Presence of Foreign Atoms

Our considerations up to now have been limited to aluminum of highest purity in which almost only aluminum atoms were present. Normally, commercial purity aluminum contains a few tenths percent iron and silicon or different alloying elements. This brings up the question of what happens to the foreign elements during solidification.

In the melt, the situation is simple. The foreign atoms that are soluble, according to the phase diagram, move in and out between the aluminum atoms without special order. However, in solidified metal one must differentiate between homogeneous and heterogeneous structures depending on the arrangement of the foreign atoms.

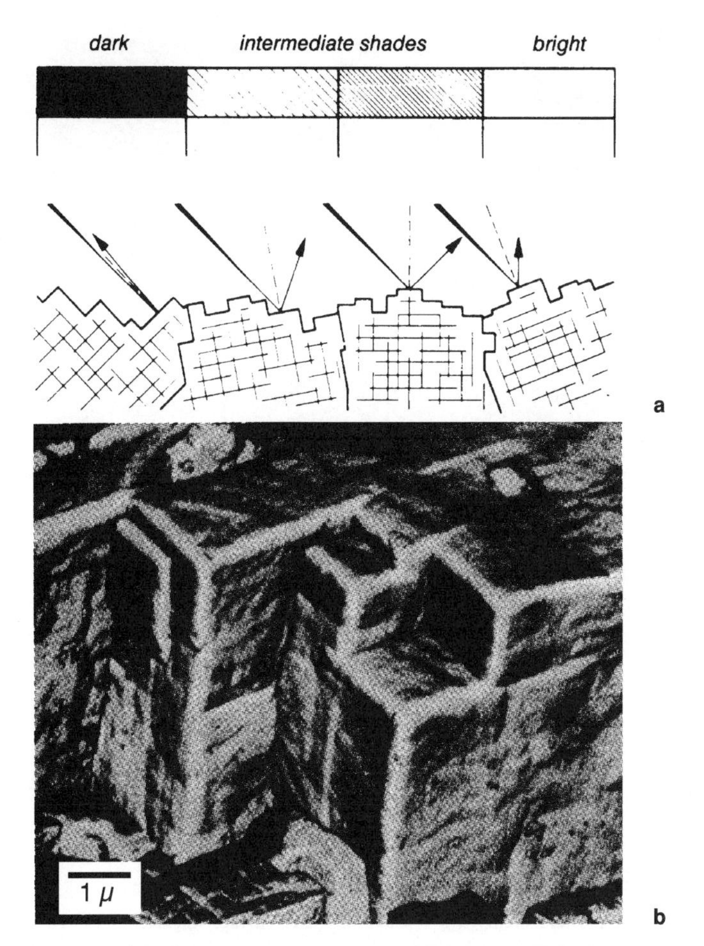

Fig. 4.6: (a) Reflection of incident light by etched grains. The observer sees from above "steps" in the crystal structure revealed by etching. In the plane of observation, each slope in the crystal surface reflects more or less light into the eye of the observer. The crystal on the far left reflects the light directly back at the source and, therefore, seems dark. Whereas, the crystals on the far right reflect the light directly in the eye of the observer and appears bright. The schematic section, perpendicular to the etched surface, shows four crystals with their crystallographic planes parallel to the page. Angle of incident light: 45° (M. Schenck). (b) Electron micrograph of an etched aluminum surface (99.99% Al). One can see that the etchant removed layers of atoms, one after the other, preferentially from crystallographic planes parallel to the faces of the elementary cube.

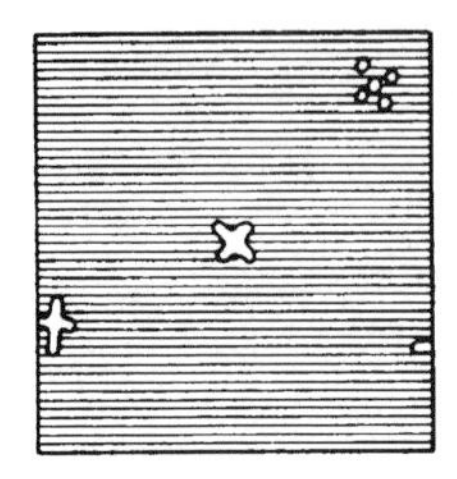
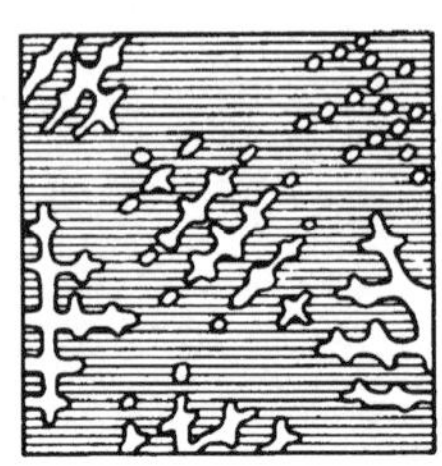
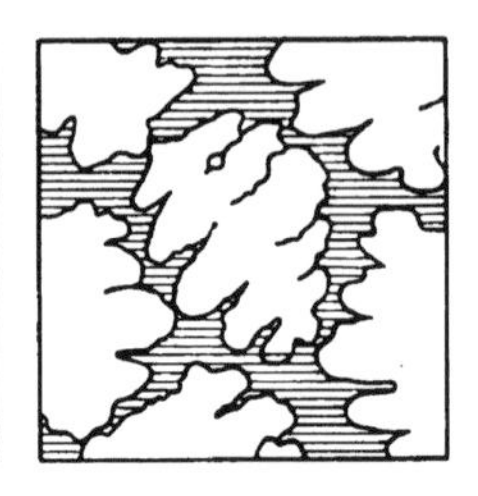
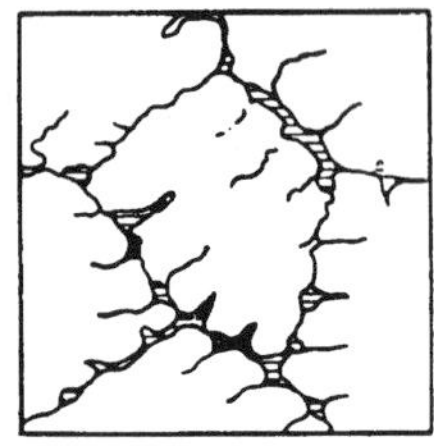

Fig. 4.7: Schematic representation of dendritic crystallization of a pure metal. (G. Sachs)

4.5.1 Homogeneous structure

A homogeneous structure is characterized by the fact that the atoms, whether host or foreign atoms, are uniformly mixed with each other even in the smallest crystals. Such a structure is to be found in super-purity aluminum (over 99.99% Al). If one etches a cross-section of super-purity aluminum, a faint line can be seen at the grain boundary since the etchant takes the metal atoms into solution more easily at lattice imperfections present at the grain boundaries (Fig. 4.11).

If one adds a metal that is soluble in the aluminum to super-purity aluminum—for example, 2% magnesium—the picture does not change much. The structure is still homogeneous since the magnesium atoms and the aluminum atoms have formed a solid solution that appears to be of one crystal type in the microsection. The individual atoms cannot be seen due to their small size so that even observation with a microscope cannot resolve the atoms in solution. The formation of solid solutions from the melt can be compared to the solidification of colored water in which the colored particles freeze in their random position in the ice.

Two different types of solid solutions are illustrated in Fig. 4.8.b and Fig. 4.8.c. In substitutional solid solutions (Fig. 4.8b), the aluminum atoms in the lattice are replaced by foreign atoms without basically changing the lattice structure. Substitutional solid solutions occur preferably if the foreign atoms have a similar diameter to the aluminum atoms. In addition to this, the chemical affinity as well as valence of the metal plays a part. Examples of this type are copper, silicon, or magnesium in solution in aluminum.

Another type of solid solution is the interstitial solid solution (Fig. 4.8.c) in which the foreign atoms are located in the interstices formed by the aluminum atoms in their normal lattice position. One of the requirements for this arrangement is that the foreign atoms have a very small diameter. The only really important example of this type in aluminum technology is the solubility of hydrogen atoms in aluminum.

4.5.2 Heterogeneous structure

Many metals are not soluble at all in solid aluminum, while others are soluble only in trace amounts. For example, iron or titanium are soluble only to about 0.03% maximum. In a melt with a temperature above 700°C, these atoms may be intimately mixed with the aluminum atoms, but in the instant when solidification starts they disturb the lattice

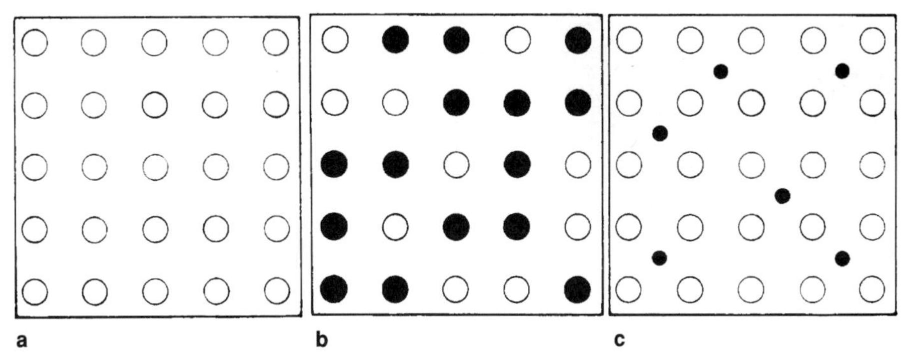

Fig. 4.8: Atomic arrangement in a homogeneous structure (schematic). (a) pure metal, (b) substitutional solid solution (example: copper or magnesium dissolved in aluminum), and (c) interstitial solid solution (example: hydrogen dissolved in aluminum).

structure of the aluminum and are rejected by the aluminum atoms, so to speak. The foreign atoms (for example, iron atoms) form a separate crystal with aluminum atoms in which three aluminum atoms join together with each iron atom. This precipitated intermetallic compound has the composition $FeAl_3$. The atomic arrangement of a heterogeneous alloy is shown schematically in Fig. 4.9. In other cases, the foreign elements precipitate alone (silicon), or two foreign elements precipitate together as is the case with magnesium and silicon.

In a heterogeneous structure, small foreign crystals are located within and between the aluminum grains. The origin of a heterogeneous structure may be compared to the crystallization of a concentrated solution of salt in water. As is known, the solution freezes below 0°C, which means that the freezing point is lowered by adding salt to water. The same is true in the case of aluminum with the addition of other metals. For example, an alloy of aluminum with 5% magnesium solidifies at less than 580°C instead of the 660°C at which super-purity aluminum solidifies. This reduction in the solidification point holds true for all common aluminum alloys whether they are homogeneous or heterogeneous.

The cast structure of commercial aluminum alloys usually consists of solid solutions with embedded heterogeneities so that both crystal types almost always exist together. Fig. 4.10 shows a heterogeneous structure, Fig. 4.11 shows a predominately homogeneous structure.

4.6 A Closer Look at the Cast Structure

4.6.1 Thermal analysis

In a thermal analysis, the temperature is measured as a function of time during cooling of the melt up to the point of complete solidification. This provides information on the progress of solidification and the solubility of different elements in aluminum. With the aid of supplementary metallographic and physical testing methods, this information is used to construct "phase or equilibrium diagrams."

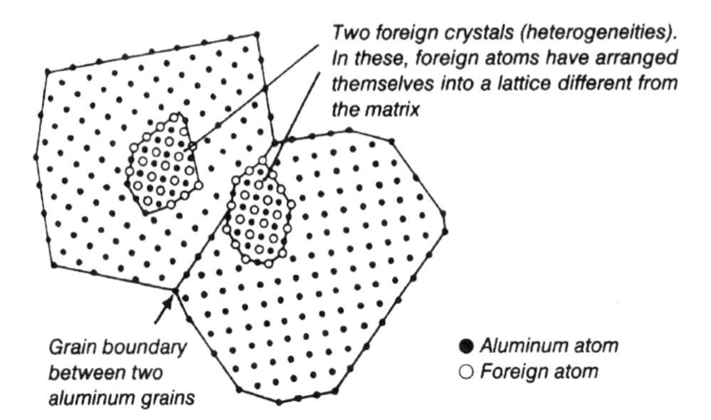

Fig. 4.9: Atomic arrangement in a heterogeneous structure (schematic).

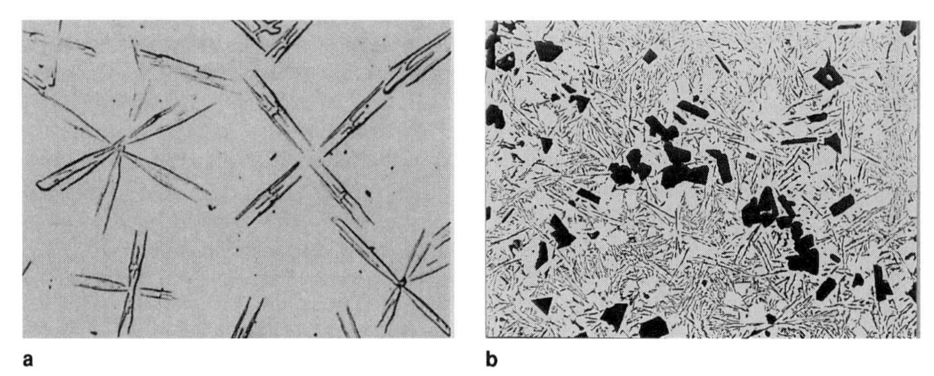

Fig. 4.10: (a) Cross-section through an aluminum-1% titanium alloy. A second, titanium-rich crystal is embedded in the aluminum matrix giving a typical heterogeneous structure. Microsection through an aluminum-titanium hardener (Alusuisse). (b) Casting alloy with 18% silicon and grain refining with phosphorus addition. The large precipitates are primary silicon crystals; the smaller are of the aluminum-silicon eutectic. (Alusuisse)

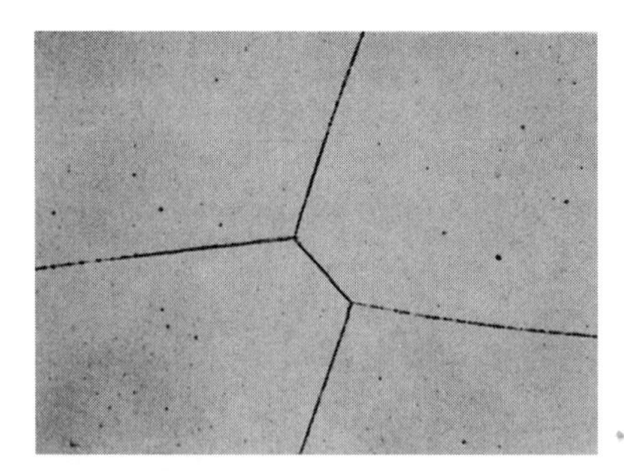

Fig. 4.11: Cross-section through a homogeneous cast structure, etched to show grain boundaries (99.99% Al).

4.6.1.1 *Thermal analysis of an alloy that solidifies homogeneously*

It has already been seen that alloying additions lower the melting point of aluminum significantly in commercial alloys. The crystallization of super-purity aluminum and an aluminum alloy differ especially in the temperature-time curve during the solidification process. Such cooling curves are simple to plot (Fig. 4.12). As an example, if super-purity aluminum is heated to 700°C, all the metal is liquid (A). The bath is allowed to cool and the temperature measured with a submerged thermocouple as a function of time. As soon as the bath has cooled to 660°C, solidification begins (B).

As the heat of solidification is released, there is a delay in the cooling. In the case of super-purity aluminum, the temperature remains absolutely constant from B to C at which time the melt has completely solidified. After that, the solidified metal continues to cool as shown in the C–G curve.

Conditions are different in alloys. Consider an alloy of 5% magnesium produced with a super-purity aluminum base (dashed curve in Fig. 4.12). Solidification begins at point D, which is lower than the solidification temperature of super-purity aluminum. The solidification of an alloy takes place over a temperature range. How broad this freezing range is depends entirely on the alloy in question. The temperature at which solidification begins is called the liquidus temperature, and the temperature at which solidification is complete is the solidus temperature.

Unalloyed aluminum solidifies in such a manner that there is an abrupt transition from liquid to solid. In most alloys there is a mushy zone between the solid structure and the liquid metal. The thickness of this zone is determined by the solidification range (Fig. 4.13). The eutectic alloys are an exception. (See Section 4.6.8 for a discussion of eutectic alloys.)

4.6.1.2 *Equilibrium Diagrams*

In aluminum alloys, the range of solidification is dependent on the type and amount of the foreign atoms. The addition of different alloying elements has a widely varying effect on the solidus and liquidus temperatures. This variation can be seen by viewing an equi-

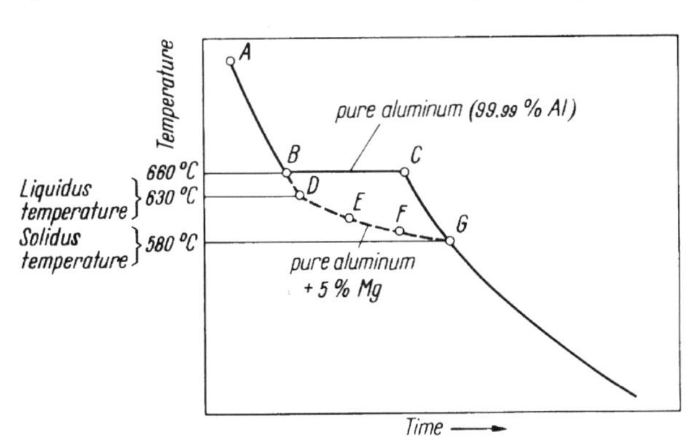

Fig. 4.12: Cooling curve for pure aluminum and an alloy containing 5% magnesium. Line B–C = solidification of pure aluminum. Curve D–G = solidification of the 5% magnesium alloy, for which the liquidus and solidus temperature is given.

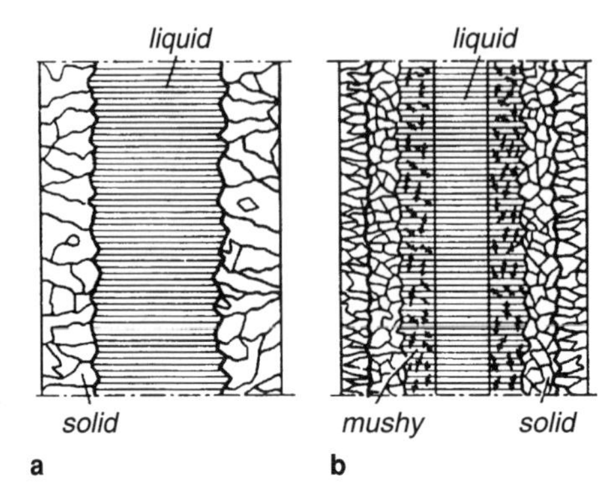

Fig. 4.13: Two types of solidification fronts: (a) Pure aluminum (99.99% Al), or eutectic alloy, and (b) alloy with solidification range.

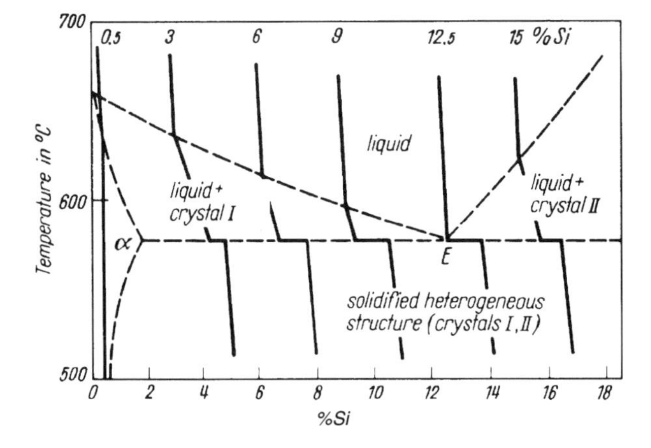

Fig. 4.14: Equilibrium diagram constructed from cooling curves for the aluminum-silicon system. E = eutectic point. Crystal type I = aluminum-rich solid solution with Si atoms in solution ("α-solid solution"). Crystal type II = silicon crystals (containing small amounts of Al atoms in solution). α = homogeneous solid solution (crystal type I).

librium diagram obtained with the aid of numerous cooling curves. In Fig. 4.14, the aluminum-silicon system shows schematically how the solidus and liquidus points—and, subsequently, the equilibrium diagram—are obtained from the temperatures at which a delay or a change in slope occurs in the cooling curve. It can be seen that, with increasing silicon content, the liquidus temperature decreases to a point and then starts increasing, while the solidus temperature remains constant. It can be seen from the equilibrium diagram that a maximum of 1.65% Si is soluble in aluminum at 577°C and that this solubility of silicon decreases in the solid phase with decreasing temperature.

The equilibrium diagrams are valid for equilibrium conditions (i.e., for crystals that have uniform composition throughout the entire cross-section). Such an ideal formation of the crystals is not obtained by commercial casting techniques since the movement of the atoms is not sufficient to form, in the short time available, crystals that have a uniform concentration of foreign atoms throughout the cross-section. This leads to the subject of "grain segregation" (Fig. 4.15 and Fig. 4.16).

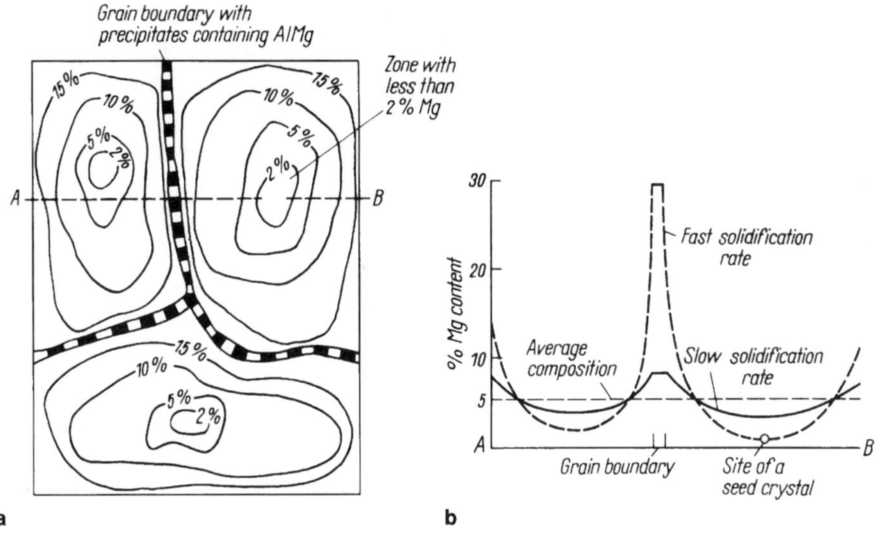

Fig. 4.15: (a) Section through the cast structure of an aluminum-5% magnesium alloy with grain segregation. The rings show areas of equal magnesium concentration. The center of crystallization lies in the middle of the innermost zone (schematic). (b) Influence of solidification rate or homogenization on grain segregation. Dashed line = A–B in Fig. 4.15.a. Solid line = magnesium distribution in the same alloy when the solidification rate is slower (example: permanent mold or sand casting) or after a rapidly solidified structure has been given a homogenization treatment to equalize the distribution of foreign atoms.

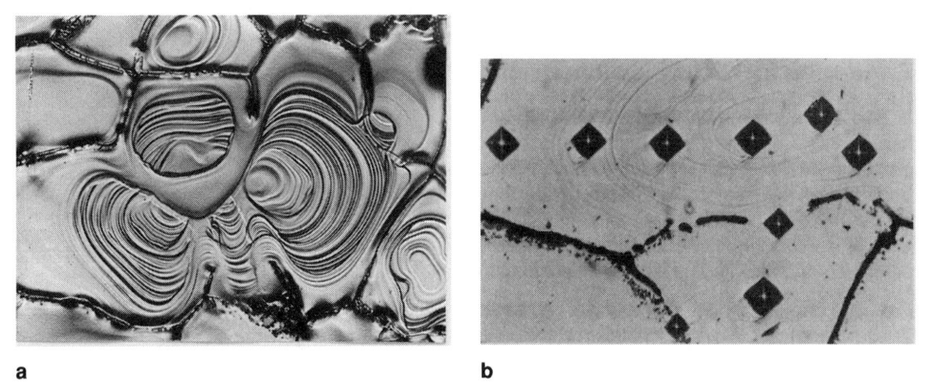

Fig. 4.16: (a) Coring in 99.5%-purity aluminum. Micrograph through a DC cast ingot. Residual melt veins (black) can be seen around the dendrite arms. Coring actually indicates a short interruption of solidification (Kostron). (b) Grain segregation in the region of neighboring dendrite arms made evident through microhardness tests. The size of the impression is a measure of the concentration of dissolved foreign atoms. Note: Two neighboring dendrite arms ("cells") are partially divided by residual melt veins and partially by a homogeneous structure area with a high concentration of foreign atoms.

4.6.2 Grain segregation

For the present, the discussion will be limited to relationships in the cast structure. The relationships obtained in the newly solidified structure are largely maintained since the alloy is usually cooled from the solidus temperature to room temperature as rapidly as possible. Such rapid cooling suppresses structure changes that could occur if the material in the solid phase were held for longer periods of time at elevated temperatures, for instance, above 400°C.

If the solidification of an alloy is investigated, the solidifying crystals have a different composition than the melt from which they are solidifying. If, for example, an alloy of aluminum and magnesium is observed, the phase diagram shows that 17.4% Mg is soluble in aluminum at the eutectic temperature of 449°C, as seen in Fig. 4.17.[1]

Consider more closely the formation of the cast structure of an aluminum alloy containing 5% magnesium. As the crystals begin to grow, they contain only approximately 1% magnesium. During the further process of solidification, the melt is enriched in magnesium, which steadily increases the magnesium content in the crystal as it grows.

In Fig. 4.12, the crystals at point D have an average of only 1% Mg; at point E, 3%; and at point F, approximately 5% Mg. The material left to be solidified at the grain boundaries (point G) can have a significantly higher magnesium content, even greater than the solubility limit even though that limit is quite high.

This non-uniform distribution of the alloying elements in the structure is called grain segregation. This is represented in Fig. 4.15.a. One can see that at the grain boundaries, a second type of crystal that is rich in magnesium forms so that there is a heterogeneous structure at that point. The extent of the grain segregation depends mainly on the rate of solidification. The slower an alloy solidifies, the smaller is the grain segregation.

The influence of the solidification rate or an homogenization treatment is represented in Fig. 4.15.b. The grains containing segregation show "coring," similar to the growth rings in a cut tree, since the content of foreign atoms changes in concentric rings (Fig. 4.15.a and Fig. 4.16).

Even in the solid state, the atoms have a great deal of mobility at high temperatures. That means that at sufficiently high temperatures, the atoms can change places in the solid structure. This process is known as diffusion. This means the grain segregation can be eliminated, thereby achieving a uniform distribution of the foreign atoms in the structure. This process requires a certain amount of time. If the newly solidified structure cools very slowly, or if an alloy in the solid state is annealed long enough at temperatures just under the solidus temperature, the variations in the content of the foreign atoms within a grain may be entirely eliminated.

[1] The source for this diagram is Hansen and Anderko. Other investigators, Willey, Dix, and Keller, report the solubility of magnesium in aluminum as 14.9% at a eutectic temperature of 451°C.

4.6.3 "Non-equilibrium" in commercial alloys

In Fig. 4.17.a, the appearance of a second magnesium-rich crystal type in the structure of aluminum with 7% magnesium caused by grain segregation can be clearly seen. This crystal type melts at 449°C. Such a phenomenon makes grain segregation very significant for commercial alloys since homogeneous, segregation-free alloys begin to melt at temperatures about 100°C higher. (See the right side of Fig. 4.17.a). In the copper-aluminum system, the maximum solubility of copper in aluminum is 5.7% at 547°C, nevertheless a second copper-rich crystal type appears already at 1% Cu (Fig. 4.17.b).

In commercial casting processes, solidification rates and cooling of the solidified structure are closely tied together by the rapid heat removal per unit of time. Especially in direct chill or strip casting techniques the solidification and the cooling after solidification is so rapid that, in addition to the grain segregation, a considerable supersaturation

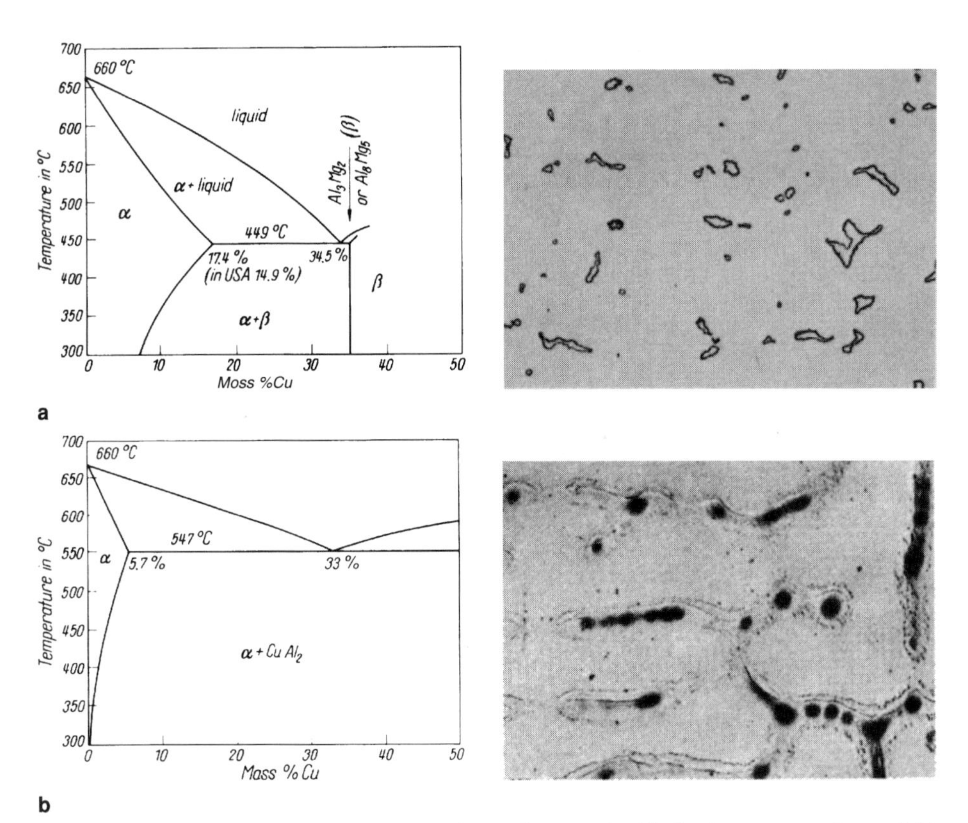

Fig. 4.17: Grain segregation images and equilibrium diagrams for (a) aluminum-magnesium and (b) aluminum-copper. The micrograph of Fig. 4.17.a shows the cast structure of an aluminum-1% magnesium alloy. Although the solubility limit lies at 17.4% magnesium, magnesium-containing inclusions appear, due to strong grain segregation. (Hanemann-Schrader) The micrograph of Fig. 4.17.b shows the cast structure of an aluminum-1% copper alloy. A heterogeneous structure through grain segregation, in equilibrium, 5.7% Cu is soluble in aluminum. (Hanemann-Schrader)

of the alloying elements occurs. According to the phase diagram, a maximum of 1.8% manganese is soluble in aluminum, but with rapid cooling (e.g., splat cooling), 9.2% manganese may be retained in solution. Of this amount, 7.8% is in supersaturated solution at the solidus temperature. During homogenization of the cast structure, the supersaturation is removed to a great extent by precipitation processes. Direct chill cast material is often fabricated without homogenization. Therefore, due to the history of the cast structure, one must expect a "non-equilibrium state" in semifinished material because of grain segregation, supersaturation, or precipitation of alloying elements. The equilibrium conditions will be described first.

4.6.4 Solidification under equilibrium conditions—the Lever Rule

With the aid of the equilibrium diagram, it is possible to follow the solidification of a homogeneous alloy in Fig. 4.18.a. It is assumed that the solidification takes place so slowly that equilibrium conditions can be maintained during solidification. With a melt composition S_1, the first crystal to form at the start of solidification has the composition K_1, which is poorer in the alloyed metal than the average composition. The concentration of the melt continues to move to the right in the process of crystallization (from S_1 to S_4).

When the alloy reaches temperature A, a melt with composition S_2 is in equilibrium with crystals of composition K_2. The amount of each phase can be read from the lever rule in the phase diagram. The ratio of the length of the line K_2A to the length of the line K_2S_2 is the same as the ratio of the liquid to the solidified portion of the melt. Based on the lever rule, 80% of the alloy is solidified at temperature B.[2] The crystals of composition K_3 are in equilibrium with the remaining 20% liquid, which has the composition S_3. At point K_4, the line of the average composition of the alloy crosses the solidus line of the phase diagram. At this temperature, the remainder of the melt solidifies and the resulting crystals have a composition corresponding to the average for the alloy, namely K_4. These crystals are a solid solution with the alloy metal dissolved in the crystal lattice of the base metal. The aluminum-rich solid solution, which solidifies first, is labeled with the Greek letter alpha (α), as shown in Fig. 4.17.a. The next solid solution is called the beta (β) phase. It has a different crystal structure. Normally, the solid solution that exists at the highest temperature is called alpha and the next solid solution ranges are called beta, gamma, delta, etc. in order of lower equilibrium temperature. However, this rule is not always followed.

Consider more closely from Fig. 4.18 the formation of the cast structure of an aluminum alloy. As the crystals begin to grow, they contain only a small amount of the alloying element. During the process of further solidification, the melt is enriched in the alloying element that steadily increases the alloy content in the crystal as it continues to grow.

It represents quite a mental exercise to imagine how a crystal zone of the composition K_4 can form from the last liquid component of the melt of the composition S_4. This may be explained by the fact that in attaining equilibrium during solidification, the foreign atoms at the surface of the newly solidified crystals may diffuse into the center of the crys-

[2] Percent of solid phase + $(BS_3/K_3S_3) \times 100$; percent of liquid phase = $(K_3B/K_3S_3) \times 100$.

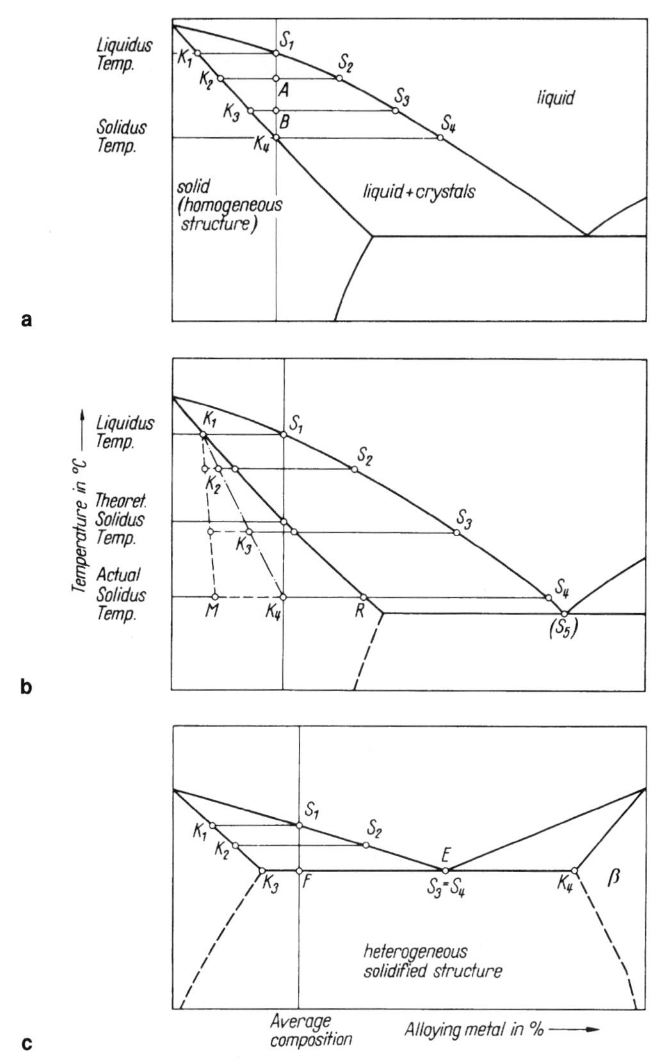

Fig. 4.18: Observing crystallization with the aid of an equilibrium diagram. (a) Solidification of a homogeneous alloy under equilibrium condition. (b) Solidification of a homogeneous alloy under non-equilibrium conditions. (c) Solidification of a heterogeneous alloy under equilibrium conditions. The β crystals are solid solution crystals, rich in the alloying metal.

tals where there is a deficiency of such atoms. From this, the significance of the equalization of the concentration during and after solidification can be seen.

4.6.5 Non-equilibrium in hypoeutectic alloys

Equilibrium conditions are not attained in commercial casting techniques as mentioned before. The homogeneous alloy outlined in Fig. 4.18.a was crystallized with a relatively slow solidification rate of 5–10 cm / min, which corresponds to the conditions during DC casting to a certain extent.

Based on Fig. 4.18.b there is not enough time to achieve equilibrium through diffusion of the atoms. Because of this, melt compositions and crystals (S_3, S_4, K_2, K_4) are formed that are not even possible according to the equilibrium diagram. The composition of the center of the crystal changes very little during solidification because there is not enough time available to allow the foreign atoms to reach the center of each crystal by diffusion (Curve K_1–M). The growth zone, which is the outer shell of the crystal, is correspondingly enriched during solidification (Curve K_1–R) with alloying elements that greatly exceed the average concentration of the alloy. It is only after the average (Curve K_1–K_4) between the concentration in the central and the outer edge of the crystal has reached the composition of the alloy (at K_4) that the solidification at a temperature lower than the theoretical solidus temperature is complete.

If the crystal segregation is increased somewhat by more rapid cooling, a melt of composition S_5 with a heterogeneous structure would subsequently be formed. See Fig. 4.17 for a practical example. A heterogeneous structure would also result if the addition of the alloying metal were increased in Fig. 4.18.a or Fig. 4.18.b. This leads to a discussion of the solidification mechanism of a heterogeneous alloy.

4.6.6 Solidification of alloys with heterogeneous structures

The term "heterogeneous alloys" is used here to designate those in which the content of foreign atoms is greater than the maximum solubility in the solid state as shown by the equilibrium diagram. With heterogeneous alloys, in every case an additional crystal type besides the aluminum matrix occurs, while in "homogeneous alloys" formation of a second crystal is dependent on the degree of crystal segregation, supersaturation of the alloying elements, and the heat treatment.

In this book, the expression crystal segregation is often used as an overall term, but actually it can be divided into two parts. It can be seen from Fig. 4.18 that a more dilute alloy (hypoeutectic) with deviating composition splits off during solidification from the original melt. Both change their composition as solidification proceeds. In the strictest sense, only the changes in concentration of a crystal are to be designated as crystal segregation while the enrichment of the alloying elements in the residual liquid is partially an independent process that is influenced indirectly by grain segregation. The solidification of the residual liquid results in heterogeneous particles in the structure, which are not in agreement with the phase diagram, due to conditions of rapid solidification (Fig. 4.17). The expression "crystal segregation" is used to describe the entire process.

In order to determine solubility of silicon in aluminum, a series of alloys with increasing silicon content was produced. In segregation-free microsections, which have been annealed for long times at temperatures just under the solidus line and then cooled rapidly, it can be recognized that if the silicon content is above 1.65%, small silicon-containing embedded foreign crystals form for the first time. This indicates that the maximum solubility of silicon in aluminum is 1.65%, as shown in Fig. 4.14. The heterogeneous alloys always have at least two recognizable crystal types when observed under a microscope, as shown in Fig. 4.10. The amount of embedded second phase in the cast structure is further increased by crystal segregation.

Finally, it must be remembered that most alloys are based on commercial purity aluminum, which always solidifies with precipitates containing iron and silicon. The iron content of commercial purity aluminum exceeds the solubility limit significantly. In general, the small iron and silicon contents are ignored in the classification of an alloy. For example, an alloy of 5% magnesium in aluminum is classified as a homogeneous alloy even though there are small precipitates containing silicon and iron, and, if crystal segregation is sufficient, there are also magnesium-rich particles.

4.6.7 Equilibrium diagram for a heterogeneous alloy

The solidification of a heterogeneous alloy under equilibrium conditions may be followed with the aid of Fig. 4.18.c. At the beginning of the solidification of the melt S_1, an α-solid solution crystal K_1 forms, which is poor in the alloying element. Such a crystal is called a primary crystal. In this case, the solidified structure is still homogeneous when half of the melt is solid (composition K_2). If the melt has attained composition S_3, two different crystals—K_3 and K_4, which are significantly different in composition—form simultaneously. In the present example, crystal K_4 is the p-solid solution that contains considerably more of the alloying metal and has a different crystal structure than the α-solid solution. In this way, a "heterogeneous structure" is formed. In this case, it is called eutectic. Since the composition of the melt lies between the two crystal types K_3 and K_4, the "eutectic" composition of the melt does not change any more ($S_3 = S_4$) during the course of the solidification.

The alloy depicted in Fig. 4.18.c with an average composition that lies to the left of the eutectic point E is called a "hypoeutectic" alloy. If the composition is the same as point E, an alloy is designated eutectic; if the content of the alloying element is higher than E, it is called "hypereutectic."

4.6.8 What is eutectic?

In order to answer this question, it is necessary to consider the cooling characteristics of a heterogeneous alloy in detail. One such example is an alloy of 5% silicon in aluminum, since the solubility limit of 1.65% Si in the solid state is exceeded. This approximates the composition shown in Fig. 4.18.c.

The temperature of the melt as a function of time is measured during slow cooling. The results are shown schematically in Fig. 4.19. At time A, the crystallization of an aluminum-rich primary crystal (α-solid solution) begins.

From A to B to C, the silicon content of the α-solid solution increases constantly in agreement with the phase diagram. At C, the growth zone is so highly enriched with silicon that the solubility limit is exceeded. Silicon crystals begin to precipitate. This means that from time C, two different crystal types form that, in total, have the same average composition as the melt from which they form. If, during solidification, a second crystal type occurs along with the α-solid solution there are three possibilities. The second crystal type may be a β-crystal in Fig. 4.18.c, an intermetallic compound (for example, $CuAl_2$ as in Fig. 4.17.b) or crystals of the alloying metal. The latter case occurs in the aluminum-silicon system. A closer look shows that in the solidification of aluminum-silicon alloys, the silicon crystals that form contain a very small number of aluminum atoms in solution, which means that they could also be identified as a β-solid solution. The concentration of the liquid and solid phases does not change from time C on.

The mixture of two crystal types that solidify simultaneously is called "eutectic" (from the Greek word *eutektos*, which means easily melted or fused). The prevailing temperature at time C is called the eutectic temperature. The figure shows that the temperature remains constant to time D (i.e., until the metal is completely solidified). After that, the normal cooling of the solidified cast structure takes places.

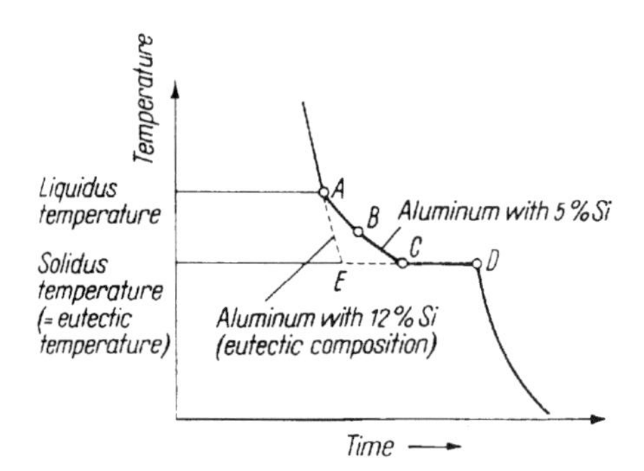

Fig. 4.19: Solidification curves for aluminum with 5% and 12% silicon content (compare Fig. 4.14). A–C: Beginning of crystallization. Growth of aluminum-rich primal crystals (α-solid solution). C–D: Simultaneous growth of two crystal types, formation of a eutectic network around the primary crystals. E–D: Pure eutectic crystallization. D: End of crystallization.

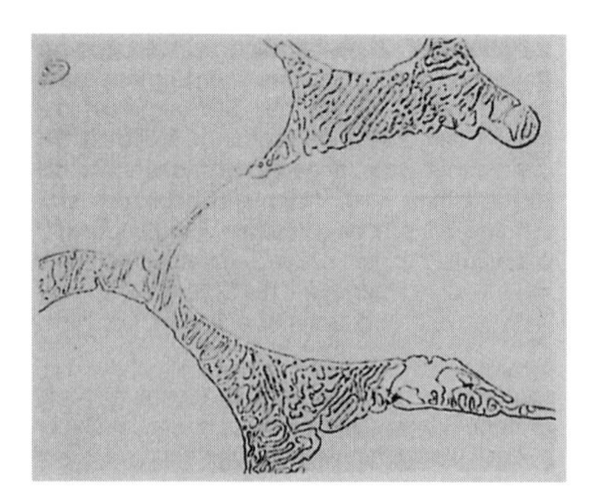

Fig. 4.20: Structure of a hypoeutectic alloy. An aluminum-10% copper alloy, eutectic around larger aluminum-rich primary crystals (α-solid solution). (Hanemann-Schrader)

In a micrograph, a eutectic structure is very often fine grained. This can be demonstrated by the structure of an alloy of aluminum and 10% copper as presented in Fig. 4.20. In this "hypoeutectic" alloy, the volume percent of the eutectic is relatively small. The eutectic consists of small elongated $CuAl_2$ crystals enveloped in an α-solid solution. The $CuAl_2$ crystals may occur as primary crystals or in the eutectic. In Fig. 4.21, micrographs of the aluminum-silicon system are presented. (See Fig. 4.14 for the aluminum-silicon equilibrium diagram.)

As shown in Fig. 4.21, the proportion of aluminum-rich primary crystals in the structure becomes less with increasing silicon content. At silicon contents above around 12%, silicon crystals form as primary crystals that are embedded in the eutectic structure.

As can be seen from Fig. 4.14, the liquidus temperature and the temperature range over which solidification takes place decreases continuously with increasing silicon content. The eutectic composition is attained at 12.5% Si [i.e., from the beginning both crystal types—I (aluminum-rich α-solid solution) and II (silicon crystals)—crystallize at the same time]. This solidification takes place at a constant temperature of 575°C, which is the eutectic temperature (dashed curve in Fig. 4.14). The simultaneous crystallization of both crystal types without the formation of primary crystals leads to an especially fine-grained eutectic structure.

A purely eutectic structure has special commercial advantages for castings. But just by casting an alloy of eutectic composition, there is no guarantee that the desired fine-grained structure will be achieved. Often one crystal type dominates in the eutectic crystallization, which causes the formation of a "divorced" eutectic. The eutectic crystallization of alloys with approximately 12% or 13% Si can be greatly influenced by minute sodium additions. In this way, a "modified" eutectic structure is obtained (Fig. 4.22.a).

The processes involved in the modification through sodium additions have not been conclusively explained. Apparently, the sodium promotes a certain supercooling of the melt which leads to a fine-grained eutectic crystallization. The sodium addition shifts the eutectic point somewhat (to 13% Si). Fig. 4.22.b shows an unmodified alloy containing primary silicon crystals while the Fig. 4.22.a shows a modified alloy with no large primary silicon crystals.

In hypoeutectic or hypereutectic alloys, the eutectic almost completely surrounds the primary crystals, which solidify first (Fig. 4.20 and Fig. 4.21).

What has been said up to now concerning alloys of two metals is valid in principle for three or more alloying components. A ternary eutectic temperature can be considerably lower than the eutectic temperature of the binary systems of the metals involved.

4.7 Constitutional Undercooling[3]

As we have seen, solidification under non-equilibrium conditions means that there is insufficient time for diffusion to even out different concentrations of the alloying ele-

[3] In collaboration with K. Buxmann.

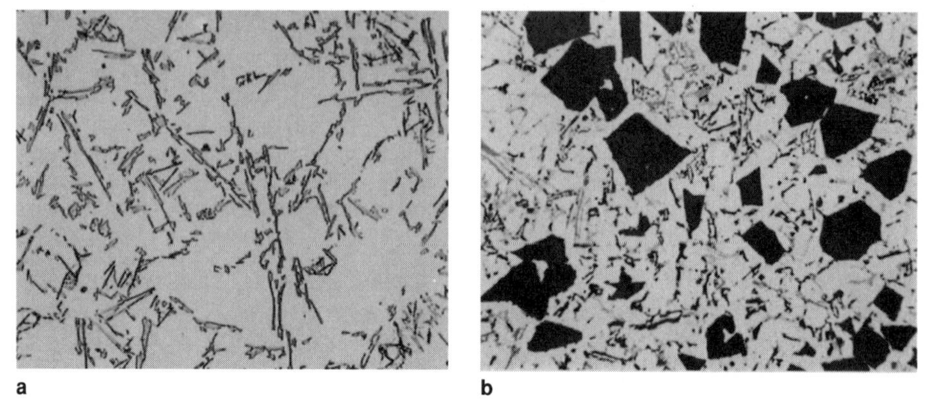

Fig. 4.21: Structure of sandcast aluminum-silicon alloys. (a) Aluminum alloy with 8% silicon hypoeutectic alloy with aluminum-rich primary crystals. (b) Aluminum alloy with 20% silicon hypereutectic alloy with primary silicon crystals. See Fig. 4.22 for eutectic alloy structure. (Alusuisse)

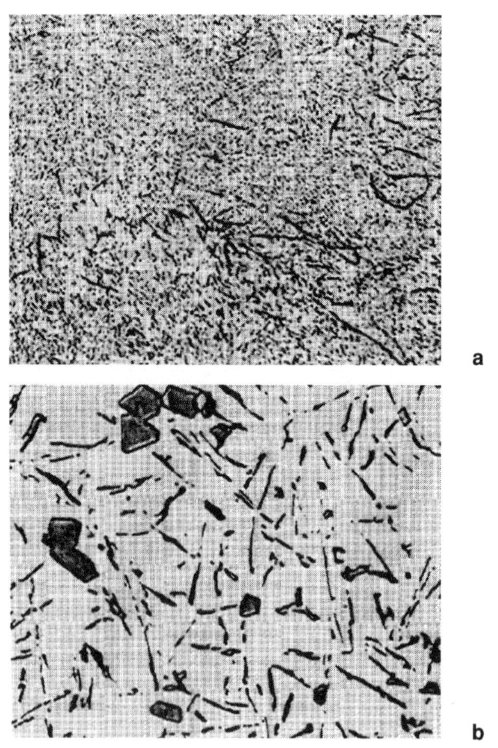

Fig. 4.22: Microstructure of an aluminum alloy with about 12% silicon. Eutectic alloy: (a) modified; (b) unmodified.

ments that develop within and between the liquid and the solid phases. Whereas such concentration differences in the solid phase can be easily identified by microprobe analysis as grain segregation in the cast structure, we can calculate the concentration of an alloying element in the melt immediately in front of the advancing solidification front provided that its diffusion constants are known. A key relationship is shown in the phase diagram Fig 4.18.b. If a concentration K_1 of a given alloying element is found in the solid phase immediately behind the solidification front, then the corresponding concentration in the melt before the front must be S_1 and similarly S_2 for K_2 and so on. The result of such a calculation is shown in Fig 4.23.

In a layer just a few microns thick immediately ahead of the solidification front, the amount of a particular alloying element can be enriched to several times its nominal level. At an advancing solidification front, the local temperature, T_q, must increase along a gradient from the solid to the liquid, as shown in Fig. 4.23. According to the phase diagram, the higher the concentration of an alloying element, the lower is the actual liquidus temperature, T_l. Fig. 4.23 plots both these temperatures against the distance from the solidification front. The shaded area between these two temperature curves reveals a zone ahead of the solidification front where the actual melt temperature is, in fact, lower than the liquidus temperature; this zone is undercooled to the extent of the difference T_l minus T_q.

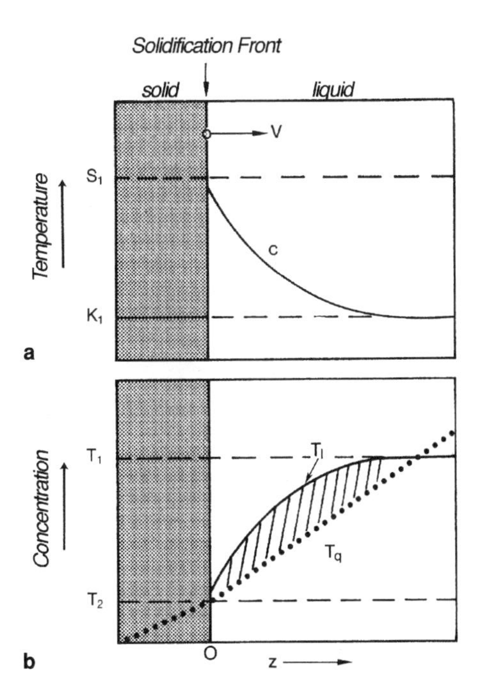

Fig 4.23: The concentration (c) of an alloying element and the temperature in the region of a uniform solidification front advancing at speed v in the solid and liquid phases (after Kurz & Fisher). The shaded area in Fig. 4.23.b shows the undercooled melt, its upper boundary being the liquidus temperature in accordance with curve c in Fig. 4.23.a. v = speed of solidification; c = concentration of the alloying element in solution; z = distance from the solidification front; T_1 = liquidus temperature; T_2 = solidus temperature; T_l = liquidus temperature according to concentration c; T_q = actual temperature.

4.7.1 Cellular and equiaxed solidification

As constitutional undercooling develops, a uniform solidification front becomes unstable. Should the enriched layer at a particular place be somewhat thinner because of some slight irregularity, then locally solidification can proceed more quickly, causing the solidification front to bulge out. The result is that the alloying atoms diffuse sideways, and the bulging accelerates until the front breaks through the undercooled zone. So, when a uniform solidification front grows out into the melt, a thin enriched layer is formed that causes constitutional undercooling, which, in turn, destabilizes the solidification front. Finger-shaped protuberances, called cells, develop and grow out into the melt. The alloying elements are pushed mainly into the spaces between neighboring cells, which explains why in a cast structure the cell boundaries show up by their higher content of alloying elements. If the content of alloying elements is high enough and the solidification proceeds relatively rapidly, cell head surfaces themselves become unstable and form

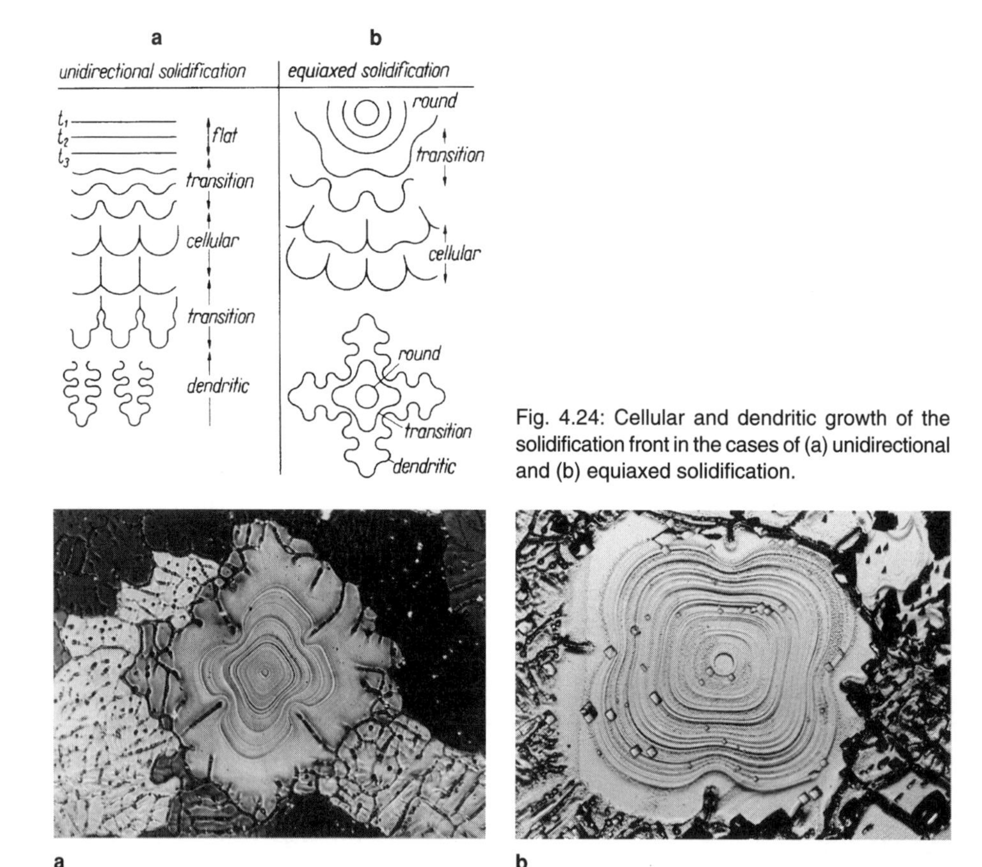

Fig. 4.24: Cellular and dendritic growth of the solidification front in the cases of (a) unidirectional and (b) equiaxed solidification.

Fig 4.25: Metallographic evidence of an early stage of cellular growth recorded in "growth rings" due to periodic variations in the concentration of foreign atoms in the constitutionally undercooled layer. (VAW)

branches called secondary dendrite arms. These usually take a fixed direction relative to the main axis of the dendrite (Fig. 4.25). Fig. 4.24 shows how the shape of the solidification front develops with increasingly marked constitutional undercooling.

Indigenous solidification means that crystals form freely floating in the open melt, as distinct from exogenous solidification, which proceeds from a cooled surface. Indigenous cellular solidification leads to a branched structure as shown schematically in Fig. 4.24.b. This cellular and dendritic growth also shows up in microsections of solidified samples. Cell size is an important quality criterion, because it determines how finely or coarsely the alloying constituents are distributed in the cell boundaries. The cells are defined by the irregular pockets of residual melt trapped between them. Cells within an as-cast grain may also be identified by the regularly spaced pattern they create in microsections. In cross-sections through dendrites, the cells can, therefore, be recognized either by their regular, branched patterns, or by the less regular pockets of residual melt trapped between them. Free-floating crystals can be observed in DC-cast structures of wrought alloys as well as in casting alloys. Fig. 4.25 well illustrates the growth of such free-floating crystals. Etching reveals concentric growth rings marking fluctuations in the content of alloying elements. These rings record periods of local undercooling, when melt temperature and composition allowed the floating crystal to grow. Fig. 4.25 illustrates the transition from early globulitic solidification to dendritic growth and the formation of cells according to Fig. 4.24.b. "Indigenous growth" results in a crystal that has roughly equal diameters in two main crystallographic directions and is, therefore, called an "equiaxed grain."

The photographs of Fig. 4.25 demonstrate the enormous difference between slowly growing free-floating crystals and the surrounding fine cells. The floating crystal grows slowly because it is out of contact with the main solidification front through which the heat extraction takes place. Local differences in heat extraction and cooling rate always create corresponding differences in cell size (Fig.4.27).

Constitutional undercooling occurs even in the solidification of DC-cast high-purity Al (99.99%), resulting in a cellular structure. However, at this level of purity, solidification nuclei are not able to grow in the constitutionally undercooled zone; hence, the grains grow in columnar form toward the center of the ingot. Al of 99.9%, on the other hand, produces enough constitutional undercooling so that suitable seeding of the melt with grain refiner can produce a completely globular or equiaxed grain structure. The nuclei form in the constitutionally undercooled zone and grow uniformly in all directions. Continuous cast aluminum and most aluminum alloys typically have a cellular structure, in which the branching of cells, unlike dendrites, is not characterized by a fixed angle (e.g., 90°). The following discussion will therefore be confined to cellular solidification.

4.7.2 Cell structure due to heterogeneous solidification

For hypoeutectic alloys (all conventional wrought alloys), the micro-segregation within grains is the result of alloying elements being depleted near the surface of the growing grain and finally accumulated at the grain boundary. We can now refine this description in the light of our understanding of cellular and dendritic structures. Only a small pro-

Fig. 4.26: Structure of 99.5% aluminum, DC cast with grain refiner. Grain size is smaller than 0.5 mm; cell size is from 50 microns to about 300 microns. The residual melt has solidified mainly in the cell boundaries. (VAW)

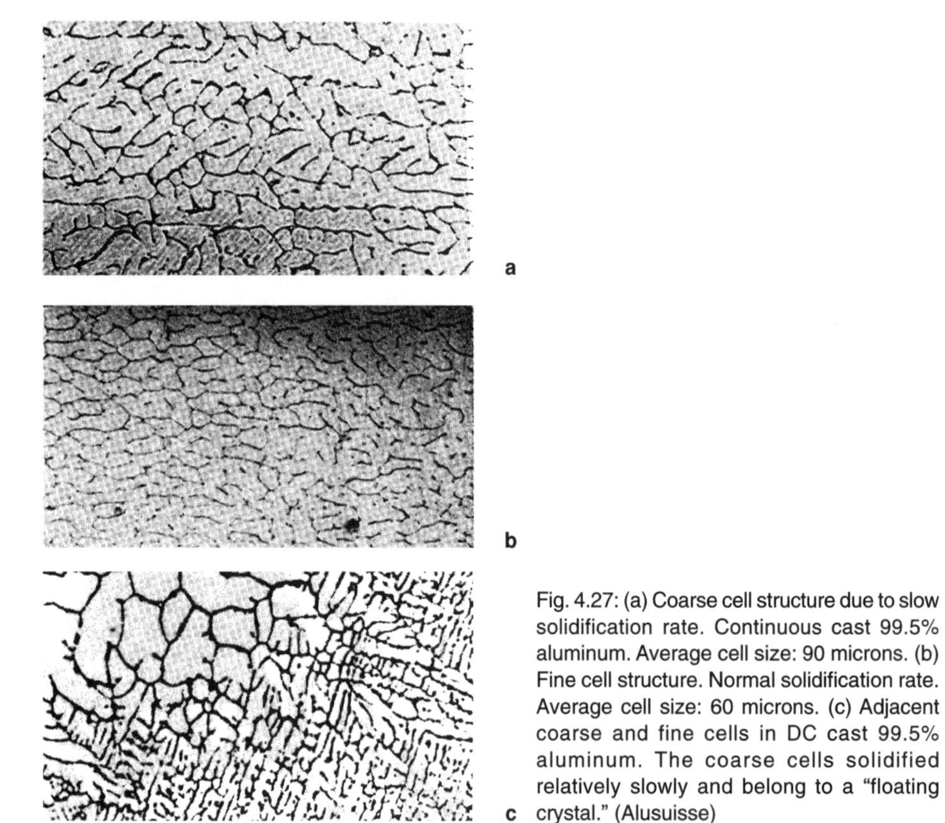

Fig. 4.27: (a) Coarse cell structure due to slow solidification rate. Continuous cast 99.5% aluminum. Average cell size: 90 microns. (b) Fine cell structure. Normal solidification rate. Average cell size: 60 microns. (c) Adjacent coarse and fine cells in DC cast 99.5% aluminum. The coarse cells solidified relatively slowly and belong to a "floating crystal." (Alusuisse)

portion of the alloying elements appear at the true grain boundary. Cells created by dendrite arms push aside most of the impurities and alloy elements, which then form an enriched zone between the cells. There, they become part of the residual melt, which solidifies last. When pure aluminum or an alloy contains more than 0.03% iron (which is the case with aluminum of 99.9% or lower purity), then the intercellular residual melt concentrates so far that it finally reaches the eutectic composition, and it solidifies with eutectic phases in the cell boundary. This explains why the globular cast grains shown in Fig. 4.26 are divided into cells. Various intermetallic phases mark the cell boundaries. More rapid solidification produces finer cells, and, conversely, slower solidification leads to coarser cells. This means that the intermetallic phases in the structure are also distributed more finely or more coarsely according to the solidification rate. Fig. 4.27.c shows in DC cast aluminum a progressive transition in cell-size from coarse at the top left, which could be indigenous (and, therefore, slow) solidification, to fine at the lower right.

Later we shall see that such a solidification pattern with its large variations in cell size can cause surface streakiness in anodized semifabricated products. A fine and uniform distribution of intermetallic phases in the cast structure, therefore, requires a fine and uniform cell structure. These phases from the cast structure are inherited in the extruded, rolled, or forged product, where fine and uniform intermetallic phases are often an important quality requirement. We can influence grain size and structure by later forming and heat treatment, but solidification conditions permanently define the distribution of many intermetallic phases. For this reason, later chapters will again refer to cell structure.

Chapter 5. Melt Quality and Treatment[1]

5.1 *Quality of the Aluminum Melt*

By definition, quality means a product's suitability to fulfill explicit or implicit requirements. Anyone who wants to supply high-quality aluminum must, on the one hand, understand its condition from the inside and, on the other hand, thoroughly understand the customer's requirements. Every process step in making and fabricating aluminum fixes some quality characteristic once and for all. For instance, freshly tapped liquid potline metal contains various dispersed particles and dissolved impurities that can be only partially removed by melt treatment in the casthouse. Certain microstructural properties of the product are fixed during solidification in the casthouse so that subsequent operations, such as heat treatment and working, can have only a limited effect on them. In contrast to some other metals, therefore, what happens in the casthouse has a major influence on the final quality of the product.

5.1.1 Impurities in the untreated melt

The quality of intermediate and end products is effected by unwanted impurities. These impurities occur mainly in melts from freshly tapped potroom metal or in secondary aluminum from old scrap.
We can distinguish three categories of impurities:

- Dissolved alkali metals (mainly sodium)
- Particles floating in the melt (oxide or insoluble intermetallic compounds such as titanium diboride)
- Dissolved hydrogen (mostly supersaturated)

We will deal with these three categories of impurities in turn, although several melt cleaning methods remove or reduce all three types of impurity.

5.1.2 Soluble impurities

Soluble impurities in potline metal are mainly substances introduced with the electrolysis feedstock: alumina and anodes. Dissolution of steel components can also significantly increase iron contamination. The standard purity of internationally traded aluminum ingot is 99.7%, determined mainly by the two chief impurities: iron (typical range, 0.1–0.2% Fe) and silicon (typical range, 0.05–0.1% Si). Special precautions in the smelter operation can raise purity up to 99.95%. These include, for example, using higher purity alumina and anodes and selecting metal from the best pots. Applications needing yet higher purities, electronics for example, require costly additional processes, usually in dedicated plants. These processes include three-layer electrolysis, gravity segregation during partial solidification, and zone refining techniques (See Chapter 2).

[1] By K. Buxmann.

As already mentioned in Chapter 2, certain elements like titanium and vanadium reduce the electrical conductivity of aluminum even at quite low levels of about 50 parts per million (ppm) (i.e., 0.005%). If such metal is to be used for conductors, then a melt treatment with boron can improve electrical conductivity at moderate cost. This generates highly insoluble borides TiB_2 and VB_2 that settle out of the melt.

Freshly tapped potline metal contains between 20 ppm and 80 ppm of sodium from the cryolite, and if the latter contains lithium, then the primary aluminum could also contain 2 ppm to 10 ppm of lithium. Both traces are extremely harmful for many semi-finished products, so that quality assurance directives usually require lowering Na and Li to about one tenth of the initial levels mentioned above.

5.1.3 Non-metallic insoluble impurities

5.1.3.1 Examining the melt for non-metallic inclusions

Melts from in-plant process scrap and from old scrap (secondary aluminum) are as a rule less pure, depending on the purity of the input metals and the condition and degree of contamination of the recycled products (see Chapter 2). When charging old scrap, it is particularly difficult to keep down contamination by iron, magnesium, copper, and zinc. Traces of lead and tin are also damaging in certain applications.

Non-metallic particles dispersed in the melt require particular attention. Often having a similar density to aluminum, these particles may not settle out, but become inclusions cast into the ingot, causing diverse problems when they reappear in subsequent processing and in the product. Metal quality relative to such dispersed particles can only be described accurately by distribution curves like Fig. 5.1, which show the size and frequency of each type of particle as determined by an intensive study.

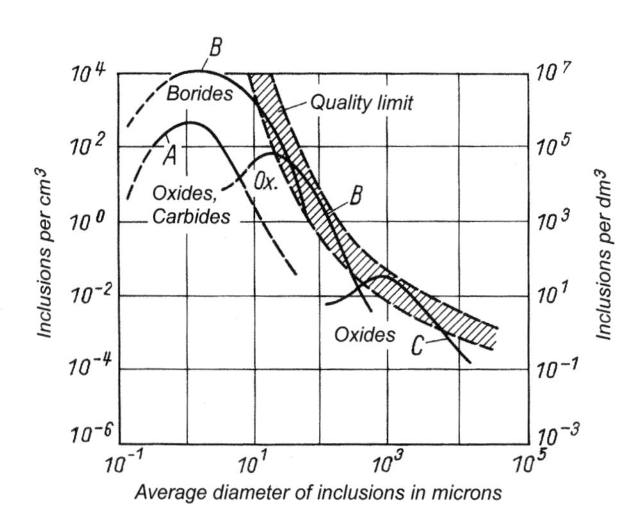

Fig. 5.1: A graphic for quantitatively describing inclusions in cast aluminum. A—inclusions from the electrolysis process; B—inclusions from melting and melt treatment including grain refinement; C—inclusions caused by casting, for example, by faulty melt transfer during DC casting.

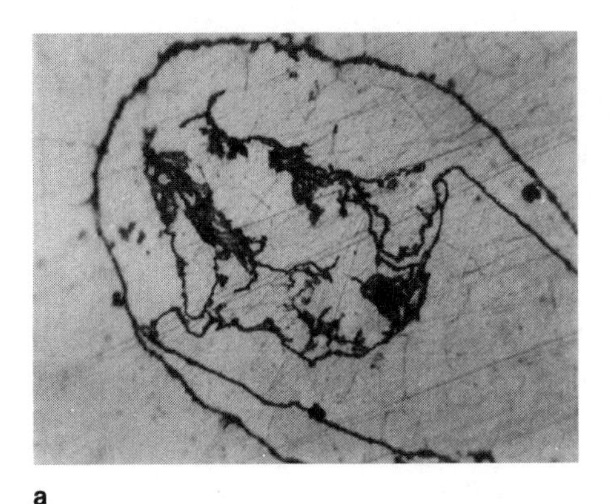

a

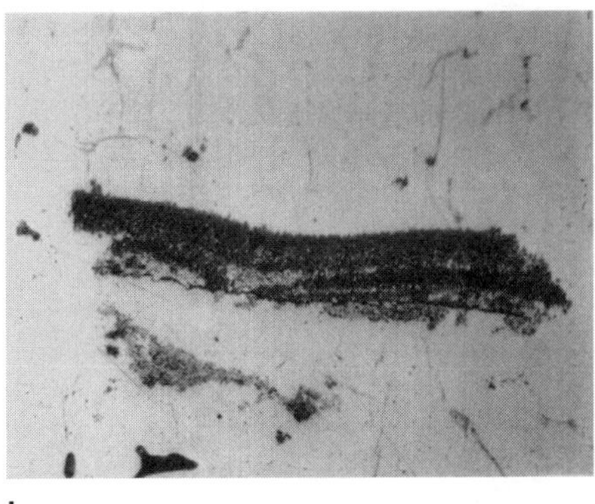

b

Fig. 5.2: Oxide inclusions in aluminum located by diamond milling and anodizing, then conventionally repolished to show detail (magnified about 400×). (a) An oxide skin inclusion and (b) an oxide flake inclusion—a stirred-in, furry oxide flake from the surface of an aluminum-magnesium alloy melt.

Freshly tapped potline metal generally contains dispersed particles that are mostly finely distributed in the form of oxides, borides, fluxes, and carbides. However, exceptional cases of large particles or clusters can occur, which are difficult to detect in practice. For instance, freshly remelted secondary metal from old scrap usually contains many oxide inclusions.

Even one inclusion of 0.1 mm diameter per cubic centimeter of aluminum can noticeably affect further processing or even the quality of the final product. However, it is difficult to test aluminum for such inclusions. To determine metallographically the shape, size, and distribution of inclusions in cast ingot requires preparing about 100 cm^2 of microsection and carefully polishing the surface, preferably by diamond milling. Subsequent anodizing reveals inclusions embedded in the transparent anodic film, where they are easy to see and evaluate. Fig. 5.2 shows oxide inclusions that have been located in this way.

The type, size, and distribution of inclusions in the melt can be determined by rapidly solidifying a melt sample on a copper block and then preparing a metallographic section from it as described above. There are some widely used methods of concentrating inclusions from the melt so as to evaluate them in a small microsection. The usual method is to squeeze the melt through a fine filter plate about 10 mm across and examine the filter cake. If it is important to find even the smallest inclusions, which normally escape through the filter plate, then they can also be sedimented by centrifuging a molten sample.

More recently, physical measurements have gained in importance for evaluating inclusions while the metal is still liquid. The most notable are ultrasonic inspection and electrical resistance measurements along capillary tubes through which the melt flows. Unfortunately, none of these methods is able to measure simultaneously all of the parameters needed for a full description of the inclusions. We must, therefore, use various different processes if we wish to obtain a comprehensive view of the diverse types of inclusions in a melt and their shape, size, and distribution.

5.1.3.2 Oxide skins and their properties

Oxides and other non-metallic inclusions in the structure of castings and wrought products constitute breaks in the metal matrix with all of their consequent disadvantages, especially for mechanical properties.

When liquid aluminum comes into contact with air it instantly forms an oxide skin that is normally much less than a micron thick. Yet this skin has extraordinary strength and a special type of toughness that can be described as superplastic. This can be seen as the oxide skin is stretched, for example, by causing a float to rotate on the surface of the melt in a laboratory crucible. On exceeding a critical deformation, the skin develops minute cracks, but heals them immediately without tearing widely or losing its resistance to further deformation. The oxide skin also shows this peculiar behavior when aluminum is poured slowly from a ladle. The aluminum flows under a compact skin that slightly tarnishes the liquid metal's metallic sheen. When a foundryman casts the metal into molds by hand, he or she practices pouring through this oxide sleeve, and takes care that it does not tear as the mold fills. If the metal flows much faster, the oxide skin appears to follow it in jerks. Thus, at first people mistakenly thought that aluminum had a high viscosity, whereas, in fact, it flows as easily as water. Similarly, the conspicuously rounded meniscus at the edge of a pool of aluminum reveals the oxide skin's resistance to stretching, which masks the liquid's own surface tension.

At temperatures below 750°C the oxide skin on liquid pure aluminum has an amorphous, non-crystalline structure (i.e., like glass). This extremely thin layer is so compact that oxygen from the air can hardly diffuse through it. What limits the reaction is the very low diffusion coefficient of oxygen in the oxide skin. Diffusion controls all oxidation processes where a barrier layer forms. Thus, liquid aluminum, unless atomized to minute droplets, cannot burn. Water vapor deserves specific attention, as it can react with liquid aluminum (see the formulae in the passage "Hydrogen content" in Section 5.1.4).

Certain impurities tend to concentrate in the oxide skin, especially alkali metals like sodium and lithium and earth alkali metals like magnesium. They make the layer much less compact, thus facilitating further oxidation and hydrogen penetration. Experienced foundrymen recognize melts containing too much sodium and lithium by their reduced metallic gloss as well as by their stiffer flow from a ladle, both of which correspond to a thicker, stiffer oxide skin.

Oxide layers on alloy melts can have a completely different composition and structure than those on pure aluminum and can be correspondingly thicker and less compact. For instance, the oxide layer on an aluminum-magnesium melt consists mostly of pure magnesium oxide or of aluminum-magnesium spinel. If this layer is held for a few hours at a temperature higher than 750°C, then it ceases to protect against oxidation. The metal no longer oxidizes in layers, but rather the oxide penetrates into the metal as dendrites, forming a thick, black, furry oxide crust (Fig. 5.2). This process can continue up to the point where all of the aluminum is oxidized.

5.1.3.3 *Melt preparation and its effects on metal quality*

Melt preparation varies according to starting material and product requirements: initial metal quality, the customer's needs, and the above metallurgical conditions all affect the operations of charging, melting, alloying, and the various melt cleaning stages.

Modern casthouses in smelters and semi-fabricating plants cast ingots, slabs, and billets from furnaces holding 25–100 tonnes of liquid metal. These furnaces are heated either directly by flame or indirectly by electric induction. In smelters, pure aluminum at 950°C is siphoned out of the pots into transport crucibles, generally holding 5–10 tonnes, to be taken to the neighboring casthouse.[2] In the casthouse, the aluminum at 750–850°C is tipped into the batch furnace, like water from a bucket. As it cascades, the liquid metal mixes with air, immediately forming compact and mechanically strong oxide skins as described in the previous section. The melt's turbulent flow intimately mixes it with oxide skins and gas bubbles, which float to the surface as dross when the melt settles in the furnace. In many smelter casthouses, liquid metal is siphoned from crucible to batch furnace so as to minimize this contamination. Fig. 5.3 shows a section through a solidified lump of dross.

Dross also forms when in-plant process scrap and old scrap are melted. The major factor here is the oxide skin that covers every solid piece of aluminum and remains mechanically intact even after the metal inside it has melted. For instance, a piece of aluminum foil held over a flame soon melts, but the liquid metal cannot run out because of the surrounding oxide skin. It is easy to understand that more dross can form if the scrap has a greater surface area in relation to its volume. Therefore, melts from thin scrap, such as foil or used beverage cans, need vigorous stirring to break the oxide so that drops of molten metal can escape and flow together. This stirring effect can be achieved electrically, as with an induction furnace, or mechanically, as with a rotary furnace or metal pump. When loading runaround scrap into a furnace hearth, one should always try to put the most massive material in the hottest part of the furnace. It is also important to

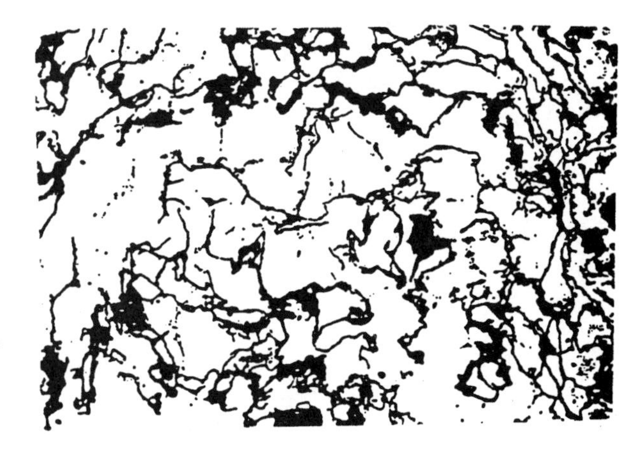

Fig. 5.3: A microsection through a solidified lump of aluminum dross. It consists of an intimate mixture of oxide skins (lines), metal (pale), and pores (dark). Magnification about 300×.

[2] Sometimes liquid aluminum is transported over longer distances to casthouses, so as to save remelting solid metal.

note that alloy scrap forms much more dross than pure aluminum. Finally, molten flux can reduce oxidation of the metal and is used especially in rotary furnaces. There, the salt flux agglomerates small melt droplets, dissolving the oxide films. Fortunately, by using all of the available techniques, one can achieve metal yields of 95–99% when remelting solid scrap. In spite of this, throughout Europe, aluminum casthouses produce 250,000 tonnes of dross each year, which requires costly disposal. Specialized dross treatment plants yield about 50% as secondary metal.

Particles of oxide skin tend to remain in suspension, being bulky and similar in density to the melt. Some of these skins could become non-metallic inclusions. Therefore, melt treatment and transfer need to be optimized to achieve the required metal quality, taking into account the oxidation characteristics of the alloy and the melt's self-cleaning capacity during settling.

The potline metal or the remelted scrap usually does not have the composition required for the product and, therefore, needs the addition of alloying elements. Elements that easily dissolve in the melt, such as copper, magnesium, silicon, and zinc, can be added in the pure form, whereas those elements that are slow to dissolve, like iron, manganese, titanium, and chromium, are added in the form of pre-alloys (i.e., more concentrated alloys with aluminum). To save time, it is desirable to obtain early confirmation from the analysis laboratory that the melt sample has the correct composition. However, before taking this sample, one must first be reasonably sure that all alloying additions are uniformly mixed in the melt.

Here too, precautions are needed to avoid introducing harmful contaminants together with alloying elements. Magnesium addition is especially critical, because it is lighter than aluminum and tends to float on the melt surface. This magnifies the risk factors for dendritic oxidation, producing oxide flakes that can be dragged into the melt by any movement at the surface. Oxide flakes are the most common and most damaging inclusions in alloys containing magnesium.

It is important to realize that any material that comes into contact with the melt might react with it, introducing dissolved or dispersed contaminants. Alloy melts are often chemically more aggressive than pure liquid aluminum. Furnace, crucible, and transfer channel linings thus need to be selected carefully. Metal tools used for melt treatments and taking samples need prior coating with a wash of parting compound. The quality of this coating needs to be carefully specified.

5.1.4 Hydrogen content

5.1.4.1 Sources of hydrogen

The most important source of hydrogen in aluminum is the reaction between the liquid metal and water that is usually present as moisture in the atmosphere. Moisture can also come from combustion gases as well as the remelting of oily, wet, or corroded scrap (the corrosion product of aluminum is a hydrated oxide).

Fig. 5.4: Hydrogen content in aluminum as a function of temperature with equilibrium values under a pressure of one atmosphere.

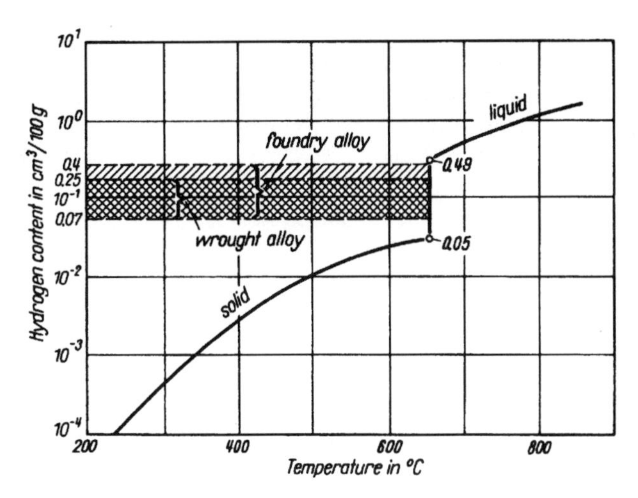

The reaction between aluminum and moisture proceeds by the equation:

$$2Al + 3H_2O \rightarrow Al_2O_3 + 6H$$

Hydrogen generated by this reaction is dissolved in atomic form in the liquid or solid aluminum. The supersaturated dissolved hydrogen has a strong tendency to precipitate as H_2 molecules in solid metal, forming secondary porosity.[3] Hydrogen gas precipitation during solidification or during high temperature annealing of solid metal can cause a number of problems. For example, it can embrittle thick plates of high-strength alloys, lowering fracture toughness and fatigue strength. To prevent this, hydrogen must be reduced to sufficiently low levels in the melt.

The higher the water vapor pressure above the surface and the higher the temperature of the melt, the greater the hydrogen content in the melt. The influence of temperature on the equilibrium solubility of hydrogen in aluminum is presented in Fig. 5.4. Since freshly tapped primary metal has a high temperature of 900°C it reacts readily with moisture and usually has a relatively high hydrogen content (under some circumstances up to 1.0 cm^3/100 g aluminum). In Fig. 5.4 we see an abrupt drop in the solubility of hydrogen as aluminum solidifies.

The hydrogen contents that usually are present in aluminum products are shown as shaded areas in Fig. 5.4. The solubility of hydrogen in aluminum in the solid state (at the solidus temperature) is ten times less than in the melt at liquidus temperature. This means that the hydrogen present in the melt is trapped in a supersaturated condition during the solidification process or to an even greater extent in cooling the solid aluminum. Fig. 5.4 is valid for a hydrogen pressure of one atmosphere. The proportions in practice may vary. The reaction with moisture results in a higher hydrogen pressure. With the rapid solidification rates attained by commercial casting techniques, hydrogen does not have enough time to escape from the solidifying structure. Thus, hydrogen in solid aluminum is always supersaturated to a certain extent.

[3] See Section 5.3 for primary and secondary porosity.

The equation for the reaction between aluminum and water vapor shows that in addition to hydrogen, aluminum oxide also forms. The aluminum oxide collects mostly in layers at the surface of the melt. This oxide skin can be drawn into the interior of the metal during remelting as well as by movement or flow of the melt. The oxide is then present as an inclusion (Fig. 5.2).

5.1.4.2 Industrial methods to determine melt quality

Quantitative measurements to evaluate melt quality thoroughly are very troublesome and time-consuming. Routinely, people responsible for quality control in the casthouse generally need only a fast, relative test for overall quality. One such method often used consists of letting a melt sample solidify under reduced pressure (e.g., 50 millibars) in a coated crucible. The sample is subsequently sawed lengthwise and visually examined for coarse porosity. Coarse pores are a sign of either excessive hydrogen content or medium hydrogen content with too many inclusions. In this case, hydrogen gas provides the energy for the porosity, and inclusions provide nuclei for the hydrogen bubbles.

Other casthouses check their metal quality using variants of this principle. In one common and rapid method of estimating hydrogen content, a sample of about 200 g of the melt is kept liquid in a vacuum chamber. The operator carefully watches the liquid surface through a window as the pressure is slowly reduced until, at about 0.05–0.10 bar, small bubbles of hydrogen appear at the surface. At the first bubble, the operator notes the pressure, from which he or she can deduce the approximate hydrogen content via a calibration curve to decide whether the quality is satisfactory. At other casthouses they let the sample solidify at normal atmospheric pressure and then examine it in sections for microscopic defects. In any case, it is important to have the result before casting begins or at least early during casting, so as to limit the damage in case of a bad result.

Such empirical methods enable the casthouse to guarantee the customer's quality requirements. However, the information they provide is only global and relative. Spot checks with more detailed examination will remain essential for the more demanding products such as plate and forgings in high-strength alloys for aerospace applications.

5.2 Melt Treatment

5.2.1 Removal of alkali and earth alkali metals

Here we will take the example of sodium removal, because this is the usual target metal. Sodium is particularly harmful, because it can cause tearing during hot working and damage the surface properties of certain products. Like hydrogen, sodium diffuses into bubbles during gas rinsing; however, its partial pressure is low and in inert gas, the effect is too slow to be useful. Continuous treatment with argon or nitrogen in a unit like that in Fig. 5.5 would decrease the sodium level at best by half, which is often not sufficient.

Since sodium often needs to be reduced to less than a fifth of its initial level, a reactive gas, usually 1–6% chlorine, is mixed with the carrier gas. This immediately reacts with

Fig. 5.5: Degassing an aluminum melt by the SNIF process. (a) A diagram showing the principles of operation and (b) a general view. (photo Praxair)

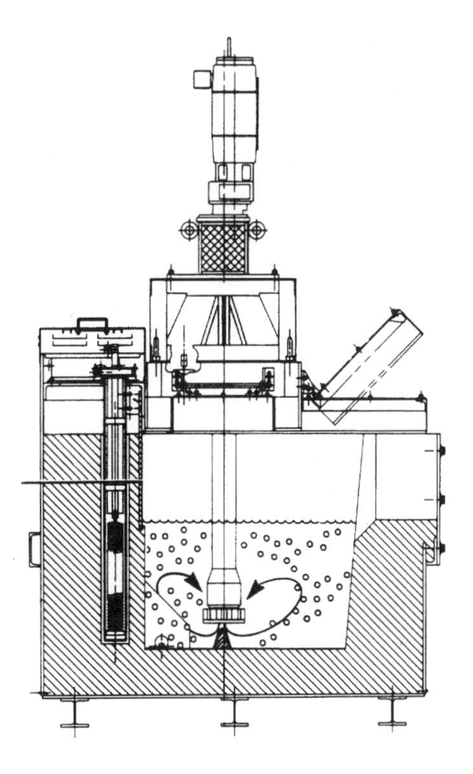

a

b

the sodium as it enters the bubbles, producing NaCl. By reducing the partial pressure of sodium in the bubbles, the chlorine allows more sodium to enter them until the chlorine is used up.

Sodium also tends to be lost from the melt by various "self-cleaning" mechanisms. For instance, if a melt sample is held for a few hours in a new laboratory crucible of graphite or ceramic, its sodium content falls to a fraction of the initial level as the sodium atoms diffuse into the crucible material. Industrial furnace, crucible, and channel lining materials show a similar capacity to absorb sodium. When potline metal is transferred by cascade, a large part of the sodium disappears, also by evaporation and inclusion in the oxide skins. Some active melt cleaning processes also use absorption of sodium into solid materials with a large contact surface. For instance, passing the melt through a bed of petroleum coke granules about 10 mm across can reduce the melt's sodium content to a third or a fifth. Another industrially used process involves stirring or blowing aluminum fluoride granules (AlF_3) into the melt. These granules react on the surface with sodium to form cryolite ($NaAlF_4$).

To prevent sodium from building up in refractory linings throughout the casthouse or foundry, it should be removed as early as possible. This means treating the metal in the transport crucible on the way from the potline. Here, rotors are used to inject either a rinsing gas or aluminum fluoride granules.

In regard to removing other alkali and earth alkali metals, smelter managers are tackling the problem at its source by preventing them from getting in. Fortunately, they abandoned the once popular addition of lithium to cryolite several years ago, because lithium contamination is particularly harmful. It is a problem for recycling and will discolor aluminum foil even in small quantities. They also avoid calcium contamination by carefully selecting the materials that the aluminum melt touches.

5.2.2 Removal of non-metallic inclusions

5.2.2.1 Settling and gas rinsing

The most important requirement for a modern melt cleaning system is to remove non-metallic inclusions as completely as possible. However, it is equally important to avoid introducing inclusions during other melt treatments. This involves attention to various metallurgical relationships that were already mentioned in previous chapters. The simplest and most traditional method of removing inclusions is to hold the melt in the casting furnace without disturbance for 1–4 hours. The heavier inclusions tend to settle to the bottom and form clumps there. The less dense inclusions and those that adhere to bubbles tend to float to the surface where they stick to the oxide skin.[4]

Another cleaning method uses flotation: if an inclusion in the melt touches a bubble of injected rinse gas, surface tension attaches it to the bubble, which carries it to the surface. Thus, the gas rinsing methods described above serve not only to remove hydrogen and

[4] During holding, the sodium content of the melt also goes down (self-cleaning mechanism).

alkali metals, but also reduce the content of non-metallic inclusions to a third or a fifth of the prior level.

Here again, chlorine additions help to make the bubbles smaller only in simple, static injection systems. By contrast, the modern rotor systems disperse the rinsing gas finely enough without chlorine.

5.2.2.2 *Filtration*

Filtration is being increasingly used as an additional stage of melt cleaning. The filter bed can consist either of a ceramic foam plate or a bed of granules. Ceramic foam plates are generally manufactured from a very open-pored plastic foam, which is then saturated with a suspension of alumina and binder. Calcining evaporates the plastic and consolidates the alumina as an inverse foam. Such plates are light and sufficiently strong. As shown in Fig. 5.6, they are used as inserts in filtration units placed between the casting furnace and the casting machine. Before casting begins, the filter plate is made red hot with a gas burner, so that the liquid metal passes easily through it. As non-metallic particles in the melt come into contact with the alumina surface, they can adhere or become mechanically trapped.

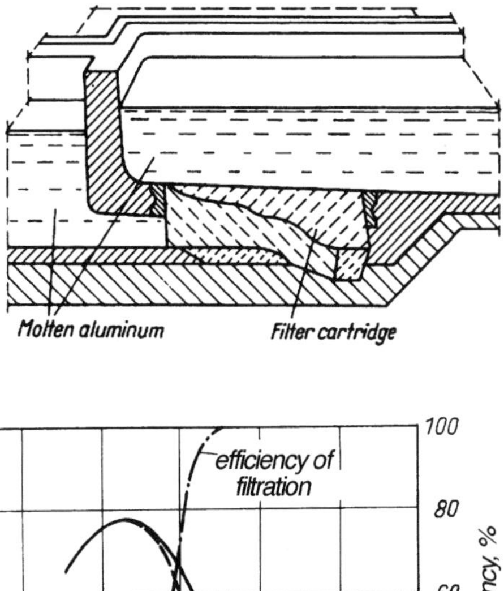

Molten aluminum
Filter cartridge

Fig. 5.6: A replaceable ceramic foam filter insert in a continuous filtration unit. A continuous casting line requires only minor modifications to incorporate such a unit.

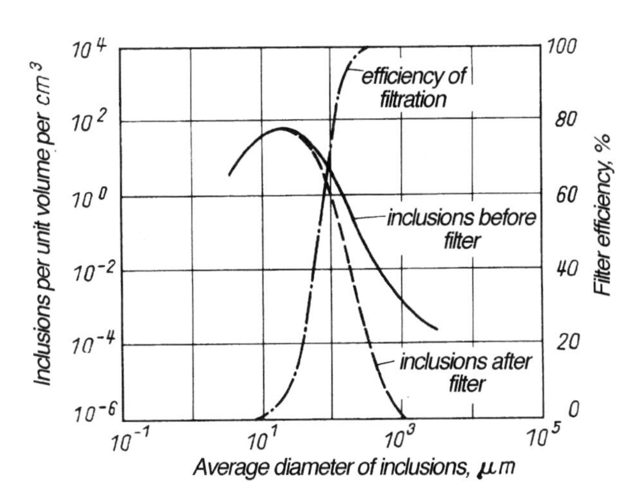

Fig. 5.7: A graphic representing the efficiency of a continuous filter as a function of inclusion size, in this case, for inclusions of type B (magnesium oxide or titanium diboride) from Fig. 5.1.

Bed filters are beds of melt-resistant granules about 1 cm across, like petroleum coke or special ceramics, contained in filter units for continuous filtration. These filters are much coarser, but also an order of magnitude bulkier so as to obtain a similar melt contact time. Bed filters work purely in depth: no filter cake can form at the surface. Whereas ceramic foam filters normally need changing after each cast, bed filters can remain in use for up to several weeks before needing renewal, which is, however, a correspondingly bigger task.

As with inclusion removal by settling during holding, continuous filtering methods have different efficiencies according to the type, size, and shape of inclusion or flake considered. The efficiency of continuous filters can be experimentally measured by taking melt samples at the filter entry and exit and then evaluating these samples according to Fig. 5.1. Comparing the two size distribution curves for a particular type of inclusion (as in Fig. 5.7), we can calculate a "filter curve" that describes the filter's efficiency as a function of inclusion size.

The curve shown in Fig. 5.7 is typical for flake inclusions of magnesium oxide or titanium diboride. It represents the filtration efficiencies to be expected from filtering with rinsing gas and rotor, ceramic foam inserts, or granular beds. Quantitative comparisons have so far shown no significant difference in efficiency between these three types of continuous melt cleaning. They are all essentially better at taking out oxide skins than compact particles and flakes, because flexible skins floating in the melt stick more easily to rinse gas bubbles or filter surfaces.

Therefore, selecting the right type of melt cleaning system does not depend on fundamental, metallurgically proven differences in efficiency, but on criteria such as reliability, ease of use, and cost. It is a matter of choosing the right combination of cleaning stages in order to achieve the degree of melt cleanliness needed both reliably and at an acceptable cost.

5.2.3 Removal of hydrogen /"degassing"

The three central tasks of melt treatment are to achieve sufficiently low levels of hydrogen, alkali metals, and non-metallic inclusions. The term "degassing" tends to be used loosely to include many processes designed to remove other impurities as well as hydrogen. Unalloyed liquid aluminum contains between 0.5 cm^3/100 g and 1.2 cm^3/100 g of hydrogen, corresponding to its solubility at temperatures of 660°C and 800°C, respectively. But as soon as the metal solidifies, equilibrium hydrogen solubility is only 0.05 cm^3/100 g, even at solidus temperature. In industrial practice there is too little time for supersaturated hydrogen to escape from the melt as bubbles. Instead, it remains supersaturated in the solid metal, where it can cause various quality problems. Therefore, it is usual to lower the hydrogen level of the melt to well below 0.25 cm^3/100 g and to below 0.1 cm^3/100 g for heat-treatable alloys (see Fig. 5.4).

Excess hydrogen in semis ingot is highly undesirable because it leads to porosity in the cast structure and blisters during heat treatment of wrought products. Conversely, as Fig. 5.4 shows, a moderate hydrogen content is acceptable or even desirable in casting alloys, especially in relatively slowly solidifying sand castings. There it serves to compensate for solidification shrinkage with relatively fine porosity rather than coarse shrinkage voids (as further described in Chapter 6).

In theory, the equilibrium diagram of the aluminum-hydrogen system shows that for most applications aluminum should reach a sufficiently low hydrogen content if held in a vacuum furnace at a pressure below 0.01 bar. One problem is that 40 cm under the liquid surface the pressure on a bubble is not 0.01 bar but 0.11 bar due to the weight of metal above. Furthermore, to nucleate a new bubble against surface tension requires overcoming a substantial extra energy barrier. For these reasons, vacuum degassing is seldom used industrially to remove hydrogen from aluminum melts.

More usual are processes that flush out hydrogen by injecting bubbles of another gas. If very fine bubbles of an inert gas, preferably argon, can be uniformly distributed in the melt, they will provide stable nuclei into which hydrogen can diffuse until it reaches a partial pressure in equilibrium with the hydrogen still dissolved in the melt. These bubbles saturated with hydrogen rise to the surface, where the hydrogen burns.

This principle of rinsing with another gas is dominant industrially and has many variants differing in the gases used and in the way the gas bubbles are injected and distributed in the melt. Earlier processes used mainly chlorine, either pure or diluted with argon or nitrogen as the carrier gas. Chlorine reacts violently with liquid aluminum, releasing so much energy as it forms aluminum chloride gas that this reaction product and the carrier gas are dispersed and rise as fine bubbles. Rinsing with a chemically reactive gas like chlorine removes hydrogen quite effectively, even when injection takes place through a simple submerged tube.

By contrast, if chemically inert gases are simply injected through a tube, aluminum's high surface tension (two orders of magnitude higher than water) results in fist-sized bubbles that escape rapidly. These conditions do not meet the requirements for hydrogen removal: dross forms, but the hydrogen content stays the same. However, modern melt cleaning processes succeed in finely dispersing inert gas (often with a small amount of added chlorine) in the melt using various techniques, especially injection through rotors. In this way, hydrogen and if chlorine is used, sodium are effectively removed in an environmentally acceptable way.

Fig. 5.5 illustrates the SNIF process, which is normally placed between the casting furnace and the casting machine. In continuous operation (at normal casting rates), it can reduce hydrogen levels to half their initial value. Processes with other trade names, such as ALPUR and RDU, work on similar principles and with similar capacity.

5.3 Origin of Porosity in Castings

Three different types of cavities can occur in the cast structure: shrinkage cavities caused by the reduction in volume during the transformation from liquid to solid (6–7% in commercial purity, less for most alloys); air bubbles resulting from air being introduced into the metal during casting and not having time to escape before solidification; and hydrogen pores in which the precipitated hydrogen agglomerates during solidification or in

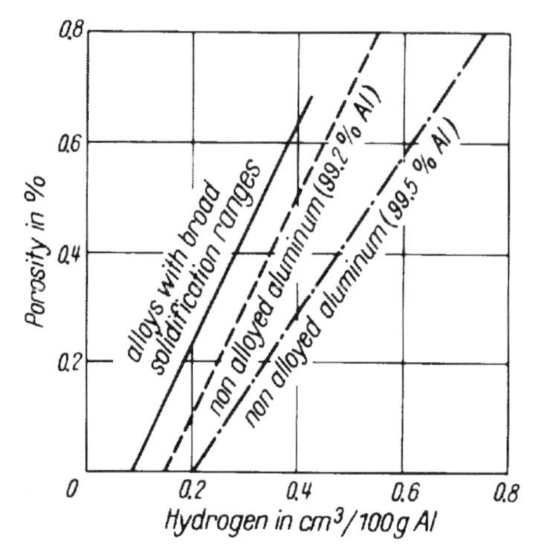

Fig. 5.8: Primary porosity in a DC cast ingot as a function of hydrogen content.

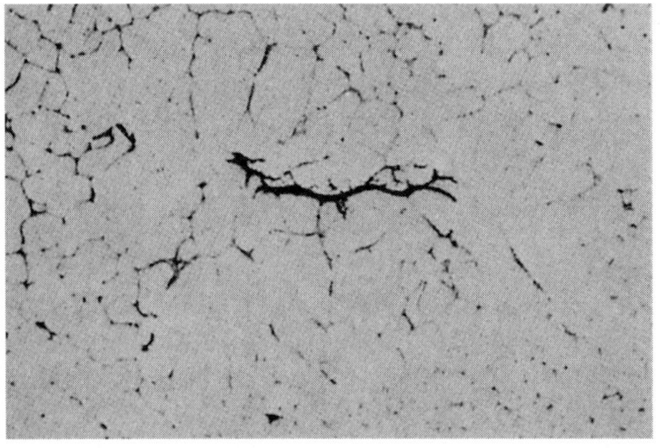

a

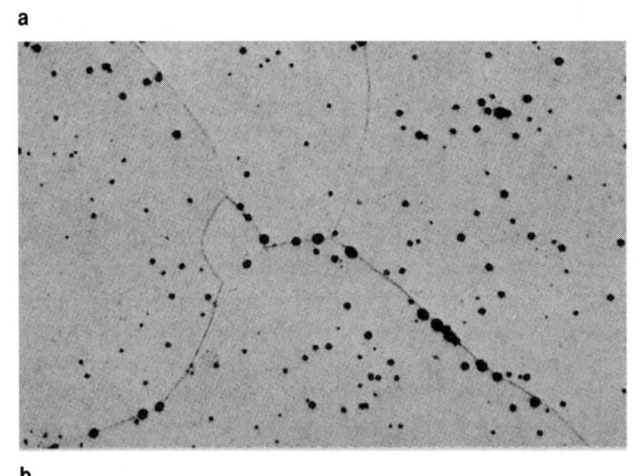

b

Fig. 5.9: Two different forms of primary porosity in DC ingots caused by hydrogen precipitation during solidification. Magnification (a) 140:1 and (b) 450:1

the solid state. The first two types are important in mold casting and are discussed in the next chapter.

Hydrogen precipitation in cast structures occurs in the form of relatively fine pores with a diameter of 0.001–0.5 mm and plays a role in DC ingot casting as well as in the production of castings and wrought semis. These pores may originate either during solidification or afterward and are accordingly identified as primary or secondary porosity depending on the time the precipitation took place. Higher hydrogen contents in the melt and slower cooling rates favor the formation of primary porosity.

In Figure 5.8 the amount of primary porosity versus hydrogen content in DC cast ingot of pure aluminum and alloys with broad solidification ranges is presented. Figure 5.9 shows the appearance of primary porosity.

The primary porosity is usually nonuniformly distributed in the metal. In commercially pure DC cast ingot with a normal hydrogen content of 0.1–0.2 $cm^3/100$ g, the primary porosity is less than 0.1 vol.% and is not detrimental to this material or the wrought alloys of medium strength. Higher hydrogen contents give rise to primary and secondary porosity, which can cause cracking during hot rolling and blister formation during annealing. In DC casting of high strength alloys, the hydrogen content of the melt must be significantly lower. It should be maintained at less than 0.08 $cm^3/100$ g because, on the one hand, the broad solidification interval and the normal dendritic crystal growth of these alloys promote the precipitation of hydrogen as primary and secondary porosity and also because the high-strength alloys are more sensitive to faults in the structure than relatively softer materials.

The secondary porosity consists of very fine pores normally only a few microns in diameter (0.001–0.01 mm). These pores may form or become enlarged during annealing of billets and semi-fabricated products. Secondary porosity is uniformly distributed and has generally been considered relatively harmless compared to primary porosity.

Somewhat different considerations govern the production of castings. In high-strength casting alloys, the hydrogen content is kept as low as possible (less than 0.1 $cm^3/100$ g aluminum) in order to attain the highest strength. This is especially true for sand castings because of the slower solidification rates, since there is more time available for precipitation of the hydrogen compared to mold casting. On the other hand, alloys that have been degassed to this extent tend to show more shrinkage cavities and impose very exacting requirements on casting, gating, and risering technology. As a result, one employs somewhat higher hydrogen contents in the melt (0.3–0.4 $cm^3/100$ g) for castings of uniform wall thickness with medium strength requirements. The small amount of primary hydrogen porosity that this causes is considered to be favorable in hindering the shrinkage behavior without appreciably affecting the mechanical properties in medium-strength alloy castings.

Excessive gas contents must be prevented in DC casting as well as in mold casting. High gas contents lead to extensive primary porosity, reducing the density of the structure, the strength, and especially the elongation at rupture.

Chapter 6. Industrial Casting Processes

The basic processes for the solidification of aluminum as described in Chapter 4 are of fundamental importance for all types of aluminum casting. There are three main groups of industrial casting processes, viz.:

- Casting of ingots for remelting
- Billet, fabricating ingot, sheet ingot, bar and strip-casting for further working into extrusions, forgings, wire, rod, bar, and a variety of rolled products
- Casting by various processes into shapes suitable for final use without further deformation

Unalloyed remelt ingot is often cast immediately after the smelting of primary aluminum in the smelter casthouse. Primary metal may be cast directly from transfer crucibles or transferred to an alloying and pretreatment holding furnace and then poured from this furnace into an ingot-casting line. This is also where the various master alloys are cast for further alloying in mold-casting and in making semifabricated products. Instead of primary metal, aluminum scrap is often used on a large scale as input material for all the processes and products mentioned above.

6.1 DC Casting

Direct-chill (DC) casting is a semicontinuous casting process, in which water-cooled molds initiate the first part of solidification. Thereafter, water sprays impinge on the shell of solid aluminum enclosing the still liquid core. DC and its variants are practically the only casting method used for making rolling ingot and extrusion ingot. (Hereafter, this book refers to extrusion ingot as "billet," which is the popular usage. Sheet ingot is often called "rolling ingot" or "rolling slab.")

Vertical DC (VDC) casting proceeds vertically downwards into a pit and can produce the widest wide range of alloys and cross-sections.

Horizontal DC (HDC) has proved useful for:

- Sheet ingot for commodity products as end use
- Extrusion billet up to about 280 mm in diameter in commonly used extrusion alloys such as 6061 and 6063
- Remelt ingot in rectangular or more complex sections (mainly foundry alloys)
- Electrical conductors (bus bars)

In the USA, HDC is dominant for billet casting on cost grounds, since it offers more continuous operation, facilitates process/handling automation and reduces end-related scrap while not requiring such deep foundations or high buildings as does VDC.

In Europe, where casthouses produce a wider variety of alloys and ingot sizes in smaller batches, VDC is dominant.

Fig. 6.1: Vertical direct casting (VDC) of rolling slabs. (Gautschi)

Wrought alloys can be cast in the form of the following DC ingot types:

- Extrusion billets are cast horizontally in any length or vertically in "logs" up to ten meters long. The long extrusion billets are then usually cut into lengths suitable for the extrusion press. Billet cross-sections are:
  - Generally circular with diameters of 100–550 mm.
  - Occasionally cast with two parallel faces and two semicircular sides, to make extra-wide profiles from specially modified presses.
  - Occasionally hollow, in the form of thick-walled tubes, for extruding seamless tubes.
- Rolling ingots, from which plate, sheet, and foil are produced by hot- and cold-rolling. These are generally of rectangular cross-section and can be up to 1300 mm thick and more than 2800 mm wide. They are typically cast vertically up to about 10 m long, or horizontally cast continuously and cut with mobile ("flying") saws to the desired length for subsequent hot and cold rolling.
- Forging stock or wirebar in various shapes such as round, square, or rectangular.

Fig. 6.1 shows a vertical DC casting unit in use. Fig. 6.2 is a schematic representation of the process, showing one strand of a vertical DC casting machine just after the start of a casting operation, or "drop." Above the casting pit, fixed to the casting frame, is the water-cooled mold in the form of a ring whose inside dimensions correspond to the cross-section of the ingot to be cast. Before casting begins, a hydraulic cylinder raises a false bottom (stool-cap) into each mold. Its sides are sprayed with cooling water from nozzles built into the bottom edge of the mold. In conventional DC casting, the liquid metal is fed into each mold through a casting nozzle, below which a deflecting device distributes the

Fig. 6.2: Schematic representation of a vertical DC caster: 1—feed trough; 2—nozzle; 3—float-valve; 4—distributor; 5—mold; 6—solidifying ingot; 7—sump; 8—water-cooling sprays; 9—stool-cap; 10—stool.

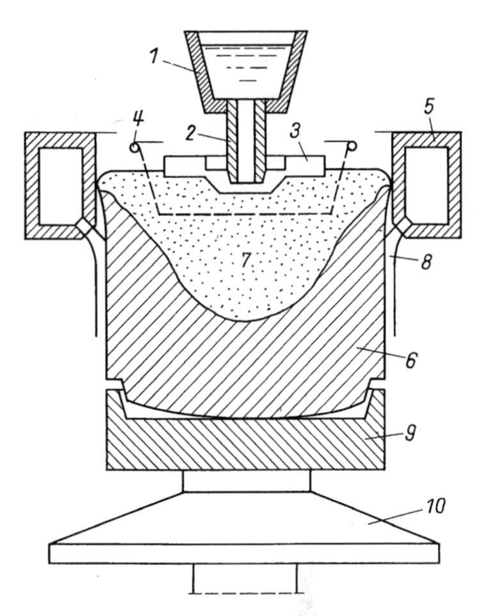

liquid metal stream in the mold.[1] Usually, a number of such molds are mounted together in one frame and are fed from one casting furnace.

At the start of the drop, liquid metal flows from the holding furnace via a trough to the nozzle and then into the hollow formed by the mold and the stool-cap. As soon as the melt surface in the mold reaches a predetermined level, the stool begins to descend slowly and steadily on the hydraulic ram. The water sprays of the mold now no longer strike the stool-cap, but play directly on the thin outer solid skin after contact with the mold. This direct-chill water efficiently removes most of the heat of solidification. As the stool descends with the solidifying ingot, the float under the nozzle maintains the liquid level, allowing a constant flow of liquid metal to enter the mold. The metal distribution system further ensures an optimal distribution of the melt, for safety and for uniformity of the ingot cast structure.

The hydraulic cylinder allows the solidifying ingot to descend steadily into the pit until it reaches its full length, at which point the casting stops.

In a modern DC casting installation, up to a hundred extrusion billets or ten rolling ingots can be cast simultaneously. This is only possible if a high degree of automation is used. Accurate control is vital for both quality and safety. Uncontrolled mixing of water with liquid aluminum can under certain conditions cause an explosion. Precautions include careful control of the melting and casting processes, and special coatings to prevent contact with potential initiation sources.[2]

[1] For hot top casting, level feeding is used instead of a nozzle.

[2] While thousands of tonnes of molten aluminum are produced and handled without incident every day, explosions do infrequently occur. Because of the chemical reactivity of aluminum, an explosion, when it does occur, can be violent. The industry, through the Aluminum Association and through the efforts of individual companies, has spent the last 45 years studying these phenomena and how they may be prevented.

6.1.1 Grain refining of wrought aluminum materials

A regular, equiaxed grain structure with grain diameters in the range 0.2–1.0 mm is an important quality requirement for DC ingot. An ideal structure with a fine uniform grain size can be seen in Fig. 4.26. To achieve this, a grain-refining master alloy in the form of a rod about 10 mm in diameter is fed into the trough or into pretreatment systems during DC casting. Between 0.2 kg and 1.0 kg of grain refiner is used for each tonne of aluminum. The usual master alloy for grain refining consists of aluminum with 5% titanium and 1% boron.[3] Since both these elements are only slightly soluble in aluminum, the master alloy contains intermetallic phases. Fig. 6.3 shows a longitudinal metallographic section of such a grain refining rod. We recognize larger light grey crystals of titanium aluminide ($TiAl_3$) embedded in the paler aluminum matrix and also numerous small particles of titanium diboride (TiB_2). Examination at substantially higher magnifications with an electron microscope shows that the TiB_2 particles have deeply fissured forms or shapes, comparable to ice crystals in snowflakes.

The grain-refining wire penetrates only a few centimeters into the liquid metal before becoming so hot that the aluminum matrix dissolves and disperses the two different intermetallic phases into the melt. The $TiAl_3$ particles immediately begin to dissolve in the melt and after about 30 seconds have disappeared. The TiB_2 particles on the other hand, being practically insoluble, travel in the melt as far as the solidifying ingot, in which they become embedded. Those TiB_2 particles that were initially embedded in the $TiAl_3$ play an especially important role in the grain refining mechanism. During their passage down the trough, they are to a large extent washed free of their surrounding $TiAl_3$. Because of their fissured shape, their crevices can conceal some remaining $TiAl_3$, which is thereby able to "survive" right up to the solidification front of the ingot. When such $TiAl_3$-loaded TiB_2 particles cool to the solidus temperature, a peritectic reaction causes the $TiAl_3$ to surround itself spontaneously with a layer of solid aluminum. This crystal can grow very quickly in an undercooled melt zone. The more such solidification nuclei ap-

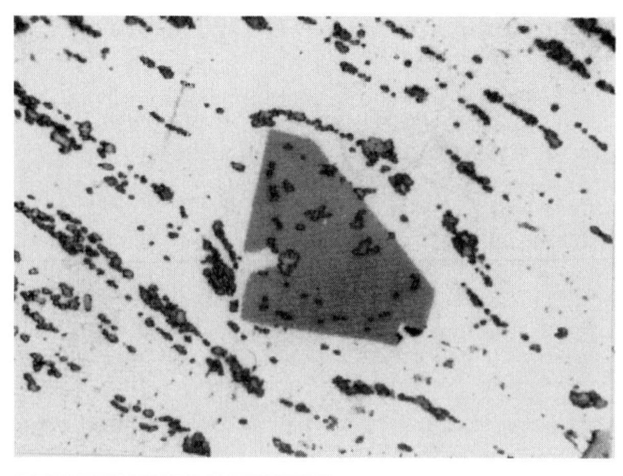

Fig. 6.3: Lengthwise section of grain refiner wire made of aluminum with 5% titanium and 1% boron (after Kiusalaas).

[3] Frequent variants are 5:5, 5:2, 3:1, etc.

pear per unit volume of the solidifying melt, the finer will be the grain of the resulting cast structure. It is important to note that TiB_2 never nucleates a grain—$TiAl_3$ does.

The $TiAl_3$-containing particles tend to agglomerate into clusters as large as 50 μm in diameter, a condition to be avoided. Such nonmetallic inclusions are very harmful to quality, especially in products for decorative anodizing or where machining is required.

Titanium carbide particles can play the same role as $TiAl_3$ in nucleating aluminum crystals, so that rod made from an aluminum-carbon master alloy has occasionally been used for grain refining.

The grain refining mechanism described above explains the phenomena observed in practice and is supported by evidence from laboratory investigations. The technical literature nevertheless proposes various other possible mechanisms, which will not be discussed here.

6.1.2 As-cast structure

Only in high-purity aluminum is the cast structure free of intermetallic phases. Metal direct from the potroom and alloys made from it contain, in contrast, iron- and silicon-bearing intermetallic phases that form by eutectic reaction. These phases decorate the cell boundaries in the cast structure, as already explained in Chapter 4. A primary task of casthouse practice is to ensure that the cast ingot has as fine and uniform a cell structure as possible. This in turn ensures that after further working by extrusion, rolling, forging, or drawing, the product has a correspondingly fine and uniform distribution of the intermetallic phases. This is especially important for extruded and rolled semifabricated products, which are anodized. Thus, Fig. 6.4 shows a section of a cast structure with non-uniform structure and the corresponding sheet surface, both after anodizing. It can be seen how variations in structure lead to a streaky surface appearance. At any given point in the solidifying ingot, the more rapidly the melt solidifies the finer is the cell structure. Structures that have solidified at slower rates display a coarse cell structure.

It is possible to measure the "local solidification time," which determines the extent of this cell coarsening, by allowing small thermocouples to freeze into the metal during DC casting. In this way, the liquidus and solidus surfaces in the solidifying ingot can be pinpointed. Their relative positions are illustrated schematically in Fig. 6.5 through the thickness of the ingot. The vertical thickness (D) of the so-called "mushy" zone located between the liquidus and solidus surfaces is directly proportional to the local solidification time. The cell size (s), also shown, is proportional to the cube root of the local solidification time.

The conventionally cast ingot cross-section often consists of a 5–20 mm thick coarse-grained shell, an underlying zone with fine grain, and a somewhat coarser-grained center. The coarse-grained outer zone can cause problems. It is least pronounced when the ingot is cast with the lowest possible metal head, so that the point of impact of the cooling water and the point where the freezing shell moves away from the mold are as close as possible to each other. Recently, it has been possible to eliminate this shrinkage gap altogether by

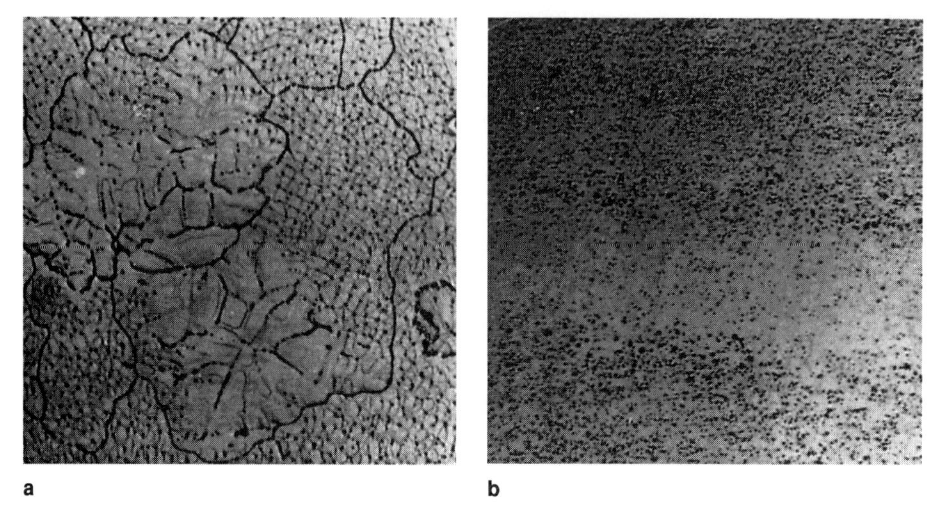

Fig. 6.4: Non-uniform cell size (a) in a metallographic section of a rolling ingot and the resulting streaky surface (b) of a sheet rolled from the ingot. Both the microsection and the sheet were anodized to highlight the defect.

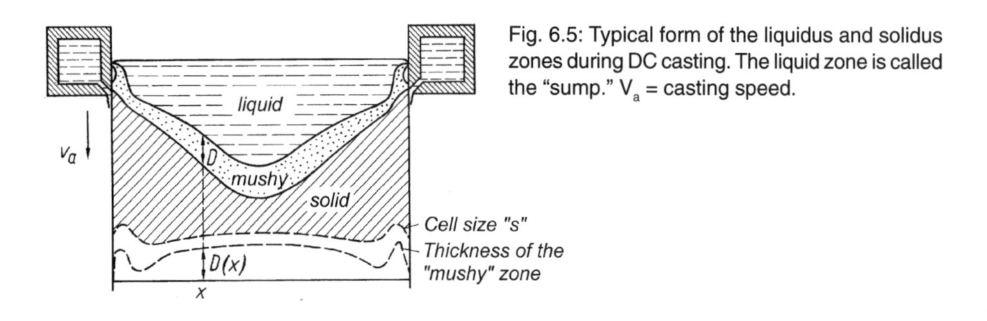

Fig. 6.5: Typical form of the liquidus and solidus zones during DC casting. The liquid zone is called the "sump." V_a = casting speed.

using electromagnetic molds. With this method, inductor coils carrying high-frequency AC current at several thousand kHz are built into the mold. These induce eddy current forces in the melt that repel it from the mold wall and thereby prevent its coming into direct contact with the wall. The heat is then extracted entirely by the direct chill of the water, resulting in an extremely smooth surface, uniform subsurface structure, and an absence of segregation zones that can occur in conventional DC casting.

6.1.3 Surface quality of the cast ingot

The surface of most DC-cast ingot looks rough and uneven, like the bark of a tree, as can be seen in the typical examples shown in Fig. 6.6. These surface defects ("bleed bands" or periodic surface segregation) must often be removed by machining for quality reasons, increasing the cost of the product. Rolling ingots have to be milled ("scalped") and billets for direct extrusion must be turned before further processing. There are several reasons for these surface characteristics:

- Cold shuts occur when the melt meniscus freezes against the mold wall and fresh molten metal overflows the frozen crust without fusing to it.
- Surface segregation occurs when molten metal "bleeds" through the partially solid shell, driven mainly by the metallostatic pressure of liquid metal above (Fig. 6.7).

"Solidus" and "liquidus" in Fig. 6.7 and in the following text refer to the start and end positions of solidification in the mushy zone rather than to temperatures in the equilibrium phase diagram.

Cold shuts occur mainly in pure aluminum and in alloys that solidify within a short freezing range. They can generally be avoided by increasing the casting temperature or the casting speed ("drop rate") or by restricting the flow of heat through the mold wall (e.g., by fine vertical grooves). Pronounced surface segregation is, on the other hand, characteriztic of more highly alloyed metal with a long freezing range. In this case, the liquid metal can "bleed" through the frozen skin to form a layer as much as 1–2 mm thick and containing alloying elements at more than double their normal concentration.

In conventional DC casting, surface segregation can be reduced by shortening the depth of the frozen crust in contact with the mold. Such a short crust is less susceptible to leaks and the metallostatic head stays low. The hot liquid metal flowing into the mold should

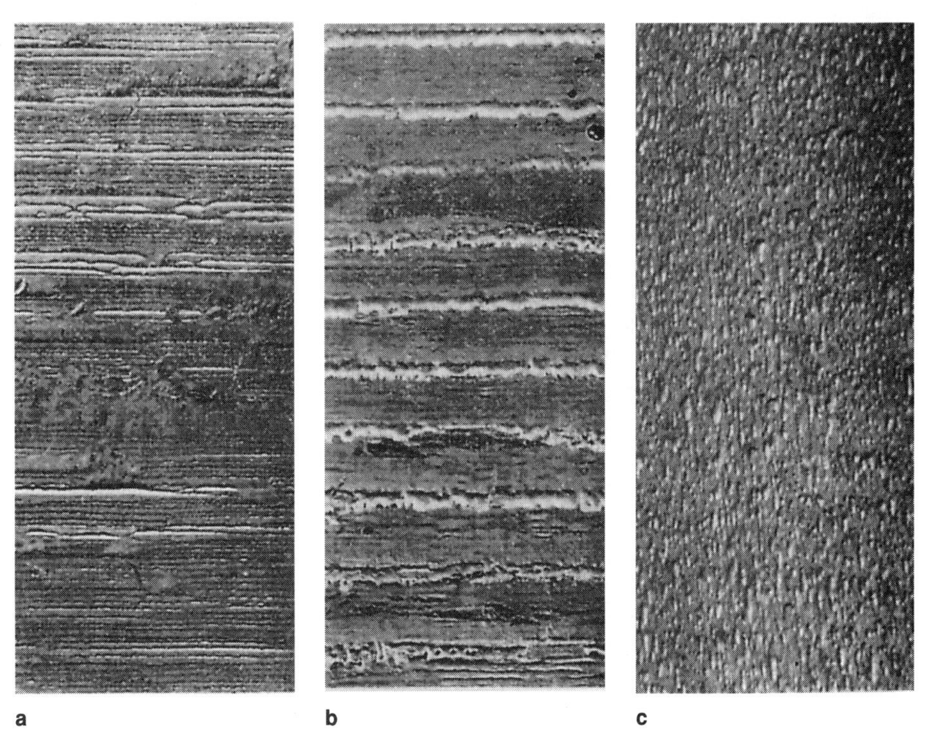

Fig. 6.6: Typical surface appearance of DC cast ingots: (a) surface with cold shuts; (b) surface with periodic liquation; (c) uniformly rough surface with gas blebs and contact lines.

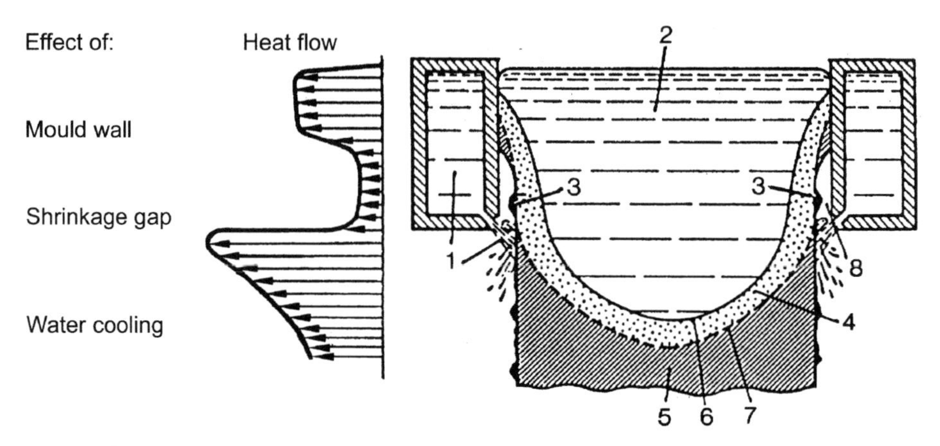

Fig. 6.7: Surface segregation due to the formation of a shrinkage gap during DC casting. Owing to the reduced heat transfer, molten metal can bleed through the partly solidified shell under the metallostatic head. 1—water cooling; 2—liquid metal (the "sump"); 3—surface segregation or bleed bands; 4—mushy zone; 5—solidified ingot; 6—liquidus surface; 7—solidus surface; 8—shrinkage gap.

be prevented from flowing directly onto the solidifying crust. It is also helpful to avoid too high a casting temperature. The most thorough way of avoiding these surface defects is, as mentioned before, to use electromagnetic casting (EMC), which eliminates the causes of cold shuts and bleed bands. The segregation layer in EMC ingot hardly exceeds 70 μm for any alloy. The improved surface quality often eliminates edge-cracking during hot rolling.[4]

6.1.4 Porosity in the ingot

There are many and varied sources of unsoundness in DC ingots, which can cause difficulties in their further processing. Source often encountered are: an accumulation of supersaturated hydrogen, or internal voids, or oxide inclusions. In the following, a few mechanisms for forming porosity are described, and some instances are given where it must absolutely be avoided.

As mentioned elsewhere, the metal volume decreases by 7.5% between the solidus and liquidus surfaces due to shrinkage during solidification. This volume deficiency has to be made up by fresh molten metal, which seeps along the channels between the dendrites. The internal flow resistance to this interdendritic "feeding" is particularly high in the narrow channels in the neighborhood of the solidus surface. Too little feeding causes voids of primary porosity between the dendrites. The feeding conditions are favorable in DC casting because of the relatively flat sump and the high temperature difference. In the microstructure of pure aluminum, there is almost no such porosity.

[4] Recently, the uniqueness of EMC cast surface quality has been challenged by sophisticated VDC technology, which has almost completely eliminated the air gap. ("Low head composite" casting by the Wagstaff company)

On the other hand, problems with porosity can occur in the high-strength 2xxx and 7xxx alloys. Owing to the wide freezing range of such alloys, the solidus and liquidus surfaces are far apart, so that the feeding channels are long. Large-diameter fabricating ingot in alloys of the 2xxx and 7xxx series are often fabricated into forgings with very high strength requirements, in which case any internal unsoundness is unacceptable. These alloys must be cast slowly, so that the sump depth is reduced.

For high-strength alloys, the casting of pore-free billets of large cross-section requires a low hydrogen level in the melt, if possible below 0.10 cm^3/100 g, which necessitates appropriate methods of degassing (see Chapter 5.2.3).

6.1.5 Macrosegregation in the ingot

The same mechanism—the feeding of fresh molten metal through the mushy zone toward the solidus surface—is also the origin of the inverse macrosegregation, which is a typical feature of DC-cast ingot. It can happen that in the middle of the ingot the concentrations of alloying elements may be 10% lower than in the outer parts, or even less. This phenomenon can also be explained to some extent in terms of feeding between the dendrites. Because of the mechanism of microsegregation already described, the less residual melt is left between the dendrites, the more it is enriched with alloying elements. Thus, the further solidification progresses, the greater the enrichment. The concentrations of the alloying elements are correspondingly lower in the cells that have already solidified than in the fully liquid metal above the liquidus surface. At the bottom of the sump, in the middle of the ingot, fresh molten metal with the normal liquid concentration forces its way into the interdendritic spaces, displacing the enriched intercellular residual melt, which then seeps further in the direction of the solidus surface.

When DC ingot solidifies with a deep sump, as it does with large mold sizes and high casting speeds, this displacement of the intercellular melt is strongly directed from the inside toward the outside of the ingot, for it proceeds at right angles to the liquidus surface. Thus, the alloying elements are carried from the inner to the outer zones of the ingot, which are enriched by an amount corresponding to the alloy dilution in the ingot core. The extent of this inverse macrosegregation can be reduced by casting more slowly, because this flattens the sump. Inverse macrosegregation must not be confused with the surface segregation illustrated in Fig. 6.7, where the fresh liquid metal is displaced by the metallostatic head.[5]

6.1.6 Hot and cold cracking

If in DC casting the casting speed is increased above a certain limit, the feeding problem becomes so pronounced that the internal stresses close to the solidus surface produce cracking instead of or in addition to interdendritic porosity. Cracks propagate in the casting direction, so that by the end of the "drop" the ingot may be cracked over a considerable length. Hot cracking can also occur at the start of the drop because the impact of the cooling water on the newly solidified shell creates thermal stresses. This sort of hot crack-

[5] We have given a short and simplified analysis of the mechanism of inverse macrosegregation, compared to Aluminum Association definitions.

ing can be avoided by careful design of the stool-cap, by appropriately controlling the metal distribution, and by programming the speed and cooling during the start of the drop. In particular, the direct chill by the water must be reduced at the start, for example by mixing the water with carbon dioxide or air, or by applying them in pulses. Good grain refining is also important because a fine-grained shell is better able to resist the stresses that cause cracking.

As long as the conditions for hot cracking are avoided, DC cast ingots of pure aluminum and soft alloys can be cooled by direct chill down to room temperature as quickly as desired, with no risk of cracking from the ensuing internal stresses. Hard alloys in the 2xxx and 7xxx series, on the other hand, develop relatively high internal stresses that can result in post-solidification cracking when cooled in the as-cast state. Big ingots in these alloys should, therefore, be directly cooled with water only down to 350°C and then dried, which can be arranged with various devices, for example with wipers or blow-offs. If such an ingot is cooled too quickly, cracking can occur suddenly, violently, and destructively. This represents a significant safety hazard both for people and for the equipment.

6.2 Strip Casting

In the 1930s and 1940s, just as DC casting was superseding static mold casting for wrought alloys, the development of continuous strip casting began on an industrial scale. It seems appropriate to mention the names of the three leading pioneers.

In the USA, Joseph Hunter began by developing the Hunter-Douglas process (chains of rigid molds with internal water-cooling) and subsequently the Hunter Engineering twin-roll casting process.[6] Also in the 1950s, Hunter developed the cold extrusion press, the continuous strip-painting process, and, finally, a complete system—the Hunter-Douglas process. The raw material was a scrap-based alloy of the Al-Mg-Si (6xxx series) with a very wide range of elevated copper and iron content and other impurities; it was hot-rolled immediately after homogenizing and then cold-rolled to a very high overall reduction. The end product was coiled stock used mainly for making venetian-blinds and sound-absorbing ceiling panels. Thanks to the very high degree of cold work, the variations in the alloy composition had little effect, while the thick, baked, paint film effectively limited the risk of corrosion due to the copper content. Thus, a brilliant inventor recognized the synergy between strip casting and scrap recycling, a good four decades ago.

Quite independently, William Hazelett developed a continuous casting machine using two endless steel belts. Today, a sizeable number of Hazelett machines are in regular use for casting strip in zinc, copper, and copper alloys, as well as copper bars for making wire. Apart from these, there were about ten Hazelett machines producing aluminum strip for a total of about 500 kt/y in 1995.

A third individual contributing original ideas to existing strip casting principles is William Lauener. He converted the Hunter-Douglas process into a "block caster" by reversing the heat flow, improved the automation of the process at "Roll Casters," and recently

[6] There had been some previous publications from other sources.

came up with the concept of continually tensioning the traveling thin steel belts to eliminate their warping during casting as a result of the "cold framing effect."

6.2.1 Distinguishing features of the strip casting[7] process

Unlike DC casting, strip casting can be considered as producing a semifabricated product, as can the continuous processes for casting wirebar. The process has proved itself for the production of coiled strip suitable for painting and also as a method of manufacturing foilstock, yielding strip from 3–20 mm thick and up to 2000 mm wide.

The strip casting processes have made considerable progress in the last four decades and now account for a remarkable share of the world output of rolled aluminum semifabricated coilstock. Strip casting is for many products a low-cost alternative to the rolling of relatively thick ingot in big, expensive hot mills. This is why very many strip casting installations are in newly industrializing countries. Fig. 6.8 shows a modern block caster installation.

In the 1950s and 1960s, there was great enthusiasm for strip casting, with the idea of avoiding investment in hot-rolling mills. It soon became evident, however, that continuously cast strip could not match the quality of strip produced by conventional hot-roll-

Fig. 6.8: View of a modern strip casting plant for a maximum width of 1750 mm (Caster II). (Courtesy of Lauener and Alusuisse)

[7] In this text, the term "strip casting" is often used to denote both thin strip and slab casting processes generically; twin-belt casters and block-casters produce slab, typically 20 mm thick, which can be continuously rolled into strip.

ing, especially for hard alloys and for products with high quality requirements. It is not easy to cast alloys with more that 3% magnesium nor any heat-treatable alloys by strip casting. The same goes for strip requiring a decorative streak-free surface after bright anodizing. Finally, continuous cast strip has limitations for products requiring extensive forming, such as deep-drawing or bending over a small radius. The reason for this is easily found in the relatively slight deformation, typically in rolling strip from a twin-roll caster from six to one millimeter. The cast grains in such a structure tend to show up after further working in the form of a surface texture known as the "orange-peel" effect.

6.2.2 Wide strip casting

For wide strip between 1000 mm and 2000 mm wide, we can distinguish three kinds of casters, according to the form of the cavity in which the metal solidifies: between hollow rolls, between thin steel belts, or between moving solid metal blocks.

With twin-roll casters (Fig. 6.9) the molten metal is fed horizontally (in earlier models also vertically) between two hollow rolls which are cooled internally with water. The casting speed was for a long time typically 1 m/min at 6 mm gauge. There are well over 100 such casters installed. The range of application of these machines depends in the first place upon the alloy, because they are most suitable for alloys with a short freezing range. Alloys cast in this way include pure aluminum, the 3xxx series, and the 5xxx series with up to 2.5% magnesium. Because of the overriding effect of the freezing range, the casting rate falls off very rapidly with increasing magnesium content. The solidification zone is roughly 10–20 mm long and is followed immediately by a zone of hot-rolling in the same roll gap (see Fig. 6.12). Table 6.1 shows clearly the decisive influence that the roll diameter has on the casting rate and, hence, the productivity of this type of caster. The table shows that increasing the roll diameter from 620–960 mm boosts the annual capacity by up to 50%.

Fig. 6.9: Twin-roll caster for width 1500 mm (Caster I, Alusuisse).

Table 6.1: Capacity of twin-roll casting machines

(Courtesy of Péchiney Aluminum Engineering)

Caster model	Jumbo 3C 620	Jumbo 3C 840	Jumbo 3C 960
Roll diameter (mm)	620	840	960
Strip width (mm)	1600	1800	2000
Maximum strip thickness (mm)	12	12	12
Minimum strip thickness (mm)	6	6	6
Maximum coil weight (t)	12	14	16
Production rate at 1000 mm width (t/h)	1.35	1.50	1.70
Maximum capacity (kt/y)	16	20	25

Solidifying the metal between twin rotating steel belts, as with the Hazelett machine (Fig. 6.10), produces slab from 15–20 mm thick and up to 1500 mm wide. The material is then immediately hot-rolled down to coilable gauge of 3–6 mm. The alloy range is wider than with twin-roll casting and includes, for example, alloys suitable for canstock and autobody sheet, which are now very much in focus.

To achieve consistent quality and mainly to suppress warping of the steel belts, a coating is applied to reduce the heat flow into the belt. This belt coating is the subject of continuing development for relatively wide and demanding end products like can stock and automotive body panel sheet.

In the original casters using massive moving steel blocks (the "block caster"), the heat flow was unidirectional; that is to say, the heat passed directly from the solidifying metal through the steel blocks into the cooling water. In the Hunter-Douglas process, which was used to cast up to a width of about 600 mm, steel mold-blocks were used, rotating like caterpillar tracks and cooled with water from the inside. The resulting high thermal stresses in the blocks gave rise to fatigue cracking (thermal checking) at the surface of the blocks, which limited the surface quality of the cast product.

Fig. 6.10: Casting between twin rotating belts (by Hazelett at LMG).

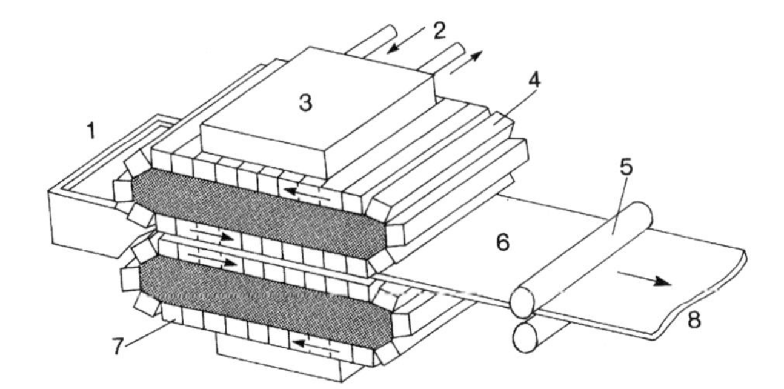

Fig. 6.11: The principle of operation of a strip casting machine with rotating steel blocks ("block-caster"): 1—metal feed; 2—upper spray-cooling; 3—water feed and takeoff; 4 and 7—mold blocks; 5—pinch rolls; 6—cast strip (15–20 mm thick by 600–1800 mm wide). (Lauener Engineering)

In all the above-mentioned strip casters, the heat flows in one direction during solidification (i.e., into the internally cooled rolls or blocks, or through the thin steel belts). In an independent development of the original block caster process, which has taken place since 1970, the steel blocks serve as heat sinks to absorb the heat of solidification, this heat being given up to the cooling water only after the blocks have lost contact with the strip and during their return to the point of contact with liquid metal (Figs. 6.8 and 6.11).

A feature of the Lauener-Alusuisse process is that the heat flow in the blocks is reversed and that, furthermore, block temperatures between 50°C and 200°C can be used. This facilitates casting those alloys that have a tendency to hot shortness when cooled too abruptly. This process has recently been used by a subsidiary of Coors for casting magnesium-manganese alloys for canstock, and in particular for recycling UBC. Casting slab at about 20 mm gauge, followed immediately by hot-rolling, it has a production rate of about 14 t/hr per m of width. Problems sometimes encountered are structural discontinuities at the site of the joints between the blocks, and an unduly short lifetime of the blocks.

In both cases (Hazelett and block casters), there are at least two hot-rolling stands in tandem with the caster. Owing to the high throughput, at least two relatively large holding furnaces are needed in order to ensure continuous operation. The annual capacity for one meter width with three shift operation can reach 100 kt/yr for a block caster, about 150 kt/yr for a Hazelett line.

6.2.3 Features of the microstructure of material from twin-rolldrum casters

Solidification is very rapid in twin-roll casting, leading to characteristic features in the microstructure, in particular a marked supersaturation of the relatively insoluble elements such as manganese, iron, and chromium. This makes it possible to push the level of manganese in supersaturated solution as high as 2%. For example, one alloy well-tried in this respect contains 1.6–1.7% Mn, 0.2–0.3% Cr, up to 1.0% Fe, and a maximum Si level of 0.25%. This and similar alloys have interesting properties:

- High strength at elevated temperatures, which is a special advantage in the market for light-gauge sheet for thermal insulation.
- A strikingly high resistance to corrosion, which makes the material particularly suitable for use in agricultural irrigation and drainage systems.

6.2.4 Further development of strip casting

Strip casting has gained a high degree of acceptance and has achieved success in terms of development of equipment, process, and products. Today, development has by no means come to a standstill; on the contrary, it has gained new impetus. This has happened because even large aluminum firms have shown new interest in strip casting, particularly because beverage can recycling is much more efficient if UBC can be transformed into canstock in decentralized locations, in so-called "mini-mills." Development of twin-roll casters has turned toward automatic control of the process via an array of sensors and actuators. Thus, for example, the temperatures of the melt, cast product, and caster rolls are automatically measured on-line and recorded, and the separating force between the caster rolls as well as the drive torque are measured and regulated. Such means have led to improvements in strip quality and at the same time to a significant increase in the productivity of strip casting plants.

Fig. 6.12 shows a cross-section of the solidification zone during twin-roll casting. The narrow band in which the freshly solidified metal is compressed by the rotating rolls is the focal point for controlling the process.

- The roll-separating force and the torque depend critically on the length of this band, and constitute key factors for automatic control.
- The degree of hot-working has a decisive influence on the quality after cold-rolling.
- The heat-transfer is at its highest in this zone of compression; this creates scope for reducing the strip thickness and increasing productivity.

Fig. 6.13, shows significantly higher productivity with decreasing thickness. For this reason, a strip gauge of 1 mm is today a declared developmental goal, with progress reported in Hunter casters.

The chief goal remains the manufacture of high-quality canstock in the magnesium-manganese alloy group. While this goal is difficult to attain by strip casting, there are claimed to be—either under development or in production—twin-belt and twin-roll casting machines suitable for producing canstock. Much effort is also being directed at expanding the range of products that can be competitively made by strip casting.

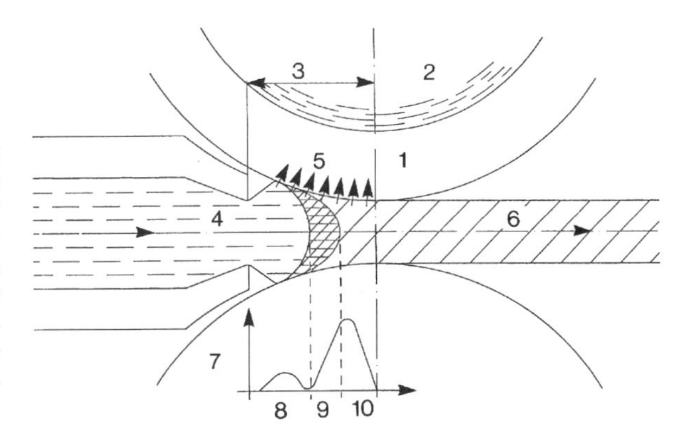

Fig. 6.12: Cross-section through the solidification zone of a roll caster: 1—caster rolldrum; 2—cooling water; 3—arc of contact from the nozzle to the nip; 4—melt; 5—heat flow; 6—solidified strip; 7—changes in heat transfer coefficient; 8—liquid; 9—mushy; 10—solidified.

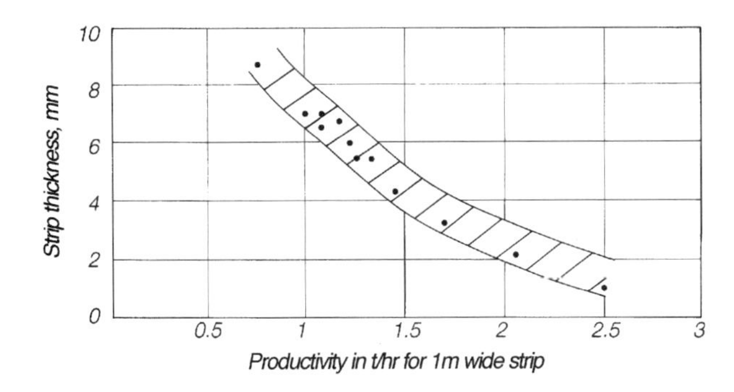

Fig 6.13: Productivity for a roll caster with change in 1145 strip thickness. (Lauener)

Reynolds, the world's biggest producer of aluminum foil, has about ten twin-roll casters for making foilstock. This proves that packaging foil in the intermediate gauge range can be produced quite satisfactorily by twin-roll casting.

Of late, alloys based on pure aluminum with high iron content (e.g., 8014 and 8079), specially tailored for roll casting, have come into use. For example 8079 has 0.7–1.3% Fe and 0.1–0.3% Si. It is claimed that with the correct choice of alloy, high-quality light-gauge foil can be made out of foilstock from the thin-strip casters described earlier. A major advantage of thin-strip casting is that it reduces the number of rolling passes: it is expected that foilstock of 0.6 mm will be cold-rolled in one pass from 1–2 mm cast gauge.

6.2.5 Other methods of strip casting

Special processes are devoted to casting narrow strip up to 300 mm wide with a combination of a water-cooled casting wheel with a rotating steel belt wrapped half-way around its circumference ("rotary machines"). Such narrow strip casters are mostly used for producing small discs, called "slugs," for impact extrusion into tubes, aerosol containers and cans.

A closely related process is the use of the casting wheel to make wirebar, the casting cavity being closed with a peripheral steel or copper alloy belt up to the end of the solidification zone. The bar is then fed immediately into a multistand rod mill, usually of about ten stands. In Europe, the Properzi machine, as shown in Fig. 6.14, is the one most often used. The usual alloys are electrical conductor grades (pure aluminum, 1350, and sometimes 6101) and the 3xxx series. Properzi rod is usually delivered at 12 mm diameter.

6.2.6 Recent versions of the twin-belt caster

There are some novel twin-belt strip casting units in pilot production or in development. One such development is the Kaiser "Micro-mill." The technical essentials of this process can be summarized as follows:

- The stock is cast up to about 500 mm wide, and the production rate relatively low, giving an annual capacity of around 45 kt/y.
- The cast strip is hot-rolled in-line, flash-annealed, quenched and cold-rolled in-line to finished canstock gauge.

Fig. 6.14: A Properzi machine for wirebar casting. (LMG)

The chief economic advantages claimed by Kaiser, as compared with the conventional process, are:

- The equipment is small-scale, relatively simple and cheap.
- The dedicated plant can be located close to the canmaking plant. This means that process scrap can be efficiently recycled, and transport and inventory costs reduced.
- The whole cycle from liquid metal to finished canstock ready for the canmaking plant takes no more than a day.
- The flash-anneal and quenching makes it possible to use a cheaper, scrap-based alloy.

Another example of the further development of strip casting is a twin-belt caster with unique features, developed by LAREX (US Patent no. 4.964.456). Bare belts of uncoated steel provide a high heat transfer, so there are no problems with coatings on the belts. The belts, of determined length, are unreeled from a coiler and are recoiled after passing through the casting zone. The belts are stretched by about 0.4% while passing through the casting zone, by which means distortion by the "cold framing" effect is avoided. The belts can be re-used up to 50 times.

6.2.7 Outlook

There seem to be a "golden rule" governing the success of strip casting installations: the wider the strip, the greater the difficulties. Only roll casters have reached up to 2 m width of strip in successful operation. For block casters, it has been proven that a width of about one-half meter is commercially manageable. Upscaling from a successful installation at half a meter width to above one and a half meters can be extremely difficult. According to industry sources, it remains uncertain whether or not continuous commercial operations are practical at widths at or above 1.5 meters. With this in mind, the concept of the micro-mill is quite logical.

6.3 Industrial Mold-Casting Processes[8]

6.3.1 Sand casting

In sand-casting, re-usable, permanent patterns are used to make the sand molds that are destroyed in removing the solidified castings.

There are two basic kinds of molding material—natural and synthetic. The former is called "green sand" and consists of mixtures of sand, clay, and moisture. The latter is called "dry sand" and consists of sand and synthetic binders cured thermally or chemically. The advantage of synthetics is that their composition can be closely controlled within tight limits. The sand cores used for forming the inside shape of hollow parts of the casting are made using dry sand components.

The extra metal needed to fill and feed the casting generates runaround scrap and causes relatively low metal yields. Sand castings require cleaning, which involves removing molding sand and cutting off the extra metal in the attached runners and risers.

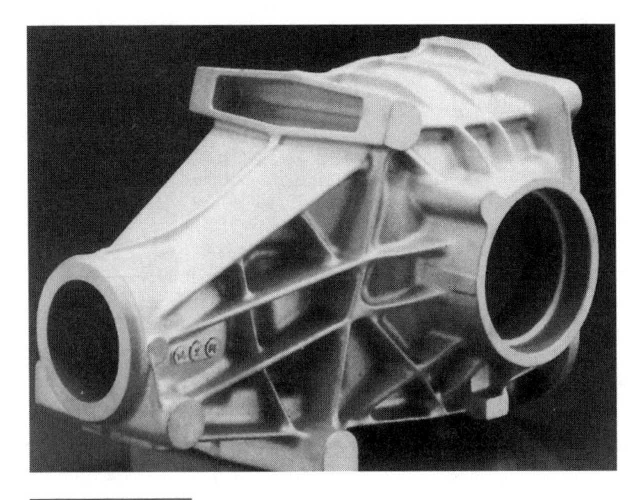

Fig. 6.15: Rear axle housing made from 380.0/AlSi9Cu3 by sand-casting. Weight: 6.8 kg. (Südalumin)

[8] By W. Schneider.

Of all the casting methods, sand-casting is the most versatile. It is not only suitable for small-scale production but can also be cost-effective where castings are required in larger numbers. Furthermore, the size and complexity of castings that can be produced in this way is almost unlimited, as can be seen for example in Fig. 6.15.

6.3.1.1 *Low-pressure sand-casting*

Low-pressure sand-casting is an innovative variant of sand-casting. The principle is shown in Fig. 6.16. The casting cavity is filled from below via a fill tube—the riser stack—and a nozzle from a crucible under low gas pressure (up to 0.7 bar). The pressure is maintained during solidification. With microprocessor control, the pressure during injection can be varied as required to ensure that the mold fills in the optimal way for its particular shape. Thanks to this technique, castings can be made with wall thicknesses as low as 2.5 mm, which is difficult with conventional sand-casting. By means of this process, other important requirements can be satisfied. These include close dimensional tolerances, good surface quality, high mechanical properties, and optimal microstructure. Productivity is higher and cleaning time lower, since the conventional sand-casting arrangements for gating and risering are not needed.

Fig. 6.17 shows a mass-produced casting for the aerospace industry, made by low-pressure sand-casting. It is a landing-flap mounting in A356.0-T6 alloy, which has the required high yield and ultimate tensile strength in the load-bearing zones. Apart from such regular mass-production, low-pressure sand-casting is especially well-suited to making prototypes of castings being developed for permanent-mold casting, more particularly for pressure die casting. This gives the designer the opportunity of making a prototype which he can then test under simulated conditions of service. As a method for making prototypes, low-pressure sand-casting most closely simulates the pressure die casting process.

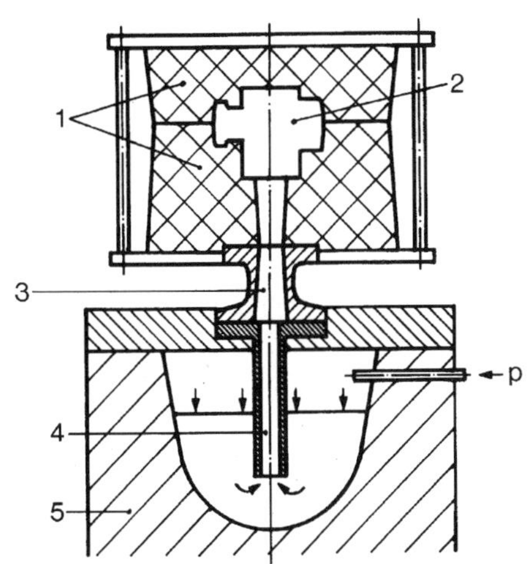

Fig. 6.16: Schematic view of a low-pressure sand-casting installation: 1—sand mold; 2—casting cavity; 3—melt injection nozzles; 4—feed pipe (riser); 5—pressure-tight crucible; p—low-pressure gas.

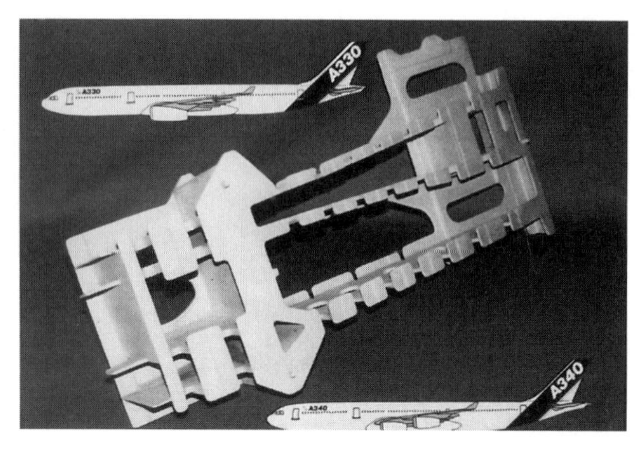

Fig. 6.17: Landing-flap mounting in A356.0 was made by low-pressure sand-casting. UTS = 330 MPa, $TYS_{0.2}$ = 270 MPa, El_5 = 5%. (Honsel)

6.3.2 Permanent mold casting

With permanent mold casting, called "gravity die casting" in Europe, the mold is not destroyed at each cast but is permanent, being made of a metal such as cast iron or steel. Coatings and parting compounds are used to prevent liquid metal from welding onto or eroding the mold and to help separate the product from the mold. Inner surfaces and non-re-entrant cavities in the casting are formed from removable mold sections and re-entrant parts by smooth-coated sand cores or by loose metal cores.

Permanent mold casting is suitable for mass production and for fully mechanized casting. Filling of the metallic die and solidification take place under gravity alone at ambient pressure, with suitably adapted gating and risering arrangements. Pouring can be done by hand when only relatively few castings are needed, or it can be fully mechanized and even automated for mass production.

Some of the mold casting processes described below under die casting (Section 6.3.3) are also sometimes described as permanent mold casting processes, as the molds are not destroyed but rather are continuously reused.

6.3.3 Die casting

6.3.3.1 Low-pressure die casting

The operating principle is virtually the same as for low-pressure sand-casting (see Fig. 6.16). The metallic die can contain sand cores. It is filled from a pressurized crucible below, and pressures of up to 0.7 bar are usual. Low-pressure die casting has proved itself especially suited to the production of components that are symmetric about an axis of rotation. Cast automotive wheels are typical of such symmetric parts.

6.3.3.2 *Back-pressure die casting*

The crucible and the die are held in separate pressurized chambers. Each chamber can be pressurized up to 10 bar. Filling the die requires a pressure difference of about 1.7 bar, which is adjusted by reducing the pressure in the die chamber. The melt solidifies under higher pressure than with low-pressure die casting.

6.3.3.3 *Pressure die casting*

Pressure die casting belongs to the family of permanent-mold casting. In pressure die casting, a molten charge is injected into a water-cooled die of heat-resistant steel by a casting piston under high pressure and at high speed. Pressure is applied throughout solidification. The process causes turbulent filling of the die, leading to gas inclusions in the casting originating from entrapped air and from the die-lubricating and parting compounds. These inclusions generally rule out subsequent heat-treatment or welding because they would cause blistering. Aluminum casting alloys are mainly pressure die-cast using horizontal cold-chamber machines. In these machines, the casting chamber, into which a measured amount of melt is introduced, is horizontal and is arranged at right angles to the die. The melt quantity is regulated automatically. Pressure die casting is suited to the production of intricate castings of large surface area and accurate dimensions and yields an outstandingly good surface quality. A selected example of such a casting is shown in Fig. 6.18.

Both the machine and its dies are very expensive, and for this reason pressure die casting is economical only for high-volume production. Complex cores cannot be used to form re-entrant inner details, limiting the range of shapes that can be cast.

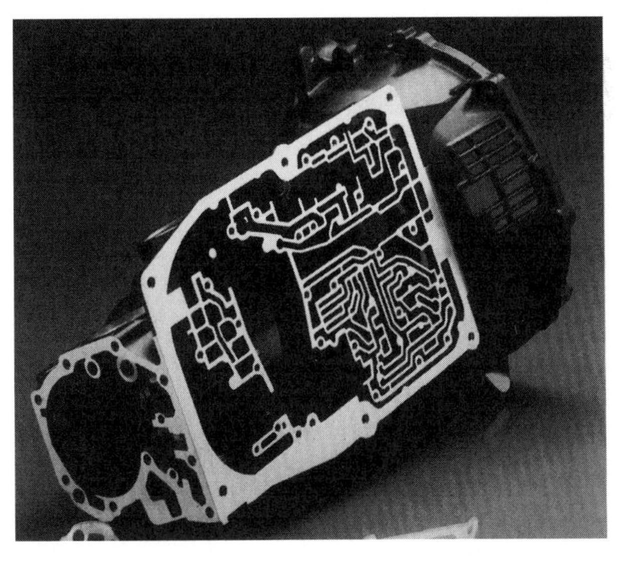

Fig. 6.18: Gearbox casing for a passenger car in AlSi9Cu3 (380.0 type), pressure die-cast. (Alumetal)

The chief objective of the newest variants of the pressure die casting process is to avoid the compressed gas inclusions that are a feature of conventional pressure die casting and which cause porosity in the cast structure and blistering during heat treatment. Avoiding entrapped gas makes the castings heat-treatable and weldable. This is achieved either by evacuating both the die and the casting chamber (vacuum die casting) or by slow filling while evacuating the die. The purpose of filling more slowly and with much reduced pressure is to minimize turbulence. Such processes need suitably designed injection systems with special controls to program the speed of the casting piston.

6.3.3.4 Vacuum die casting

The vacuum die casting process takes two different forms.

By the first method, the melt is poured into the casting chamber in the traditional way (i.e., from a ladle). The die is evacuated after the casting piston has advanced past the pouring hole.

The second method is illustrated in Fig. 6.19. The melt is drawn into the casting cavity by the vacuum itself via a ceramic tube that is attached to the casting chamber and dips into the liquid melt. This variant has the advantages that more time is available to evacuate the die and transfer is less turbulent. The pressure in the casting chamber and the casting cavity must be less than 50 millibars to ensure a satisfactory casting. Castings made in this way can be welded or undergo solution heat-treatment at temperatures above 500°C with no risk of gas blisters. Depending on the choice of alloy and heat-treatment, they can have high mechanical strength with relatively high ductility. Fig. 6.20 shows several examples of the improved mechanical properties that can be obtained by vacuum die casting after appropriate heat-treatment. The technique has made it possible to penetrate new application areas, replacing for example impact-extruded, sand-cast, permanent mold-cast, and forged parts as well as, on occasion, steel. This is especially important for mass-produced parts such as are used in making motor vehicles. Vacuum die-cast and heat-treated parts are often chosen for safety-critical components because of their favorable combination of properties.

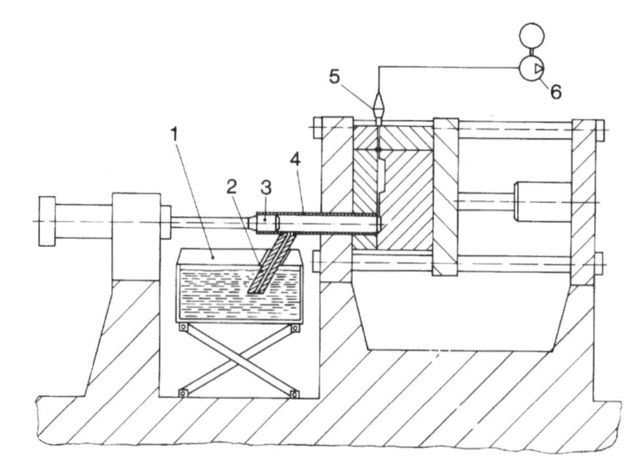

Fig. 6.19: Schematic representation of the vacuum pressure die casting process: 1—holding crucible; 2—suction pipe; 3—casting piston;, 4—casting chamber; 5—vacuum valve; 6—vacuum pump. (Vacural, after F. Kirch and W. Schneider)

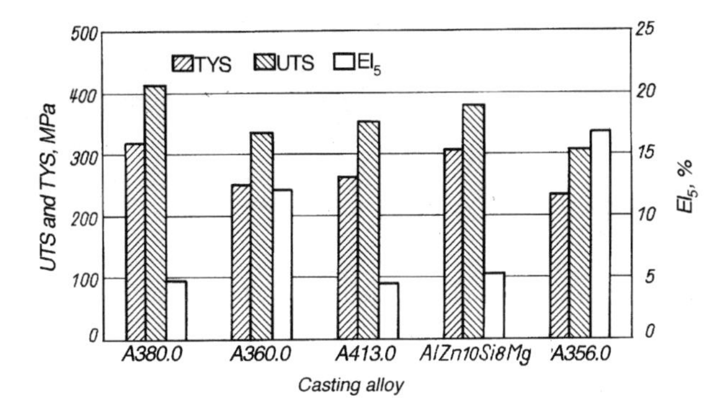

Fig. 6.20: Mechanical properties obtained by vacuum die casting of various alloys. Sample thickness: 3.5 mm, heat-treated. (VAW)

6.3.3.5 Pore-free die casting (PFDC)

In this method of pressure die casting, developed and used in Japan, the air in the casting cavity is displaced by oxygen. This oxygen reacts with the melt and forms aluminum oxides. The main purpose of the process is to deplete gas within the mold. Hydrogen plays no role. The process requires special gating to fragment the injected stream. The Al_2O_3 reaction product is very finely dispersed in the cast structure where it is said to have no negative effects. PFDC is used in Japan for producing wheel rims for cars and, recently, for utility vehicles.

6.3.4 Squeeze casting

This is another innovative die casting process that takes two forms: direct and indirect.

In direct squeeze casting, the die is filled with a defined amount of liquid metal via a trough. The die, which may have two or more parts, is closed from above by hydraulic pressure, forcing the metal to conform to the die surfaces. The pressure is maintained during solidification and can be from 500 to 1500 bar, depending on the size and wall thickness of the casting.

Unlike direct squeeze casting, the indirect version has proved itself in production and has obtained a foothold in foundries producing castings for special applications, chiefly automotive wheels and pistons. The principle of operation is shown in Fig. 6.21. It differs from the direct process in that the melt is not poured directly into the die, but into a casting chamber situated below it. The bottom of the chamber is sealed by a piston whose actuating cylinder is fixed to a pivot. After filling, the chamber is swung into position under the die and the piston is raised to fill the die. Once this is accomplished, the cylinder pressure is increased to 1000–1200 bar and held at that level until solidification is complete. The advantages of indirect over direct squeeze casting are that the die can be filled with the least possible turbulence and that horizontally or vertically closed dies can be used as well as multiple dies. The high pressures used in squeeze casting do not merely ensure optimal filling of the die; they prevent the premature separation of the solidifying shell from the die wall, which occurs with casting methods other than pressure die casting. This ensures better heat transfer, resulting in a greater rate of solidification, which, in

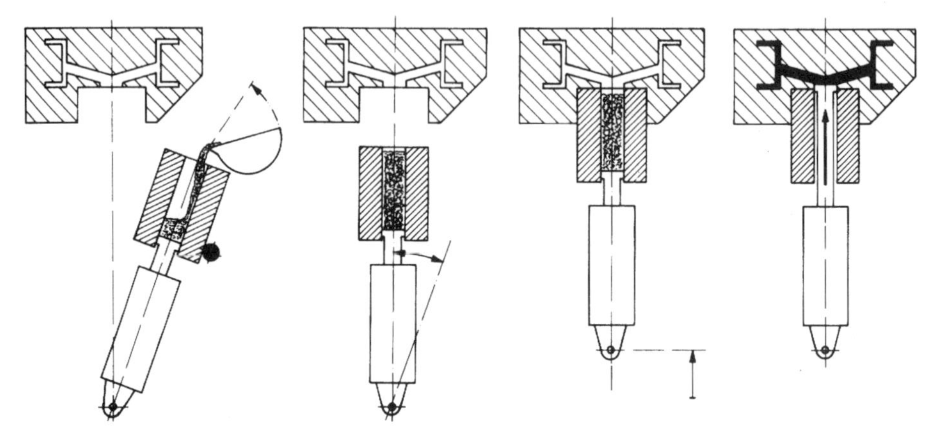

Fig. 6.21: Schematic representation of the indirect squeeze casting process. (After G.A.Chadwick and coworkers)

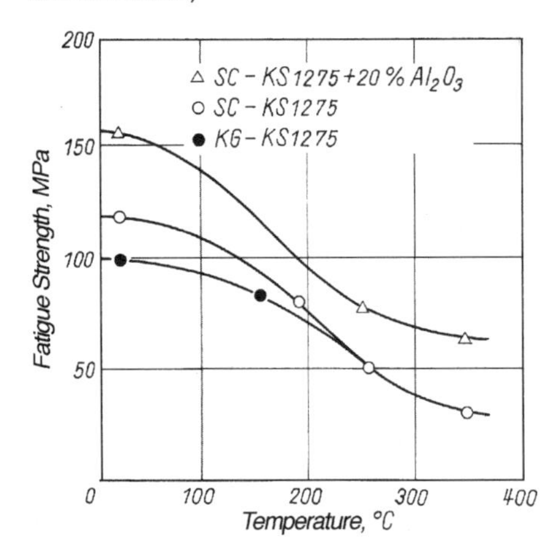

Fig. 6.22: Rotating-beam fatigue strength of G-AlSi2CuNiMg (KS 1275). KG—die-cast; SC—squeeze-cast. (Kolbenschmidt)

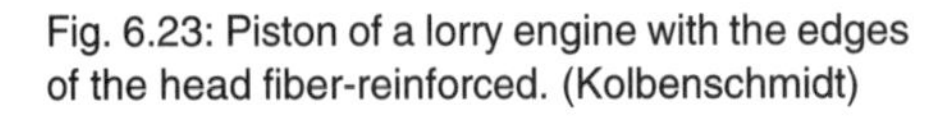

Fig. 6.23: Piston of a lorry engine with the edges of the head fiber-reinforced. (Kolbenschmidt)

turn, leads to a finer cast structure and improved mechanical properties, in particular to higher yield and ultimate tensile strengths.

Indirect squeeze casting is especially suitable for making fiber-reinforced castings. These have fiber-shaped elements (fiber cake preforms) cast into them in order to increase their strength at elevated temperatures, among other properties. The fibers are commonly made of silicon carbide or alumina. If the cast component is to be usable, it is essential for the melt to infiltrate between the fibers. This requires a low filling speed and relatively high pressure during solidification. Increased strength at high working temperatures is especially important for pistons. At temperatures above 300°C, a casting reinforced with alumina fibers has double the rotating-beam fatigue strength of the same casting without reinforcement. Temperatures as high as this occur, for example, at the edges of the piston head of a diesel engine under heavy load and in high specific output automobile engines. Fig. 6.22 shows the variation with temperature of the rotating-beam fatigue strength of a piston alloy, first permanent mold-cast, second squeeze-cast, and last squeeze-cast with fiber-reinforcement.

Under severe temperature cycling, castings with fiber reinforcement can have three to four times the strength of parts made without reinforcing additions. One of the first applications in mass production, shown in Fig. 6.23, is a lorry engine piston head with fiber-reinforced edges.

6.3.5 Thixocasting[9]

A novel casting process that has lately been the subject of much attention is thixocasting. The technique takes advantage of the thixotropic properties of metal alloys in the partly solidified state. Thixotropic behavior means that the material behaves as a solid when at rest but flows like a liquid when rapidly deformed, its viscosity falling as the stress increases. To bring this about, the metal has to be heated to a point in its freezing range between the solidus and liquidus temperatures. This point should be chosen such that for example 40% of the volume is liquid, the rest remaining solid.

Aluminum alloys having a long freezing range are suitable for thixocasting (e.g. 356.0 and 357.0). To attain a good thixotropic state, the α-crystals in the solid solution must be equiaxed in order to ensure that the liquid and solid phases flow uniformly and do not separate. Such a structure can be created by so-called "rheocasting," whereby the melt is electromagnetically stirred during solidification. This breaks off or melts off the dendrite arms, which condense to globular shapes when held just above the solidus temperature, resulting in indigenous growth of free-floating crystals with the desired equiaxed structure (see Fig. 6.24). This method is used, for example, to cast logs up to 150 mm diameter for semifabricated products. Another technique is strain-induced melt activation (SIMA), in which the material is given preliminary cold-working; during subsequent heat-treatment, the stored strain energy drives recrystallization to form equiaxed grains of α solid solution. The SIMA process is so far only suitable for producing small samples. Yet another method is mechanical agitation during solidification.

[9] Nomenclature varies; the names semi-solid forming, semi-solid forging, thixoforming, and thixoforging are also used.

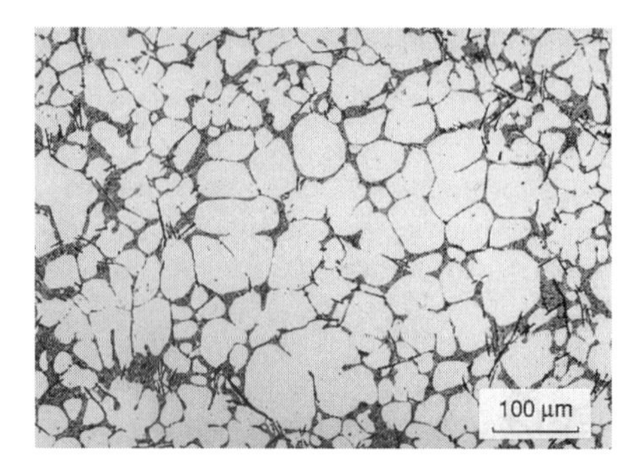

a

Figs 6.24: Commercial thixocast parts and the equiaxed development of the α-crystals in the solid solution before and after deformation (Thixo-structure). Here, the shape and size of the primary crystals remain unchanged, the solidification process being limited to the residual melt in the thin layers between them. (a) Microstructure of a log in A357.0-T6 and (b) microstructure of a landing gear component "thixoformed" in a die casting machine from the log shown in Fig. 6.39.a (SAG, Bühler). (c) 356.0 inner turbo frame for Mercendes trucks (SAG) and (d) 356.0/357.0 automotive components (Alusuisse).

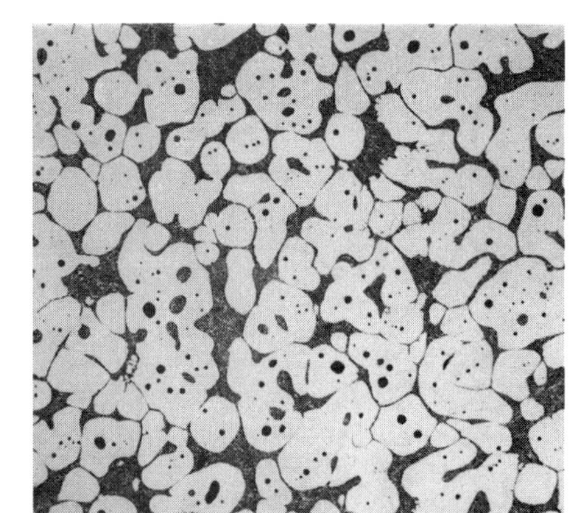

b

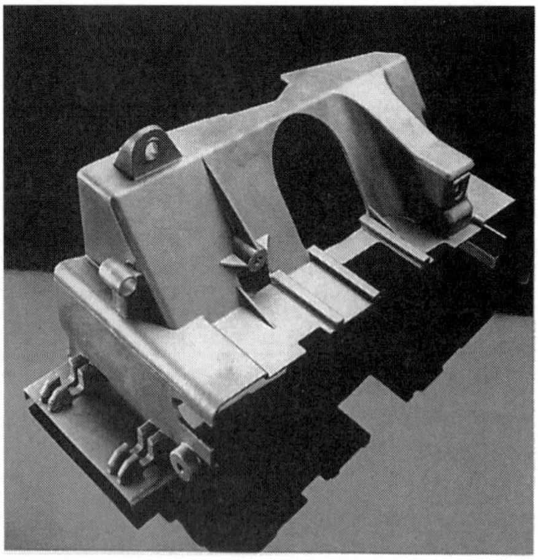

c

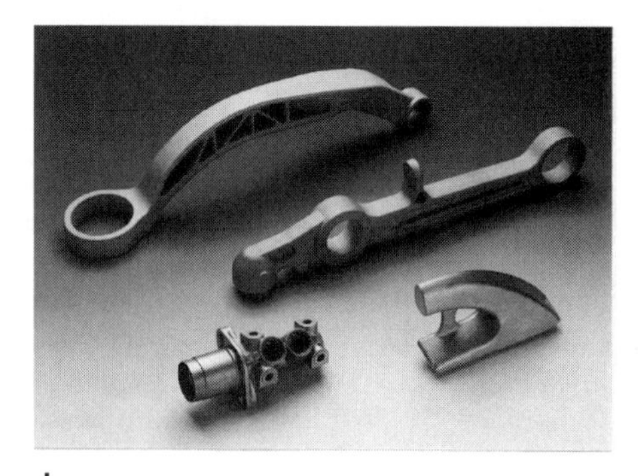

d

Current practice in thixocasting is as follows. The log is first sawn into slugs having sufficient weight for the part to be cast. The slugs are then preheated to a predetermined temperature in their melting range. This is done automatically using temperature sensors to attain a definite proportion of liquid metal, often between 35% and 50%. At this temperature, the slug is still behaving as a solid. It is placed in the casting chamber of the casting machine and is injected into the mold cavity by the casting piston. During preheating and casting, precautions must be taken to ensure that oxide skins from the surface are kept out of the casting.

A number of advantages over conventional pressure die casting methods are claimed for thixocasting, provided that key process parameters are kept under control. These advantages include a significant energy saving, including a large part of the heat of fusion and the whole of the additional heat, as well as energy for holding the melt at temperature. A further advantage is better dimensional tolerances owing to the reduced shrinkage. Since the cast part is closer to final dimensions ("near net shape"), it needs less further working and generates less scrap to be recycled. Productivity can be higher than with traditional methods of pressure die casting. Because the process temperature is about 100°C lower, the temperature cycling is less severe, lengthening the tool life. This also makes it possible to use low-iron alloys because the cooler melt has less tendency to attack the die by dissolving iron. A vital advantage is that the mold fills without air inclusions. By using a partial vacuum in the die cavity, leak-proof castings can be produced, which can be welded and heat-treated. The present-day demand for highly ductile, safety-critical components can be satisfied by rapidly solidified thixocasting alloys containing less than 0.15% iron.

The peculiar characteristics of the thixocasting process make it suitable for casting variants of the usual wrought alloys. Thixocasting is a good way of producing particle-reinforced materials because the particles have less tendency to separate out of the partially solidified melt.

Besides the classical methods of mold casting, a variety of specialized processes are also used. These include investment casting, also known as lost wax.

6.3.6 Lost-wax and lost-foam casting

In lost-wax casting, both mold and pattern are destroyed at each cast. The ceramic mold is made as a single piece. A number of patterns are prepared using injection-molding and are assembled together in a "tree," the patterns being the branches and the feed channel the trunk. The whole is coated by dipping in several baths of ceramic slurry. After these layers of slurry have hardened, the wax is melted out and the ceramic mold is fired, ready for casting, usually under vacuum, into the preheated mold. Very intricate castings can be made to close dimensional tolerances using this technique.

In lost-foam casting also, both mold and pattern are destroyed during each cast. The pattern is made of polystyrene foam, which is easily vaporized, while the mold is of sand without binder. The patterns are made by blowing the foam into specially designed molds (Fig. 6.25). After washing with parting compound, the pattern is embedded in the sand which is then compacted under vibration. The foam pattern remains in the mold and is

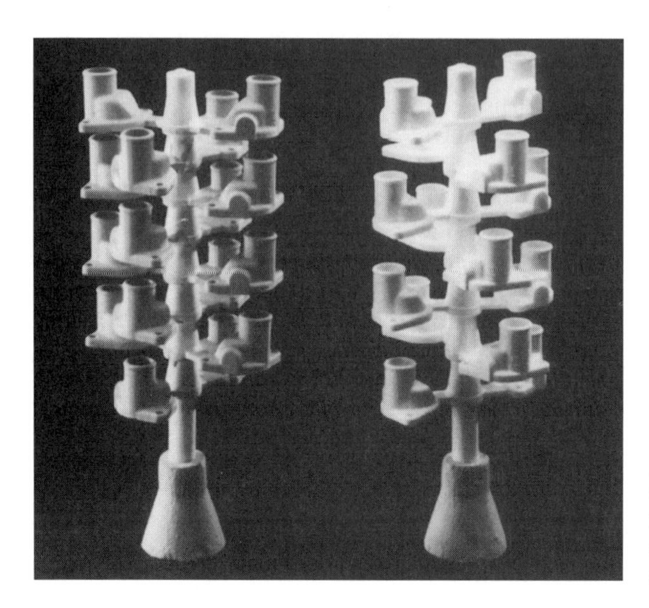

Fig 6.25: "Tree" assembled from polystyrene patterns for investment casting, before and after coating with ceramic slurry. (Honsel)

Fig. 6.26: Cast aluminum parts made by lost-foam casting. (Honsel)

vaporized as the melt fills the mold. The wash serves mainly to control the escape pressure of the vaporized foam and to support the sand in the gap between the melt and the still solid foam. The gas permeability of the wash coating is critically important: it must be regulated so as to provide a vapor pressure in the gap high enough to support the coating, but at the same time it must not seriously hinder the escape of the vaporized foam. In one variant of the process, solidification takes place under a pressure of 3–10 bar.

The chief advantage of lost-foam casting is that neither sand cores nor binding agent are needed, so that the preparation of the sand is simplified and the amount of waste sand reduced. With careful attention to practice, tolerances can be attained that are as good as those obtained by pressure die casting. The technique also makes possible constructional features that are impossible with pressure die casting or mold-casting such as, for example, precast channels or hollows without using cores. Fig. 6.26 shows a selection of components which have been made using lost-foam casting.

6.4 Quality Aspects of Mold Casting and Related Process Parameters[10]

6.4.1 Introduction and overview

The mold casting of metals represents the shortest route from primary metal to finished product, and makes possible the economical production of complicated shapes. Essentially, liquid metal is poured into a mold where it solidifies, taking the form of the cavity. As the liquid metal solidifies there is a solidification shrinkage that tends to produce porosity or larger voids. To prevent or reduce these defects, further liquid metal must feed to the casting during solidification. In gravity casting, this liquid metal comes from risers placed higher and which solidify later than the product. In pressure casting, continued liquid metal input pressure is maintained during solidification.

In gravity casting, the liquid metal is poured from a container into the funnel-shaped *sprue* and fills the mold cavity only under its own weight. There are basically two ways to fill the cavity:

1. *Top gating*: From the sprue, the liquid metal enters through *gates* sited above the mold cavity. Flow is turbulent, the liquid metal filling first the mold cavity and last the risers. Solidification tends to be from bottom to top, which favors feeding from the risers.
2. *Bottom gating*: From the sprue, the liquid metal is channelled through *runners* and enters through gates sited below the mold cavity. This reduces turbulence, but the risers receive the first-poured and subsequently cooled liquid metal. Solidification then proceeds in two directions, towards the gates and towards the risers. Fig. 6.33 schematically presents different arrangements of risers. The individual mold casting processes were described in the previous section of this chapter.

[10] By F. Kahn.

6.4.2 The quality aspect

Current processes are optimized, and new ones are being developed, with the objective of improving quality. Controlled pressure is an important aspect in this, because it enables better filling (thinner walls, more detail) as well as better feeding. The level of pressure applied varies. The mold is filled at low pressure so as to minimize turbulence and the consequent gas and oxide inclusions, whereas high pressure is used during solidification in order to reduce porosity and shrinkage voids.

Solidification creates a cast structure that largely determines the properties of the cast product. This cast structure is characterized by cast phases (e.g., aluminum-based solid solution and heterogeneities like eutectic phases), by the cast grain shape ("morphology," for example equiaxed, columnar), by segregation and structure defects such as porosity (gas pores, micro-shrinkage voids), and by impurities (e.g., oxide inclusions). The foundryman can improve the cast structure by various means. These means include melt treatment to remove impurities (degassing, filtering) and to influence grain and phase structure (adding grain refiner or modifier), the choice of appropriate casting conditions, and devices to influence solidification in the mold. Such devices can be metal chill plates in sand molds to locally encourage a pore-free zone within the casting.

After casting, the product's properties may be improved by suitable heat treatment. But to attain optimum properties, we one must first choose an appropriate alloy and casting process. Aluminum casting alloys are suitable for practically all casting processes. These casting alloys belong principally to the alloy families Al-Si, Al-Si-Mg, Al-Si-Cu, Al-Mg, and Al-Cu-Ti. The various alloys vary in their ease of casting, called castability (see Section 6.4.6).

For the best cast quality, components must be *designed with a view to the casting process* to be used. This requires close co-operation between designer and foundryman so as to avoid casting problems from the start (Simultaneous Engineering, see Chapter 17).

6.4.3 Aluminum foundry alloy types and their properties

In comparison with wrought alloys, casting alloys contain larger proportions of alloying elements such as silicon and copper.[11] This results in a largely heterogeneous cast structure (i.e., one having a substantial volume of second phases). This second phase material warrants careful study, since any coarse, sharp, and brittle constituents can create harmful internal notches and nucleate cracks when the component is later put under load. The fatigue properties are very sensitive to large heterogeneities. As will be shown later, good metallurgical and foundry practice can largely prevent such defects. Table 6.2 groups by application the most common standard casting alloys, together with their tensile and hardness values, with the casting processes used.

The elongation and strength, especially in fatigue, of most cast products are relatively lower than those of wrought products. This is because current casting practice is as yet

[11] There is no international accord on aluminum casting alloys, so both USA and ISO/DIN alloy designations are generally shown; the compositions may vary and so the properties may not always be identical.

Table 6.2: Range of representative properties for some commonly used aluminum casting alloys[12]

Application group	Alloy and Temper	Casting Process	UTS (MPa)	$TYS_{0.2}$ (MPa)	El_{50} (%)	Hardness (HB5/250)
General uses	B413.0/AlSi12-T6	S	140–200	70–100	3–10	45–60
	332.0AlSi9Cu3-T6	PM	160–240	100–160	0.5–3	70–110
	A360AlSi10Mg-T6	S	200–230	170–260	1–4	80–110
High-strength components	364.0/AlSi11-T6	PM	150–230	80–110	6–13	45–65
	A357.0/AlSi9Mg-T6	PM	240–340	190–280	3–7	80–115
	A356.0/AlSi7Mg-T6	I	230–320	190–260	3–6	80–115
	201.0/AlCu4Ti-T6	S	240–380	160–230	3–10	85–105
Especially good corrosion resistance	512.0/AlMg3-F	S	130–190	60	3.0	50–60
	355.0/AlSi5Mg-T51	PM	140–200	100	1.0	60–75
Pressure die casting	380.0/AlSi9Cu3-F	D	240–310	140–240	0.5–3	80–120
	A413.0/AlSi12(Cu)-F	D	220–300	140–200	1–3	60–100
	360.0/AlSi10Mg-F	D	220–300	140–200	1–3	70–100
	520.0/AlMg9-F	D	200–300	140–300	1–3	50–100

(1) Properties from separately cast tensile test pieces.
(2) S = sand casting; PM = permanent mold casting; I = investment casting; D = Die casting.

unable to reliably prevent casting defects. In recent years however, innovations in casting processes have brought about considerable improvements, which should be taken into account in any new edition of the relevant standards.

The first group in Table 6.2 are the general-purpose alloys, which are used to make components with moderate strength. These alloys can be die or sand cast. It is important to appreciate the difference between the properties of die cast and sand cast components, a difference that applies generally to all casting alloys. Thanks to the finer structure attained in pressure die cast components, they have better elongation, tensile, and fatigue properties than similar sand cast or conventionally die cast components.

Alloy B413.0 / AlSi12 is notable for its very good castability and excellent weldability, which are due to its eutectic composition and low melting point of 570°C. It combines moderate strength with high elongation before rupture and good corrosion resistance. The alloy is particularly suitable for intricate, thin-walled, leak-proof, fatigue resistant castings.

Alloy 332.0 / AlSi9Cu3 is one of the most frequently used aluminum casting alloys. This is because it is made almost exclusively from recycled scrap and is, therefore, relatively cheap. It has only medium strength, low elongation, and reduced corrosion resistance compared to purer, primary alloys. Its main uses are for machine components, household appliances and, thanks to its high strength at elevated temperatures, internal combustion engines.

[12] For compositions, see Appendix D.

Alloy A360.0/AlSi10Mg is heat treatable, thanks to a small magnesium addition of 0.4–0.6%. After heat treatment, its strength is substantially higher, with some loss of elongation. Otherwise, the alloy has similar properties to B413.0; it is suitable for high-strength engine components such as crankshaft casings and cylinder heads.

The second group of alloys in Table 6.2 have particularly good mechanical properties and are especially suitable for lightweight components requiring high strength and toughness.

Alloy 364.0/AlSi11 has medium strength, and, especially when die cast, it can achieve excellent elongation values. It is, therefore, suitable for complicated, thin-walled, leak-proof castings resistant to fatigue and impact loads.

Heat-treatable alloy A357.0/AlSi9Mg possesses good strength and elongation combined with excellent castability, thanks to its near-eutectic composition. It resists corrosion well because of its reduced impurity levels. It is used for complicated, thin-walled parts that require good levels of toughness. It is also used for aerospace applications.

Another heat-treatable alloy with high toughness is A356.0/AlSi7Mg, which is much used in all casting processes and is the first-ranking alloy for investment casting. Examples of its uses are aerospace castings, car steering casings, and vehicle wheel rims, as well as bicycle wheel hubs made by a special pressure die casting process.

The strongest of the common casting alloys is heat-treated 201.0/AlCu4Ti. Its castability is somewhat limited by a tendency to microporosity and hot tearing, so it is best suited to investment casting. Its high toughness makes it particularly suitable for highly stressed components in machine tool construction, in electrical engineering (pressurized switchgear casings), and in aircraft construction.

The common feature that the third group of alloys has is excellent resistance to corrosion. Alloys 512.0 and 514.0 (AlMg3 type) have medium strength and good elongation and are suitable for components exposed to sea water or to similar corrosive environments. These alloys are often used for door and window fittings, which can be decoratively anodized to give a metallic finish or in a wide range of colors. Their castability is inferior to that of the Al-Si alloys because of their magnesium content and, consequently, long freezing range. For this reason, they tend to be replaced by 355.0/AlSi5Mg, which has long been used for similar applications.

The last group in Table 6.2 summarizes the most important alloys for die casting—the method most used world-wide to make aluminum castings. By the use of die casting methods, the alloys 360.0/AlSi10Mg, 380.0/AlSi9Cu3, and A413.0/AlSi12 (already discussed above) attain significantly higher strength and hardness, but at some cost to elongation when compared with sand and permanent mold castings. For die castings where decorative anodizing is particularly important, alloy 520.0 is the most suitable.

Conventional die casting (at low pressures) tends to yield noticeably low elongation values, as outlined earlier, and is, therefore, unsuitable for safety-critical components. In recent years, higher pressure die casting (e.g., squeeze and thixocasting) has been devel-

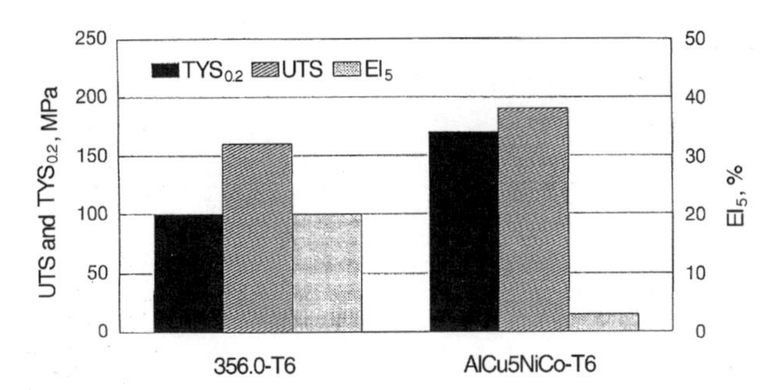

Fig. 6.27: Tensile properties at 200°C of two sand-cast aluminum alloys (best-achievable data from Honsel).

oped further, as described in Section 6.3. As a result, elongation values of well over ten percent are now attainable, together with higher strengths. This considerably widens the range of application for pressure die castings.

Besides the standard aluminum casting alloys, there are special alloys for particular components, for instance: engine piston heads, integral engine blocks, or bearings. For these applications, the chosen alloy needs good wear resistance and a low friction coefficient, as well as adequate strength at elevated service temperatures. A good example is the alloy 203.0 / AlCu5NiCo, which to date is the aluminum casting alloy with the highest strength at around 200°C. Fig. 6.27 reveals the strength of this alloy after heat treatment in comparison with that of A356.0 / AlSi7Mg at 200°C. It is used for high-strength gearbox housings.

Another interesting alloy is B390.0 / AlSi17CuNiMg, which is low-pressure cast to make engine blocks for luxury cars. The cast structure contains hard primary silicon crystals. A special treatment causes these crystals to protrude slightly from the cylinder bore and, thus, provide excellent wear resistance for the cylinders and also for the pistons, which are made of a similar alloy. Engines made using these alloys achieve considerably higher performance than is possible with conventional gray cast iron cylinder blocks.

Further development of cast aluminum alloys seems unlikely with conventional alloying techniques. Instead, new composite materials will be produced. One starting point is the long-established method of composite casting by which steel inserts are placed in the mold and become integrated in the aluminum casting (as in pistons for heavy vehicles). This leads to other methods of reinforcement, such as using high-strength fibers or suitable insoluble particles (see Chapter 14). Another possibility is to use alloys that solidify with large amounts of heterogeneous precipitates and to arrange casting conditions so that these phases grow directionally to achieve the desired increase in strength in the principal loading direction.

6.4.4 Melting methods

Like all light metals, aluminum casting alloys are highly reactive in the molten state. This necessitates special care in the design and operation of melting, holding, and casting furnaces. These are heated by either gas or oil or electrically by either the use of resistance elements or by electromagnetic induction coils. In small and medium-sized plants, the crucible furnace remains the usual melting unit. It is a fireclay / graphite or fireclay / sili-

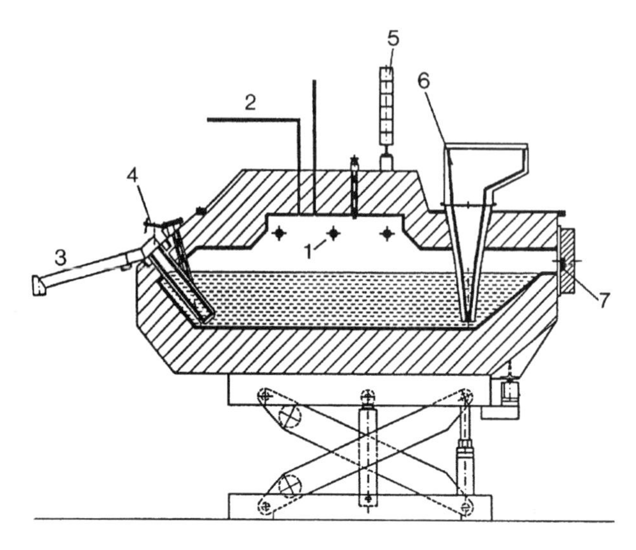

Fig. 6.28: Injection furnace for aluminum alloys: 1—electric resistance heaters; 2—gas pressure regulation; 3—casting channel; 4—riser sensor; 5—metal level indicator; 6—charging device; 7—safety plug. (Westofen)

con carbide (SiC) pot holding up to 500 kg of aluminum and designed to be easily loaded whole into an insulated furnace chamber. An alternative method offering considerable savings in energy and melting costs is to have liquid metal delivered directly from the smelter.

Casting and batch injection furnaces are most often built on to pressure die casting machines, but they have also proved their value on turntable casting machines, in permanent die casting, and in sand casting plant for high production rates. These furnaces allow faster cycle times and lower casting temperatures. The electric resistance heated injection furnace shown schematically in Fig. 6.28 is not of the crucible type but is of a more specialized design in which low-pressure gas under electronic control displaces a measured charge of liquid metal. The melt is generally subjected to most of its cleaning and other treatments before being transferred into the casting or injection furnace. To ensure uniform melt composition and temperature, all of these furnaces must provide some mixing movement by convection or by magnetic induction.

6.4.5 Metallurgical measures to assure melt quality

Two of the major problems when melting aluminum and its alloys are oxide formation and hydrogen absorption by the melt. Both can seriously affect cast quality by producing oxide inclusions or porosity. Melting practice, therefore, usually involves special measures to check for and, if necessary, eliminate them. The principles of melt treatments are described in Chapter 5. In the following paragraph, we describe melt treatment specifically for casting alloys.

6.4.5.1 *Grain refining*

Grain refiners, which are usually materials that liberate titanium, boron, or carbon, are generally added in the form of master-alloy wire. Both in wrought and in casting alloys, grain refining is a well-proven method to influence the nucleation conditions in a melt, so that it solidifies with as fine-grained and dense a structure as possible. The main advantages in mold casting are:

- Improved flow to give better castability, especially evident in better feeding of the partly solidified regions;
- Reduced tendency to hot cracking;
- Reduced shrinkage and gas porosity; and hence
- Improved strength and elongation in the product.

Hyper-eutectic AlSi alloys can be grain-refined with additions that release phosphorous, which promotes the nucleation of primary silicon.

6.4.5.2 *Modification*[13]

Modifying Al-Si alloys of eutectic and hyper-eutectic (excess Si) composition means treating the melt to hinder primary silicon from precipitating to form coarse, irregularly shaped particles. We can basically distinguish three types of structure in these Al-Si alloys: coarse crystalline, lamellar, and modified (Fig. 6.29).

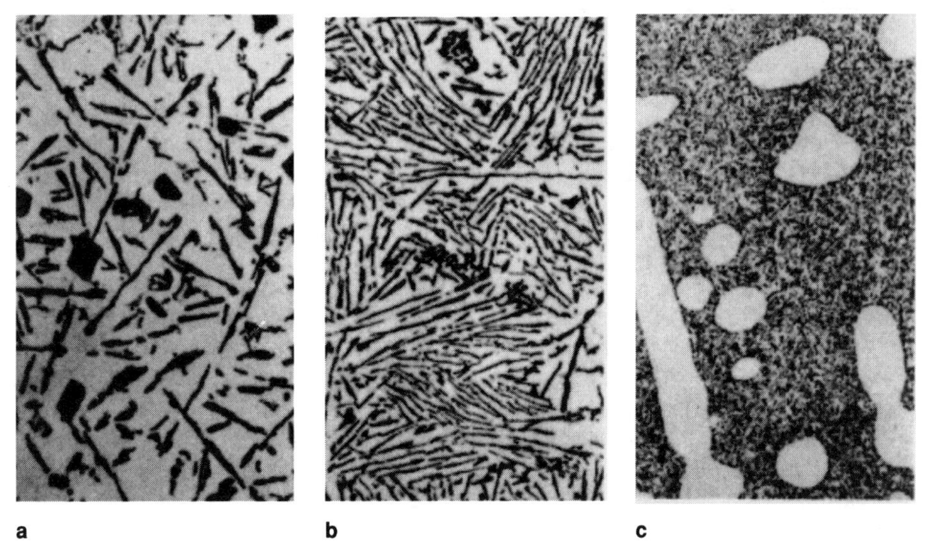

Fig. 6.29: Three types of structure in cast B413.0/AlSi12: (a) angular, (b) lamellar, and (c) modified.

[13] In the USA, "modification" refers only to hypoeutectic structures. The primary phase is called "refinement."

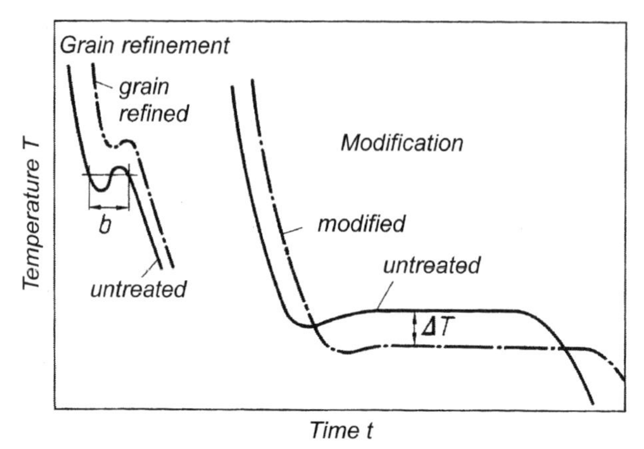

Fig. 6.30: Thermal analysis cooling curves for Al-Si alloys. Left: The influence of grain refinement on the undercooling of primary aluminum in a hypoeutectic alloy. Right: The influence of modification on the eutectic solidification plateau in a eutectic alloy.

Coarse crystalline structure is mainly characterized by the appearance of angular primary silicon particles of various sizes in a fairly coarse eutectic matrix. It occurs at higher active phosphorous concentrations (>100 ppm P) and is the usual structure resulting from untreated melts. As a rule, such a structure is undesirable from the user's point of view, since it reduces strength and elongation values.

In the *lamellar* structure, silicon forms bundles and fans of silicon plates and needles, which individually can reach several millimeters in length in the cast structure. Eutectic grains can grow to several centimeters across during solidification. In sand castings this lamellar structure can cause brittleness, whereas in die casting it tends to be modified in the course of solidification. Nevertheless, these castings are often unusable because of surface porosity or sub-surface porosity, usually called "pin-hole porosity," which is almost impossible to detect reliably. Lamellar crystallization occurs only at very low phosphorous and sodium levels, a situation that is rare in practice.

Finally, the modified structure is typified by a finely dispersed silicon phase present in the eutectic. This results in much higher tensile strength and elongation, and enabled a breakthrough in the use of AlSi castings in the 1920s, beginning in Germany. The melt can be modified by adding capsules of metallic sodium or compounds that release sodium. Alternatively, additions of strontium have proved successful in die castings in recent years. In contrast to sodium, which burns off and is lost fairly quickly, strontium lasts longer and may be included in remelt ingot. No other attempts to introduce reliable modification methods have so far succeeded.[14]

To monitor both grain refinement and also modification treatments, thermal analysis has proved a useful inspection tool in recent years. Fig. 6.30 shows schematic cooling curves for treated and untreated samples of two alloys, one hypoeutectic (less Si) and one eutectic composition. The melt samples were poured into beakers of molding sand containing thermocouples. They were then allowed to solidify slowly enough to avoid the effects of high cooling rates on crystallization. The hypoeutectic sample without grain refiner clearly shows pronounced undercooling (interval b) before showing a brief rise in temperature caused by the heat of solidification of the primary aluminum crystals. With the eutectic

[14] Antimony additions are called eutectic grain refiner, and in die-cast AlSi7Mg they produce finely lamellar eutectic silicon.

alloy, on the other hand, the modified sample shows that hindering silicon crystallization depresses the eutectic solidification plateau by an amount (ΔT) throughout solidification. This temperature difference ΔT is a measure of the degree of *modification,* and, according to practical experience, it should amount to 6–10°C.

According to current theories, sodium or strontium at levels of 100–200 ppm can cause modification by inactivating the aluminum phosphide nuclei. Without these nuclei, primary silicon crystallization and its eutectic are retarded, and its degenerate, angular form is thus suppressed.

6.4.6 Casting properties

The designer wants high-strength light alloy components with excellent dimensional accuracy and surface quality. However, he is limited not only by material properties and by the shapes that the forming process can make, but also by the alloy's casting behavior. The term "castability" groups together several important properties of melts that are very important for making a satisfactory product. Among the most important of these are:

- Fluidity and mold-filling ability
- Solidification characteristics such as the tendency to "piping" voids
- Feeding ability in the semi-solid state
- Tendency to hot tearing

A dominant factor in castability is the solidification type, or "morphology," of the cast structure, since in most industrial casting processes (except investment casting) solidification starts as the mold begins to fill. Fig. 6.31 shows the main basic forms of solidification that can occur, depending on the type and quantity of alloying elements used. Combinations of these forms tend to occur in casting production.

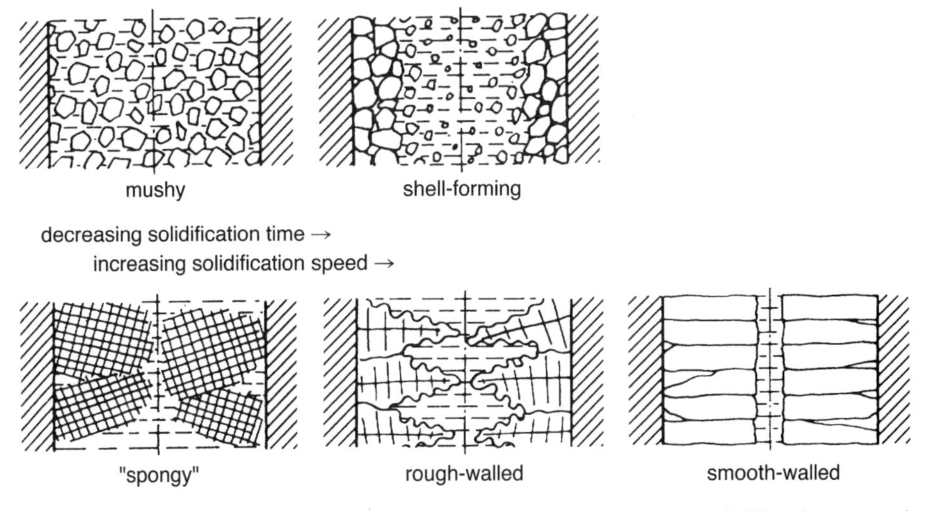

Fig. 6.31: Typical solidification types and the influence of solidification speed. Upper row: Indigenous types. Lower row: Exogenous types.

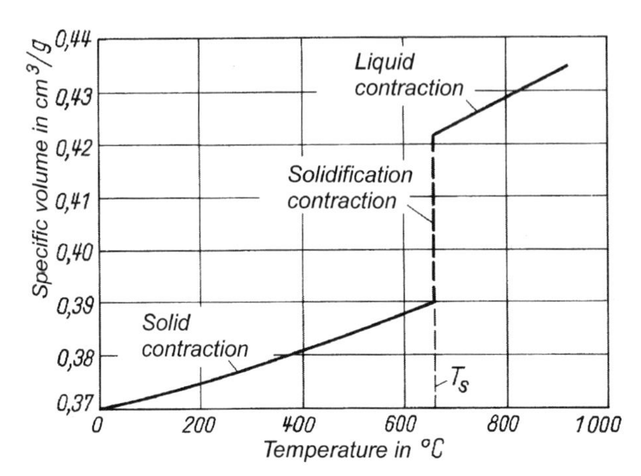

Fig. 6.32: Specific volume of aluminum as a function of temperature.

Among the indigenous types (crystallizing on free-floating nuclei) that produce equiaxed grains, we call the solidifying melt "mushy" when crystals grow fairly uniformly across the whole section being considered. When heat is rapidly extracted through the walls of the mold, a shell forms at the mold wall, and this makes for more complete filling and feeding of the mold. Also favorable are the exogenous types, where crystals nucleate on the mold walls. Best for feeding is the smooth-walled type, but this occurs only with modified eutectic castings of the B413.0/AlSi12 variety. All other alloys with a longer freezing ranges tend to be in the spongy-mushy class.

An alloy's fluidity and mold-filling ability can be measured with suitable tests, such as the cast spiral test, which is mainly used in experimental work. This type of test is too expensive and time consuming for normal production use and is generally not necessary, since the actual cast component is its own best witness.

Almost all metals shrink in volume when the atoms form regular crystallinities as they solidify. This is the cause for the occurrence of the so-called *"piping"* voids. There are three distinct parts to the curve in Fig. 6.32, showing how the specific volume of pure aluminum varies with temperature. First, the hot liquid cools and contracts linearly as far as the solidification temperature; this has little effect on cast quality. Secondly, a volume contraction occurs during the solidification phase. This carries the risk of creating voids, whose shape and size depend on the alloy's solidification type, and on the cast product and mold design. Solid shrinkage is shown in the third part of the curve. Accordingly, the designer must calculate the acceptable tolerance between the dimensions of the mold and those of the finished casting.

Solidification shrinkage amounts to 7.1% by volume in pure aluminum. This can be reduced to 4% in some alloys, usually by adding silicon, because instead of shrinking, this element increases in volume by about 10% as it solidifies.

The appearance of shrinkage voids in and on the cast product can vary greatly. They can occur as macroscopic centreline piping cavities, microporosity, surface dishing, or as punctures through to sub-surface voids. Alloys of the smooth wall, rough wall, or indigenous shell solidification types usually show a tendency to piping, whereas the spongy and mushy types tend rather to microporosity. The foundryman has a number of techniques with which he can prevent such shrinkage defects, or at least considerably reduce their severity.

Gas porosity is a closely related phenomenon. Some of it stems from lubricants or coatings used in mold casting, but the bulk results from hydrogen precipitation during solidification in the form of primary porosity with a pore diameter mostly above a few microns and even as large as one millimeter.

Evenly distributed fine gas porosity is desired during casting for a slow solidification to supply a shrinkage deficit in volume from a multitude of small pores rather than from large shrinkage voids.

All of the above refers to primary porosity. Its generation is described in Section 5.1.5. Secondary porosity is that generated by precipitation of hydrogen in solid aluminum at elevated temperatures, with pores of a diameter around one micron or less. It is generally of little or no importance for the quality of castings, the exceptions being higher strength alloy castings that must be heat treated.

The total of all voids in a casting is determined by a measurement of the density or specific weight of the casting. This is well suited for a "go-no go" quality check, but it does not distinquish among the fine, uniformly distributed porosity that may be acceptable and the macroscopic voids that are unwanted.

Hot tearing can occur as cracks between the grains in the mushy state, during solid contraction if this contraction is opposed by the shape of the mold, by friction against its surface, or by the cores used in making hollow castings. This risk is especially high with permanent molds, which cannot flex like sand molds. Cracks can heal only as long as the metal is still locally in a partially liquid state and has adequate opportunity to feed the cracked region. Alloys with a long freezing range are most at risk, being of the spongy or mushy solidification type. For critical shapes, the measures to prevent porosity and cracking should start already with the design of the shape to be cast and with the choice of alloy.

6.4.7 Mold filling

6.4.7.1 Preventing hydrogen pick-up and oxide inclusions

Because liquid aluminum readily reacts with air or moisture, special precautions are needed to prevent oxide skins and hydrogen from entering the melt as it flows into the mold. From the crucible or ladle, the melt should fall only the shortest possible distance in air before entering the mold, which must be completely dry. Entry channels (sprue, runners, and gates) should be streamlined according to hydrodynamic principles. This includes controlling the pressure by narrowing the cross-section area, and avoiding sharp corners and abrupt section changes, where turbulence can cause air to be entrapped. To reduce reaction with gases, it is an advantage to fill the mold upwards from a (first) gate at the bottom of the mold.

a Mold filled from bottom up, with a riser on top of plate standing up.	↑		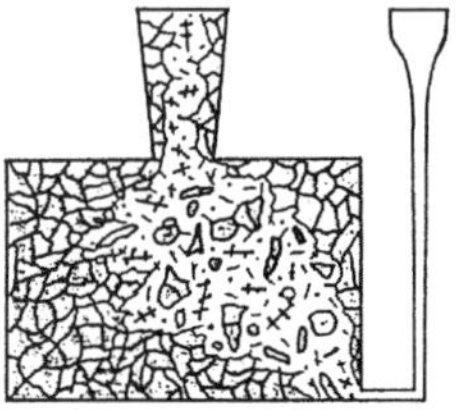
b Riser in the gate for a plate lying flat.	→		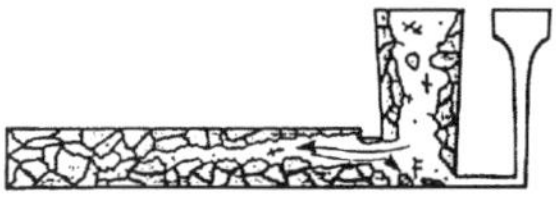
c Riser near the gate for a plate standing up.	→		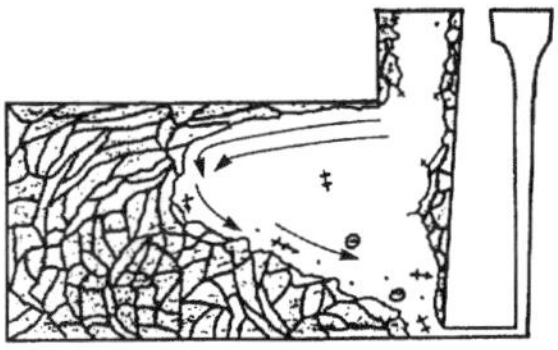
d Continuous feeding from the base in low-pressure casting process	↓		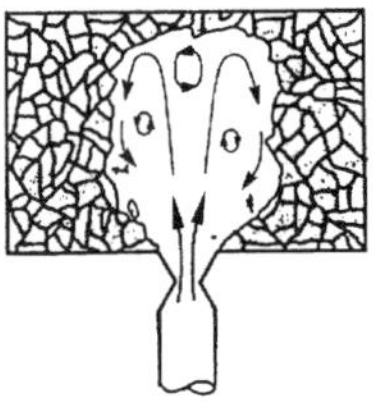
	Solidification direction	Cross-section	Lengthwise section

Fig. 6.33: Techniques of feeding to prevent shrinkage cavities: Examples of casting a rectangular plate in four different orientations, showing two cross-sections of each. Solidifying grains are shown schematically; the mold is not shown here. The melt input funnel and distribution channel (sprue and runner) join to the product at a neck (gate), which is, as a rule, at the lowest point so as to vent air upwards. The riser is an added reservoir for liquid metal to feed shrinkage cavities in the product before it finally solidifies.

6.4.7.2 Controlling directional solidification

Fig. 6.33 illustrates schematically the typical arrangements for feeding solidification shrinkage by using gating and risering techniques. The most common layout in aluminum foundries is still simply bottom gating (Fig. 6.33.a), but this has some drawbacks. Only relatively cool metal reaches the riser at the end of mold filling, and, on the other hand, the lower end of the mold and plate tend to become overheated. This causes the exact opposite of the desired directional solidification towards the riser. A better technique is to use a "layered" filling sequence from the side (Fig. 6.33.b and 6.33.c), in which the mold is filled relatively quickly, the metal starting to freeze towards the riser. With risers on the side, and especially with bottom risers (low-pressure casting, Fig. 6.33.d), hot metal flow reinforces the efficiency of the risers.

The foundryman can also influence the direction of solidification by slightly tapering the thickness, or by adding local chills or insulation to the mold. It is difficult and laborious to achieve perfect directional solidification in all parts of a complicated casting, hence it is difficult to avoid shrinkage porosity in all parts of a casting. The reason lies in the large differences in local solidification time, which is roughly proportional to the square of the thickness. How local solidification time affects dendrite arm spacing and toughness in the component will be described next.

Where, because of reduced cross-section, a part of the casting passes rapidly across the liquidus-solidus interval, a fine grain size, small dendrite arm spacing, and, therefore, high toughness will result in this region. However, compensating for volume shrinkage during casting involves feeding liquid metal from the riser to the advancing solidification front, as well as a sufficient level of superheat in the melt itself. Keeping the feeding channel toward the riser opening means imposing a progressive increase in the total time to cool from superheat to liquidus and, therefore, to cause solidification to proceed in the direction of the riser. The solidification time in general should be as short and as uniform as possible. It is technically feasible to achieve this sequential cooling fairly simply by using "active" cooling in permanent mold casting, as shown in Fig. 6.34.

In this method, the mold (7) with cooling blocks for controlled heat extraction (chills) (2, 3, 4) is filled with melt through the sprue/riser (6). Coolant then flows through the chill (2) furthest from the riser until it has removed the heat of solidification from the adjoining region of the workpiece. When this region has cooled to the solidus threshold, a signal starts the coolant flow in the next chill (3) and so on until the solidification front, driven by this sequence, reaches the riser. In each specific volume element the cooling rate can be measured and controlled by a temperature sensor.

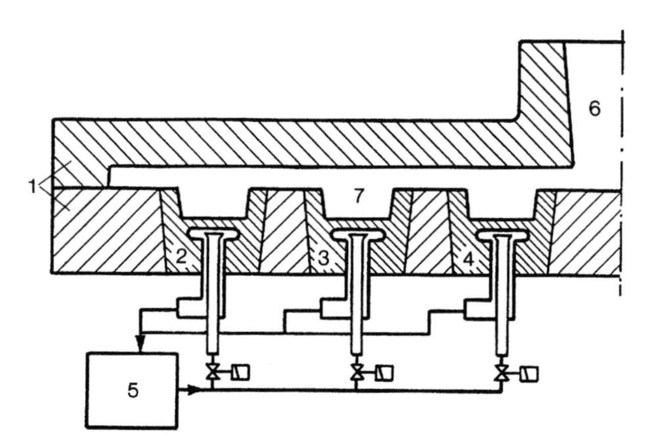

Fig. 6.34: Principles of sequential cooling for active control of solidification in a permanent mold: 1—permanent mold; 2, 3, 4—chills with actively controlled heat extraction; 5—controller; 6—entry sprue and riser; 7—component shape with wall thickness variations.

Mold filling and solidification still cause problems in conventional pressure die casting: the melt is injected with considerable turbulence into the mold cavity, trapping gas bubbles in the structure. In contrast, filling and solidification are usually satisfactory in the low pressure die casting of aluminum alloys.

6.4.7.3 Computer-aided optimization of mold filling[15]

Computer simulation of mold filling and solidification processes is today a state-of-the-art procedure for optimizing both the casting process and quality of the cast components. Even before making a mold we can visualize these critical casting processes by numerical simulation. This enables us to optimize the component geometry, the design of the gating and risering system, and the control of solidification. We can thus avoid expensive trial-and-error changes to the mold, pattern, and melt feeding system. Savings also result from avoiding time-consuming test casts and from the shortening of design and development times. Quality can be improved by anticipating casting defects as well as correcting the design of the cast part and the casting system before the first cast run.

The available mathematical models for mold filling are based on liquid flow and heat transfer equations. A grid represents the component shape to be cast. Temperature calculations based on the heat transfer equations then make predictions either for the grid intersections (finite difference method, FDM) or for the individual elements between them (finite element method, FEM). Today, powerful computers are available for the use and further development of this technique. The following section uses an actual case to illustrate how computer simulation is used to optimize casting design.

6.4.8 Computer-aided optimization of a cast component

6.4.8.1 Basic principles

Mold and process design in the foundry is still dominated by empirical methods in which experience and empirical rules are used to provide the necessary technical base. Physical and thermodynamic laws of solidification are known in a general and idealized forms, but are not applied in their full mathematical depth due to the complexity of production conditions. Since computerized simulations of pouring and solidification have become available, however, we now have an opportunity of visualizing the temperature field effects in the mold and workpiece. Prompt and rapid analysis of the effect of changing various parameters in mold construction and during the casting process allows us to dispense with costly trial-and-error methods. The main consideration here is not just to reduce the cost of tooling but to achieve an overall reduction in the cost of producing castings. Without going into the details of a specific program, the following text describes simulation calculations with a particular software package.

[15] By A. Mundl, including computer-aided optimization of castings.

6.4.8.2 Computer program

The program in question uses both the finite difference and the finite element techniques and provides results in a short enough time to be usable even for complex castings. Before beginning the calculation, the boundary conditions must be defined. These include the initial temperature and the mass flows. Temperature-dependent material properties (thermophysical data) such as thermal conductivity, thermal capacities, and heat transfer coefficients of the materials and coatings are stored in data files.

A practical example

The steering arm shown in Fig. 6.35 illustrates the modern way of developing a structural component as a die casting. Starting with a computer-aided design (CAD) model of the surface, the design was transferred by data link to the simulation program. A finite-element method (FEM) program called *ANSYS* calculated the strength and stiffness of the steering arm. A significant aspect in the economical use of FEM and its integration into CAD is the use of a precursor program to prepare the calculation. The pre-processor helps collect the extensive data necessary to set up the calculation model (Fig. 6.35.a). It is usual to generate the elements from the surface contours of the workpiece design and then to subdivide this shape into finite elements according to the mechanical loads on them. The principal economy lies in the automatic calculation of the grid points, compared with manual methods.

The completed geometrical grid with the defined boundary conditions and external loads is introduced into the main computer program, which then calculates the internal stresses. The computer can display the results of this calculation graphically, showing stresses, strains, and deformations from various viewpoints and in various relationships. Representing the results in color makes it easy to judge the effectiveness of the design. Although not shown here, color coding the values at the grid points makes areas of high stress immediately recognizable, and they are easy to record and compare with subsequent designs (Fig. 6.35.b).

To obtain the most efficient use of material, the component to be cast is subjected to a parameter optimization program. For this calculation, the various performance parameters must first be weighted for importance and combined to form an overall performance criterion. The program then varies the design parameters so as to maximize this overall performance criterion. The optimized component shape is then transferred into the CAD system.

The CAD design data are then transferred as a data file to the mold maker to help him decide the manufacturing procedure. For subsequent manufacturing of the pattern by numerically controlled (NC) machining, the design data must be completed by adding the necessary shrinkage allowances and by designating the parting line. Gates and risers have to be added to the CAD design of the mold, the data for these elements being generated by the use of a solidification simulation program.

Besides calculating the heat balance for the mold, the solidification program can also simulate mold filling over several casting cycles (Fig. 6.35.c). As well as temperature distributions in space and time, the program can display on screen the temperature gradients, phase boundaries, and important casting process parameters, such as tendency to hot cracking.

To validate the model's predicted temperatures, thermocouples should be placed in the mold at points defined in the simulation model. Predicted and measured temperatures should then be in good agreement.

In a typical case, one begins by studying mold design to simulate heat balance in the mold. Logically pursuing this approach, one finds solutions that combine the designer's experience with that of the more production-oriented foundryman. Over a period of time, one can accumulate considerable experience of design and manufacture in this way.

There is a promising future for computer-aided development of molded parts in combination with optimization of the casting process.

To achieve broadly applicable and realistic results with simulation, it is necessary to have reliable values for the thermophysical properties of mold and workpiece materials. These must include specific heat, thermal conductivity, and, especially, the heat transfer coefficient at the interface between workpiece and mold. Our knowledge of these important material data is still limited, and there remains a great need for more measured data.

Table 6.3: Integrated casting technology for high-quality castings

1. **Melt (stable influence)**
   Alloy composition
   Grain refinement
   Heat content (latent heat and superheat)
   Fluidity and mold filling capability
   Solidification type

2. **Mold (stable influence)**
   Dimensional tolerances
   Mold material properties: physical, chemical, geometrical
   Gate and riser geometry

3. **Casting process (dynamic influence)**
   Mold filling, casting speed, melt reactions, hydrodynamic effects:
   Requirement: Control the mold filling process
   Solidification sequence, fluctuating heat flow, convective flow, formation of air gaps, local solidification times
   Requirement: Control the solidification process

4. **Mechanical and thermal treatment of castings (stable influence)**
   The result is a workpiece with:
   - Exact surface dimensions
   - Dense, fine-grained structure
   - Optimal properties

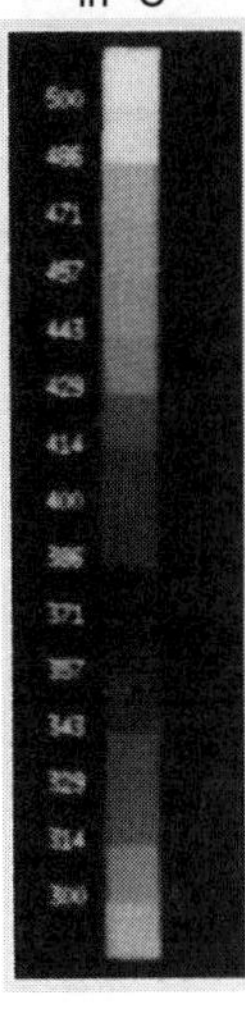

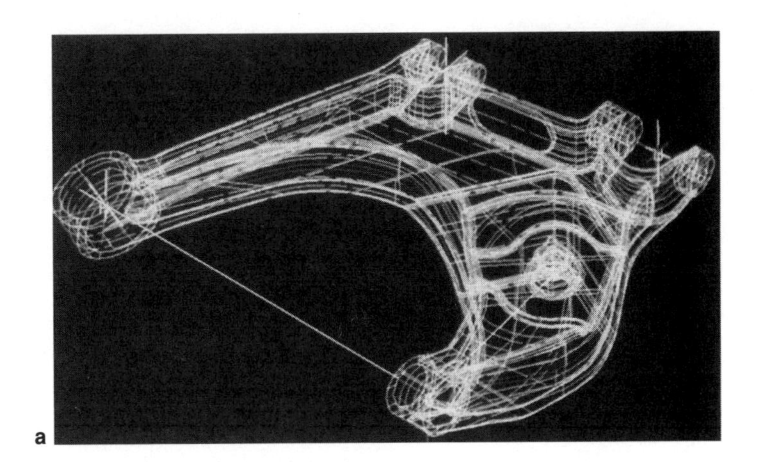

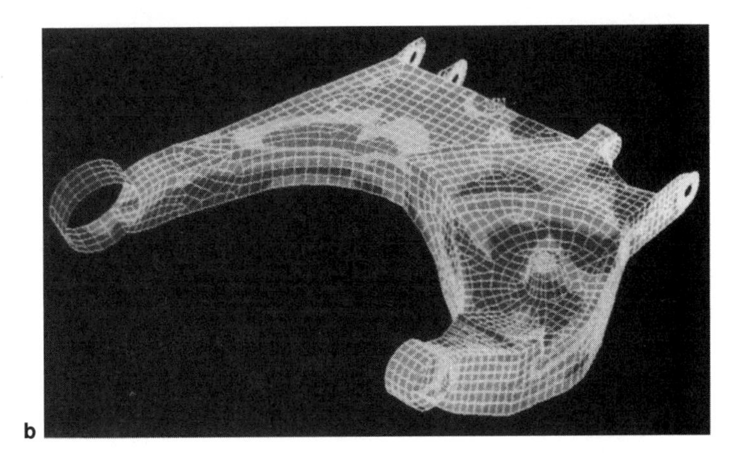

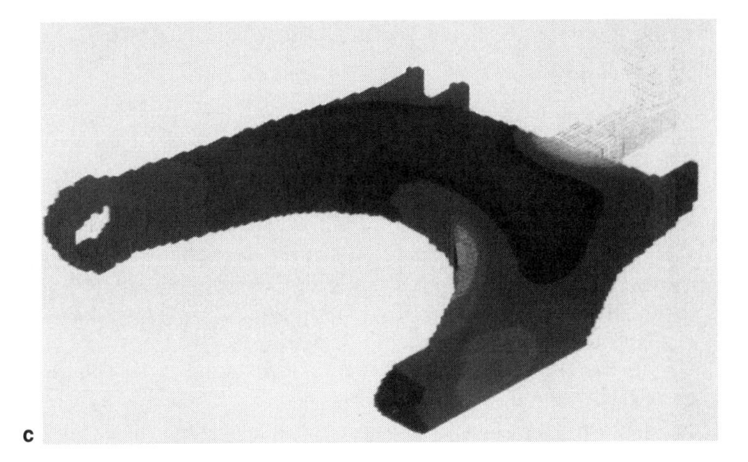

Fig. 6.35: (a) CAD model of the surface shape of the steering arm. (b) stress distribution in the steering arm by FEM analysis. (c) Temperature distribution in a solidification simulation as represented by different shades. A color version of this figure appears on page 426.

6.4.9 Quality assurance management

6.4.9.1 Integrated casting technology

A porosity-free and fine-grained structure is of vital importance for high-quality cast products. Form casting is a very direct route from liquid metal to almost finished component and requires considerable effort to ensure high quality standards. Problems in controlling the casting process arise from the fact that a very short time interval occurs in which the product takes on its final solid form, during which a whole host of influences act on it. These influences mostly act in concert, and they often interact. Table 6.3 summarizes these influences and groups them into four stages in the casting process.

The first two stages concern the separate but parallel operations of preparing the relevant alloy melt, and manufacturing the mold. Parameters involved here are considered to be stable because they have long been verified and optimized, and they can be held in a generally constant state until casting begins. The third stage is casting proper, with the actual mold filling and component solidification processes. A host of time-dependent parameters come into play at this stage, which can interact with each other and react with the above stable parameters. It is vitally important to achieve high process reliability and reproducibility. The fourth stage is cleaning and finishing the workpiece.

Two aspects are vital at the outset: the right choice of alloy and the right design of the component. In turn, these factors interact with the casting process itself.

A new design for a cast product, therefore, requires early and close co-operation between the metallurgist, designer, and production engineer (see also "Simultaneous Engineering" in Chapter 17).

Future applications for high-value aluminum castings will lead to even stricter quality requirements. Because suppliers are now legally liable for their products, a company's quality assurance procedures have become more important than before. New development of high-performance aluminum castings will depend more and more on computer simulations and on test rigs for comprehensive performance checks. To keep the variability in product properties as low as possible, advanced process control techniques will be called into play.

6.4.9.2 Testing the cast structure

Classical microscopy is still the most important technique used to provide guarantees of material quality. A significant auxiliary measurement technique is chemical analysis in the sample cross-section. Quantitative measurements of the metallographic structure over extensive regions of the workpiece, formerly impossibly time-consuming, are now feasible thanks to computerization (quantitative microscopy). This technique, however, is relatively seldomly used in general foundry practice. Dendrite arm spacing (DAS) within cast grains correlates well with strength and toughness, so DAS could be considered a reliable way of judging cast quality.

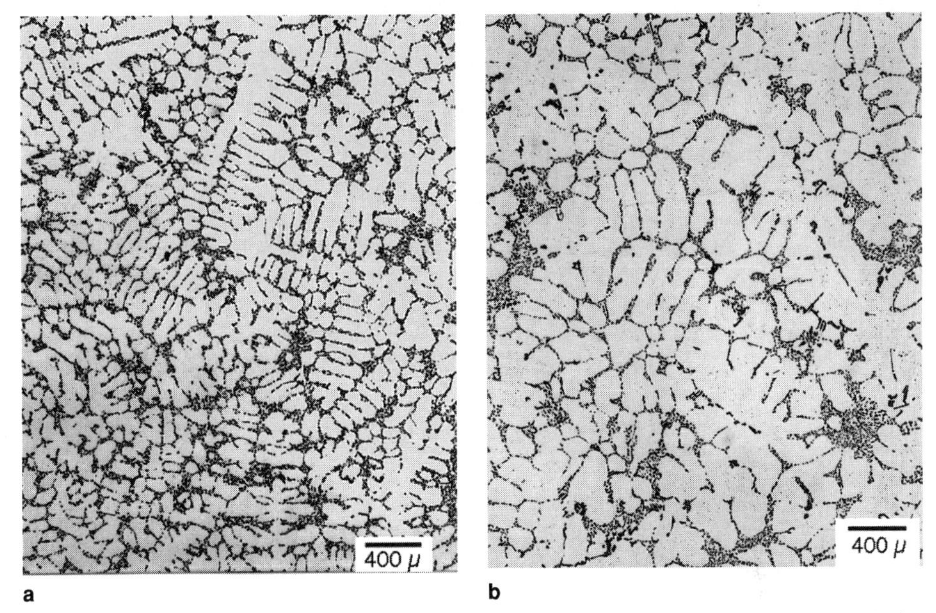

Fig. 6.36: Comparison of structure fineness using dendrite arm spacing (DAS). Two structures in the hypo-eutectic alloy A356.0-T6: (a) DAS = 20 microns; (b) DAS = 40 microns. (Audi)

Fig. 6.36 compares two microstructures in A356.0-T6. As in all hypo-eutectic alloys, DAS serves as a measure of the fineness of the structure. Fig. 6.37 shows the strong dependence of mechanical properties on DAS, average pore size (Pm), and magnesium content. With higher DAS and Pm values, tensile strength and elongation values are much reduced; and these are the usual criteria for judging a material's strength and toughness.

The possibilities for examining fracture surfaces have improved enormously in recent decades thanks to the technique of scanning electron microscopy (SEM). SEM images have a very large depth of focus. Fig. 6.38 shows the dendrites exposed by connected microporosity, leaving only minimal connections to provide the load-bearing function. A

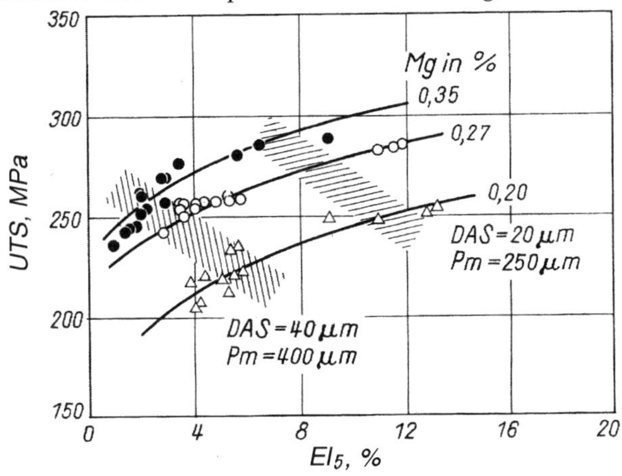

Fig. 6.37: Effect of tensile strength and elongation on average dendrite arm spacing (DAS), average pore size (Pm), and magnesium content in A356.0-T6.

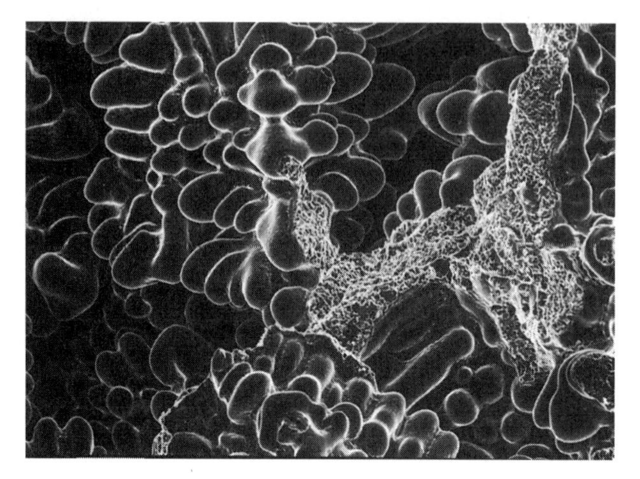

Fig. 6.38: Scanning electron microscope image of the fractured surface of a cast sample of A356.0-T6, with microporosity exposing the bare dendrites. Only on the right is there a small area (appearing fibrous) of ductile fracture where there had been cohesion. Dendrite arm spacing: 50 microns; porosity: 3-4%; elongation at rupture: 1%. (Audi)

new development is the "acoustic microscope," which uses the echo of an ultrasonic pulse to construct a view of sub-surface porosity, making it complementary to X-ray inspection. Safety-critical cast aluminum components are today X-ray inspected in large numbers (100% of car wheels, for example). Pores down to one millimeter are easily detectable by this method. Limits of acceptability must be matched to the service loads. With an X-ray microscope much finer porosity down to about 10 microns can be detected.

6.4.9.3 Mechanical testing

The classical tensile test is as important as ever in developing components. Impact tests and fatigue tests are other methods that can help to judge how well suited a casting is to the needs of its application in terms of reliability and toughness. Much depends on surface quality, which requires a dense, fine-grained structure without cracks, and a defined roughness tolerance. Sophisticated fatigue testing, simulating the whole lifetime of the component, is often the decisive factor in whether an automotive firm will accept cast components (see Chapter 7).

6.5 Outlook For the Future[16]

Novel variants of mold-casting will enable aluminum castings to penetrate markets which have been hitherto inaccessible. We may distinguish three main fields of activity in this regard:

1. Further improvement of the already-established processes by avoiding unfavorable microstructural features.
2. Development of new processes for producing castings for pre-production trials.
3. Invention of novel or fundamentally better compound materials by incorporating into the casting abrasion- and / or temperature-resistant elements as, for example, by reinforcing the metal matrix with particles or fibers.

[16] By W. Schneider.

Fig. 6.39: The incorporation of high-quality castings in the crash-resistant front end of the Audi A8.

Further penetration by aluminum in automobile applications, the single most important market for castings, will most likely occur through improved pressure die casting processes.

Minimum porosity is essential in castings for automobile parts with high mechanical property requirements, especially for those that are considered to be safety-critical, as in the Audi A8 front end in Fig. 6.39. Today the key technology in this area is vacuum die casting. This technique produces castings which can be can be welded and heat-treated. However, several other novel methods are already in use, for example squeeze casting, with the object of satisfying the ever-increasing demand for better quality. There is intense development activity in the field of thixocasting,[17] a technology that avoids the disadvantages of the traditional pressure die casting processes and could well take over new application areas.

The development of pressure die casting prototypes is made easier by using low-pressure die casting, with either sand or permanent-molds. By low-pressure die casting, a component suitable for mass-production use can be designed in the shortest possible time without incurring the high costs of making one or more experimental pressure die casting molds.

The casting operation itself must be perfected in order to improve both the quality level and its consistency. Process automation systems can be still further improved, using computers fed with more information from on-line sensors. In this regard, computer simulation, which has already been successfully applied, can and will be further developed for products and processes.

In the whole of the transport sector there is unremitting pressure to save energy by reducing the weight of vehicles while at the same time reducing manufacturing costs and improving performance. The continual improvement of existing casting methods and the development of new processes will ensure that aluminum castings continue to play a vital role in this critical area of the modern industrial economy.

[17] Nomenclature varies; the names thixoforming and thixoforging are also used.

Chapter 7. Properties of Aluminum under Mechanical Stress and During Deformation[1]

It is well known that there are two basic types of deformation: elastic and plastic. If a bar is bent only slightly and then released, it returns to its original position, demonstrating a "spring-back" effect. The bar has been deformed elastically. A well-known example of a material that behaves elastically is rubber. With more severe bending, the metal bar remains crooked and does not completely return to its initial shape. This is known as "permanent" or "plastic" deformation. A typical example of a plastic material is plasticine.

7.1 *Elastic and Plastic Deformation*

Metals behave both elastically and plastically under deformation. The kind of behavior we observe during and after a given deformation depends entirely on the type of metal and its metallurgical state. Lead, for example, has almost no elastic formability, and any working causes plastic deformation. Spring steel or hard aluminum sheet, on the other hand, will show extensive elastic deformation before deforming plastically. If the same aluminum sheet is softened, however, for example by annealing at 400°C, its elastic formability is much reduced, and it can be plastically deformed with significantly lower force than when it was in the hard condition. Experience shows that elastic and plastic deformation can take place at the same time. After severe bending, a metal bar will spring back when released, but not return to its former shape; the permanent deformation is plastic.

A designer must ensure, for instance by use of computer-aided design techniques, that throughout the structure any deformation remains within the elastic region, even when the structure is subjected to maximum service loads. For this, data are needed on the properties of the material under various kinds of loading (e.g., tension, compression, bending, and torsion). Data on material properties are obtained by testing laboratory samples. In such testing, the material is loaded in various ways. The three main loading patterns are monotonically increasing load (e.g., tensile test), constant load (e.g., creep rupture test), and fluctuating loads (dynamic tests) at different temperatures. Short-term tests are most frequently used. Since a material's behavior under short-term conditions only partly reflects its behavior under load over long periods, long-term tests are also used. Typical of such tests are standardized fatigue tests and creep tests.

One of the most important and most often-used tests is the tensile test, lasting about a minute or two at room temperature. This test will be described in some detail.

7.2 *Tensile Test at Room Temperature*

The tensile test (ASTM method E8) allows us to determine various aspects of the behavior of a metal under deformation. First, a sample must be taken from the material to be tested; it may be in sheet form or cylindrical, dependent upon the product being tested. The test piece, after machining to standardized dimensions, is fixed in a tensile testing

[1] In collaboration with W. Kühlein.

machine by clamping both ends, which have an increased cross-section. Fig. 7.1 shows a universal tensile testing machine. The machine applies a steadily increasing measured load to the sample while allowing changes in thickness, width, and especially elongation to be measured. In modern machines, the test is carried out fully automatically and pro-

Fig. 7.1: A universal tensile testing machine for determining mechanical properties.

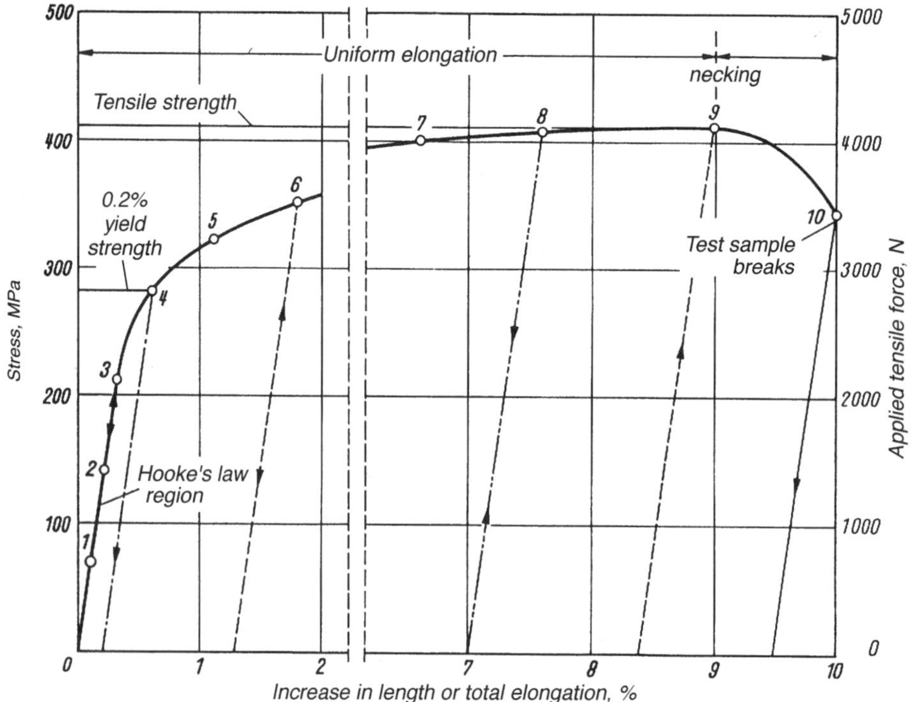

Fig. 7.2: A generalized stress-strain diagram for the tensile test.

vides a print out of material properties and the stress-strain diagram (Fig. 7.2). The latter provides a convenient way of visualizing the behavior of the metal under tensile stress.

7.2.1 Elastic elongation: "Hooke's law"

If a sample with a cross-section of 10 mm² is loaded with a force of 700 N,[2] the deformation is represented by point 1 on the graph. Under this applied tensile stress of 70 MPa,[3] the sample is 0.1% longer than it was before loading. If the load is now removed, we observe that the deformation was elastic,[4] because the sample returns to its original length. The deformation is purely elastic up to point 3; the sample behaves like a rubber band. We notice furthermore a very important feature: the elongation is proportional to the applied stress (i.e., tripling the stress triples the elongation, and so on). This is Hooke's Law, named after the physicist who discovered it. If the applied load is removed at any point between the origin and point 3, the sample springs back to its exact original length (i.e., the plastic deformation is zero). This phenomenon is indicated by the arrow pointing downwards from point 3. The slope of the Hooke's Law region of the stress-strain diagram is known as the Modulus of Elasticity,[5] or Young's Modulus, and is a very important characteristic of the material. It corresponds to the stress that would, in theory, be necessary to double the length of the sample (if the material could deform elastically so far). The modulus of elasticity of common wrought aluminum alloys is not significantly changed through alloying additions.

7.2.2 Plastic elongation after passing the yield stress

If the load is further increased to 2800 N (point 4), the sample stretches a further 0.3%. If the load is now gradually reduced, the sample shrinks back along the broken line, essentially on the elastic modulus line. When the load reaches zero, we see that the sample is 0.2% longer than it was initially; it has undergone permanent (plastic) elongation.[6] This is the yield stress or yield strength, also called the proof stress outside of the USA. The point where the permanent elongation reaches 0.2% is called the 0.2% offset yield stress or yield strength[7] or, in symbols, $TYS_{0.2}$. In the example, it is 280 MPa. A permanent elongation of 0.2% is a small but easily measured offset, and is the basis of the standard measurement of yield stress or yield strength. In a more precise version of the stress-strain diagram using a sensitive extensometer, the 0.02% yield stress may be determined, as in Fig. 7.3. Such a low permanent deformation is often important for constructional details. For most purposes, however, the more easily measured 0.2% yield strength is sufficient.

---

[2] 700 N corresponds approximately to a weight of 70 kg; more exactly, 9.81 N is the weight of a mass of 1 kg.

[3] Tensile stress (in MPa) = load (in N)/initial cross-section (in mm²).

[4] Elastic deformation (in mm/mm, usually expressed in %) = [length under load—initial length (in mm)]/initial length (in mm).

[5] Modulus of elasticity (in MPa) = stress (in MPa)/elastic deformation (in mm/mm).

[6] Total elongation is calculated as (final length –initial length)/initial length × 100 (in %).

[7] The term "yield point" came originally from the steel industry. Steel deforms elastically up to the yield point and then suddenly begins to deform plastically. Yield point is often used also in the aluminum industry, although this is not generally correct. The transition from elastic to plastic deformation is usually gradual in aluminum. This is why the yield point is specified as the stress at 0.2% elongation in the aluminum standards. Aluminum alloys that often show a yield effect are the Al-Mg (5xxx) series. Here, magnesium atoms "pin" the dislocations, which break free suddenly during yielding.

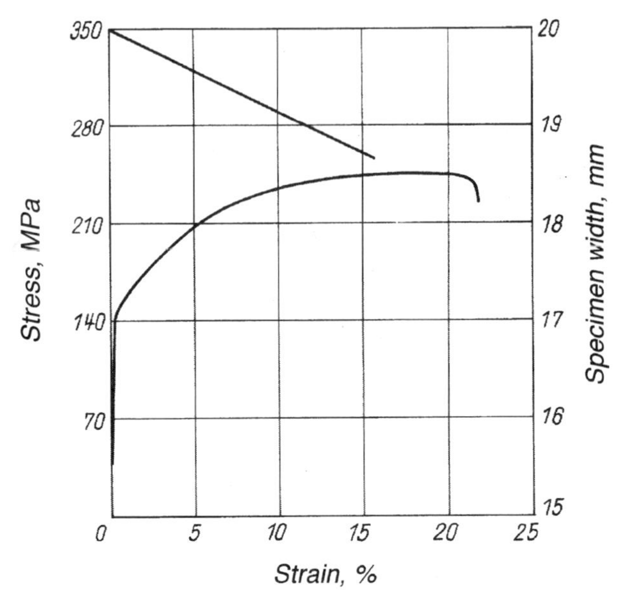

Fig. 7.3: Tensile test report from a computer-controlled testing machine. Lower curve: stress-strain diagram, upper curve: test piece width as a function of the elongation. (AMAG)

If the load is further increased (points 5 to 8), we see that an ever-smaller increment of load is required to bring about a given increment of elongation. At point 6, the total elongation is 1.8% with a load of 3500 N. If the load is then removed, the sample shrinks by the amount of the elastic deformation, as shown by the broken line. A permanent deformation of 1.3% remains, which means that the elastic deformation for this load was 0.5%.

7.2.3 Work hardening by plastic deformation

Fig. 7.2 shows clearly that with increased plastic deformation the modulus of elasticity remains unchanged, but the elastic formability is increased. At point 6 the load on the sample reaches 3500 N. If the sample is then unloaded, it shrinks elastically along the broken line by 0.5%. With renewed loading, the sample can take up to 3500 N with purely elastic deformation; as the load increases from 3500 N to 4050 N, it undergoes further elastic deformation. Now it can be released and reloaded, only beginning to deform plastically again when the load surpasses 4050 N. The plastic deformation from the start up to point 9 has made the material stronger and harder. The process is known as work-hardening or cold-working. We shall look into this in more detail later.

7.2.4 Tensile strength, total elongation, and reduction in area

The last stage of load increase brings us, at point 9, to the highest stress level at about 420 MPa; this is the tensile strength, also known as the ultimate tensile strength, UTS. The corresponding plastic elongation is 8.4%, and is known as the uniform elongation, since the lengthening of the sample is uniform over its whole parallel length right up to point

9. On passing point 9, localized necking occurs, usually near the middle of the sample. This greatly reduces the cross-section, and is accompanied by further localized elongation. It also results in a reduction in the nominal stress to about 340 MPa calculated on the original cross-section (point 10), although the true stress on the reduced cross-section would be higher.

The total elongation,[6] usually designated by El with a subscript indicating the gage length, is measured by putting together the broken surfaces and measuring the plastic deformation that occurred during the test; this is 9.5% in our example. The total elongation depends on the shape of the sample, but the uniform elongation does not. For round samples as a rule, total elongation is measured over a gage length of four or five times the diameter of the test section of the tensile specimen (4D or 5D; El_4 and El_5, respectively). However, a fixed gauge length of 2 in. (El_2), 50 mm (El_{50}), or 80 mm (El_{80}) is used in testing flat products less than 3/8 in. or so in thickness.

The reduction in area,[8] designated by RA, is measured as the permanent change in cross-sectional area at the narrowest part of the sample, and is expressed as a percentage of the initial cross-sectional area. Total elongation and reduction in area together give an indication of the extent to which the material can be worked plastically. This provides a criterion for judging whether it is suitable for a given fabricating technique, as well as for predicting its behavior when used in critically loaded structural components. Modern tensile testing machines are controlled by computer, with the tensile force and the changes in length, width, and thickness being recorded in a data-acquisition system. The mechanical properties are calculated automatically and are printed out with the stress-strain diagram. Fig. 7.3 shows a typical tensile test report.

7.3 Hardness

The temper designations "annealed" (O temper), "half-hard" (H14 temper), and "hard" or "full hard" (H18 temper) are used to characterize rolled semi-fabricated products such as sheet, plate, and bar. The hardness of aluminum can be measured directly, the hardness values corresponding roughly to the tensile strength. There are various measurement methods, for example the Brinell test (ASTM Method E 10). Here a steel ball, usually 2.5 mm in diameter, is pressed into the surface of the aluminum sample under constant force (e.g., 153 N) for 30 seconds. Then the diameter of the dent left in the surface by the ball is used to give a measure of the hardness; the softer the material, the greater is the diameter of the dent. Hardness measurements offer a quick approximate relative strength estimate, but are only a limited guide to the deformation behavior of a metal.

Hardness values determined by other methods, for example the Vickers or Rockwell tests (ASTM standard methods E92 and E18, respectively) are less frequently used for aluminum. The Vickers microhardness test has a special purpose. It is used to measure the hardness of individual constituents of the microstructure or to test anodic films. For aluminum, the Vickers and Brinell hardness values correspond relatively closely except for the hardest alloys.

---

[8] RA = (initial cross-sectional area – final cross-sectional area)/initial cross-sectional area × 100 (in %).

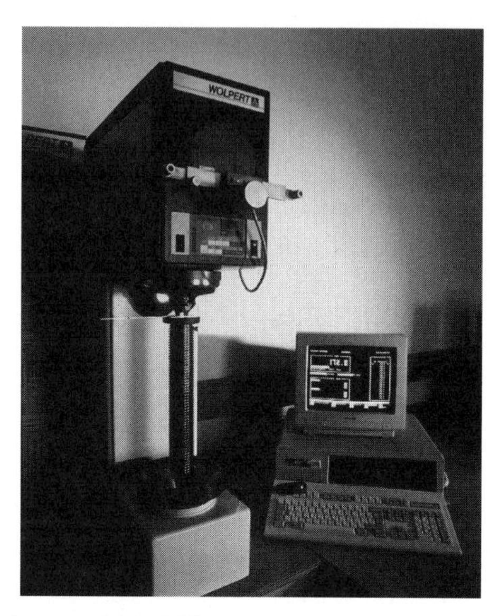

Fig. 7.4: A universal hardness-testing machine.

Fig. 7.4 shows a universal hardness testing machine. As with modern tensile testing machines, the evaluation of the test results is done by automated computer analysis.

7.4 Macro- and Micro-Notches

Fig 7.5 shows schematically a standardized macro-notch in a test sample. Similar stress concentrations occur at many different macro defects such as rivets, weld seams, or severe scratches on the surface. But much more frequent are micro-notches, which are internal discontinuities caused by large precipitates or other heterogeneities such as oxide inclusions, sizable gas porosity, or shrinkage voids. Evidence of defects such as these can be detected even in the standard tensile test, mainly by a reduced plasticity visible on the surface of the fractured sample, as well as by a lower elongation. We shall see later that macro- and micro-notches may significantly reduce fatigue strength and fracture resistance and are, furthermore, detrimental to crashworthiness (for instance, of automobile components).

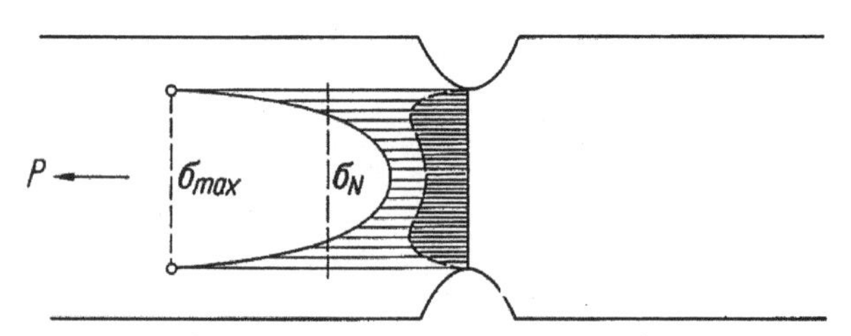

Fig. 7.5: A notched sheet sample showing stress distribution under tensile loading. σ_N = nominal average tensile stress; σ_{max} = maximum local stress at the base of the notch in the direction of load; the shaded area indicates the transverse tensile stress.

7.5 *Ductility and Toughness*[9]

Ductility is defined as the ability of a material to deform plastically rather than crack and fracture, especially in the presence of micro- or macro-cracks or notches that locally increase stresses; this characteristic is also called toughness. Elongation and reduction of area from the tensile test are relative indicators of ductility. Notched bar impact resistance, notch toughness, and fracture toughness are more specific measures, which are obtained from more specialized tests (see below) in which notches or cracks are present during the loading.

Tensile strength and elongation are significantly reduced only by larger defects, such as macroscopic porosity. Tensile test pieces usually have only microscopically small defects, such as inclusions, which have relatively little effect on standard tensile properties. In contrast, even microscopic cracks and notches may lower fatigue performance. Surface finish, as has been mentioned earlier, also may play a significant role.

7.5.1 Notch tensile test

Tensile tests of notched specimens (ASTM standard methods E338 for sheet-type and E602 for cylindrical specimens, respectively) are sometimes used to measure the sensitivity of alloy specimens to such stress raisers. The ratio of the tensile strength of the notched specimen (the notch-tensile strength) to the tensile yield strength of the material (known as the notch-yield ratio [NYR]) is the most useful measure of notch toughness from notch tensile tests.

A stress raiser can occur singly or simultaneously at many places in the cross-section of an aluminum material. A singular stress raiser is typically a macro-notch (see Fig. 7.5), but small micro-notches such as inclusions, detrimental to elongation and/or fatigue strength, could occur in many places in a test sample or a component part. This would be the case if the cross-section had many oxide inclusions or if small or medium-sized heterogeneities had been formed by some inappropriate production practice.

7.5.2 Notched bar impact test

In the tensile test, the behavior of the material is tested under gradually increasing load (monotonic loading) without the presence of local stress raisers like a notch. In practice, however, materials are often subjected to sudden impact loading in the presence of local discontinuities. The notched-bar impact test (ASTM Method E23) simulates both of these aggravating loading conditions, and also provides a measure of the behavior of an actual component under shock loads. A pendulum hammer strikes and breaks a test piece into which a notch of standardized dimensions has been cut.

The magnitude of the overswing is a measure of the energy remaining in the pendulum. The bigger this is, the less the energy was absorbed by the test piece. Dividing the energy absorbed by the area at the smallest cross-section (where the bar breaks) gives a measure

[9] With J.G. Kaufman.

of notch impact resistance. This is a number that describes the behavior of the material under sudden loading and is a measure of the material's toughness. The higher this value, the better the material resists brittle fracture.

The disadvantage of using absorbed energy as a measure is that the single value describes both strength and ductility. Two materials having the same notch impact resistance or notch toughness can have completely different strengths and formabilities. Also, for quite ductile materials like most aluminum alloys, the specimens simply bend plastically and do not break, so the energy value has little significance.

With computerized data acquisition, the force applied to the sample can be recorded over very short time increments to determine the force-time curve. Such an "instrumented" notch impact resistance test enhances the ability of the simple test to describe the behavior of a material under impact loading and gives information on elastoplastic formability as well as the tendency to form and propagate cracks. Fig. 7.6.a shows a typical impact force-time diagram, and Fig. 7.6.b shows a typical industrial test installation.

7.5.3 Fracture toughness

In general engineering practice it is wrong to assume that materials are completely uniform and free from micro- and macro-notches and cracks. Every component contains some small defects, and it is appropriate in evaluating the expected performance of any component to assume that it may contain the largest crack or stress raiser that can not be reliably found by inspection.

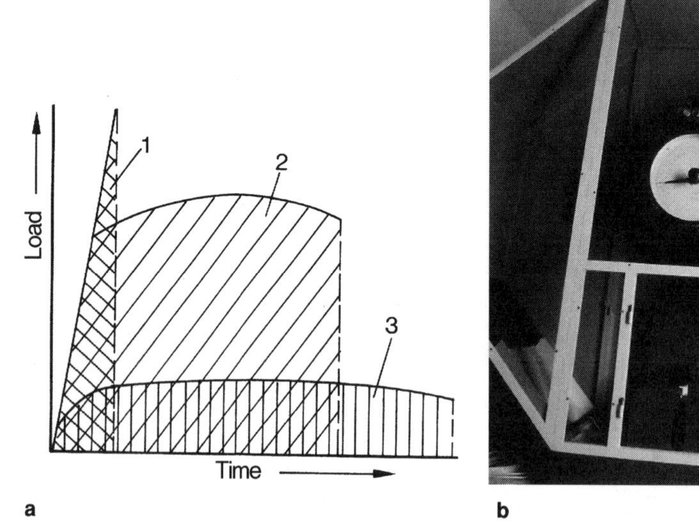

Fig. 7.6: (a) Typical impact force-time curves shown schematically. 1—brittle behavior, 2—ductile behavior, 3—tough behavior, strength, and ductility in optimal proportion (after Blumenauer). (b) Instrumented notch impact resistance test rig. (AMAG)

In recent decades fracture mechanics has been developed as a new branch of materials science. With its concepts and techniques we can quantify the toughness of materials in terms of the combinations of stress and flaw size that may lead to low-ductility fracture or brittle fracture. This measure of a material's strength under plane strain conditions is called its plane strain fracture toughness K_{Ic} and is measured by ASTM standard method E399.[10]

Given a component containing a crack of a certain size, the fracture toughness value allows us to estimate the stress needed to cause the crack to propagate in an unstable manner, leading to low-ductility fracture.[11] Thus, we have a quantitative relationship linking stress in the component to crack size and the material's resistance to crack propagation. Although the field of fracture toughness characterization is still developing, there are already a number of design standards, especially in the aerospace industry, that require dimensions to be calculated on the basis of fracture toughness criteria.

High values of K_{Ic} mean high toughness, which is especially desirable if yield and tensile strengths are high. With increasing tensile strength, toughness tends to become relatively smaller, but there seems to be no clear correlation between the standard tensile test results and fracture toughness. It is well known that high strength tends to go together with low elongation at rupture. One goal of material development is, therefore, to design materials with both high strength and high fracture toughness. Fig. 7.7 gives an overview of typical K_{Ic} values of various alloy types in relation to 0.2% yield strength. The tough-

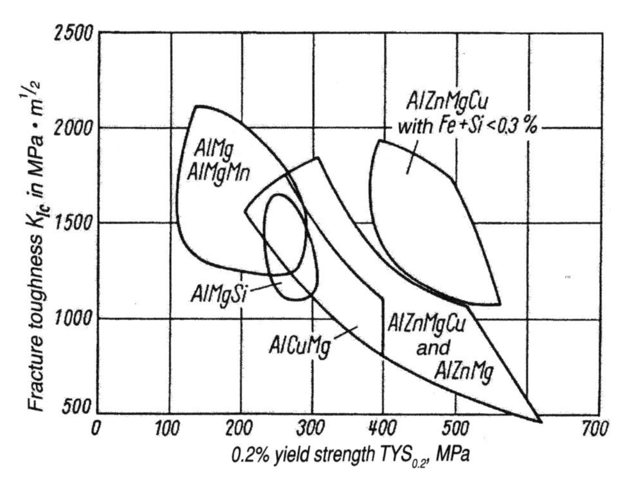

Fig. 7.7: Overview of the levels of fracture toughness of various aluminum alloy types. (from Aluminum-Taschenbuch, 14th edition)

[10] Measuring fracture toughness K_{Ic} is accomplished with ASTM method E399. K_{Ic} is derived from the evaluation of the load/strain diagram of a pre-cracked test piece. Moderately large test pieces are needed to measure K_{Ic}, which requires plane-strain conditions, meaning that the plastically deformed zone just ahead of the crack tip must be small in comparison to the crack length, the remaining thickness, and the test piece thickness. Only then is the toughness measurement independent of the test-piece geometry. Without plane strain conditions, it is still possible to use other measurements (J-Integral, ASTM Method E813, or critical crack opening, ASTM Method E1290) to describe toughness, but these are more time-consuming and not always applicable.

[11] The units of K_{Ic} are MPa.$m^{1/2}$, which is an unusual dimension.

ness values for each material depend strongly on chemical composition (where insoluble phases promote the initiation and propagation of cracks) and the microstructure and its orientation (e.g., whether the sample was aligned in the rolling direction, transverse to it, or through the thickness).

7.6 Mechanical Properties under Special Conditions

The usual mechanical properties, as given in standards and in lists of alloys, are the results of ordinary tensile tests as described above. In using these values to design structures it is important to bear in mind the test conditions under which they were obtained and to check carefully that they are adequate for the practical working conditions under consideration. The tensile test is most often carried out at room temperature, and the load is monotonically increased to the breaking point, with the whole procedure lasting typically only a few minutes. These conditions may correspond well enough to many practical situations but there are a number of important exceptions:

- Loading at high or low temperatures
- Loading over a protracted time, especially at elevated temperatures
- Alternating or frequently repeated loading

If these conditions apply, the structural design criteria must involve other mechanical properties, which are obtained from tests that more closely approximate actual service conditions.

7.6.1 Influence of temperature

Aluminum alloys cover a range of service temperatures of several hundred degrees Celsius. However, some specialized knowledge is needed to guarantee long service life of certain alloys outside the usual temperature ranges, especially in applications where safety is paramount.

7.6.1.1 Properties at low temperature

Fig. 7.8 shows the effect of low temperatures on the mechanical properties of 6061-T6. It can be seen that the tensile strength and the 0.2% yield strength increase with decreasing temperature. The elongation, on the other hand, remains roughly constant, so that aluminum retains its toughness down to the lowest temperatures, even to liquid helium temperature (–270°C), unlike carbon steels or zinc alloys that tend to become brittle. For this reason, aluminum is the preferred material for building low-temperature installations such as liquid air plants.[12] While at room temperature aluminum alloys may have lower notch impact resistance than steel, at low temperature the reverse is true; between –50°C and near absolute zero, aluminum has substantially better ductility, notch impact resistance, and toughness than most commonly used steels.

[12] The only exceptions are the 7000 series alloys, which are not recommended for service at extremely low temperatures.

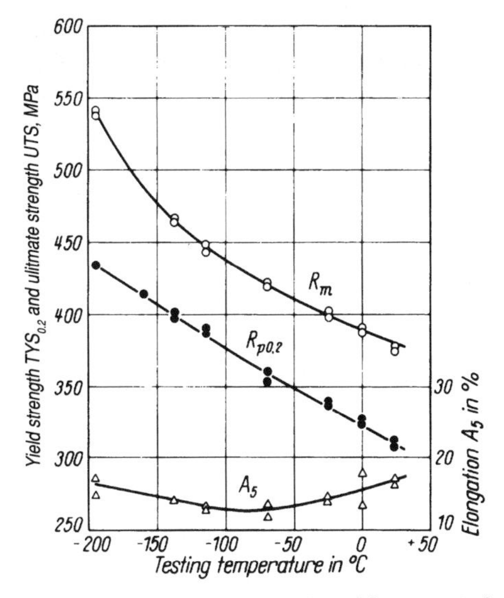

Fig. 7.8: Tensile test properties of heat-treated 6061-T6 at low temperatures. (after Mori)

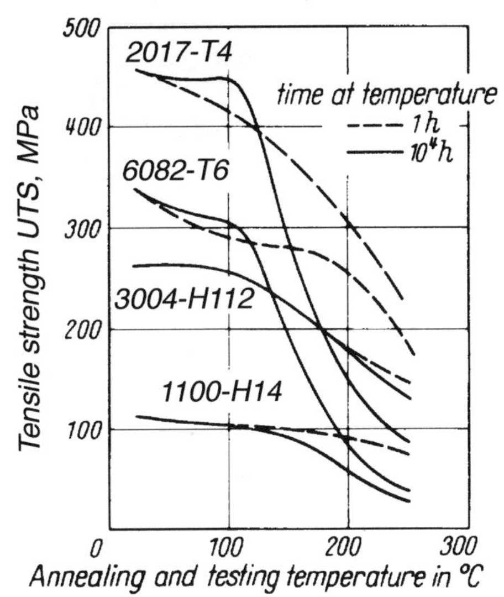

Fig. 7.9: The tensile strength of various aluminum alloys at high temperature.

7.6.1.2 Properties at elevated temperatures

With increasing temperature, metals including aluminum alloys lose strength and gain formability. This is turned to good practical account in hot rolling, extruding, or forging, for example. If a tensile test is performed at high temperature (ASTM standard method E129), the strength depends not only on the temperature of the sample at the time of testing but also on the length of time for which the sample was held at high temperature before and during testing, often referred to as the "soaking time." Fig. 7.9 demonstrates this clearly for different alloys, both heat-treatable and non-heat-treatable. The tensile strength is plotted against testing and soaking temperature for very short and very long soaking times. The graph shows two clearly distinct influences at work. First, there is the general loss of strength with increasing temperature, seen in its pure form in the two softer alloys. Second, for the cold worked or heat-treated alloys, we observe the superimposed effect of the change in microstructure resulting from the extended soak time at high temperature. In general, the material gets even softer as the duration of the soak increases until annealing and complete recrystallization occurs and the properties reach a minimum.

Temperatures between 50°C and 150°C

The effects of these slightly elevated temperatures vary significantly from one alloy family to another. Non-heat-treatable alloys (i.e., alloys designed to achieve their mechanical strength by cold working) undergo merely a slight reduction in work hardening by recovery. Heat-treated alloys begin to experience a more significant drop in strength with temperatures above about 100°C.

An additional effect of temperatures in this range is noted for aluminum-magnesium alloys with more than 3% Mg, especially those with more than 4% Mg, such as 5083. When heated above 60°C, and more particularly above 80°C, supersaturated magnesium precipitates on the grain boundaries, substantially weakening the material. This precipitation by warming, called sensitizing, forms a grain boundary layer that is electrochemically much more reactive than the matrix. Corrosion can then develop rapidly along these sensitized grain boundaries, and the material can thereby become susceptible to stress-corrosion cracking.[13]

Temperatures between 150°C and 300°C

Certain Al-Cu alloys, such as 2219 and 2618 used in aerospace applications and castings of some Al-Si-Cu alloys, such as 390.0, in internal combustion engines have to withstand higher working temperatures for long periods. Among the most resistant casting alloys are the piston alloys, which contain up to 20% of silicon (highly hypereutectic composition). The high-temperature strength of aluminum alloys containing copper has been tested and proved in this respect over several decades for such applications and is well documented.

Temperatures above 300°C

Aluminum alloys made by thermo-mechanical alloying (i.e., by mixing metal and ceramic particles that are then pressed and sintered to the required form) are very strong and fairly ductile at temperatures up to 500°C.

7.6.2 Creep resistance

The high-temperature strength characteristics described earlier were measured under the standard tensile test conditions, in which the load is monotonically increased until fracture occurs. If we apply a constant load for some period of time (hours, days, or months), we observe—particularly at higher temperatures—that the metal slowly deforms.[14] The phenomenon is known as creep and is graphically illustrated in Fig. 7.10, which shows how a sample held under constant load for an extended period of time gradually increases in length (ASTM standard method E119). The strain rate[15] is initially high, but it falls off and remains almost constant until the point of rupture approaches. Before rupture, the strain rate increases rapidly until the sample fails.

The microstructural changes taking place here can be qualitatively understood in terms of the competing interactions of work hardening and thermally induced recovery or softening. Initially, hardening is the predominant factor; then an equilibrium develops in which the effects counterbalance each other. Finally, defects appear in the material (porosity, cracks, necking) that accelerate the loss of strength and, hence, the strain rate, leading to rupture. The higher the temperature and the load, the more rapidly the material

[13] The risks are less in soft than in work-hardened conditions, for which 240–250°C desensitized tempers H116/H323/H343 have been developed.

[14] The phenomenon of creep becomes noticeable typically at temperatures higher than one-half of the absolute melting temperature T_m.

[15] Strain rate = [elongation (in %)]/[time unit (seconds or days)].

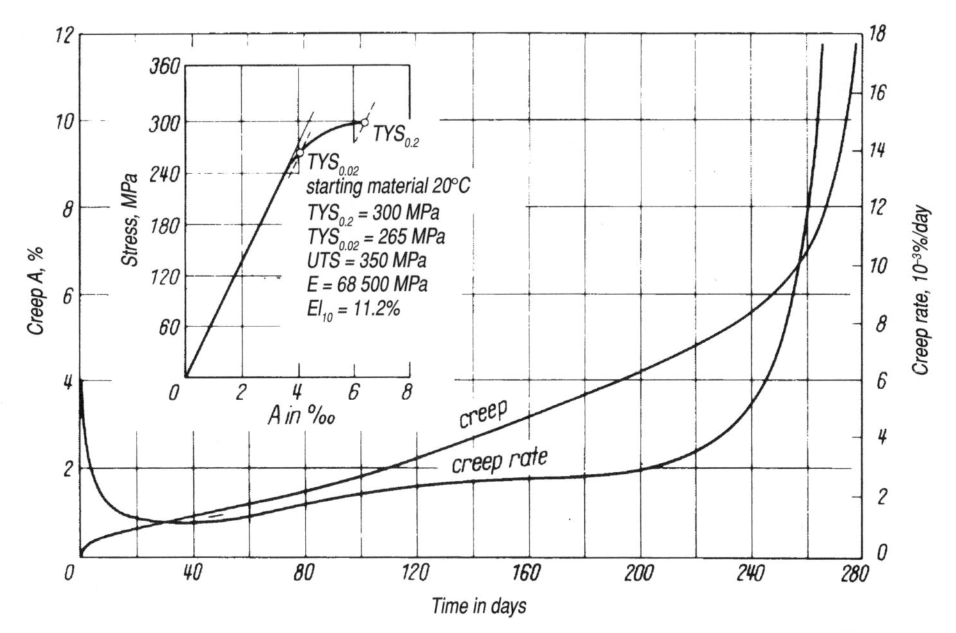

Fig. 7.10: Curves of elongation and creep rate vs. time for 6061-T6 under a stress of 180 MPa and at a temperature of 130°C. (Alusuisse)

creeps, and the shorter is the time to failure. Long-term tests are needed to establish the relationships between stress, temperature, and the time to failure. Fig. 7.11 shows the results of such a test carried out on the work-hardening alloy 5083-O. Curves of this type can serve as a basis for structural calculations, giving the maximum working stress for a given temperature and working life. Pressure vessels and other equipment for the chemical industry are frequently designed using pure aluminum or Al-Mg alloys for service lives of over ten years.

For aerospace applications, special alloys are used that have both greater high-temperature strength and greater creep resistance. Irrespective of alloy, however, their creep behavior under protracted loading must be taken into consideration when designing for service. Overland gas pipelines, overhead transmission cables and their supporting elements, and cylinders for compressed gas or hydraulic brakes are other typical applications where creep behavior is critical. For all such functions, extremely tight tolerances are set for levels of trace elements with low melting points, such as lead and bismuth. Even quite low levels of these elements dramatically increase the material's susceptibility to creep, even at only slightly elevated temperatures. Since these elements are almost insoluble in aluminum and have melting points at quite low temperatures, a low melting-point eutectic forms as a liquid film between the grains, which can lead to failure of a component under sustained loading at ambient temperature even at stress levels below the yield point (low strain rate cracking or creep cracking).[16]

[16] Load sustained over a very long time has, in the past, caused catastrophic failures in the forged clamps holding the cables in overhead electrical transmission lines.

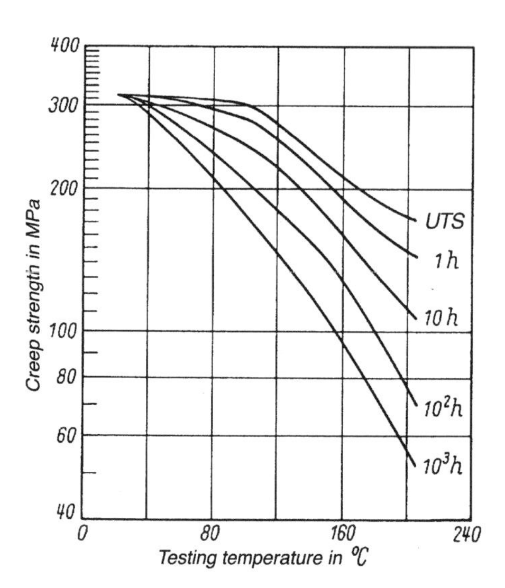

Fig. 7.11: Creep test results for 5083-O. (ASTM)

7.6.3 Fatigue strength[17]

When considering the room-temperature tensile test (Fig. 7.2), we saw that if a test piece is repeatedly loaded to a level slightly beyond its yield strength, it deforms plastically and work-hardens, and with further repeated loading and unloading it behaves largely elastically. Very precise measurements of elongation show that unloading involves a small plastic contraction, and reloading involves a similar small plastic extension, particularly at sites of stress concentration. If the applied stress is high enough and the loading and unloading cycle is repeated often enough, work hardening accumulates, caused by this alternating plastic deformation. Eventually the metal's plasticity is exhausted at some site, and a tiny crack appears there. This crack grows gradually from cycle to cycle, until the test piece finally fails. Therefore, components that are subject to repeated service loads, if they were incorrectly designed, manufactured, or employed, can break unexpectedly even at loads below the 0.2% yield strength. This phenomenon is called a fatigue failure.

Fatigue strength measurement (ASTM standard methods E 1150) involves repeatedly loading the test piece in specially designed machines. Generally the test piece is subjected alternately to specified upper and lower load limits. Fig. 7.12 represents several variants of this test using different levels of average load, in tension or compression, relative to the alternating load. The ratio of minimum to maximum load in a fatigue test is called R, the fatigue stress ratio. Two standard cases occur: first, when the load oscillates between equally large positive (tension) and negative (compression) loads (R = –1) and second, when the load alternates between tension and zero (R = 0).

Fig. 7.13 shows how the results of a series of fatigue tests are usually presented in the form of a plot of the applied stress as a function of the number of cycles of repeated load (S-N curve, sometimes called the "Wöhler curve" after one of the pioneers in such work). Below a certain stress, about 70 MPa here, there are practically no more fatigue failures

[17] With the participation of W. Kühlein.

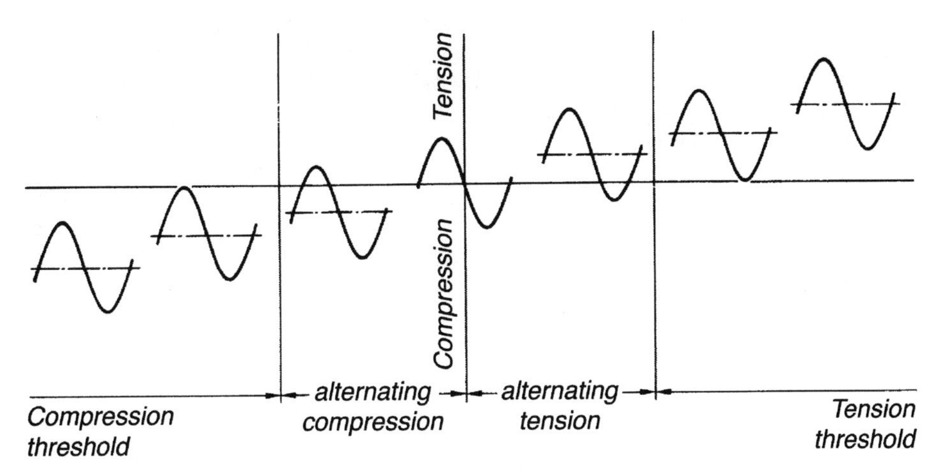

Fig. 7.12: Types of fatigue test load cycles showing variants using different levels of average stress, in tension or compression, relative to a constant alternating stress.

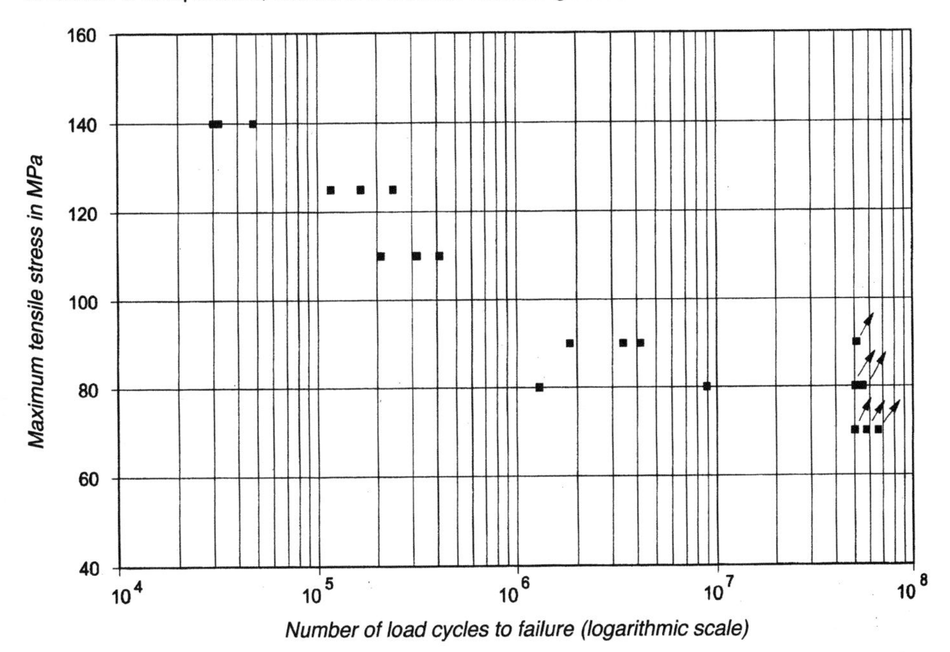

Fig. 7.13: S-N fatigue curve, measured on 7 mm cylindrical test pieces of cast 380 under alternating load. R = –1; frequency, 30 Hz; 0.2% yield strength, 85 MPa; the measurement points with an arrow are run-out survivors, still not cracked when the test ended.

even after testing for a very extended period. This stress is called the fatigue threshold or the fatigue endurance limit.[18] Alloys with higher tensile strengths tend to have higher fatigue thresholds, although the gain is less than proportional to their tensile strength.

[18] The fatigue threshold or endurance limit for aluminum is often defined as the stress that allows 10^8 cycles without failure. In practice, tests are carried out for 5×10^7 cycles.

When designing components, it is not always the fatigue endurance limit that is most significant. Of greater importance may be the fatigue life at a specific applied stress for the application under consideration (i.e., the number of cycles of loading that the material will withstand without failure).

When designing a component to resist fatigue loads, it is most important to minimize stress concentrations, local stress raisers of the type illustrated in Fig. 7.5. These can significantly reduce fatigue life, and unfortunately are always present to some degree. In the design, they occur most often at joints, notches, holes, and abrupt changes in cross-section. In the material itself, stress concentrations arise at voids, inclusions, and coarse intermetallic phases.

The influence of stress concentrations on fatigue life is so great that surface quality is even more important than choice of material. Fatigue strengths quoted in the literature are usually for ideal surfaces polished to a fine finish, although tests are often made of specimens with notches to represent stress raisers. Untreated, "mill finish" surfaces contain tiny, incipient cracks and are substantially less resistant. Scratches and corrosion may occur in service. Even protecting the surface by anodizing can reduce fatigue strength, because the anodic film contains micro-cracks. All of these notches and micro-cracks can cause stress concentrations, which could nucleate fatigue cracks and lead to premature failure.

The surface can be largely protected from fatigue cracks by pre-loading it with compressive stresses to offset potentially harmful tensile stresses. A well-proven method is peening, which means covering the surface with hammer marks. This is done industrially by shot peening (i.e., blasting the surface with hard steel or ceramic balls). The fine dents create surface compressive stresses that tend to prevent superficial micro-cracks from forming.

7.6.4 Energy absorption in crash testing

As strength increases, elongation tends to decrease. An optimum combination of these two properties is important in structures that must be designed to absorb a maximum amount of energy, for instance, in the event of a crash. This combined parameter is the rupture energy. In the simple case of the tensile test, this is graphically represented by the area under the stress / strain curve. This is similar in appearance to Fig. 7.6.a, where schematic impact force / time plots illustrate how different materials can absorb energy.

A component's ability to absorb energy depends both on its shape and on the properties of the material. This shape controls how it will deform, whereas the material's strength determines the stress level for deformation and the component's energy absorption capacity. Aluminum is an excellent material for making automobile crash beams, fenders, and bumpers. Extrusion provides practically unlimited ways of adapting the cross-section to the needs of the application. With steel, two half-shells must usually be welded together, but such joining problems are avoided with aluminum. Calculations and practical tests have shown that an optimized aluminum extruded shape, no larger overall than a nominally equivalent steel crash beam, can absorb more energy at a more uniform load. The beam's weight can be reduced by 50%.

Designs in aluminum have different shapes and cross sections compared to designs in steel. The engineer can use aluminum's special characteristics, such as its lower elastic modulus, to produce structures with higher energy absorption capacity combined with lower weight.

7.7 Differences in Elastic Deformation of Steel and Aluminum

The elastic range and the beginning of plastic deformation in a tensile test for steel and an aluminum alloy is represented in Fig. 7.14.

If one looks closely at the stress-strain diagram of a rolled mild steel, it can be seen that after the yield point is reached there is a short period in which the deformation resistance is reduced so that one finds an upper and lower yield point. Moreover, steel has a definite flow region in which it can extend several percent without any increase in stress (see Fig. 7.14). This yield point behavior is found in aluminum only in isolated cases, principally in Al-Mg alloys and even then only to a very slight extent.

Of utmost importance is the fact that the initial part of the stress-strain curve corresponding to the elastic range is three times steeper for steel compared to aluminum. The modulus of elasticity determines the slope of the initial linear part of the stress-strain relationship. Thus, the modulus of elasticity is three times higher for steel (210,000 MPa) compared to aluminum (70,000 MPa). At a given stress, an aluminum alloy shows three times as much elastic elongation as steel.

At a given elongation B, an aluminum alloy could be still in the elastic range of deformation, while the steel at the same elongation has undergone plastic deformation. Another factor of note is that for ordinary structural steel, the yield strength is less than that of a high-strength aluminum alloy.

Fig. 7.14 is worth examining to clarify typical differences between steel and aluminum. However, in the comparison of the two metals, it makes a difference whether we start with a given stress or with a given strain.

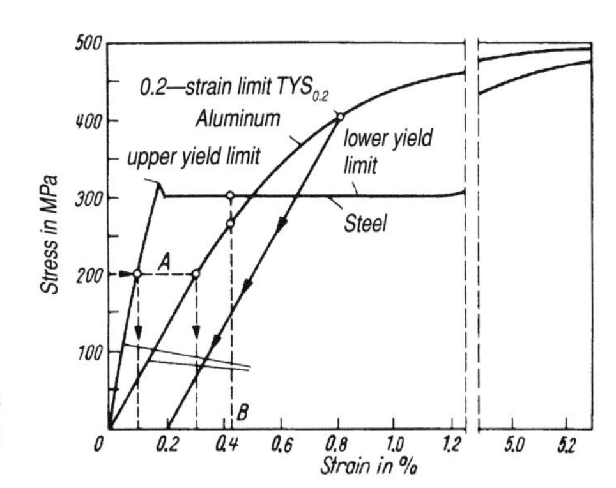

Fig. 7.14: A comparison of stress-strain diagrams for structural steel (St 37) and a high-strength aluminum alloy (7005-T5).

7.7.1 Conditions under a given stress

Imagine two bridges of the same dimensions with a specific load resting in the middle. One bridge is constructed from normal structural steel and the other of 7005-T5 aluminum alloy. Both have the same guaranteed yield strength of approximately 250 MPa. The basic difference between steel and aluminum lies in the modulus of elasticity, which is three times greater for steel than aluminum. This means that, at the same stress, aluminum has three times more elastic elongation than steel. In the case of the bridges, this means that with the same dimensions, the load-carrying members of the aluminum bridge bend or deflect three times as much as the bridge constructed of steel (see Fig. 7.15). This must be taken into account in designing a bridge, since difficulties can arise if the deflection is excessive.

The foregoing can also be seen in Fig. 7.14 from a section of the stress-strain diagram. If one follows the broken line from left to right, it can be seen that at a given stress, for example 200 MPa (Point A), aluminum shows three times as much elastic elongation as the steel. For this reason, in any substitution of steel by an aluminum alloy, the dimensions of the load-bearing members must be changed, usually by increasing the depth, in order to offset the additional potential deflection of the aluminum members.

Steel is three times heavier than aluminum of the same geometric dimensions. Because of the lower modulus of elasticity, however, it is often necessary to increase the cross-section of the aluminum part as compared to steel, so that the full advantage of the lower density of aluminum cannot be utilized. In most cases, a weight savings of around 50% can be achieved by replacing steel with aluminum.

As the designer knows, with load-bearing elements, the selection of shapes made from aluminum alloys allows the use of cross-sections with a high moment of inertia and low weight per unit length. This may be accomplished, for example, by locating as much of the available mass as possible on the periphery of the profile cross-section in the loading direction. This is shown schematically in Fig. 7.16. Aluminum extrusions of this type can be produced in one operation with relatively little cost.

In Fig. 7.17, cross-sections of rectangular shapes for the metals with similar bending strength and their respective nominal properties are presented. The product of the modulus of elasticity and the moment of inertia [(E) (I)] is called stiffness and determines the deformation resistance to a bending load and the load-carrying capability of a member before buckling. The moment of inertia can be increased much easier without a detrimental increase in weight when the density of the material is lower.

Fig. 7.15: Illustration of the difference in behavior between steel and aluminum during elastic deformation. Under a given load G (or in general: a force) an aluminum section of equal dimensions will undergo three times as much elastic deformation as steel.

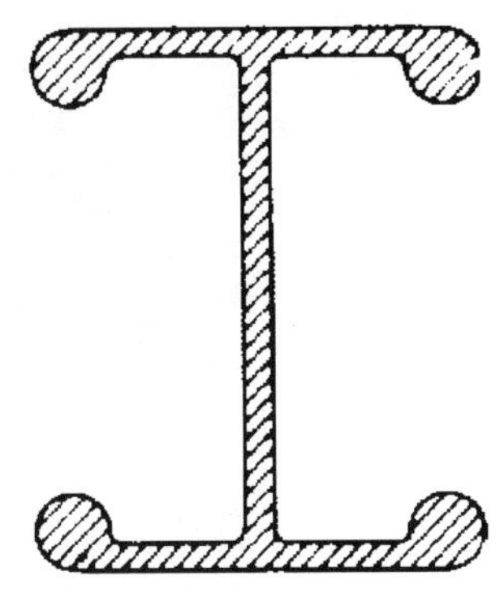

Fig. 7.16: Extruded profile with high moment of inertia.

	Carbon steel	Aluminum alloy	Magnesium alloy
Specific weight	7.85	2.7	1.75
Modulus of Elasticity "E", MPa	210 000	70 000	46 000
Height "h" for the section with equal base dimension "a"	100	144	166
P = Weight	100	50	37
I = Moment of inertia	1	3.0	4.7

Fig. 7.17: A comparison of rectangular bars from carbon steel, aluminum, and magnesium on the basis of weight P and moment of inertia I. Between the dimensions of the bars cross-section (a, h) and the moment of inertia I exists the following relation: $I = ah^3/12$. The three illustrated cross-sections have equal stiffness (E I).

The moment of inertia I for the cross-sections shown in Fig. 7.17 increases with the third power of the height or depth of the beam. The lower modulus of elasticity of aluminum or magnesium compared to steel can be compensated for by increasing the cross-section in the loading direction without making the structure too heavy.

This means that in spite of the disadvantage of a low modulus of elasticity, aluminum can be used because of its low density and good formability (for example, extruded shapes) to obtain structural members that are light and rigid by increasing the moment of inertia. It is by no means necessary that the entire cross-section of a structural member be made of solid metal such as pictured in Fig. 7.17. In many cases, it is sufficient to form the outer zones that carry most of the tensile and compression loads (see Fig. 7.16). Fig. 7.18 shows different methods of increasing the moment of inertia by introducing ribbed patterns.

A composite with a plastic core is illustrated in Fig. 7.19. For short time loads the composite has almost the same stiffness as a solid aluminum sheet of the same thickness, but with only approximately half the weight of the sheet.

A bending test comparing a composite and a solid sheet is illustrated in Fig. 7.20. From the illustration, it can be seen that aluminum can be replaced with a lighter and / or cheaper

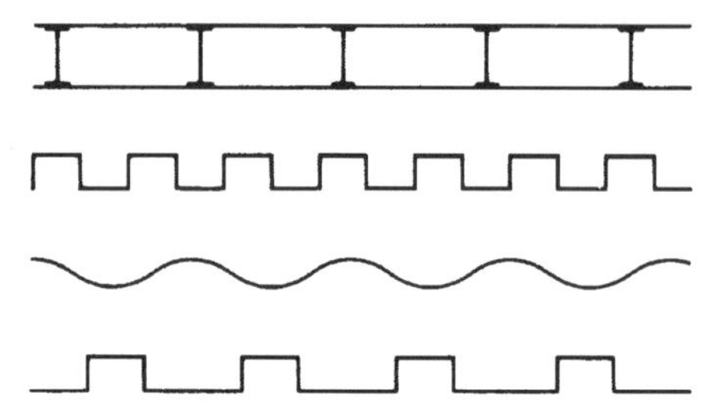

Fig. 7.18: Ribbed sheet and sheet with spacers for increasing flexural stiffness.

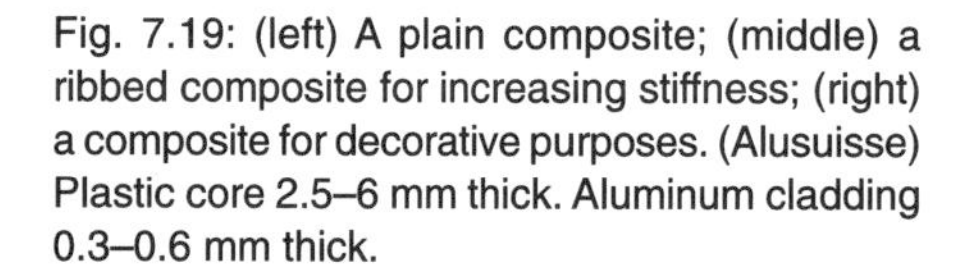

Fig. 7.19: (left) A plain composite; (middle) a ribbed composite for increasing stiffness; (right) a composite for decorative purposes. (Alusuisse) Plastic core 2.5–6 mm thick. Aluminum cladding 0.3–0.6 mm thick.

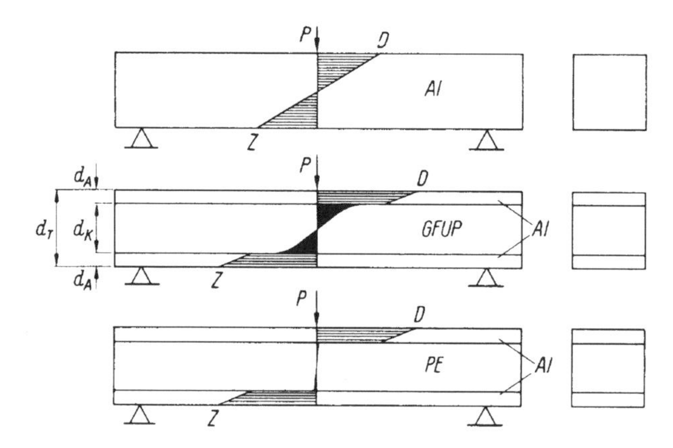

Fig. 7.20: Schematic representation of the stress distribution under elastic bending between solid Al sheet and a composite sheet (under equal load P). Typical thickness for Al-PE-Al composite: dT = about 3–8 mm; dA = about 0.2–0.6 mm; dK = 2–7 mm. Z = tensile stress; D = compression stress; GFUP = glass fiber reinforced polyester; PE = polyethylene.

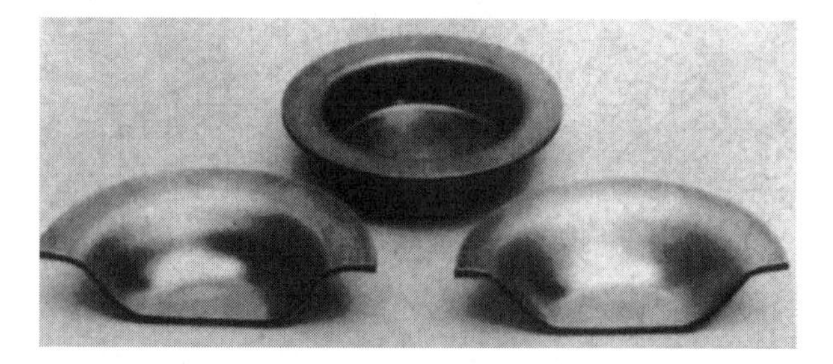

Fig. 7.21: Deep-drawn composite. (Alusuisse)

material in the neutral axis, since the maximum tensile and compression stresses in a bending load lie in the outer fibers. As shown in Fig. 7.18, the situation is analogous to that in which the solid metal in the area of the neutral zone (during bending conditions) is replaced by periodic spacers and air. However, the stiffness as shown in Fig. 7.18 is obtained in combination with limited plastic formability compared to composites with plastic cores.

In Figures 7.19 and 7.21 it is shown that composites can be worked like a solid metal by suitable material selection and joining of the components, by corrugating or deep drawing, etc., for example. In this way it is possible to obtain the maximum stiffness combined with minimum weight. Despite certain advantages of composites, they are currently applicable only for special applications.

7.7.2 Conditions under a given strain

One may consider the opposite case, in which a specific deflection in a structure should be kept in the elastic range if possible. In such cases, aluminum is superior to steel even for similar cross-sections since an aluminum alloy with similar yield strength can absorb three times more elastic deformation before a permanent (plastic) deformation starts.

This can be seen by following the line starting at B vertically in the stress-strain diagram (Fig. 7.14). At the applied elongation B, the aluminum is still in the elastic range, while plastic deformation is already taking place in the steel. Therefore, an aluminum structure is less sensitive to an applied strain so long as the force remains in the elastic range of the aluminum structure (Fig. 7.22).

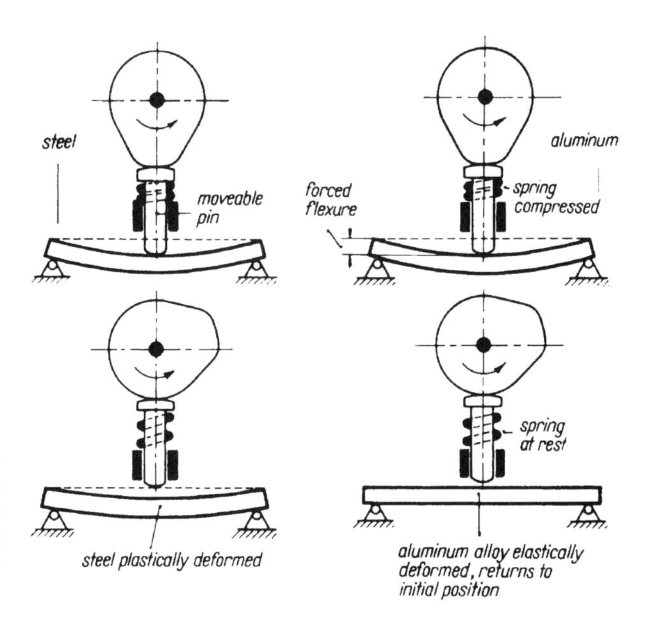

Fig. 7.22: A comparison of steel and aluminum under given deformation. Aluminum often comes out better when subjected to a specific deformation, because as long as the load does not exceed the elastic limit, aluminum can undergo three times as much deflection as steel without taking a permanent set.

7.8 Aluminum Data for the Designer (Typical Characteristics)

As far as the designer is concerned, some guidelines concerning the mechanical properties of aluminum alloys mentioned above may be summarized as follows.

A high tensile strength may be useful only if combined with a high ductility (elongation and reduction of area). The combination of these properties is important in structures that must withstand catastrophic conditions so that they are resistant to destruction. In this case, the so-called toughness or fracture energy plays a significant role. This energy is given by the area under the curve in the stress-strain diagram.

For guard rails, certain military applications, and most aircraft construction, one strives for alloys having a high total elongation along with a high ultimate tensile strength or high fracture toughness. These properties can be best achieved with high-toughness heat-treatable alloys like 7050 or 7075 or high percentage Al-Mg (Mn) alloys like 5083.

The 0.2% offset yield strength is a useful limiting factor for structures that must carry static loads. Structures are normally designed with a certain safety factor with regard to the 0.2% yield strength or the ultimate tensile strength. For example, the designer may base his or her design on 70% of the yield strength.

Aluminum's relatively low modulus of elasticity must be accounted for in the design of structural members, for example, in load-bearing beams or high facade components where the necessary limitation of elastic bending, within acceptable limits, requires that the stiffness of the cross-section be increased by increasing the height or depth (increasing the moment of inertia).

The low modulus of elasticity of aluminum may be advantageous in some cases, especially for automotive construction. The unevenness of road beds forces distortions in the automotive frame, which raises the stresses. These stresses are three times smaller in an aluminum structure than for the same steel structure. This inhibits not only premature permanent deformation, but, in addition, has a very favorable influence on durability (fatigue). As an illustration, a large all-aluminum trailer for goods transport turned over in an accident and twisted in the lengthwise direction. The experts familiar with comparable vehicles made of steel were of the opinion that the unit was badly deformed. However, when the vehicle was uprighted, to their surprise there was no permanent deformation since the aluminum structure could elastically deform three times as much as a similar steel unit and escaped without damage.

Even though aluminum has double the heat expansion of steel, stresses generated by temperature differences (thermally induced stresses) in a fixed construction element of aluminum are only 2/3 of those in steel under similar conditions. This is because of the differences in the modulus of elasticity.

Chapter 8. Fundamentals of Cold Working and Forming

8.1 Basics from Physical Metallurgy

8.1.1 Lattice imperfections

As we all know, no man is perfect. The same holds true for metals. The crystal lattice of metals and alloys contains structural imperfections. These imperfections play a role in the plastic deformation of a metal. Thus, knowledge of the lattice imperfections is necessary for a basic understanding of the properties of metals. The more important lattice imperfections are schematically shown in Fig. 8.1.

8.1.2 Vacancies

The term vacancy refers to the absence of an atom in the metal lattice. Vacancies play an important role in the movement of atoms in solid metals, which is called diffusion. To draw a comparison, imagine a parking lot completely filled with cars. None of the individual cars can be moved unless there is one free space available (i.e., a vacancy). If one such vacancy exists, the cars can be moved around so as to rearrange any individual car to a different location. By analogy, vacancies in the metal lattice make possible the movement (the diffusion) of the metal atoms within the lattice. Diffusion is important in the solution-heat treatment of heat treatable alloys or in billet homogenization in order to uniformly disperse the atoms in solution in the crystals. This takes place by diffusion jumps via vacancies (Fig. 8.2).

Vacancies originate most often in one of two ways, either by cold deformation or through heating. The latter, which are called thermal vacancies, are of prime importance. In Table 8.1 the number of occupied lattice points per vacancy is given for various temperatures. It can be seen that just before melting, one vacancy is present for just over 1000 atoms. Upon melting of the aluminum, the number of vacancies increases greatly, which is partially responsible for the increase in volume upon melting.

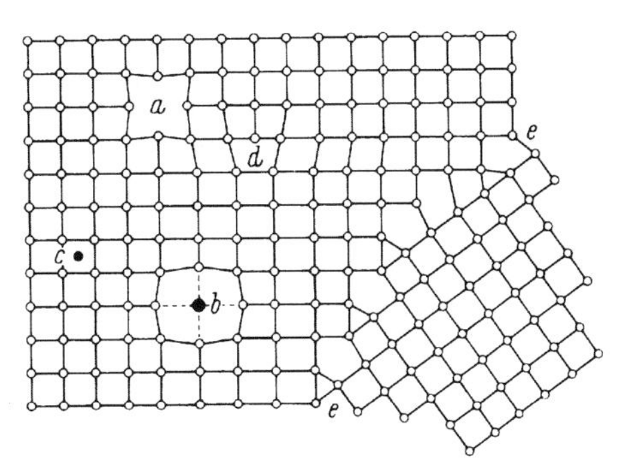

Fig. 8.1: A schematic showing principal lattice imperfections. Point imperfections (a) vacancy, (b) substitutional foreign atom (e.g., Mg or Cu), and (c) interstitial foreign atom (hydrogen atom). (d) Section through a linear lattice defect (dislocation); (e) Section through an area lattice defect (grain boundary = surface of a grain).

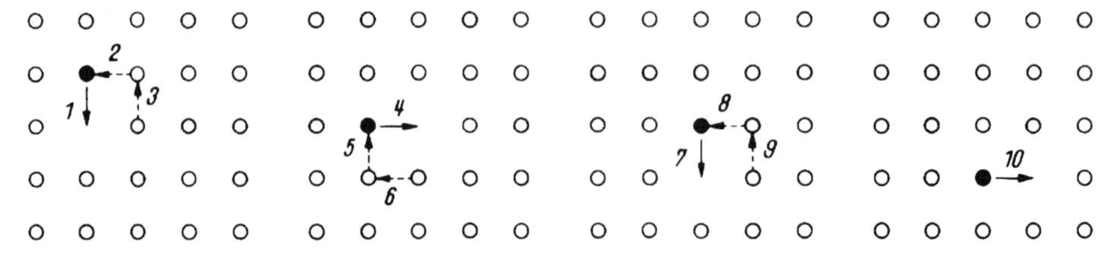

Fig. 8.2: Diffusion of a foreign atom over vacancies in ten successive steps.

Table 8.1: Occupied lattice sites per vacancy at various temperatures

Temperature (°C)	No. of occupied lattice sites per vacancy
20	10 000 000 000 000
200	12 250 000
300 (solid)	500 000
400	52 000
500	9 800
600	2 660
658	1.440

As mentioned before, vacancies can also be produced by cold deformation; however, they disappear in a fraction of a second into so-called "sinks" (for example, grain boundaries) while the vacancies in thermal equilibrium remain (Table 8.1).

8.1.3 Foreign atoms

In solid solutions, the foreign atoms may replace aluminum atoms in the lattice or may be located between the lattice points (Fig. 8.1). The arrangement the foreign atoms assume in the lattice depends on the ratio of their atomic diameter to that of the matrix atom. In addition, the chemical affinity of the element in question plays a role in the tendency of the foreign atoms to precipitate. Precipitates can also be listed as disorders in the lattice structure, but are not shown in Fig. 8.1. (See Chapter 9 for this phenomenon.)

8.1.4 Dislocations and plastic flow

Dislocations are the basis of plastic formability of crystalline materials. In Fig. 8.3 the formation and subsequent displacement of a dislocation during plastic deformation is schematically illustrated.

The dislocation lies at the point at which two atoms of the undisturbed lattice adjoin three atoms in the neighboring row of atoms. The lattice distortion is concentrated at this point. In the metal crystal, one must think of the atomic arrangement in Fig. 8.3 being extended perpendicular to the plane of the page. This means that the dislocation as a lattice disorder extends through the crystal in the form of a line.

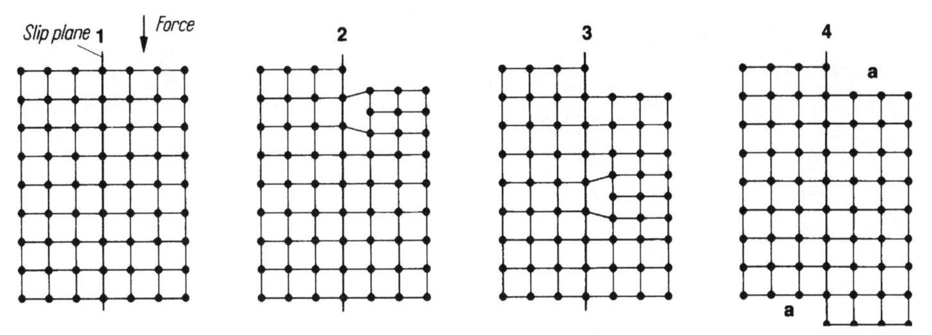

Fig. 8.3: Displacement of atoms, migration of an edge dislocation along a slip plane with plastic shear deformation; a = one interatomic distance. 1. Undeformed condition. 2. Forming a dislocation. 3. Migration of the dislocation along a slip plane. 4. Deformed condition.

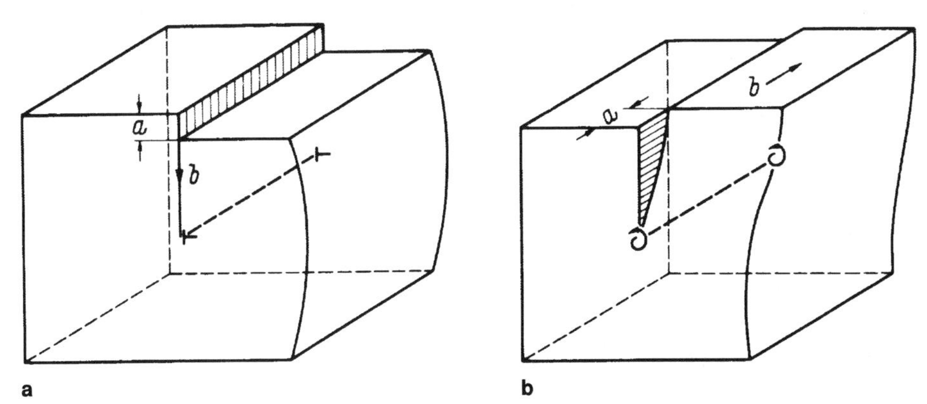

Fig. 8.4: Schematic representation of an edge dislocation A and screw dislocation B. a = one interatomic distance, b = direction of slip. The slip vector is at right angles to the dislocation in an edge dislocation while a screw dislocation is parallel to the slip vector.

There are different kinds of dislocations. Two of the most important types are illustrated in Fig. 8.4. Imagine a cube of rubber that has been sliced through halfway. In Case A, one half of the cube is pushed perpendicular to the dislocation line, and in Case B, in the direction of the dislocation line. The displacement in direction b is approximately equal to one interatomic distance. The atomic arrangement in the vicinity of an edge dislocation, Case A, corresponds to stage 3 in Fig. 8.3. This may be illustrated in the plane as an additional row of atoms forced into the lattice or in three dimensions as an inserted plane. In Case B, the neighboring atoms in the slip plane are shifted in the direction of the dislocation line by the passage of a screw dislocation through the lattice. If one traces the dislocation line from a lattice plane a few interatomic distances, the path would be similar to a winding staircase in which the difference in height between each turn is equal to one interatomic distance.

Each dislocation line may extend over hundreds or thousands of neighboring atoms. If a cold-worked piece of metal is sectioned, approximately 10^{12} dislocation lines would be intersected per cm^2 area—in a slightly cold-worked structure, approximately 10^{10} and in

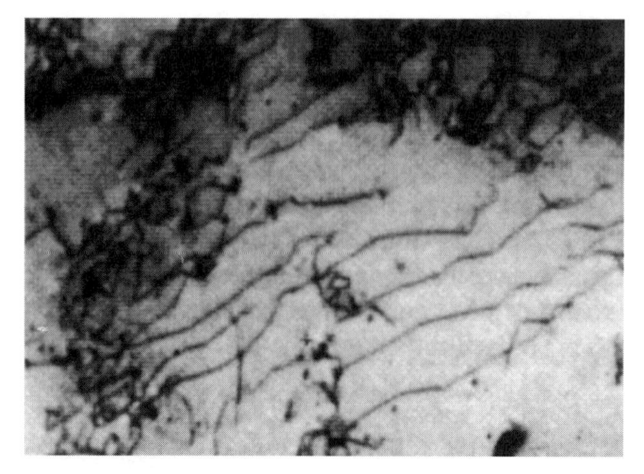

a

b

Fig. 8.5: Dislocation lines in 99.0% aluminum. (a) In the lower part of the picture are elongated dislocations that lie in the same slip plane. In the upper part of the picture are tangles of dislocations as a sign of greater lattice disorders.(b) Accumulation of dislocations due to the strengthening effect by dissolved Mg atoms in Alloy 5050. (Magnification 10,000:1.) (VAW)

the cast structure, 10^8. The latter means, for example, that in a small cube of 1 mm per side, one would find 1 km of dislocation lines each with an average spacing of approximately 1 μm. Fig. 8.5 shows an electron micrograph of dislocation lines. In Fig. 8.6, the presence of dislocation lines, even in the as-cast structure, is recognized.

In plastic deformation, the dislocations play a part in these ways:

- The movement of dislocations along the slip planes make plastic deformation possible.
- During deformation new dislocations are formed continually.
- The congestion caused by the dislocation pile-ups makes further deformation more difficult and, thus, leads to strain hardening.

In Fig. 8.7 the increase in dislocation density during cold working is presented schematically. With greater than 30–40% cold work, the congestion is not limited to the slip planes,

Fig. 8.6: Cellular dislocation arrangement in DC cast 99.0% aluminum. (Alusuisse)

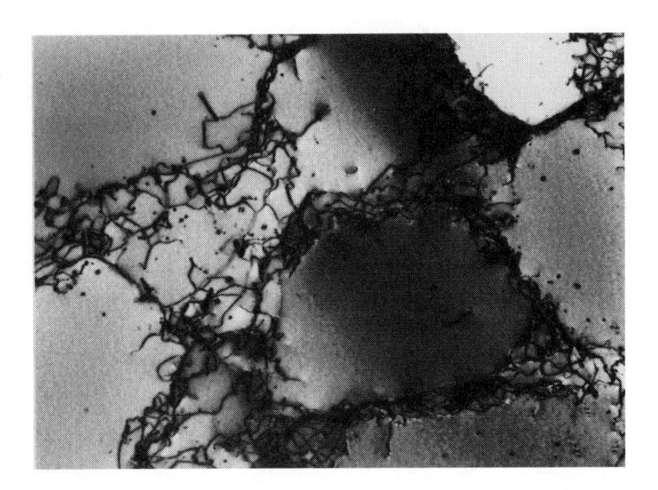

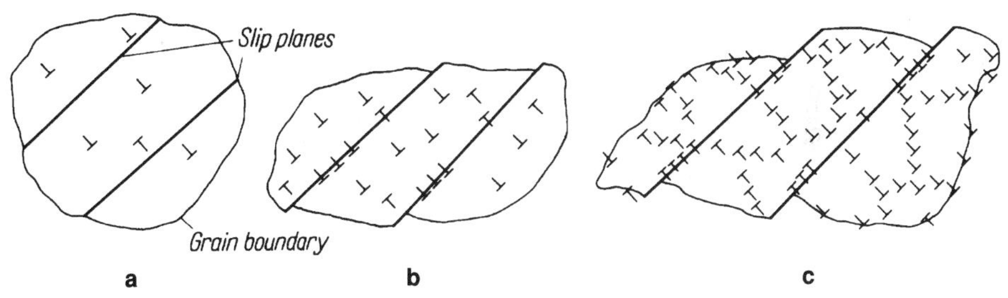

Fig. 8.7: An increase in dislocation density by cold work, schematic showing a grain with two slip planes. T = Symbol for a dislocation. (a) Before deformation. (b) After 10–20% cold work. The slip planes can still move easily, disregarding influence of neighboring grains, therefore, it is easily formable or low strength. Increase of dislocation density, especially on slip planes. (c) After 60–80% cold work. Further blocking of the slip planes through pile-up of dislocations. Accumulation of dislocations inside the grain in "honeycomb" form (indistinct subgrain boundaries). High "deformation strengthening," therefore, difficult to deform or high strength.

but dislocations may accumulate within the crystal in a honeycomb pattern similar to that shown in Fig. 8.6.

This accumulation of dislocations at the slip planes in honeycomb form is retained in the cold-worked lattice and results in strain hardening by a general distortion of the lattice. The rearrangement of dislocations by heat treating the cold-worked structure will be described later.

The foreign atoms present in the aluminum lattice can restrict the movement of dislocations significantly, which makes the mechanism of strengthening by alloying understandable on an atomic basis. (See Chapter 11 for details.) The resistance that dislocations encounter in the lattice as they move is expressed directly in the yield strength.

The discussion of the origin and function of dislocations has been limited to a few fundamentals. There are a great number of interactions between the lattice imperfections presented in Fig. 8.1.

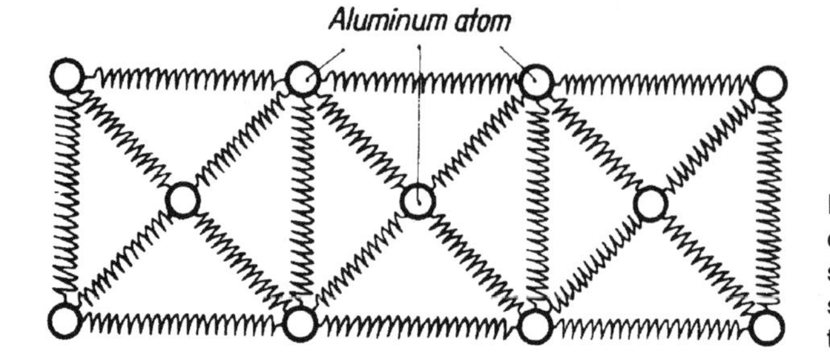

Fig. 8.8: Illustration of atomic cohesive forces with helical springs. They permit a comparison with elastic deformation of the metal lattice.

8.1.5 Elastic deformation

If the elastic behavior of metals is viewed from an atomic standpoint, one must start with the concept that the metal consists of atoms arranged regularly in a crystal lattice. The cohesion within the crystal is produced by forces of attraction that the individual atoms exert on their neighboring atoms. The attraction force is in equilibrium with a repulsion force, which is effective when the outer electron shells of neighboring atoms touch each other. This interplay of attracting and repelling forces can be represented by springs that hold the neighboring atoms in place, as shown in Fig. 8.8.

In this model, one can imagine elastic stretching such that the atoms are pulled slightly apart by an external force. If the external force is released, the atoms return to their equilibrium position under the influence of the springs.

The force required for elastic deformation is different for every metal, depending on the attraction between the atoms. Each metal has a different modulus of elasticity that is a measure of this force. It can be seen that the modulus of elasticity of a metal is not appreciably changed by small alloying additions since the foreign atoms are so greatly outnumbered that they do not significantly change the attraction forces between the atoms of the base metal.

8.1.6 Plastic deformation

During elastic deformation, the atoms leave their lattice position only to a limited extent, and upon release of the load, take their original position; in plastic deformation, the atoms are displaced over larger distances. The displacement is retained after removal of the deforming force. This process is known as slip.

Fig. 8.3 shows an example of the origin of a dislocation and its movement through the lattice in the process of plastic deformation. Planes that have the highest possible number of atoms serve as slip planes in metals. If one imagines a plane through the aluminum lattice that cuts through the highest number of atoms, this is the plane along which the atoms slip especially well by movement of the dislocations. As in the case of all face-centered cubic metals, aluminum has a total of 12 different slip possibilities in its atomic structure, which causes it to be especially formable.

Fig. 8.9: Slip lines and slip bands on aluminum. (580×) (Alusuisse)

Fig. 8.10: Rough surface on two deep-drawn cups. Grain size of the Al sheet: left = medium, right = large, thereby accenting the orange peel.

8.1.7 Formation of slip lines and bands

An unevenness can often be seen on the surface of plastically deformed metal. This is where one part of a crystal has slipped relative to another. Only certain directions of slip are possible, corresponding to certain planes of atoms in the crystal, called slip planes. They can produce visible slip lines and tend to accumulate in bands (Fig. 8.9).

If a coarse-grained aluminum sheet is bent or deep drawn, the surface becomes typically rough and dimpled, an effect known as "orange peel" (Fig. 8.10). This roughness is more pronounced as grain size coarsens. At the metal surface, these grains are not constrained and may deform more easily according to basic slip mechanisms that produce varying amounts of deformation, depending on the orientation of the grain. The anisotropic nature of the strain within the grain is responsible for the orange peel effect. Sometimes, clusters of small grains with similar orientation will deform like coarse-grained metal and produce a similar condition.

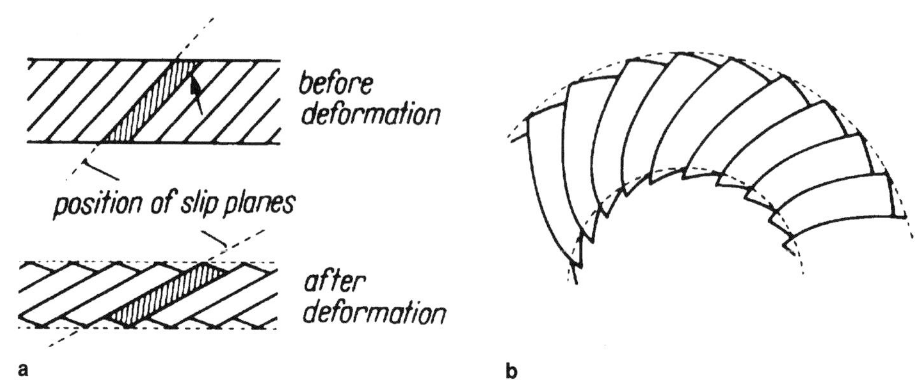

Fig. 8.11: Schematic appearance of a crystal in which slip has occurred. (a) Blocks of material are displaced relative to each other along slip planes (homogeneous deformation). (b) Bending a single crystal. Inhomogeneous deformation, the outside is in tension and the inside is in compression.

8.1.8 Behavior of a single crystal during plastic deformation

Slip can best be studied with a metal sample that consists of one crystal. By special techniques, single crystals several centimeters in length can be produced from almost any metal. Such a crystal formed from a melt has no internal grain boundaries. A test with an aluminum single crystal of approximately 10 mm diameter and 100 mm length is very informative.

This undeformed single crystal is almost as soft as cookie dough, and, in this condition, can be easily wrapped around a finger. One is astounded by how much greater force is required to straighten the single crystal thus deformed. Because of the plastic deformation that occurred in the crystal, the crystal lattice has been appreciably strengthened.

The plastic deformation of a metal is often compared with kneading of cookie dough. This comparison is not applicable, however. With cookie dough, the individual particles can slip past each other during deformation and move around indiscriminately. This is possible for metals only in the liquid state. For solid metals (i.e., in crystals), the atoms can move only in defined lattice planes.

During the deformation of a single crystal, the slip bands mentioned before can often be seen with the naked eye (compare Fig. 8.11). This indicates that during plastic deformation individual parts of the crystal slip past each other. This takes place in the so-called slip planes. In a coarse-grained structure, the slip lines can be seen as plainly as in a single crystal (see Fig. 8.9). The origin of the visible slip lines or bands is the slip of the atoms within their lattice formation. This can take place only along a few exactly defined slip planes within the lattice.

8.2 Work Hardening of Polycrystalline Material

It will be recalled that plastic deformation strengthens the metal. This was shown in the example of the single crystal that was as soft as cookie dough until bent, but then became

Fig. 8.12: Lattice of two neighboring grains in cold-worked metal. G = position of two slip planes (actually a series of parallel slip planes lie inside a grain).

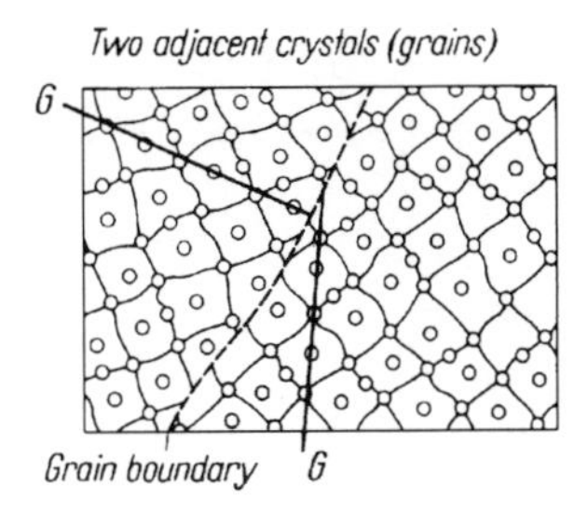

so hard that it was difficult to straighten afterward. This is not completely obvious from Fig. 8.11.b, since it would be expected that the slip of the atoms through the reaction to an applied force, as required to induce initial deformation, could be reversed. This is not the case, however, because the plastic deformation causes a permanent strengthening of the lattice, so that in the process of subsequent deformation the force required continues to increase.

The work hardening is considerably higher for polycrystalline structures compared to single crystals. The smaller the grain size, the stronger is this effect. The reason for this is that the slip planes change directions at the grain boundaries, so that the adjoining grains interfere with the slip in the neighboring crystals, since the structure maintains its solid state during plastic deformation.

Fig. 8.12 shows a simplified schematic of two neighboring grains with divergent orientation. The dislocation movement is inhibited at the grain boundaries. It can be seen from the foregoing that the plastic deformability of polycrystalline material is less and the strengthening much higher than that of single crystals. A coarse-grained material lies somewhere between a single crystal and a fine-grained structure. This means that the individual crystals of a coarse-grained structure inhibit the slip only slightly, so that a coarse-grained material has a lower strength than a fine-grained piece of the same material. The semi-finished product of aluminum or aluminum alloy should have as fine a grain as possible in order to achieve high strength values and to suppress the formation of "orange peel" during further fabrication, since this requires additional finishing.

8.3 Effect of Cold Working on Properties

The mechanical properties of aluminum (i.e., strength and the behavior under mechanical stress) can vary within wide limits depending on alloy, degree of cold working, and heat treatment. Cold deformation will be examined closely in this chapter. Table 8.2 shows a few typical values for pure aluminum and an aluminum-2.5% magnesium alloy sheet with different degrees of cold work.

Examining the values in Table 8.2 for 1100 alloys shows that the 0.2% yield strength is increased very sharply with increasing degrees of cold work. At the same time, the tensile strength and hardness increase, while the elongation decreases significantly.

The strength values behave the same way for the aluminum-magnesium alloy, with the difference that the strength is significantly higher in the soft condition for the alloy and

Table 8.2: Tensile properties of aluminum sheet

Alloy and Temper	Thickness (mm)		Tensile strength (MPa)			Elongation (%) Min. 50 mm
			Ultimate		Yield	
	Over	Through	Min.	Max.	Min.	
1100-0	1.20	6.30	75	205	25	30
1100-H14	1.20	6.30	110	145	95	5
1100-H18	1.20	3.20	150	—	–	4
5052-0	1.20	6.30	170	215	65	19
5052-H34	1.20	6.30	235	285	180	6
5052-H38	1.20	3.20	270	—	220	4

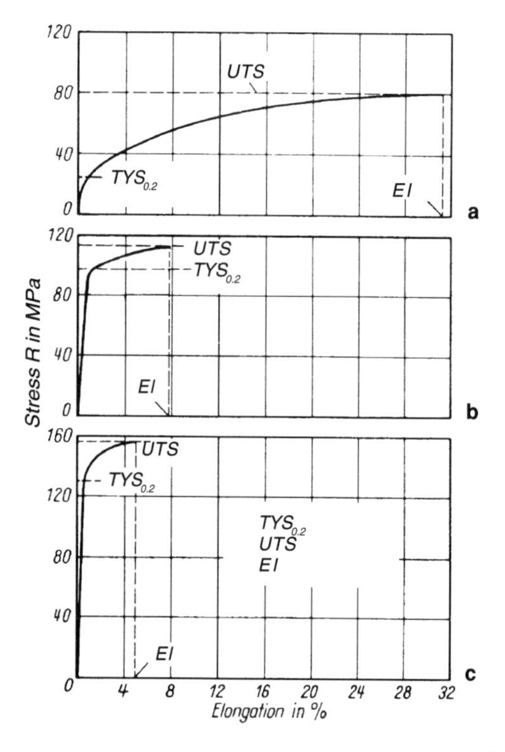

Fig. 8.13: Stress-strain diagram for three 1100 alloy sheet samples of different temper (schematic), (a) O, annealed (no cold work); (b) H14 (30% cold work); (c) H18 (70% cold work).

remains correspondingly higher. The increase in strength through cold work is called strain or work hardening, as opposed to alloy hardening. The value for alloy hardening is given by the difference between 1100-0 temper and 5052-0 temper in Table 8.2.

In Fig. 8.13, three stress-strain diagrams are given schematically for 0 (annealed or soft), H14 (half-hard), and H18 (full-hard) 1100 alloy sheet. The strong influence of the work hardening on the shape of the stress-strain curve is clearly recognized. It can be seen that the 0.2% yield strength and tensile strength are nearly identical with high degrees of cold work, while the elongation has decreased to a low value. For practical purposes, this means that severely cold-worked metal has no elongation reserve. A sheet cold worked to a maximum degree generally will fracture with a slight bend or other deformation.

Table 8.2 does show, however, that cold rolling can produce significantly higher work hardening in aluminum sheet than stretching in the tensile test.

8.4 Resulting Structural Features

A special case exists if a sheet, even though fine grained, has a so-called texture. A sheet is said to exhibit a texture if the individual small crystals are not randomly oriented, but are mostly of nearly the same orientation. One can easily imagine that subsequent forming processes of such textured sheet may result in undesirable conditions, such as Lüders lines, looper lines, and earing.

8.4.1 Lüder lines

During forming or stretching of some aluminum alloys, surface markings resulting from localized flow may appear. These are called Lüder lines or stretcher strains. They lie approximately parallel to the direction of maximum shear stress at about 45 degrees to the direction of the applied stress. When forming in tension, they occur as depressions and when in compression, as elevations. Lüder lines are usually associated with the yielding and yield-point elongation of annealed or solid solution alloys such as aluminum-magnesium. The formation of stretcher strains or Lüder lines is generally associated with stretch-forming or drawing operations where some areas of the sheet receive little or no deformation.

Aluminum-magnesium alloy (5xxx alloy) sheet is particularly susceptible to Lüder line formation, and the severity increases with increasing magnesium content. Precipitation hardenable aluminum alloys are also susceptible to the formation of Lüder lines, but normally only in the freshly-quenched condition. Alloys like 1100 or 3003 do not form Lüder lines.

Generally, Lüder lines are undesirable because they result in uneven and roughened surfaces. Most Lüder line problems arise during forming of annealed aluminum-magnesium sheet. Using strain-hardened sheet avoids the problem. However, the lower ductility of sheet that has been strain-hardened may result in reduced formability. Another method by which Lüder lines can be avoided is to form or work the material at temperatures above 150°C.

8.4.2 Looper lines (roping)

Looper lines are a form of surface roughening that is sometimes encountered during deep drawing of aluminum sheet. They take the shape of loops (ropes), hence the name. The most common cause of looper lines is a striated grain structure. This irregularity is normally due to coarse grains formed during hot rolling or intermediate annealing. Then, during temper rolling, the grains become elongated. Looper lines may also form because of segregation in the ingot, especially coarse dendritic segregation.

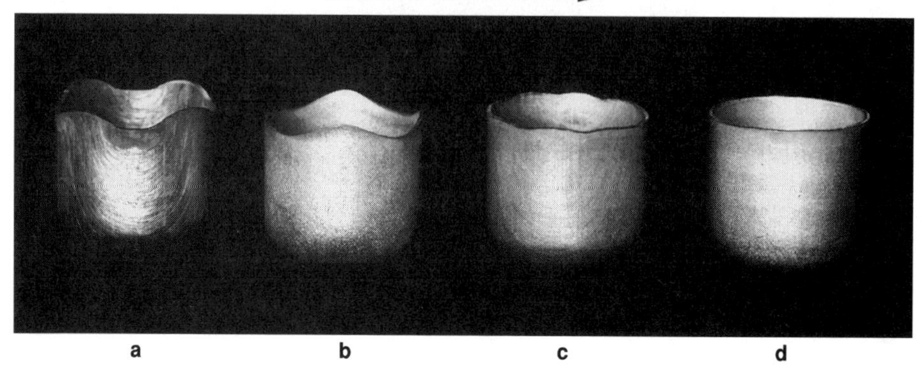

Fig. 8.14: Cup samples for investigation of earing. Rolling direction: perpendicular to the plane of the picture. (a) earing at 45° to the rolling direction (rolling texture); (b) earing at 0° and 90° to the rolling direction (recrystallization texture); (c) eight ears: four each at 45° and at 0° and 90° to the rolling direction. (mixed texture typical for canstock); (d) no earing at all (rather exceptional). (Alusuisse Singen)

8.4.3 "Earing formation" as a function of texture

Even in a casting, the crystals are not randomly oriented, but are arranged to a certain degree in an as-cast texture with numerous grains parallel to each other in the direction of heat removal. Rolling also produces a very strong texture in the finished sheet under certain circumstances.

The presence of a texture in a soft sheet can be easily determined if a small cup is drawn or pressed from a blank. The upper rim of the cup is not flat, but usually has "ears," as shown in Fig. 8.14.b.

The more pronounced the similar orientation of large groups of crystals in the sheet, the higher the ears will be. Usually there are four ears, which means that the deformation of the sheet takes place preferentially along planes perpendicular to each other in the sheet. A similar orientation of a significant portion of the grains with large areas of the structure may originate not only from the cast structure, but also by rolling, extruding (deformation texture), or annealing (recrystallization texture). The recrystallization that occurs during annealing will be discussed later. The different textures in a sheet can be determined by the position of the ears. The rolling texture (deformation texture) has its four ears inclined 45° to the rolling direction. The cubic texture that occurs during annealing under certain conditions has its ears in the rolling direction and 90° from that direction. It is possible by suitable intermediate annealing to play off these textures against each other. One can obtain a mixed texture that has eight ears in which all have about the same height (four ears each at 45° and at 0° and 90° to the rolling direction). The occurrence of eight ears is the best compromise available in deep drawing thin, soft sheet produced from DC cast material. With only four ears, considerable scrap must be removed after deep drawing since the upper edge of the deep-drawn part must be trimmed considerably more than with right ears.

Furthermore, if the earing is excessive, blanks must be correspondingly larger so that the "earing valleys" are sufficiently high to make the final part after trimming. By precise selection of the rolling and annealing conditions, earing formation can be almost completely inhibited.

In practice, this means that in highly-developed forming processes that are already close to the limits of the material, severe disruption or even a breakdown in production can result from a change in earing behavior from one rolling batch to another. One such critical operation is drawing and ironing (described in Figs. 9.28 and 9.29), typically used to make beverage cans from alloys of the type 3004. It is performed in mass production lines, each line making several hundred cans per minute. In such lines, it is considered a major risk to change from one aluminium strip supplier to another, whose strip is liable to have a rather different texture. Then, the production line will require time-consuming readjustments, such as changing the drawing tools. If the ears are very long, they will jam the trimmer, causing excessive downtime.

Over the last few decades much has been published about earing. At the beginning, many correlations of the type A-C were reported, such as the dependency of earing on factors such as:

- Iron to silicon ratio (for AA1000 series alloys)
- Casting method
- Hot rolling conditions
- Amount of cold rolling

But, above all, earing depends on the total thermal cycle that the material undergoes.

These observations concerning earing illustrate how important it is to use the overall A-B-C concept of defining the relationships between process conditions, microstructural characteristics, and the properties of the end product. More recently, studies have posed the more general and fundamental question: is it possible to define a rigorous, standard procedure to guarantee absolutely reproducible formability in deep drawing and ironing? However, empirical development based on the principle of jumping straight from process conditions to the properties of the product cannot answer this question. Even spot-checks to observe precipitates under the microscope cannot provide any overall prediction of texture.

Thus, development depends more and more on x-ray analysis of texture by goniometry across a sufficiently large cross-section. But even this considerable effort cannot guarantee the earing levels in an industrial drawing process. To master critical drawing operations, it is essential to follow the logical chain A-B-C from cause to effect in order to ensure uniform formability within the scope of a standard fabricating practice. This must include full-scale forming tests on the end product.

Chapter 9. Industrial Forming Processes[1]

9.1 Fundamentals

9.1.1 Comparison of hot and cold working

Both hot and cold working operations are used industrially in the shaping of metals. For wrought aluminum materials, hot working is generally carried out at temperatures between 350°C and 530°C. Cold deformation takes place at room temperature, though the material itself may reach temperatures up to 150°C during the process. Fig. 9.1 shows that the pressure for a 50% flattening of different metal cylinders, including aluminum, decreases with increasing temperature. This is one of the reasons why heavy forming operations are carried out mostly in the hot working range.

There are a number of reasons why the final phases of mechanical working are carried out at room temperature. The most important specific advantages of cold deformation are:

- Increased strength by strain hardening.
- Improved surface quality compared to hot working.
- For thin cross-sections, increased deformation for each operation compared to hot working.

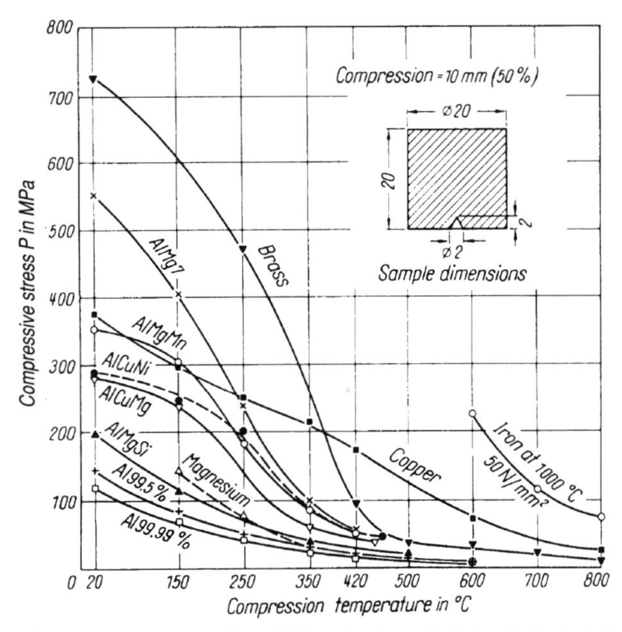

Fig. 9.1: Compressive stress necessary for a 50% reduction of initial cylinder height as a function of temperature. (Alusuisse)

[1] By G. Scharf.

The latter point may require some explanation. For example, in hot rolling aluminum, a water-based soluble oil emulsion is used as the lubricant. Thicknesses less than about 6 mm and sometimes as little as 2 mm (canstock; see 9.2.2) are cold rolled. For cold rolling, an organic rolling oil or emulsion is used that has much better lubricating properties compared to the emulsion normally used in hot rolling. However, the cold-rolling oil is flammable at temperatures above 100–150°C. When the surface area per unit weight of the material exceeds a certain value, the high heat losses and poor lubrication make further hot rolling impractical versus cold rolling.

Thus, cold rolling and the better lubrication of roll surfaces facilitates metal flow over the roll surface and permits greater reductions, improved surface finish, higher rolling speeds, and more accurate gage control. In addition, less scrap is generated, a flatter product can be produced, and the strength of the metal is increased. Additional detail on cold rolling is given in 9.3.1.

9.1.2 Thermal treatment before hot working

Rolling ingot and extrusion billet are usually cooled to room temperature in the casting process. (The hot deformation of continuous cast sheet and plate of about 10–30 mm thickness and wire rod stock of 30–80 mm diameter present a special case in which the hot rolling may be a continuous operation using the heat of casting.) They are reheated prior to hot working, for two purposes:

- Modification of the cast structure
- Facilitation of deformation

There are two different types of thermal treatment of ingot and billet prior to hot deformation, depending on the effect desired.

1. Homogenization of the ingot and billet with subsequent hot deformation at a somewhat lower temperature.
2. Preheating only to the temperature for hot working.

These two steps are often combined into one thermal cycle. In order to explain the purpose of this thermal treatment it is necessary to consider a few structural changes in the solid state and how they can be influenced by this type of thermal cycle.

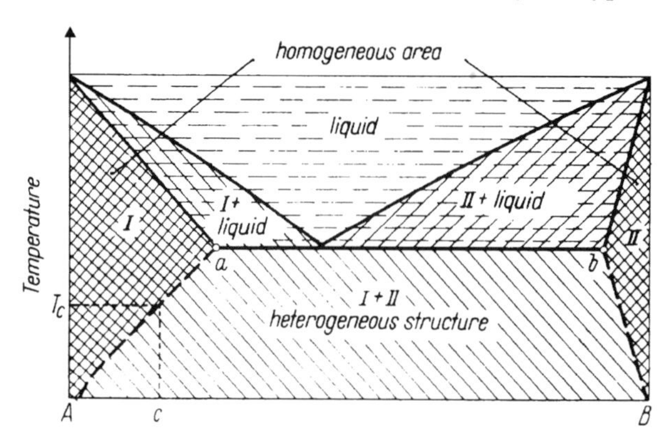

Fig. 9.2: A binary eutectic equilibrium diagram with two solid solution areas. A = aluminum, B = alloying metal or intermetallic compound.

9.1.3 Solubility in the solid state

A binary phase diagram typical of numerous wrought aluminum alloys is presented in Fig. 9.2. For our basic explanation it is unimportant whether B represents a pure metal or an intermetallic compound. In any case, there is an exactly defined crystal lattice for both the unalloyed pure metal A and the metal (or compound) B. However, it can be seen from Fig. 9.2 that both lattice types are, to a certain extent, "tolerant" in that they house foreign atoms. By the formation of solid solutions, homogeneous regions may arise in the phase diagram, which cover a range of compositions and temperatures, according to the alloy system. In the example from Fig. 9.2, crystal type I is more tolerant toward B atoms than crystal type II is toward A atoms. This depends mainly on the respective atomic radii and, thus, how much room is present in the "host" lattice. (Whether an intermetallic compound develops with its own defined system of crystallization depends on the geometric relationship as well as the chemical affinity of the partner atoms.)

The solubility is greatest at the eutectic temperature in that solid solution I accepts atoms of composition B up to point a, and solid solution II accepts A atoms up to point b in a homogeneous structure.

In the case of commercial aluminum alloys, the behavior of the aluminum-rich crystal type I α solid solution is of prime importance. With decreasing temperature, the thermal movement of the atoms is reduced. And, as a result, the lattice structure becomes tighter and more B atoms will be squeezed out of the lattice of solid solution I and precipitate in the B-atom rich crystal type as inclusions if there is sufficient time. In the case of composition c, a heterogeneous structure of a mixture of crystal types I and II can arise below temperature T_c.

The precipitation process as well as the opposite process of dissolving precipitates requires a certain mobility of the atoms that is present only at sufficiently high temperatures. Between 200–400°C, the mobility of most of the atoms in the aluminum lattice is sufficient so that they can arrange themselves into an equilibrium condition if the time at the elevated temperature is long enough. The principle is equally valid whether the shift in equilibrium results from a decrease of the temperature (heterogenization) or through increasing the temperature (solution or homogenization). However, there are some alloying elements that diffuse very slowly in an aluminum matrix. For example, iron does not diffuse to any significant extent below 400°C, and manganese and chromium diffuse only above 500–550°C.

9.1.4 Homogenization through a high temperature anneal or solution anneal

The following discussion centers on approaching equilibrium in accordance with the phase diagram.

The term homogenization is used for the elimination of supersaturation and grain segregation from the cast structure as well as for solution heat-treatment, that is, for dissolving a precipitated alloying element within the homogeneous region of the phase diagram. For both processes, the atomic picture is very similar.

It is the purpose of a homogenization anneal to take precipitates into solution and distribute the atoms uniformly in the lattice. In every case there must be sufficient time at the proper temperature to allow the atoms to arrange themselves in an equilibrium state by movement through the structure. In the case of grain segregation and supersaturation, the atoms are frozen in a nonequilibrium state. After solidification, they were not allowed sufficient time at a temperature that favored diffusion. For DC cast material, they were exposed to a favorable temperature for less than 1–3 minutes, which is too short to allow the atoms to move a distance of 0.01–1.0 mm that is necessary to compensate for grain segregation.

This is especially true of the as-cast structure. After deformation of the cast structure, the conditions for the rapid displacement of the atoms are considerably more favorable. In the as-cast structure, the heterogeneities are larger and the grain size is often above 1 mm, which means the diffusion paths are relatively long.

The heterogeneities are reduced in size by working, and zones with different contents of atoms in solution are brought closer together, which greatly accelerates equilibrium by shortening the diffusion paths.

As mentioned earlier, a pronounced grain segregation as well as a supersaturation of the alloying elements is present in rapidly solidified DC cast structures. This means that the distribution of foreign atoms in the structure is not according to equilibrium. For the production of semifinished products that must fulfill stringent requirements of formability, fine grain size, and a structure free from streaks, rolling ingots and extrusion billets are subjected to a high temperature soaking at temperatures between 550°C and 630°C prior to hot working. This serves to bring the alloying elements into equilibrium and, especially, to uniformly distribute these elements within the structure. Precipitates not in solution are spheroidized by homogenization, that is, they lose their sharp-edged shape, becoming compact and rounded off, which improves formability significantly. Veins of melt residual, which are enriched in elements difficult to take into solution (for example, iron), are not normally completely dissolved during homogenization. Thus, a homogenization does not eliminate, for example, the existence of widely varying cell sizes within the structure as shown in Fig. 4.27.c. The heat treatment time depends on grain size and the diffusion rate of the alloying components.

For many alloys, the type of cooling after homogenization is also of importance since finely dispersed precipitates may form during cooling.

If an alloying constituent is not completely soluble in aluminum at the maximum permissible homogenization temperature, according to the phase diagram, a finely divided precipitate forms both within the grain and at the grain boundaries. In this case, a high temperature anneal results in a heterogenization. (Therefore, the expression "homogenization," which is often used incorrectly, should be replaced with "ingot high temperature anneal.") A typical heterogeneous alloy is aluminum with an addition of approximately 1% manganese. In Fig. 9.3 the DC cast structure of an aluminum-manganese alloy is shown before and after high temperature anneal. In the as-cast state, the aluminum solid solution is highly supersaturated in manganese. One can see that during the high temperature anneal two different processes take place:

Fig. 9.3: Structure of a DC-cast rolling ingot. 3003 alloy: 1% Mn + 0.67% Fe + 0.16 Si (Alusuisse). (a) As-cast. Angular precipitates of the aluminum-manganese-iron phase in the cast grains and at the grain boundaries. 860×. (b) Rolling ingot heat-treated for 72 h at 600°C then quenched. Through diffusion processes, the precipitates have grown and rounded off (spheroidized). 860×. (c) Rolling ingot heat-treated for 6 h at 600°C, then furnace-cooled for 15 h to 450°C. A fine AlMnFe precipitate originated from the supersaturated solid solution due to the slow cooling. At the same time, the precipitates from the cast structure spheroidized (less than in Fig.9.3.b due to the shorter heat treatment). Mag. 860×.

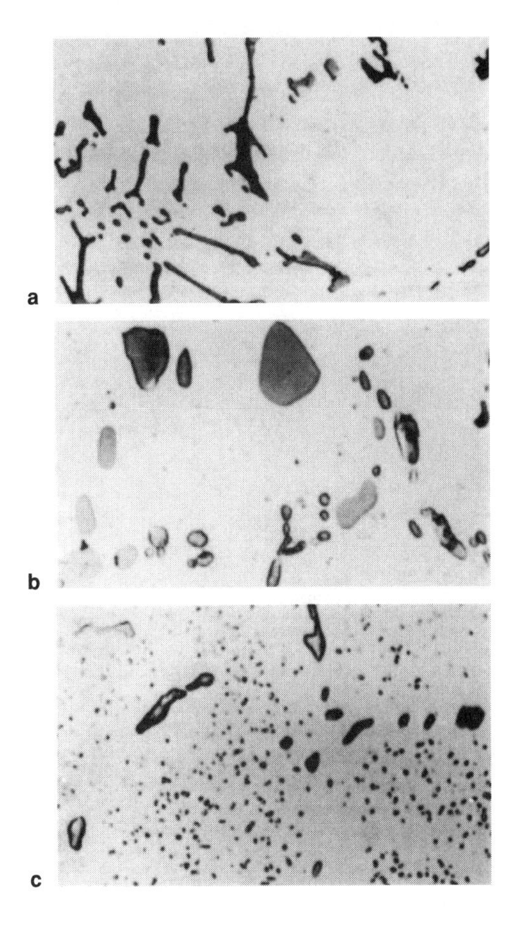

- Coarsening and rounding off of the relatively large heterogeneities from the cast structure.
- Formation of fine precipitates from the supersaturated solid solution.

In Fig. 9.4, the distribution of the foreign atoms after an ingot high temperature anneal is presented schematically. One sees that by slow cooling after an ingot high temperature anneal, the finest precipitates occur, the majority of which are not visible using an optical microscope. Finely dispersed accumulation or precipitation of foreign atoms have a special significance for the structural modifications that take place in subsequent deformation or thermal treatment (Fig. 9.5). This subject will be discussed later in connection with recrystallization and strengthening of heat-treatable alloys.

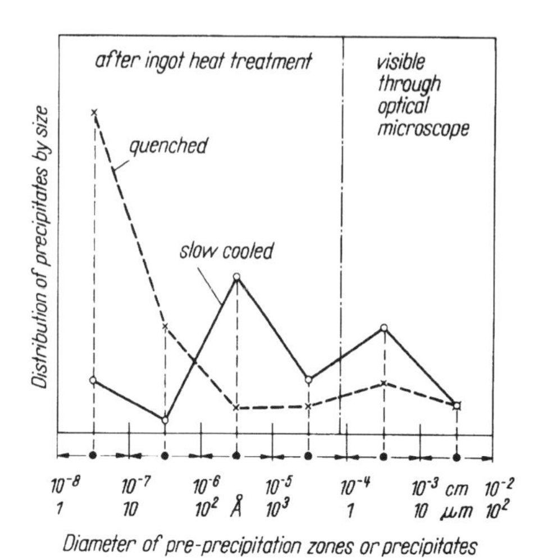

Fig. 9.4: Arbitrary distribution function of precipitated alloying elements (e.g., precipitate containing Mn and Fe in AlMn alloy) (schematic).

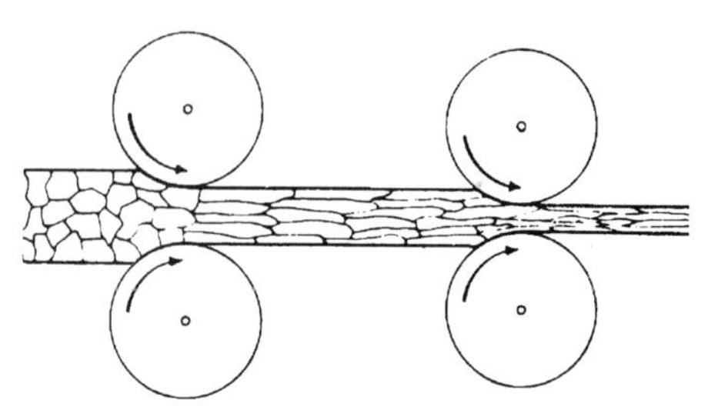

Fig. 9.5: Elongation of the cast structure by hot working. In this case, a lamellar structure is formed (deformation through two roll passes). When the deformation temperature and degree of work is high enough, the lamellar grain structure is transformed immediately into new equiaxed grains by recrystallization. This phenomenon is discussed in more detail in the section "Recrystallization" (see Chapter 10.5).

9.1.5 Heterogenization

During heterogenization or precipitation heat treatment, the atoms in solution are given the opportunity to precipitate, attaining equilibrium in accordance with the phase diagram.

If the lattice has been cold deformed prior to the heat treatment, a softening or, at higher temperature, a recrystallization takes place simultaneously with heterogenization. These events are influenced by existing precipitates and those in the process of formation. The inclusions (precipitates) that occur during precipitation heat treatment are significantly finer than those in the cast structure (compare Fig. 9.3, for example). Size and distribution of the precipitates may be influenced by the annealing time and temperature. At higher temperatures or longer times, the precipitates are larger, but less numerous.

9.1.6 Properties of hot-worked structures

During hot rolling and extruding, the grains of the cast structure are elongated (Fig. 9.9). For instance, they are increased in length 20 fold with 95% deformation. During lengthening, the inclusions present are broken up and distributed. This is shown in Fig. 9.6. The structure of a commercially pure aluminum billet is shown in Fig. 9.6.a and Fig. 9.6.b. The inclusions containing iron and silicon are present in the structure mainly at the grain and cell boundaries. After the as-cast ingots are hot rolled from approximately 200 mm to 10 mm (approximately 95% reduction), the inclusions are broken in small pieces and finely distributed.

The fragmentation and distribution of the inclusions is the decisive action in the transformation from a cast to a wrought structure. This gives the wrought structure its ductility.

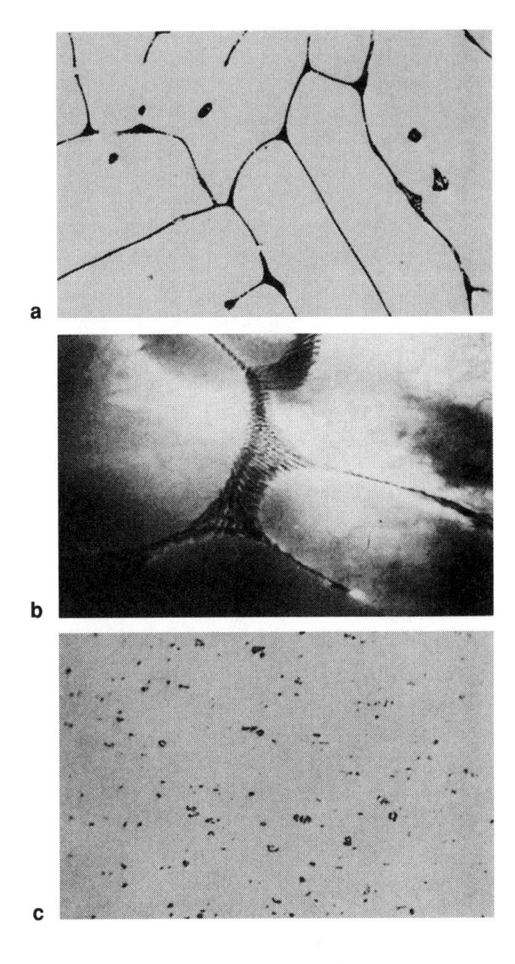

Fig. 9.6: Fragmentation of heterogeneities in the as-cast structure through hot and cold rolling. (Alusuisse). (a) Structure of DC-cast 99.5% aluminum. Solidified residual melt veins enriched in Fe and Si at the grain boundaries and partly in the grains. 450×. (b) Enlarged section from Fig. 9.6.a. The coarse residual melt veins lie at the grain boundaries (= dendrite arms). 1 800×. (c) Structure after hot rolling from about 200 mm to 10 mm and subsequent cold rolling to 3 mm. 30×.

During subsequent cold working, the undissolved particles swim somewhat passively in their surrounding structure without inhibiting the deformation significantly (Fig. 9.6.c). The size of the inclusions is not altered significantly by cold rolling.

9.2 Hot Working Processes

9.2.1 Short survey of measures and their effects

Let us see how the measures taken during hot rolling relate to transformations in the structure and the properties of the final product. This is a complex problem area. Only a few individual examples are mentioned in the following.

During hot working, the following measures are of prime importance:

- Hot deformation process, degree of deformation.
- Thermal treatment prior to hot deformation.

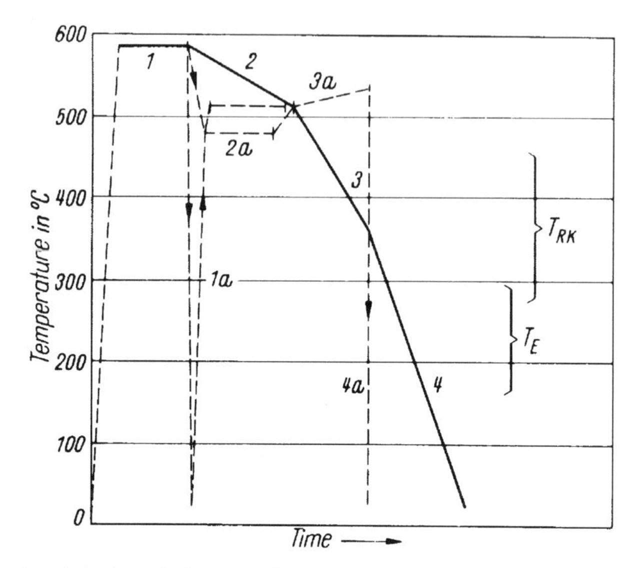

Fig. 9.7: Thermal cycle before, during, and after hot deformation (schematic). Solid line: typical cycle for hot rolling; dashed lines: variants of hot extrusion. 1: ingot thermal treatment at high temperature. 1a: heating to hot deformation temperature without homogenizing. 2: heterogenizing the ingot through slow cooling. 2a: Heterogenizing by annealing at lower temperature. 3: hot deformation, typical for extrusion. 3a: typical for extrusion. 4: air cooling after hot deformation. 4a: water quenching. T_E–T_{RK}: temperature regions for recovery and recrystallization.

- Temperature at the beginning and end of the deformation.
- Cooling after hot deformation (especially important for heat-treatable alloys).

The last three steps are grouped together under the term "thermal cycle" (see Fig. 9.7).

Among structural changes, the following are of special interest.
- Softening and recrystallization during or immediately after hot deformation.
- Precipitation or solution of foreign atoms during or immediately after hot deformation.
- Welding of casting pores during hot deformation.
- Occurrence of cracks (undesirable).
- Uniformity of hot deformation over the cross-section of the structure.

The above operations and structural changes influence a great number of the properties of the hot-worked semifinished products. Final products, annealed or not, that were cold worked after hot working are, in turn, affected. The most important properties that are influenced by the hot-working cycle are grain size, strength, formability, and texture as well as the decorative appearance of the surface (freedom from streaks).

The total thermal cycle as outlined in Fig. 9.7 has an especially strong influence on recrystallization and softening after subsequent cold work. The distribution of the foreign atoms that occurs during the important phase of hot working often does not change during recrystallization or stress relieving since the temperature ranges T_E and T_{RK}, as shown in Fig. 9.7, are generally much lower than that required for hot working.

In summary, it could be said that hot deformation and the distribution of the foreign atoms resulting from it exert a dominant influence on the properties of the semifinished material. The chain with the interconnecting measures and structural changes reaches from the cast structure to the final product. In this chain, hot deformation plays an especially important role.

9.2.2 Hot rolling

The raw material for fabricating rolled products in aluminum is rectangular DC cast ingot. Modern rolling mills use ingot sizes 200–600 mm thick, 60–2200 mm wide, and 4500–9000 mm long, usually referred to as sheet ingot. Such ingots weigh 1.5–30 tonnes. As a result of the cooling conditions in DC casting, sheet ingots have an irregularly solidified surface characterized by a heterogeneous structure with surface segregation. For this reason, it is usually necessary to mill away the cast surface, a process commonly referred to as scalping. Often the head and foot of the ingot are also sawed off in order to assure the surface quality of the product.

To improve thermal efficiency, ingot homogenizing and preheating to rolling temperature are usually combined in one operation. The hot-rolling temperature is mostly in the range 400–500°C and is usually somewhat lower than the homogenizing temperature, which means that the ingots have to be cooled after homogenizing.

Modern hotlines usually have a reversing roughing mill, in which the ingot is first rolled down with heavy reductions between 15 mm and 60 mm per pass to a slab gauge of 15–35 mm.[2] Hot-rolling then continues in a multistand tandem mill, down to 2.5–8 mm. Tandem mills can usually coil common alloys at 12–14 mm, and canstock is sometimes hot-coiled as thin as 2 mm.[3] Fig. 9.8 shows a typical modern hotline for wide strip.

Hot-rolling thoroughly works the structure, largely eliminating the original cast structure. A typical development of the grain structure during hot-rolling is shown in Fig. 9.9. Basically, the grains are elongated in the rolling direction. Depending on the alloy and on the temperature and reduction per pass, these grains undergo static and dynamic recovery as well as partial recrystallization during hot-rolling. If enough deformation energy remains in the crystal structure at finished gauge and the finishing temperature is high enough, then the grains may undergo static recrystallization in the hot coil after rolling.

An understanding of the structural changes and especially of how hot-rolling conditions affect the distribution of the alloying elements, is fundamental to planning the subsequent cold-rolling and intermediate annealing operations in order to attain the desired properties in the end product. The key element is always whether the hot-worked material was recrystallized and / or recovered.

---

[2] Hot-rolling is, however, not only done in hotlines with a rougher and tandem finisher. Worldwide, there are a great many hot-rolling installations in other mill configurations.

[3] To enhance a low-earring texture at final gauge and temper.

Fig. 9.8: A modern hotline for wide strip. (VAW/Alcan)

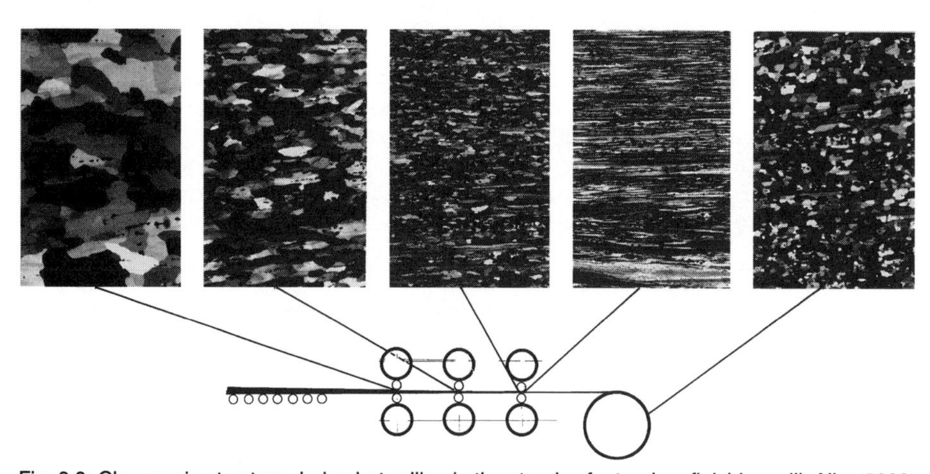

Fig. 9.9: Changes in structure during hot-rolling in the stands of a tandem finishing mill. Alloy 5083, coil gauge 4.5 mm, coil temperature 315°C. Magnified 45×. (VAW)

9.2.3 Extrusion[4]

Extrusion is second only to rolling for making semi-fabricated products. The process consists of subjecting a confined billet (usually preheated up to about 450–550°C) to high pressure, thus forcing the metal through an opening in a steel plate—the die—so as to form a profile with a constant cross-section. Extrusion is used for fabricating various metals. For aluminum it is remarkable because it combines high productivity with an unparalleled variety of extremely complex shapes that could hardly be made by any other method.

Apart from rods, bars, and tubes, the more specialized forms are interchangeably called shapes, sections, or profiles. Within wide limits, it is possible to produce almost any cross-sectional shape that the structural designer may desire (Fig. 9.10). Furthermore, an appropriate choice of alloy and extrusion conditions can result in an optimum combination of properties for the particular application. Such properties may include tensile strength, toughness, formability, corrosion resistance, and machinability.

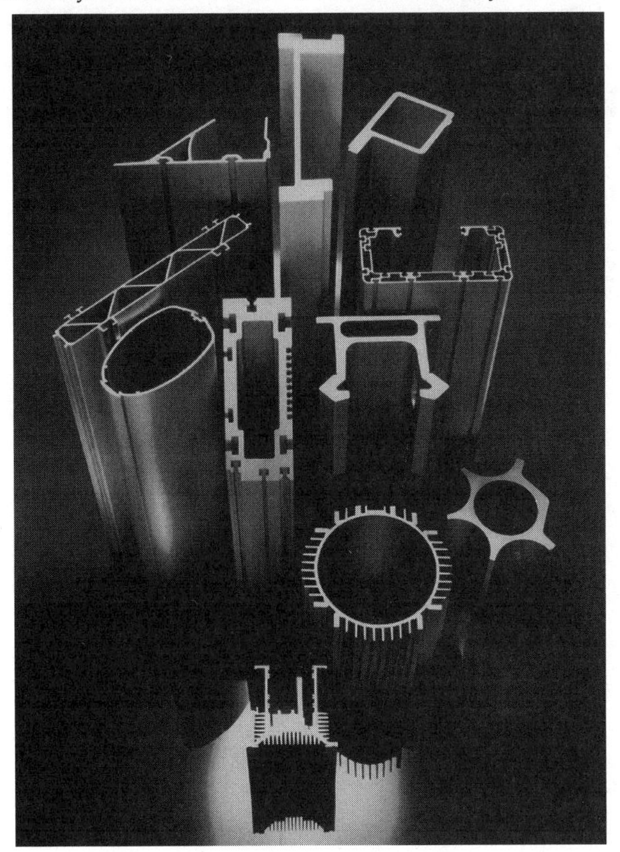

Fig. 9.10: Selected examples of typical extruded sections.

[4] By W.-D. Finkelnburg.

Fig. 9.11: Porthole die for extruding aluminum.

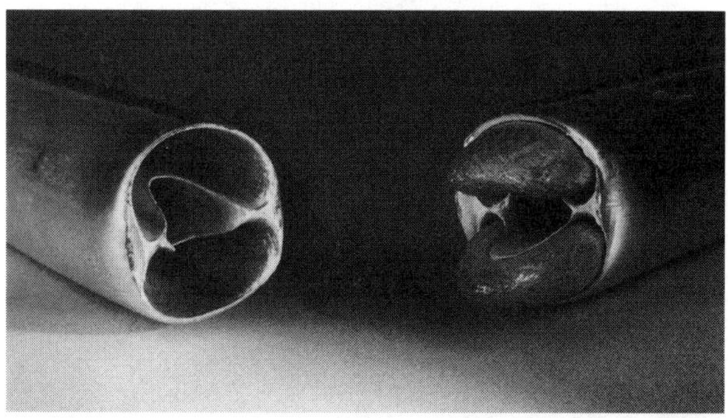

a

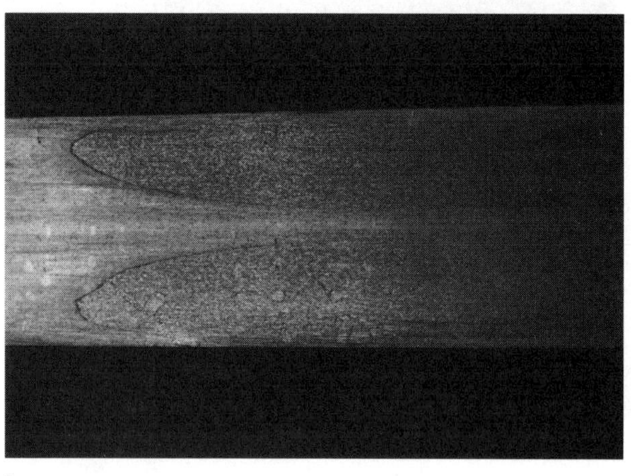

b

Fig. 9.12: Development of a tongue during the transition from one strand to the next. (a) Ends broken at the join. (b) The join in longitudinal section.

Fig. 9.13: The biggest horizontal extrusion press in Europe for forward and backward extrusion of large sections. Press force 90 MN (forward extrusion) or 100 MN (backward extrusion); maximum billet weight 1.6 t; limiting circle size[5] 800 mm; maximum extruded length 42 m. (Mannesmann-Demag, SMS-Hasenklever, Alusingen)

If the designer and the extruder work closely together from the outset, the sequence of fabrication steps from raw material right through to the finished part should be one integrated process. Cross-section design must take into account the fabrication conditions applying to the various manufacturing operations, which together affect the properties of the final product (the chain of cause and effect A→ B→ C, described in general terms in Chapter 3). Only then can the many advantages offered by the extrusion process be selectively employed to make an optimal product. This holds true for all design solutions, whether it is a matter of choosing to make a large extrusion with a big width to height ratio, such as is used in building railway wagons; putting together several smaller sections using one of the available, well-proven joining methods; or using a complex section to avoid the need to join smaller sections, which is, as a rule, necessary when using other fabrication processes and other materials.

The extreme variety in the shape of extrusions is a direct consequence of the technology of die-making. A die is basically a thick piece of steel of a grade suitable for use at high temperature, in which an opening corresponding to the form of the desired section is made by drilling, milling, and / or spark erosion. Die design is fairly straightforward for solid sections (e.g., round or flat bars) or open concave shapes (e.g., horseshoe-shaped sections). For making tubes, a mandrel, which forms the hollow center, can be inserted into the billet from the back, with the die in this case having the same appearance as a die for extruding solid round bar. This is known as seamless tube extrusion over a mandrel.

Many production extrusions are, however, complicated sections with one or more hollow chambers. In such cases, the mandrels are supported on bridges that span the die opening. The die may be of the "bridge" type (projecting into the billet chamber) or "porthole" type (Fig. 9.11). The billet is divided into several moving streams of metal that unite afresh in the hollow under the bridge before being formed finally in the die bearing. The seams or joins in the cross-section, where the separate strands flow together, are called extrusion welds. They merit particular attention in quality control, for they often stand

[5] There is a maximum possible profile size for a given extrusion press and billet size. The "limiting circle" is the circle into which all profiles must fit.

out through differences in structure and, if the process is not well regulated, through differences in mechanical properties compared with the base material (Fig. 9.12).

A whole range of different process variants and equipment types have been developed in order to make extrusion workable in everyday production. Only two of these are significant in making semi-fabricated products: forward and backward extrusion. Other processes, such as hydrostatic extrusion, continuous extrusion processes under commercial names like Conform and Castex, and cable sheath extrusion are used to a limited extent for special products. Fig. 9.13 shows a modern extrusion press suitable for both forward and backward extrusion.

9.2.3.1 *Forward extrusion*

Worldwide, the overwhelming majority of extrusion presses work on the principle of forward or direct extrusion. A hydraulic ram forces a preheated billet held in a fixed, preheated container against a die and squeezes the metal through the die opening (Fig. 9.14). The billet surface sticks to the container wall and absorbs a part of the ram force, with the resulting energy being converted to heat in a shear zone inside the billet. The ram force needed to push the aluminum through the die is proportional to the extrusion ratio, which is defined as the billet cross-section divided by the total cross-sectional area of the die openings. Table 9.1 gives an overview of typical dimensions and working conditions met in industrial practice.

Friction along the wall of the container is the cause of the material flow pattern, where the middle of the billet flows quickly as compared to the shell (Fig. 9.15). This typical material flow must be taken into consideration if a high-quality extrusion is to be produced to close dimensional tolerances. A great many defects are caused by inadequate consideration of the effects of this faster core material flow. Toward the end of the extru-

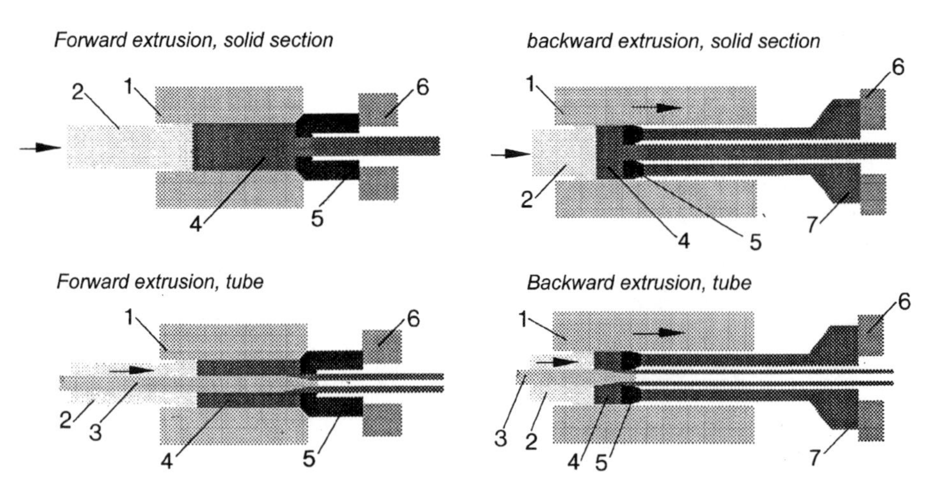

Fig. 9.14: Principles of forward and backward extrusion of solid sections and tubes over a fixed mandrel. 1: container, 2: ram, 3: mandrel, 4: billet and extruded product, 5: die, 6: backup die, 7: hollow stem carrying the die.

Table 9.1: Typical operating conditions in the forward extrusion of aluminum

Press force	10–100 MN
Container diameter	100–650 mm (occasionally rectangular, up to 250 mm × 700 mm)
Flow stress	300–1500 N/mm^2
Billet length	300–1300 mm
Billet preheat temperature	450–550°C
Extrusion (product) speed	
Alloys that extrude easily	25–100 m/min.
Alloys that are difficult to extrude	0.5–10 m/min.
Extrusion ratio	20:1 to 100:1

Fig. 9.15: Material flow in forward extrusion. (Section through a partly extruded model billet made with disks of alternate colors of plasticine).

Fig. 9.16: Display of the flow of the last part of the billet into the extruded product ("Back-end-defect"). A similar technique was used as in Fig. 9.15: the back end of the billet was covered with a thin slice of an Al-Cu alloy, which turns black when etched.

sion cycle, the notorious "back end defect" corresponds to a shortage of metal at the core. As result, metal with different structures from the back surface, billet skin, and container wall and, possibly, also dirt and impurities flows toward the middle of the billet. This leads to an inhomogeneous structure, which can easily show up as defects after decorative anodizing and also create zones of serious mechanical weakness.

Whereas removal of the unextruded billet residue tears most of the pockets of aluminum out of bridge dies, porthole dies trap a lot of aluminum in the die chamber. At the start of the next cycle in a porthole die, the old metal mixes with that from the new billet, but without adequately welding across the contaminated interface. This produces a transition zone in the extrusion that contains structural defects and, as a general rule, should be scrapped, not delivered with the product. Fig. 9.12.a shows the typical fracture form of such a billet junction. Another problem often (if not always) encountered is due to the build-up of frictional heating in the billet, as its temperature rises continuously during the extrusion cycle. This often causes changes in structure along the length of the extrusion and carries the risk of making the surface rougher toward the end of each cycle.

9.2.3.2 Backward extrusion

The conditions in backward or indirect extrusion are fundamentally different. The die is mounted on a hollow ram or stem. The container and the billet do not move relative to each other, but are pressed together against the die by the hydraulic system. The hydraulic pressure causes the container to slide slowly over the stem that carries the die (Fig. 9.14). In this process, friction between container and billet does not occur; thus, the whole of the hydraulic force is available to deform the metal. For this reason, backward extrusion is used above all for extruding high-strength aluminum alloys that are difficult to deform, particularly the Al-Zn-Mg-Cu and Al-Cu-Mg alloy groups. The higher available press force for the same press size leads to further advantages compared with forward extrusion:

- The billet preheat temperature can be lower, allowing a higher extrusion speed.
- The absence of wall friction means that longer billets can be used.
- The different material flow results in a more uniform microstructure from the beginning to end of the extruded product.

Backward extrusion has, however, significant disadvantages compared with forward extrusion. One of these is that the section has to pass through the hollow stem, which means that for the same billet diameter, the limiting circle size is smaller than in forward extrusion. The material flow is completely different from that in forward extrusion, so that the billet surface becomes the surface of the extrusion. Therefore, in general, the billets have to be turned before extrusion (scalped) to remove the cast surface skin.

9.2.3.3 Extrusion in practice

It must be emphasized yet again that an extrusion is more than a semi-fabricated product made to certain dimensions. The properties of the finished extrusion depend on the interactions between several influences: an appropriate choice of alloy, the quality of the billet, its heat treatment before extruding, the exact design of the die, the process conditions during the actual extruding cycle, and the treatment after extrusion is completed. The difficulty in practice lies in suitably matching together the individual process steps. If 6063 billet is slowly cooled after homogenizing, the resulting coarse precipitates of Mg_2Si can be put back in solution before extrusion by rather long preheating, for instance in a gas-fired furnace (C2 in Fig. 9.17.a). On the other hand, if the press is equipped with an induction furnace that heats the billet rapidly, it is absolutely necessary to cool the billet quickly after homogenizing (C1 in Fig. 9.17.a); otherwise, the precipitates will be coarse and take too long to redissolve, resulting in inadequate strength after artificial aging. Differently again, the high-strength Al-Zn-Mg-Cu and Al-Cu-Mg alloys, whose extrusion temperature is too low to be combined with solution treatment, undergo a separate solution heat treatment in an air furnace or salt bath, immediately followed by water quenching. Here, a goal of billet homogenizing is to create a microstructure most suitable for extruding, in which no low melting point segregates from casting lead to partial melting and tearing as the extrusion emerges from the die.

A different example of the need for properly tuning together several process steps concerns the effects of cooling at the press immediately after extruding. With Al-Mg-Si alloys, which are to receive surface treatment by chemical or electrochemical means, one must not only watch the levels of iron and other trace elements such as manganese, chromium, and zinc; it is particularly important to quench the material rather rapidly with water sprays in order to attain a highly reflective surface. One would normally quench only as fast as absolutely necessary in order to avoid distortion due to internal stresses created by varying strains resulting from the non-uniform temperature during the quench. Of late, quenching has been developed using computer-controlled jets of water or mixtures of water and air so that thicker parts of the section are cooled more and thinner parts less intensely (see Chapter 13.2). Similarly, with regard to corrosion resistance, which is particularly important for certain applications, quenching must be done rapidly for some alloys and may be done more gently for others. In the following section we shall examine the effects of these various process treatments on the microstructure.

9.2.3.4 Properties resulting from preheating and extrusion[6]

Today, the overwhelming majority of extruded profiles are made from easily extrudable 6xxx-type (Al-Mg-Si) alloys. These alloys belong to the precipitation hardening alloys, that is, to those which develop much higher strength through artificial aging. Extrudability depends on the alloy composition. The popular extrusion alloy 6063 contains barely 1% of the main constituent Mg_2Si and little or no additions of manganese or chromium. Manganese and chromium have low solubility and, thus, form fine, stable precipitates that significantly increase flow stress and, therefore, reduce extrusion speed.

To be economically viable, the production of extruded profiles must fulfill both metallurgical and forming process requirements. Both product quality and productivity depend to a substantial extent on billet structure. Of primary importance is the quality of the cast structure, which should be fine and as uniform as possible both in cell size and in grain size. This ensures that the diffusion distances are short and that the AlFeSi primary precipitates are finely distributed. However, the solidification conditions necessary for this

[6] By G. Scharf.

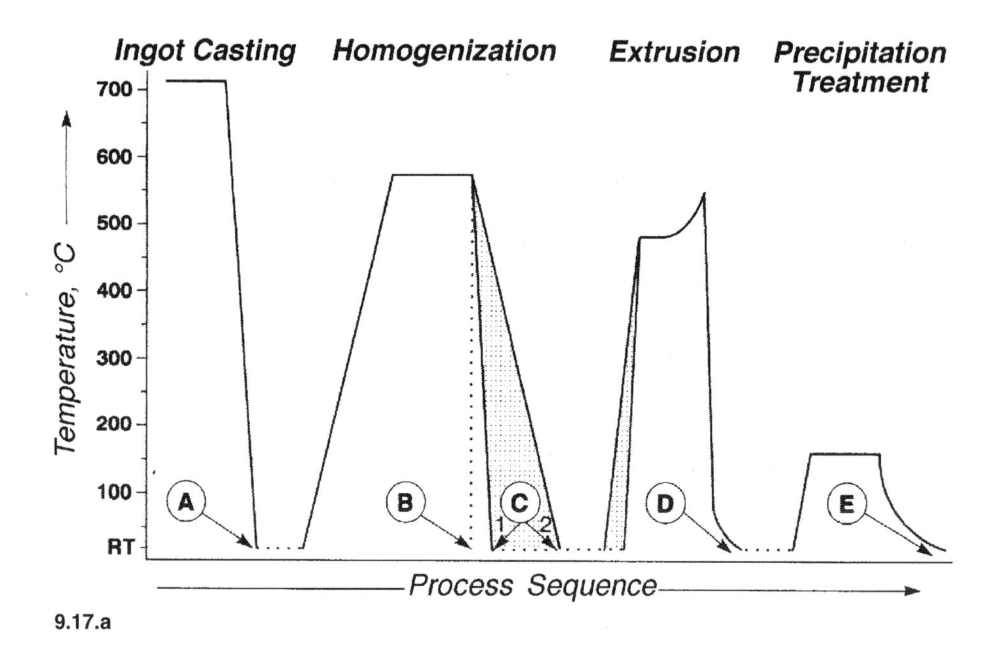

Fig. 9.17: (a) Thermal cycles in the fabrication of 6063 extruded profiles. (b) Microstructures A–E: Structure changes from ingot to extruded profile in 6063.

A: As-cast extrusion ingot: grain segregation, with veins of residual melt in the grain and cell boundaries. (Preparation: 50× magnification; transverse section Barker anodized, photographed in polarized light).

B: Extrusion ingot from A after treatment at about 580°C and quenching in water. Diffusion has largely eliminated grain segregation and substantially agglomerated the primary phases. (Preparation as A).

C1: Extrusion ingot from A, after treatment at about 580°C and cooling in air. Inside the grain are fine, secondary Mg_2Si precipitates. Agglomerated primary AlFeSi phases mark the grain boundaries. (Preparation: 110× magnification, transverse section etched 3' RT in H_2SO_4 + HF).

C2: Extrusion ingot from A, after treatment at about 580°C and slower cooling. Mg_2Si precipitates inside the grain are fine, but are coarser in the grain boundaries. Agglomerated primary AlFeSi phases mark the grain and cell boundaries. (Preparation: as C1).

D: Extruded profile, quenched in forced air (T4 temper). The matrix is a metastable, supersaturated solid solution; primary intermetallic phases are aligned in the extrusion direction. (Preparation: as C1).

E: Extruded profile, air cooled from about 580°C then precipitation heat treated (T5 temper, Mg_2Si precipitates). Inside, the grains are coherent (too fine to see with an optical microscope) but coarser in the grain boundaries, which etching reveals more clearly. (Preparation: as C1).

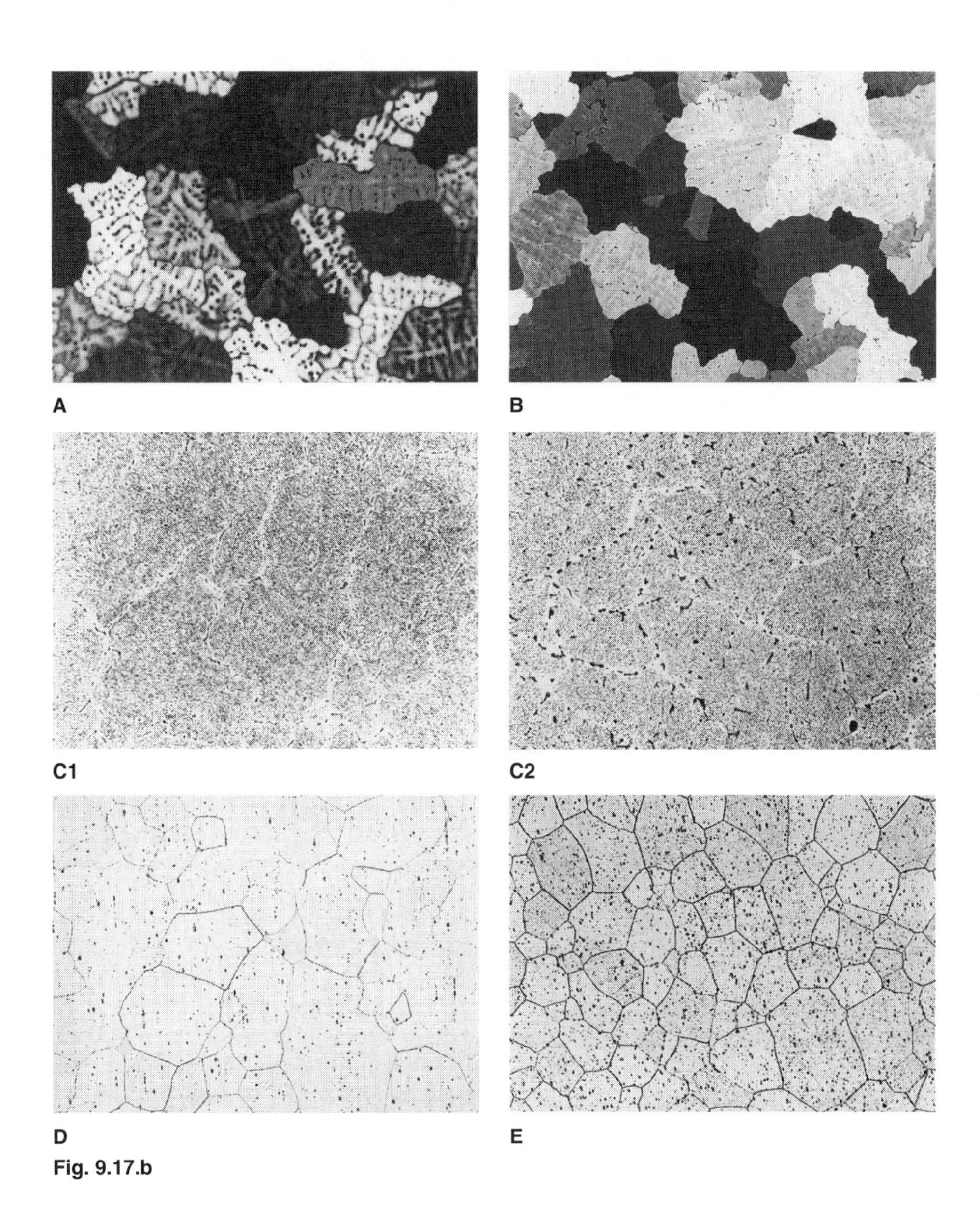

Fig. 9.17.b

also create pronounced grain segregation, partially supersaturated solid solution, and veins of residual melt. Heat treating the billets just below the solidus temperature largely removes these unwanted structures and considerably improves ductility.

Homogenization causes changes in the structure that are important for formability and later heat treatment. Looking at the thermal history through the whole fabrication process from ingot to the precipitation treated profile (Fig. 9.17.a), one sees a succession of different structures. It is important to see this sequence as a whole, since every structure change imposes changes in the conditions of ensuing process steps and the resulting properties. Only if these stages are well-matched can the end product be produced economically.

Fig. 9.17.b illustrates the microstructure changes from as-cast billet to the finished profile of 6063. The first stage is to remove the severe microsegregation (A) caused by the non-equilibrium conditions of billet solidification. At the homogenization temperature, Mg_2Si is soluble in the aluminum matrix, and Mg and Si diffuse to even out their concentrations. However, this solid solution is stable only at high temperatures. The cooling rate from the homogenizing temperature has a critical influence on structure. Mg_2Si tends to precipitate to approach equilibrium. With slower cooling, the Mg_2Si precipitates have more time to grow, and their size increases (C1 and C2). Since dissolved Mg_2Si raises flow stress, precipitating it lowers the initial extrusion pressure peak. Looking at the whole process, an optimum precipitate size would be one stable enough not to dissolve during billet preheating before extrusion. On the other hand, the precipitates must be small enough to dissolve with the extra heat and deformation passing through the die. A suboptimal microstructure could result with either type of furnace; for instance, if Mg_2Si precipitates dissolve incompletely during extrusion, then the material will not achieve full strength (T6) after subsequent precipitation heat treatment.

A crucial task for the process expert is to match the process parameters to each other. The main parameters governing Mg_2Si are billet homogenization, including cooling conditions; preheating to extrusion temperature; extrusion speed together with die exit temperature; and profile cooling conditions. By more closely examining die exit temperature, we will see how difficult it is match all these parameters. The choice of an optimum die exit temperature for 6xxx alloys is governed by two mutually contradictory criteria.

First, the die exit temperature should be as low as possible so as to maintain a smooth, bright surface on the profile. Low exit temperatures also allow high extrusion speeds, thus reducing costs. But from a metallurgical standpoint, the exit temperature must be above 500°C so as to dissolve the Mg_2Si. Without this, the profile will not later reach the required strength. The metallurgical properties for the alloy 6063 permit a particularly favorable compromise; the temperature range for hot working largely corresponds with that for solution treatment. Furthermore, this is an alloy that is not very quench-sensitive, tolerating moderately slow cooling rates, so that thin profiles can be quenched in air with little coarse precipitation and, hence, little loss of strength. Thus, solution treatment and cooling can be combined with the extrusion operation. Cooling may be with forced air or water mist sprays and must continue to below 200°C (Fig. 9.17.b, microstructure D). As a final operation, the profile develops maximum strength during precipitation treatment (E) in a furnace at a temperature in the region of 160°C to 180°C.

Fig. 9.18: A polished cross-section of the grain structure of extruded sections; magnification 15×, Barker anodized and viewed in polarized light (a) 6063, recrystallized and (b) 6082, predominantly non-recrystallized, fibrous structure with extrusion texture, but a thin, recrystallized outer zone.

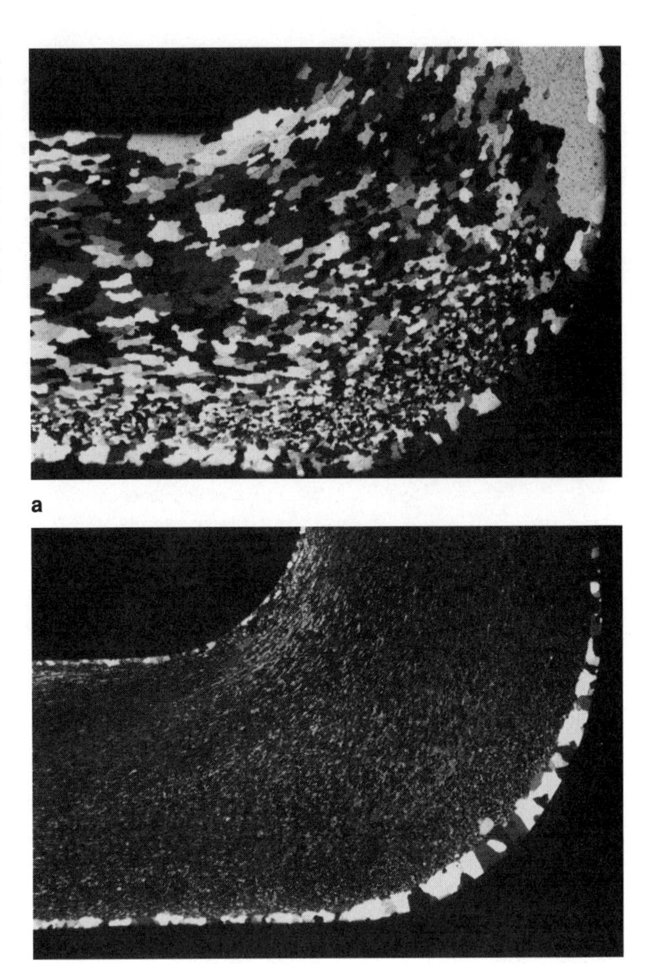

Besides the easily extrudable 6063 discussed above, the stronger alloys, 6061, 6082, or 6351 are much used in cars and trains and in demanding structural applications. To maintain good toughness as well as high strength, these alloys need not only the main Mg_2Si alloying for strength in the framework of cars, but also a suitable hot-worked structure, called extrusion texture. In AlMgSi alloys, this structure can only be achieved by alloying with recrystallization-inhibiting elements such as manganese and chromium.

Therefore, extruded sections of 6061 with such recrystallization inhibitors do not form recrystallized structures like those of 6063. Instead, the cast grains are stretched to form fibers of a hot-worked structure containing a strong deformation texture. This crystallographic texture consists of similarly oriented, fibrous grains, called the extrusion texture. Tensile strength in the longitudinal direction is considerably higher with extrusion texture than with the recrystallized structure. Fig. 9.18 compares the grain structures of a recrystallized section with a nonrecrystallized one.

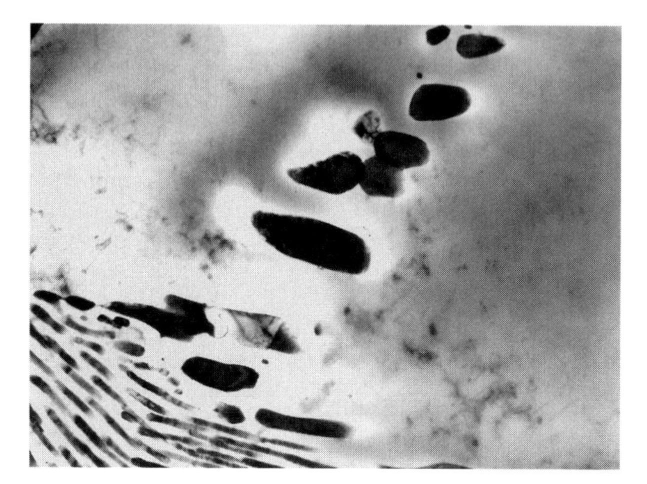
a

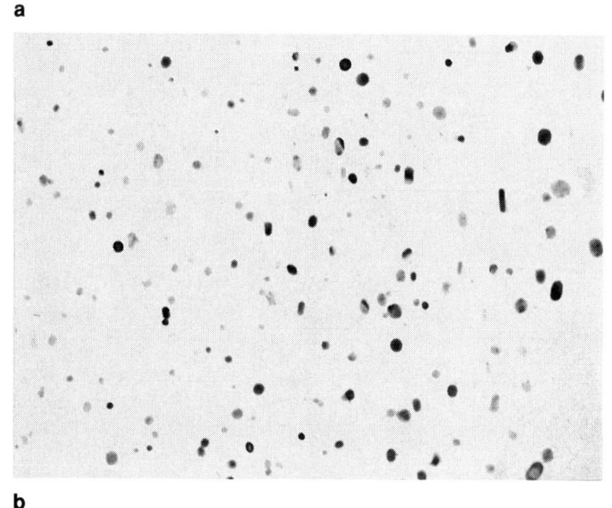
b

Fig. 9.19: Structure changes in 6082 billet. (a) As-cast DC billet; structure with primary and eutectic phases. TEM image magnified 4200×. (b) Structure after so-called "homogenization" and quench, showing heterogenized manganese-containing phases as fine secondary precipitates. TEM image magnified 8400×.

However, this alloying treatment to obtain a favorable extrusion texture brings certain disadvantages. Extrusion pressure increases, as does sensitivity to quench rate. Both effects are due to the manganese- or chromium-containing phases, which separate from supersaturated solid solution to form fine secondary precipitates during billet heat treatment (Fig. 9.19). As well as retarding recrystallization, such Mn- or Cr-containing phases also diminish extrudability (e.g., as regards extrusion pressure, speed, and surface quality). Thus, profiles will often show a mixed structure, developing a recrystallized outer band (Fig, 9.18), especially with lower billet temperature and higher extrusion speed. The increased quench rate sensitivity means that it becomes important to quench the alloy rapidly and without delay from the solution treatment temperature, since otherwise the Mg_2Si phases tend to nucleate preferentially on the Mn- or Cr-containing phases and so fail to strengthen the alloy.

In practice, a slower quench is often desirable to prevent uneven thermal stresses from causing distortion and twisting. Distortion necessitates expensive extra straightening work, which may be impossible on complex hollow profiles. Therefore, Al-Mg-Si alloy profiles are typically quenched continuously with jets of air / water mist shortly after they emerge from the press. For particularly thick sections, a standing wave (held between forward and backward-facing water jets) or a separate quench tank may be used. Warming the water to 30–50°C or using glycol additions can diminish distortion by reducing the quench rate. (For computer-controlled quenching see Chapter 13.)

For high strength alloys such as 2024 and 7075, which are the preferred alloy types for aerospace construction, much the same considerations apply as for 6061. Unfortunately, it is more difficult to reconcile the different requirements for these alloys containing more copper. Phases with a lower melting point make it necessary to lower the extrusion temperature and speed and, with insufficient quench rate, can diminish the product's corrosion resistance. Both phenomena may be mastered by precise control of billet preheating and solution treatment temperature and by immediate and rapid quenching. However, as a rule, these copper-containing alloys cannot be successfully quenched at the press, but require separate solution treatment and quench after extrusion. Nor can they be formed to hollow profiles using porthole or bridge dies because of mechanical limits on pressure and partial melting and tearing at higher temperatures.

The copper-free 7xxx alloys such as 7005 and 7020 are often used in welded constructions because they age harden after welding and can be solution treated over a wide range of temperatures. However, unsuitable metallurgical treatment can make these alloys susceptible to stress corrosion cracking. It is, therefore, essential to ensure correct metallurgical treatments throughout the critical fabrication and finishing processes. Typically these treatments will include billet homogenization (440–480°C), a strictly held slow quench rate (0.5–8°C / second) from the solution treatment temperature, and a two-stage precipitation treatment at 90°C, then 150°C. If the customer's subsequent fabrication will change the metallurgical condition (e.g., by heating or bending operations), then he or she should consult his or her supplier for further advice. If the prescribed fabrication conditions are accurately followed, then the alloy will not be susceptible to stress corrosion cracking.

9.2.4 Forging

Forging is the forming of metals by hammering or pressing, usually at high temperatures. It is an important metalworking process for making wrought products from aluminum alloys, although quantities are less than in rolling and extrusion.

There are two main types of forging machines: a forging hammer or drop hammer strikes the metal with rapid blows, whereas a forging press squeezes the metal slowly and steadily into shape.

Two types of tooling are used: first, open dies for forging between open tools, such as a hammer and anvil, and second, more expensive closed dies, which surround and shape the product.

Which types of forging machine and tooling are used depends on several factors such as the degree of deformation desired and the force needed as well as the complexity of the product, the precision required, and not least, the cost of tooling.

9.2.4.1 Open-die forging

Open-die forging uses simple tools that do not entirely enclose the billet. It typically starts by upsetting the billet, which means reducing its height, and progressively approaches the shape of the final component. This open-die forging stage particularly aims to create a fiber structure aligned with the main stresses that the component will bear in service. Although tooling costs are hardly significant, as a rule this forged pre-form will need considerable machining effort to become the final component. Open-die forging is, therefore, typically used for short production runs or for pre-production prototypes. It serves also as a preparation for closed-die forging, so as to approach the final shape before the relatively expensive closed dies complete the forming stage.

Forging billets are made either from cast (preferably DC cast) billet or cut from extruded bars or, occasionally, from hot rolled plates. Just before forging, the billets are heated to a temperature between 350°C and 500°C.

9.2.4.2 Closed-die forging

Closed-die forging uses precisely machined die blocks that completely surround the preformed blank as they close. The blank is struck or squeezed until it entirely fills the hollow of the die (Fig. 9.20), usually producing an accurate, almost finished shape. Excess metal is squeezed into the parting line between the die parts, producing a thin flange or flash on the product that must later be removed, often with a trimming die. Fig. 9.21 shows a selection of closed-die forged engineering components. The component is forged as closely as possible to its final shape, so that subsequent machining hardly cuts through the fiber flow lines or then only in low stress zones. Thus, the machining of closed-die forged parts usually amounts to little more than finishing the mating surfaces. Such machining should avoid surfaces that are stressed in fatigue, because cut fibers can become critical notches under fatigue loads. Die forged components can weigh up to several hundred kilograms.

9.2.4.3 Forged structure

Hot working in compression breaks down and compacts the cast structure, giving forged products higher strength and, in particular, higher elongation than cast products. Tensile properties parallel to and perpendicular to the flow direction of the fiber structure may be very different, as material is much weaker across the fibers, especially in fatigue; the extent of the difference is dependent upon the proportions of the finished forging. Transverse properties, especially in the short transverse direction (through the thinnest dimension), may be much lower than longitudinal ones. Because of this, the fiber direction is very important in highly stressed forged components where loading along the fibers is especially favorable under fatigue loads. Part of the art of forging is to align the fiber flow with the main stresses in components in order to best use the material's strength. Thus,

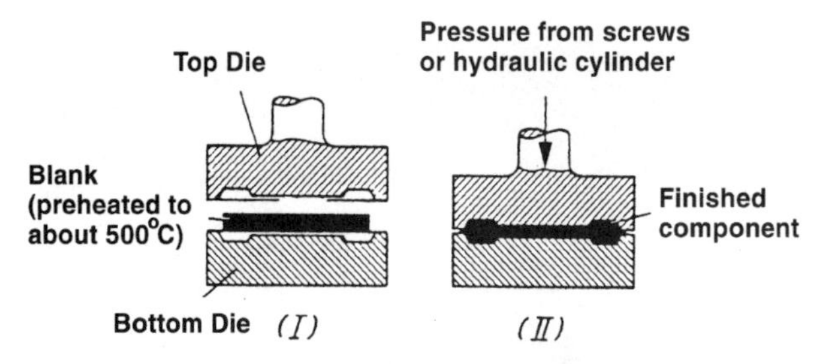

Fig. 9.20: Schematic illustration of closed-die forging before (I) and after (II) the two die halves meet.

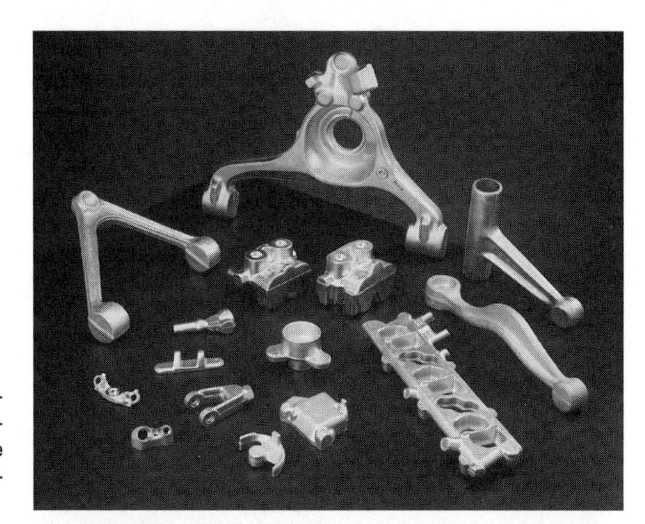

Fig. 9.21: A selection of die-forged engineering components in aluminum alloys, some of them for safety-critical components in cars. (Fuchs)

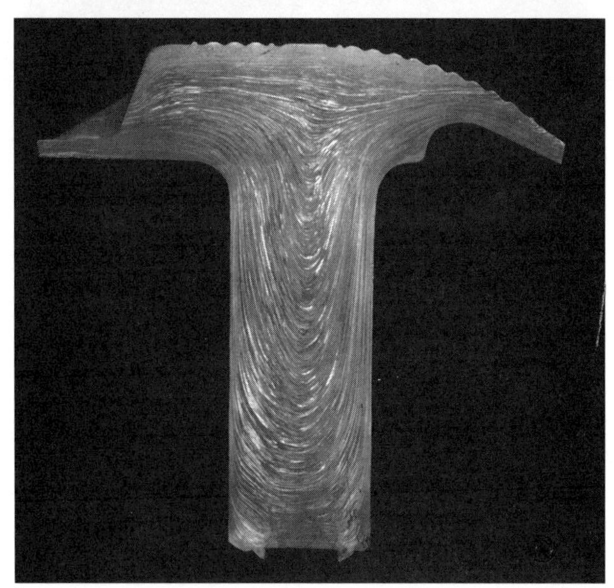

Fig. 9.22: Fiber flow pattern in the cross-section of an aircraft wheel in alloy 2014. Etching reveals variations in reflected light, corresponding to changes in grain orientation (texture) as the fibers follow the section.

Fig. 9.23: Forged car wheel of 6082. (Fuchs)

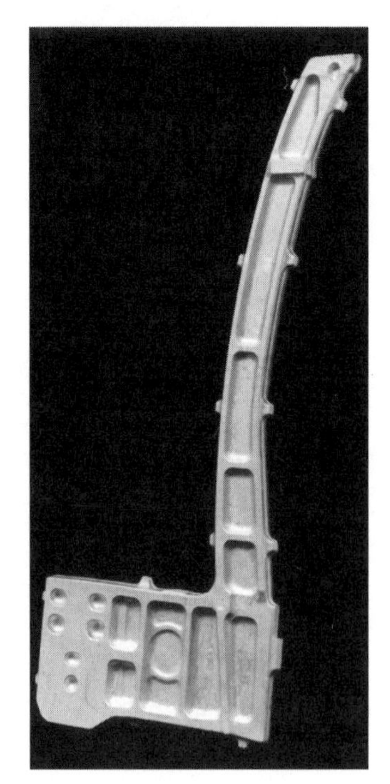

Fig. 9.24: Forged component of aluminum-lithium alloy 8090-T852 for a helicopter. Mass: 100 kg; maximum length 2240 mm. (Westland, Augusta, Fuchs)

fiber structure, shown in Fig. 9.22, is by no means a thing to avoid. Even if final machining cuts into the fibers to some degree, they keep most of their advantages. This is because the material still has optimal strength and toughness in the region of the notches from which fatigue cracks start, and in yielding it reduces stresses better than if the stresses were transverse. Fig. 9.22 shows such fiber structure in a forged aircraft wheel rim.

9.2.4.4 *Typical end products*

To supplement the examples of forged automotive products in Fig. 9.21, Fig. 9.23 shows a forged wheel often used on more expensive cars. Such car wheels are only roughly formed by forging and then roll-formed to the final shape. In the last few years forged wheels have tended to have a smaller share of the market for car wheels in face of improved low-pressure casting, which is cheaper. On the other hand, forged wheels generally have better strength and toughness and are about 15% lighter. For safety-critical components, forged aluminum components continue to dominate the market. If forged wheels can be fabricated at lower cost they will have a promising future.[7] However, improved molding processes are enabling castings to steadily replace forged components for automotive applications use.

9.2.4.5 *Alloys*

In addition to the usual Al-Cu (2xxx) and Al-Zn-Mg (7xxx) aerospace alloys, some Al-Mg-Si (6xxx) alloys and some non-heat treatable alloys of the Al-Mg (5xxx) type are also used for forgings in considerable quantities.

Fig. 9.24 shows a forged component for an aerospace application. It is made of the Al-Li alloy 8090. Its mass of 100 kg and maximum length of 2240 mm are impressive, and they underline the importance of large forged components in all landing gear for modern aircraft. Future aerospace constructions are inconceivable without large closed-die forged parts of aluminum, magnesium, and titanium. The maximum size for these parts is limited by the available press force. The larger forging presses in the U.S., France, and Russia are rated at about 50,000–70,000 tonnes force and can produce large closed-die forged parts weighing up to one tonne.

9.2.4.6 *Outlook for the future*

Today, a modern aluminum forge is in many ways a high-tech manufacturing plant, using ever more multi-stage transfer presses and automatic loading devices, including robots to manipulate parts during forging. The alloys used are very demanding in their heat treatment and in the properties to be guaranteed. Thanks to the intense working that large forges can provide, even for high-strength aluminum alloys, and their capacity to produce large and massive items in a range of forms, such large forged components of aluminum will continue to be needed in the future.

On the other hand, we must await the market verdict on small and medium-sized forged components for use in applications other than aerospace. It remains to be seen whether

[7] Development of thin wall forging has started recently (1996) with this in view.

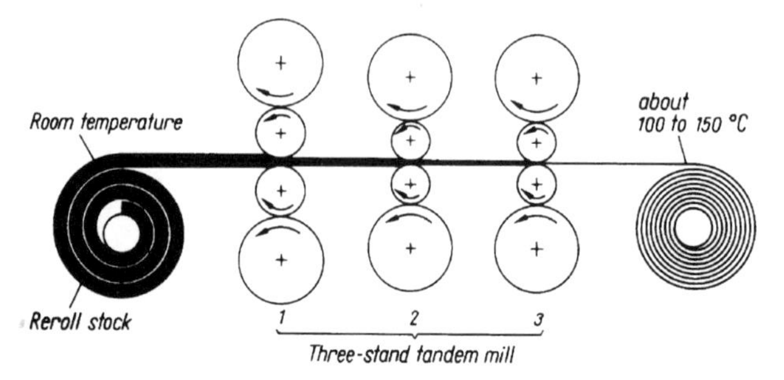

Fig. 9.25: Tandem cold-rolling (schematic). The deformation during rolling causes a temperature increase in the coil.

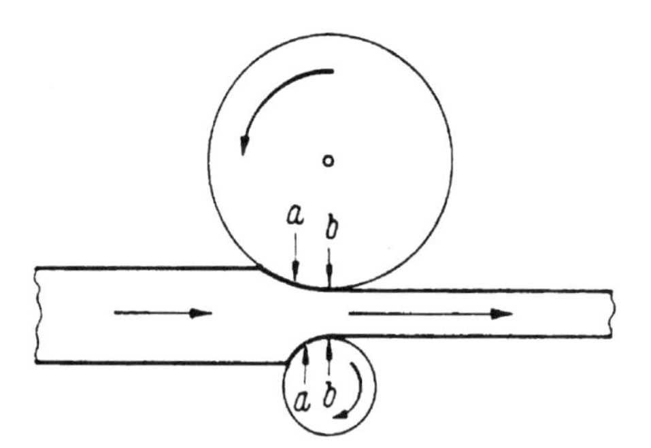

Fig. 9.26: The influence of roll diameter on friction in the roll gap. Within the arc a–b the product being rolled moves faster, relative to the roll surface. Depending on the condition of the roll, the no-slip point "a" can shift in the direction of rolling. The shorter a-b is, the smoother the surface of the rolled sheet or coil and the smaller the force and friction loss.

forgers will be able to reduce their costs so as to remain able to compete with specialized casting methods or alternative processes such as thixoforming. Another competing shaping system is computer-controlled machining of thick plates. This reduces cost and also minimizes residual stress in the finished product compared with forgings.

9.3 Cold Working Processes

9.3.1 Cold rolling[8]

After hot rolling, sheet or coiled sheet is often cold rolled. If for no other reason, cold rolling is often necessary for most applications in order to impart a certain strength to the material.

In some cases, cold rolling is carried out continuously on multiple stand mills (in a so-called tandem arrangement). Fig. 9.25 shows such an arrangement. The three rolling stands pictured are 4-highs, that is, every mill has four rolls. The small diameter work rolls are prevented from bending by the large back-up rolls. Small diameter work rolls are advantageous since they allow a large reduction of the material being rolled. The cold rolling of coiled reroll stock normally yields a rougher surface than the cold rolling of individual sheets.

[8] With R. Dean.

If highly reflective surfaces are desired, the last few passes are made with a smaller reduction that brightens the surface (see Fig. 9.26). This effect can be greatly enhanced by polishing the work rolls to a high luster.

Alloys may be cold rolled to thicknesses of around 0.05 mm. Pure aluminum can be rolled to foil down to 0.003 mm. As the degree of cold work increases, the amount of power required for further deformation increases. In order to achieve high degrees of deformation, intermediate thermal treatments are used in order to soften the cold-worked metal so it can be worked further.

Lubricants are applied in cold rolling to reduce the friction coefficient and, hence, rolling load and to protect the roll surface. They also function as roll coolants. If an oil film thickness is formed in the roll gap, which is larger than the roll roughness, a matte surface finish is produced on the strip. It is generally desirable to produce a bright finish, so there is usually some asperity contact with the smooth roll finish being imprinted on the strip. These two requirements mean that cold rolling lubrication is almost always operated in a mixed regime where there is some asperity contact but also some hydrodynamic lubrication occurring.

Rolling oils are a key element in aluminum cold rolling. For a wide range of products, mixtures using a base aliphatic mineral oil with selected load-bearing additives to improve boundary lubrication are used. Much development has been carried out on water-based cold rolling lubricants, and for a number of significant applications, most notably for canstock manufacture, they are state of the art.

Two specific products made by cold rolling, lithographic sheet and foil, will now be described in more detail.

9.3.1.1 Cold rolling of aluminum lithographic printing plates

The lithographic printing process makes use of oleophilic image areas and hydrophilic non-image areas. The process, which first appeared in the 18th century, originally used stone as the substrate; lithos is Greek for stone. Stone was replaced by various metals, and in recent years aluminum has emerged as the substrate material that most suits the

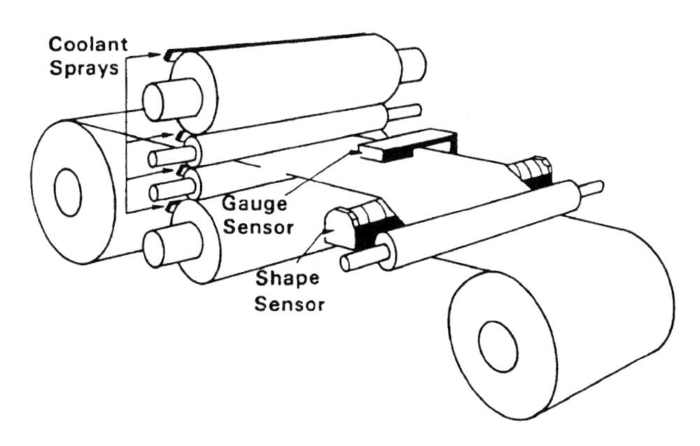

Fig. 9.27: Sketch of four-high mill with shape control.

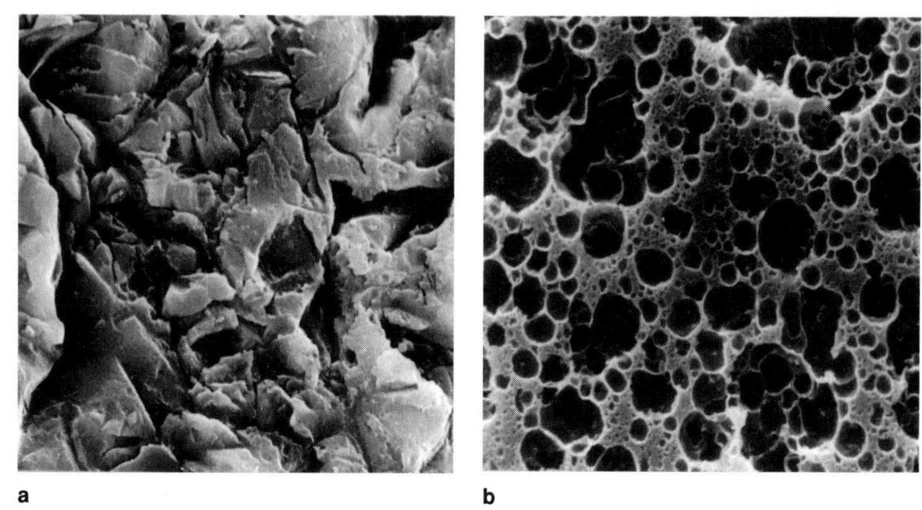

Fig. 9.28: Photographs of surface morphology of aluminum lithographic sheet material after graining. (a) mechanically grained. (b) electrochemically grained.

requirements of the printing industry. The sheet surface is coated with a light-sensitive polymer that forms the image areas on the hydrophilic aluminum. An important advantage of aluminum stems from the ease with which the metal surface can be roughened, either mechanically or electrochemically, to increase its water retentive properties. This operation is known as graining. It is frequently followed by an anodizing operation to improve wear resistance and water retention of the finished printing plate. Coils are supplied for graining in hard temper. In North America, alloys of the 3003 type are most commonly used; in Europe and Japan, 1050 is the standard for lithographic sheet.

Aluminum lithographic sheet is a difficult product to manufacture. Its surface must be blemish free and the sheet itself must be extremely flat. The production of flat sheet requires accurate control of shape and cross-section after every rolling pass on the processing route to the final product. The complete rolling sequence from hot mill to final gauge, typically 0.35–0.10 mm, must be optimized in terms of pass schedules and roll cambers to produce the required coil quality economically. It is essential to accurately control mill work roll cooling and work roll bending to maintain the roll gap geometry. Mill parameters such as tension in the roll gap, rolling load, entry and exit gauge, and mill speed must be maintained within close limits.

The development of computer-based mill control systems in the last two decades has led to considerable improvements in flatness and gauge accuracy. Efforts have been directed over the years toward developing automatic shape control systems; the successful introduction of shape measuring contact rolls has made this possible. The systems use a segmented roll at the outlet of the mill (Fig. 9.27) that measures the radial forces exerted by an arc of the web on the individual segments of the shape roll. This information is fed to the computer control system and used to adjust mill steer, work roll bending, and coolant distribution on the entry side of the work rolls. Control of coolant distribution linked with optimization of the spray system required to apply the coolant is an important method of correcting shape problems.

Gauge control is achieved by adjusting roll force, back-tension, and mill speed. With higher rolling speeds, lower response time mill control systems have been developed. Modern control systems adjust mill operating parameters so that a precise exit gauge is achieved at the highest possible speed. Accurate non-contact gauges are a key element of any control system. These devices use spectrally tuned x-ray beams that minimize sensitivity to alloy composition variations. Automatic compensation for temperature and any oil or oxide build-up on the sensor windows are built into the systems.

Frequent work-roll changes are necessary to obtain a blemish-free surface. Coils are stretch-leveled after the final rolling pass. This operation is often combined with degreasing/pre-etching to optimize the surface for the coil processing lines of the lithographic plate manufacturer.

Variation in gauge across the width combined with coiling tension are important quality parameters since creep can occur during coil storage if coil tensions are too high. Flatness of the product after sheet cutting can be badly affected.

Various mechanical and electrochemical processes have been used to produce the required topography on the aluminum surface. Fig. 9.28 shows two examples. Mechanical methods using nylon brushing with an alumina slurry are still used, but electrochemical surface preparation is the most common method.

The electrochemical processes use alternating current etching in optimized dilute acid solutions. Many acid compositions have been patented. The first electrochemical graining processes used 0.5% to 2% v/v hydrochloric acid. Variants on this basic electrolyte have included mixtures of hydrochloric and phosphoric, nitric, and various carboxylic acids, including acetic and boric acid. Many advanced etching processes have been developed using a nitric acid base. Some of these modifications in electrolytes have been introduced to improve the fineness and uniformity of the grained surface and make better print reproduction possible. With higher processing line speeds, there have been improvements in electrode design, electrolyte agitation, and alternating current wave form. Some of the latest lithographic coil processing lines combine chemical etching with mechanical and electrochemical processing to produce a surface topography optimized for the particular type of lithographic plate.

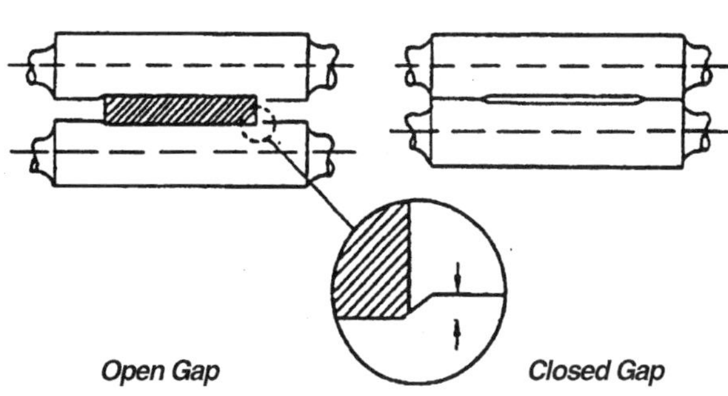

Fig. 9.29: Open and closed gap rolling.

Although a specialized market, lithographic plate manufacturers purchase approximately 250,000 tons of aluminum coil annually. The heavy investment in processing lines has meant that plate manufacturers have become very critical rolled coil customers. Line costs are too high to allow material to be rejected because of coil defects after processing to lithographic sheet.

Used printing plates are a valuable source of high purity secondary aluminum.

9.3.1.2 Cold rolling of aluminum foil

While rolling to gages greater than 0.1–0.2 mm, depending upon mill geometry, there is generally a gap between the work rolls; this is referred to as open-gap rolling. However, for lower gages, such as those for foil products, this gap is closed, and further reduced (Fig. 9.26). Closed gap rolling to the lower foil gages makes use of speed and tension control to obtain the required material thickness (Fig. 9.29).

The last foil rolling pass to final foil gage is usually performed by feeding two metal webs into the foil mill at the same time in a double rolling pass. One aluminum surface is in contact with the steel work roll, which imparts a bright surface luster, whereas the other foil surface is pressed into the other aluminum web, producing a matte appearance on that side. Welding of the two aluminum surfaces is prevented by introducing a light mineral oil between them as an interleave. At the mill outlet the two foil webs are coiled together. In the next processing step, material is edge-trimmed, separated, and perhaps slit. The individual foils are typically 0.018–0.005 mm thick.

9.3.2 Drawing

Heavy deformation can best be carried out at hot working temperatures (extrusion, forging, hot rolling). However, there are a few techniques for heavy deformation that can be carried out at room temperature. Cold rolling has already been described. Other processes include drawing of wire, rod, bar, and tube and the impact extrusion and ironing of the walls of containers (for example, seamless cans or pressure vessels). While the drawing of wire, rod, bar, and tube often takes place in a semi-finishing plant, impact extrusion and ironing of containers are usually carried out by the final fabricator.

Wire, rod, bar, and tube drawing usually takes place in the cold condition, because the drawn aluminum must transmit the deformation force. As can be seen from Fig. 9.30, the tensile force acting in the rod is converted to a compressive force in the die that reduces the cross-section of the rod.

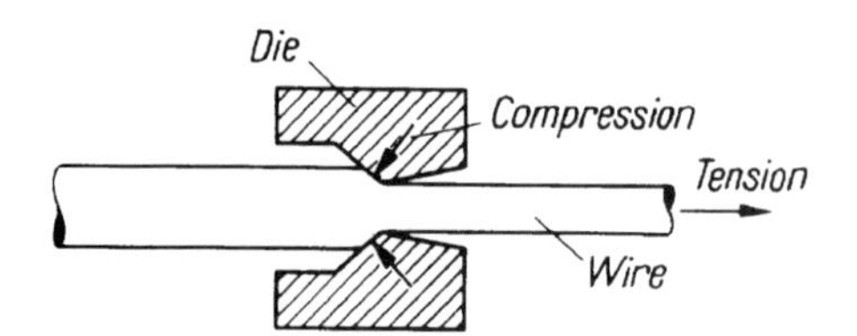

Fig. 9.30: Schematic of wire drawing operation.

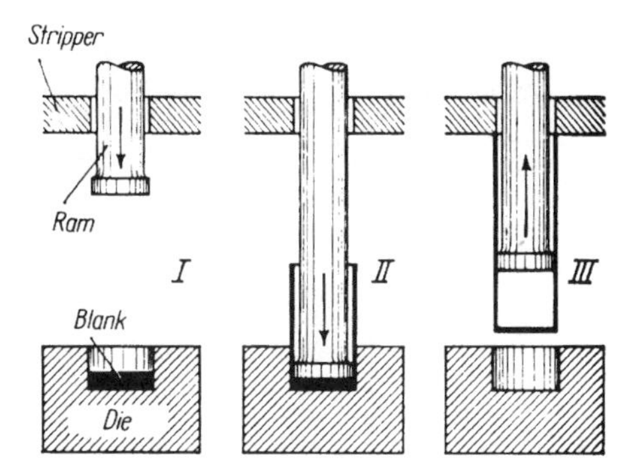

Fig. 9.31: Schematic representation of an impact extrusion. Starting material is usually a circular blank from sheet.

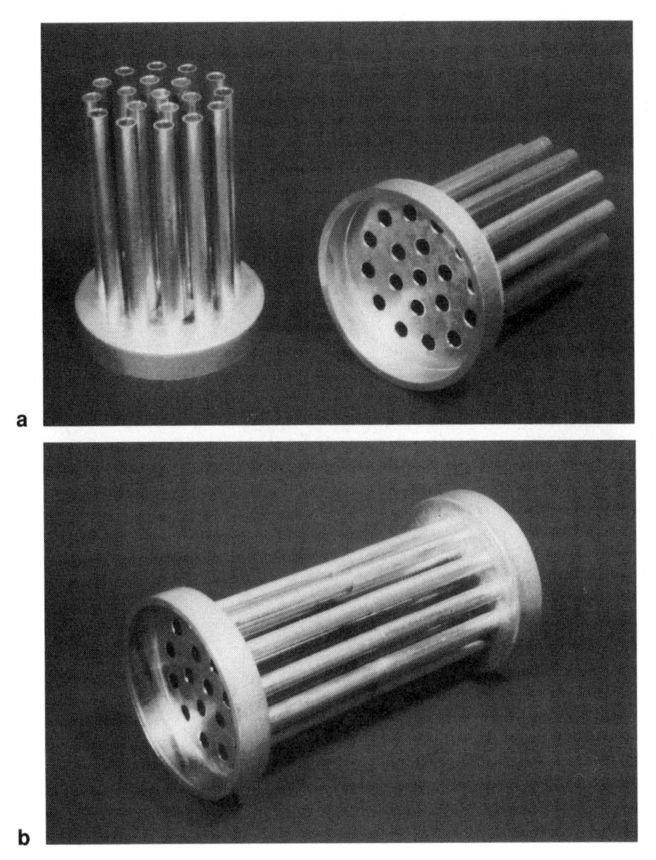

Fig. 9.32: Impact extrusions for a heat exchanger (alloy 1100). (a) Two halves of the heat exchanger before assembly and (b) the finished exchanger assembly. The weight of the starting slug was 45 g. (Alutec Pforzheim)

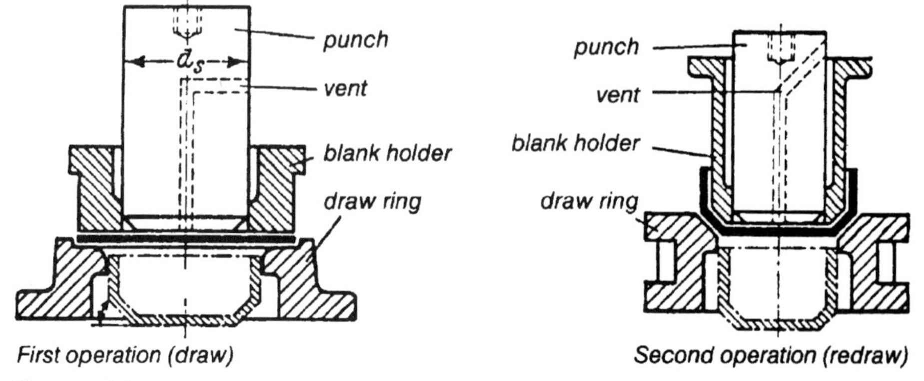

Fig. 9.33: Schematic illustration of deep drawing. For deep drawing, precision sheet (mostly circular blanks) is formed in a lubricated fixture. A blank holder prevents wrinkles from forming. For extra deep draws, the operation can be carried out in successive steps (possibly with an intermediate anneal). d_s = punch diameter. Shown in solid black are (left) the blank and (right) the semifinished deep-drawn part.

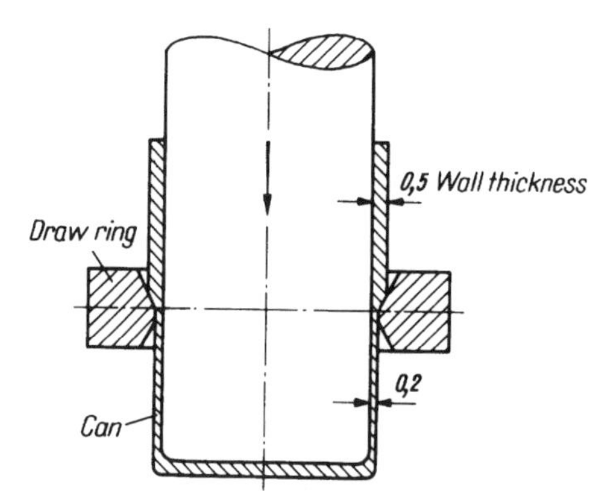

Fig. 9.34: Schematic showing wall ironing of an aluminum beverage can. Often a series of draw rings are used.

9.3.3 Impact extrusion

Impact extrusion is shown schematically in Fig. 9.31. Since some installations can operate up to 250 pieces per minute, impact extrusions are especially significant for the mass production of tubes and cans and for industrial pressure vessels and condenser housings. Important applications are also being developed for automotive components such as shock absorbers.

A typical application of impact extrusion is the heat exchanger in Fig. 9.32. The material is pure aluminum, 1100.

9.3.4 Sheet forming processes

9.3.4.1 Deep drawing

In processes applied to sheet deformation, the thickness of the starting material is not changed significantly, if at all. (Exception: wall ironing or intense stretch forming.) These processes include deep-drawing, roll-forming, corrugating, embossing, stamping, and bending, which serve to complete the fabrication of aluminum semi-finished products.

The especially important deformation process of deep-drawing is shown schematically in Fig. 9.33. Deep-drawing sheet must meet very exacting requirements: it must be fine-grained and as texture-free as possible (minimum earring). For extensive deep- drawing, the elongation as a rule should be 6–12%. A further indication of deformability is the ratio of tensile strength to yield strength.

The significance of this quotient can be seen from the following: The tensile strength is a measure of the magnitude of the deformation force, which can be transferred through the wall of the work piece in the true deformation zone (the die gap) by the punch. The larger this quotient is (i.e., the greater the difference between tensile and yield strength), the higher is the drawability, and, in general, the formability of the material.

Also of importance is the deep-drawing ratio β = blank diameter/punch diameter. The deep-drawing ratio for aluminum at room temperature may be a maximum of two. The deep-drawing of aluminum and aluminum alloy sheet is more difficult than deep-drawing of steel sheet. Steel has a generally higher elongation so that the part can be drawn to a greater degree from stretch forming with a reduction in the sheet thickness around the punch head. On the other hand, the aluminum blanks must be larger than the corresponding steel blanks, so that through the use of a blank holder or hold-down ring during deep drawing, the metal can be given a greater reduction without the formation of wrinkles. The inherent danger in wrinkling must be compensated for by a suitable fine adjustment of the hold-down ring. Thus, in the change over from deep-drawing steel sheet to aluminum sheet, definite additional precautions are often necessary.

The classic deep drawing of aluminum sheet is applied for cylindrical end shapes like pots and pans. In the automotive sector, body shapes are often formed by a combination of deep and stretch drawing.

9.3.4.2 Wall ironing

A technique for extreme deformation that has increased in importance is the ironing of impact-extruded or deep-drawn cups into thin-walled containers. It is shown schematically in Fig. 9.34. Better surfaces and thinner walls can be achieved by this method compared to impact extrusion or deep drawing alone.

Can making from aluminum for beverage cans started out with impact extruded and, later on, wall-ironed cans. By the end of the 1960s, the drawn-and-ironed (D&I) can replaced them almost completely. The drawn and ironed can is described in detail in Chapter 13.

Chapter 10. Removal of Work Hardening through Heat Treatment

10.1 Properties of the Cold-Worked Structure

A cold-worked structure can be recognized after etching by the elongated grains in the longitudinal direction (Fig. 10.1).

We have already considered the atomic arrangement of a cold-worked structure. The uniform arrangement of the atoms in the metal lattice is thoroughly distorted during the process of slip. The slip planes are blocked so that further deformation of the aluminum requires ever increasing forces and, finally, no additional deformation is possible.

Fortunately, nature in her wisdom provided two possibilities for eliminating distortions in the severely deformed lattice—either by partial annealing (or recovery) or by annealing (recrystallization). These processes will be discussed in the following sections.

10.2 Removal of Lattice Distortion

The atoms always seek to go from a high energy state to a low energy state. A part of the energy used in cold deformation is stored in the form of lattice distortions. In order to remove these, it is necessary that the atoms migrate to their regular lattice position, as described in Section 10.3. So long as the mobility of the atoms is limited (temperatures below around 100°C), the removal of the lattice distortions due to cold deformation proceeds slowly and incompletely. At temperatures between 100°C and 250°C, the mobility of the atoms increases to the point that approximately half of the lattice distortions are removed within a reasonable length of time through movements of the atoms (partial anneal or recovery).[1] At higher temperatures above the recrystallization threshold, the

Fig. 10.1: Etched structure of cold-rolled sheet. Note the elongated grains.

[1] Some significant recovery takes place in super-purity aluminum (>99.98% Al) and Al-Mg alloys even at temperatures around 50°C.

mobility of the atoms in the structure has become so great that the atoms assemble themselves in newly formed crystals of an imperfection-free lattice (recrystallization).

10.3 Recovery and Recrystallization from an Atomic Standpoint

The atoms attempt to return to their orderly position in the lattice structure, and the elevated temperature makes this possible. The rearrangement of the atoms is almost complete during recrystallization, but only partially so during recovery. Recovery does have the effect of removing a significant percentage of the blocked dislocations so that the slip planes are reactivated.

a

b

Fig. 10.2: Transmission electron micrographs showing subgrain structure in 99.99% Al, 0.1 mm thick: (a) hard rolled; (b) after recovery—two hours at 150°C. (Alusuisse)
Mag 350×

A few typical dislocation densities have already been reported:

State	Dislocation Lines/cm²
After severe cold work	approximately 10^{12}
After recovery	approximately 10^{10}
After recrystallization	approximately 10^{7}–10^{8}

During recovery, a significant number of the dislocations migrate away from the pile-ups that formed during cold working and rearrange themselves to form the subgrain boundaries. This greatly reduces the dislocation density caused by cold work.

Some electron micrographs are reproduced in Fig. 10.2 that illustrate the occurrence of subgrains in the cold-worked and recovered condition. Because the subgrains normally have a diameter of 1–5 μm, an electron microscope is preferred for their investigation.

During recrystallization, the remaining dislocation pile-ups disappear almost completely from the slip planes.

10.4 Origin of Subgrains

In order to understand the formation of subgrains, let us make a short excursion into dislocation theory.

The subgrain boundaries are formed by the accumulation of dislocations in a "dislocation wall" separating subgrains with a slightly different orientations Fig. 10.3 shows the arrangement of dislocations in such a subgrain boundary.

During recovery, the dislocations arrange themselves in relation to each other in such a way as to form subgrain boundaries. The larger the number of dislocations forming a subgrain boundary, the greater is the difference in the angle of orientation between the

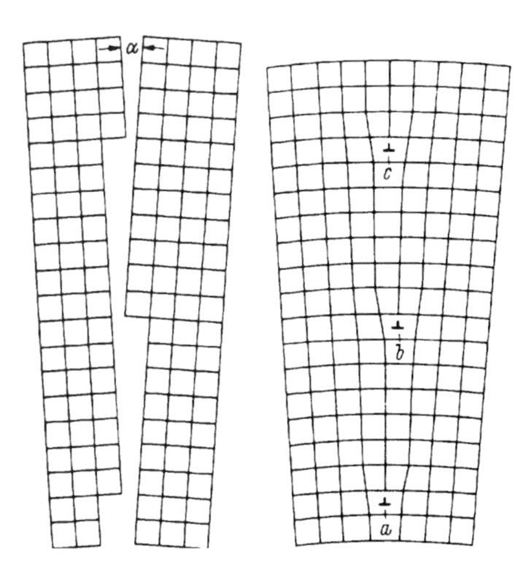

Fig. 10.3: Subgrain boundary (or small-angle grain boundary). Two grains with a mismatch angle, "a" (left), can be united by including a row of edge dislocations (⊥) at "a," "b," and "c," into the subgrain boundary (right).

neighboring subgrains. It can be seen clearly in Fig. 10.2.a that subgrains are already being formed during deformation. It is characteristic of the cold-worked condition that the subgrain boundaries are not clearly defined, and that dislocations also concentrate within the interior of the subgrains. Fig. 10.2.b shows the same sample as Fig. 10.2.a except that it has been given a recovery treatment.

There are relatively few dislocations in the interior of the subgrains since they have migrated to the subgrain boundaries. The subgrain boundaries are much more clearly defined than in the cold-worked state since the difference in orientation between neighboring subgrains has increased. In addition to this, some of the subgrains have already begun to grow at the expense of their neighbors.

If aluminum is cold worked and tested at very low temperatures (for example in liquid air at –193°C), the subgrains do not form. This shows that the migration of dislocations is facilitated by increasing temperatures, similar to the diffusion of foreign atoms. The migration rate is sufficient at room temperature to promote formation of subgrain boundaries. The difference in angular orientation between two subgrains is up to 2° in the cold-worked or stress-relieved state. After recrystallization, only extremely small differences in angular orientation, of one minute or less, remain in the subgrain boundaries. These subgrain boundaries are not visible and do not have any work-hardening effect.

10.5 Recrystallization

To produce the soft or annealed condition, a cold-worked material is brought to an elevated temperature for from a few minutes to several hours depending on furnace design and special conditions. In the case of pure aluminum, temperatures in the range of 300–400°C are usually selected.

During such an anneal, the process of recrystallization takes place—that is, a new crystal structure is formed in the solid state (compared to the cast structure that forms from the liquid metal).

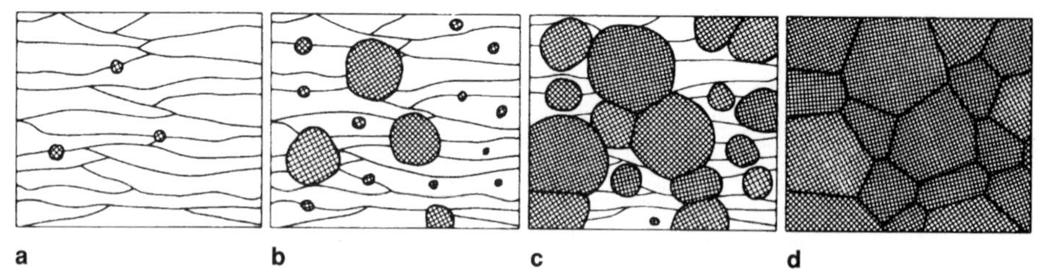

Fig. 10.4: Nucleation and grain growth during recrystallization. This schematic shows instantaneous images at constant temperature with increasing time. (a) Four recrystallization nuclei have formed in the deformed metal (shortly after reaching recrystallization temperature). (b) More nuclei have formed, and the initial ones have grown. (c) The nuclei have grown further. A few recrystallized grains are touching one another hindering their growth. (d) Primary recrystallization is complete. A recrystallized structure has grown out of the cold-worked grains.

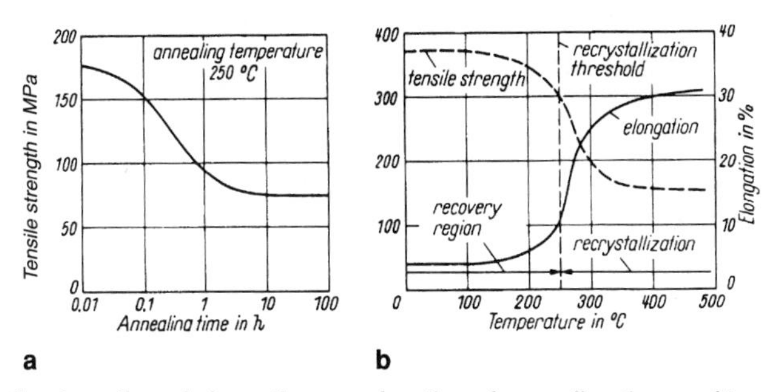

Fig. 10.5: Tensile strength and elongation as a function of annealing time and temperature. Starting material: hard-rolled commercial purity sheet. (a) At 250°C, a few hours are required to complete recrystallization. (b) Annealing time at the given temperature is five minutes. Because time at temperature is relatively short, recovery is not great. The recrystallization threshold is clearly defined and lies about 250°C. At temperatures above about 400°C, primary recrystallization is complete in five minutes.

Fig. 10.4 illustrates the process by which recrystallization nuclei become active in a severely deformed aluminum structure and grow to a relatively imperfection-free crystal structure. If the process is interrupted after a few seconds by quenching the sample, one can see that a few recrystallized grains have already grown at the expense of the surrounding cold-worked structure, while most of the structure is not yet recrystallized. With increasing annealing time, recrystallization proceeds until, finally, the entire structure has transformed into new recrystallized grains that can be clearly seen after etching.

The mechanical properties are altered significantly upon recrystallization. The tensile strength decreases to the values for the annealed condition; at the same time, the elongation increases correspondingly. Fig. 10.5.a shows the changes in tensile strength as a function of time. The transformation from the hard to the soft condition appears to be continuous, even though the corresponding structural changes are variable. If one observes a given small section of the structure during recrystallization, it is either hard (that is, most of the dislocations created during cold working are still there) or soft (i.e., recrystallized). The samples involved in the tensile tests, for Fig. 10.5.a, contained material not yet recrystallized as well as completely recrystallized material.

The recrystallization rate increases with increasing temperature due to the increased atomic mobility. From this, it can be seen that for a given annealing time, recrystallization does not take place until a definite temperature is achieved (Fig. 10.5.b). This "recrystallization threshold" is dependent on the alloy and the amount of cold work; it must be determined empirically for each specific case.

The temperature at which recrystallization begins is influenced mostly by the following factors:

1. Alloying additions: The recrystallization temperature increases mainly through the transition metals, such as Cr, Fe, Mn, V, or Zr in solution or finely dispersed precipitates.
2. Annealing time: The shorter the time, the higher the recrystallization temperature. A highly deformed pure aluminum sheet recrystallizes, for example, at 500°C in a few seconds, at 380°C in a few minutes, and at 280°C in a few hours.

Table 10.1: Temperatures required for the recrystallization of pure aluminum

(Annealing time: two hours)

Cold work reduction (%)	Temperature for complete recrystallization (°C)
5	500
20	400
40	360
80	320
98	300

3. Amount of cold work: With increasing cold work, the recrystallization temperature decreases steadily. For pure aluminum, the temperatures in Table 10.1 are required for recrystallization if the annealing time is two hours.

In general, a fine-grained recrystallized structure is desirable. A structure with 400–1000 grains/mm^2 is preferable. A fine-grained condition is obtained when the maximum number of recrystallization nuclei are active at the same time. If this is not the case, fewer, coarser grains result, which is most undesirable since this condition leads to "orange peel" during deformation, as discussed earlier.

In order to obtain a fine-grained structure, the basic rules are that the sheet have sufficient cold work (if possible, at least 30–50%), that it be heated at the maximum rate attainable, and that it be kept at maximum temperature for the shortest time possible. In most cases, this will ensure fine grain without difficulty.

An extremely large grain is usually obtained if material with the "critical" degree of cold work of 2–10% is annealed (Fig. 10.6). This danger exists in a deep-drawn part with large differences in degrees of deformation. In this case, the slightly deformed zones may be within the critical range (Fig. 10.7.)

As soon as the cold-worked structure has transformed completely into a recrystallized structure, "primary recrystallization" is complete. If the anneal is continued after this point, or if the temperature is raised, the conditions are favorable for grain-coarsening or "secondary recrystallization," in which the individual recrystallized grains grow at the expense of the neighboring grains (Fig. 10.8).

Because of the inherent variation in grain size, secondary recrystallization is undesirable and should be prevented by selection of proper annealing conditions. Experience shows that coarse grain may result during annealing under the following conditions:

a. Too little cold work
b. Heating rate too slow (especially in Al-Mn alloys)
c. Too high annealing temperatures
d. Excessive annealing times
e. Unfavorable chemical composition (e.g., insufficient iron content)

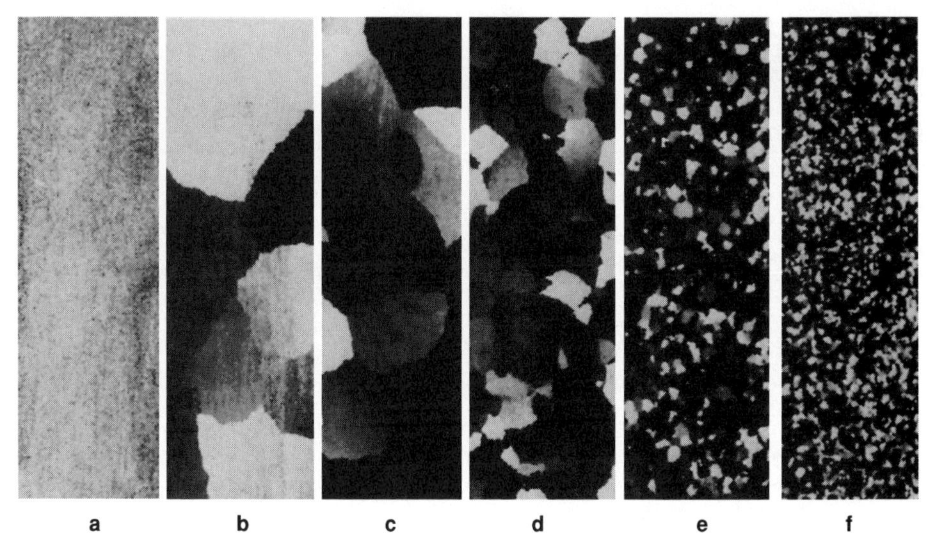

Fig. 10.6: Recrystallized grain size as a function of cold work. The following percentage numbers indicate the degree of cold work before annealing: (a) 0%; (b) 2%; (c) 4%; (d) 6%; (e) 8%; (f) 10%. (Alusuisse)

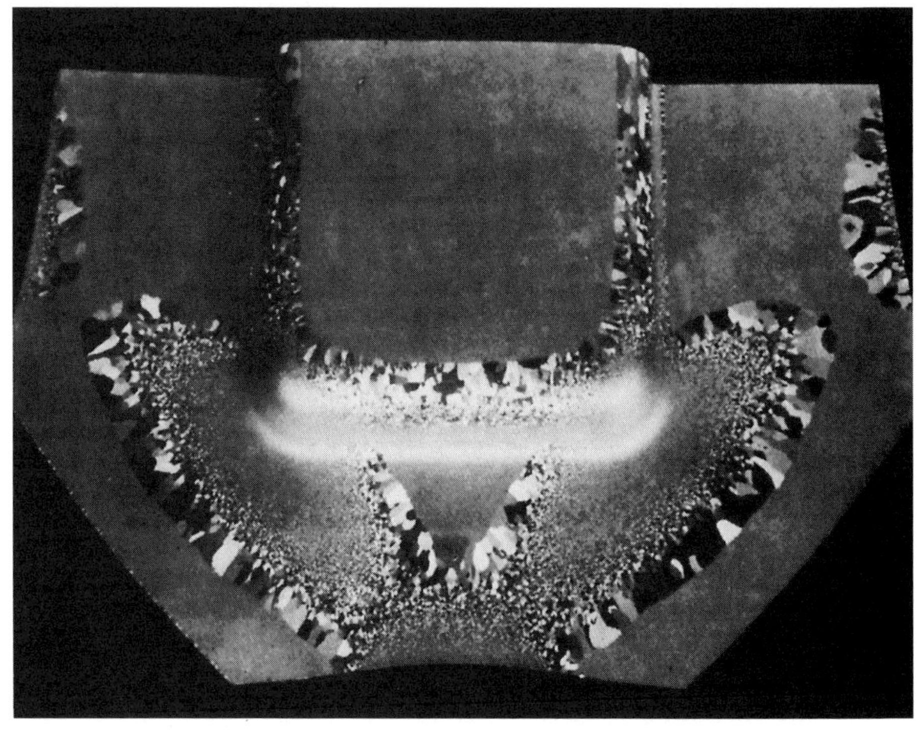

Fig. 10.7: A deep-drawn 5052 part, recrystallized during an intermediate anneal. Note coarse grain formation in regions with slight deformation.

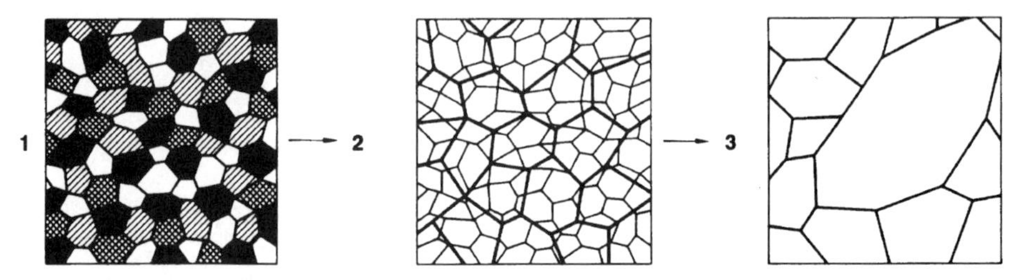

Fig. 10.8: A schematic of primary recrystallization and grain growth in cold-rolled sheet. (G. Masing)

1. Primary recrystallization is complete, a soft, recrystallized structure has formed. After etching, the approximately random orientation of the equiaxed grains is shown by shading.
2. When annealing for longer time or at higher temperature, after primary recrystallization is completed, a few grains with favorable orientation and size grow at the expense of their neighbors.
3. Secondary recrystallization can take place instead of (or after) grain coarsening, whereby single giant grains grow. Secondary recrystallization occurs especially at long annealing times at high temperature and after previously high cold deformation, notably when the heterogeneities due to precipitation of alloying elements do not dissolve simultaneously in the neighboring grains.

10.6 Removal of the Lattice Imperfections by Partial Annealing

If cold-worked aluminum is heated to temperatures below the recrystallization temperature, a continual reduction of lattice distortions due to piled-up dislocations takes place. This process, which is called partial annealing or "recovery,"[2] proceeds relatively slowly. Only about half of the hardening from cold work can be removed by partial annealing, while all of it may be overcome by recrystallization.

After severe amounts of cold work, partial annealing is carried out at temperatures of 200–300°C[3] and times up to 15 hours. The difference between recrystallization and partial annealing can be readily seen by observing the changes in the tensile properties as a function of annealing time and temperature in Fig. 10.8 (see also Fig. 10.5). In partial annealing, there is no discontinuous change of the tensile properties. There is also no rearrangement of the grain structure visible in a light microscope; the elongated lamellae of the deformed grains are retained.

As shown in Fig. 10.9, partial annealing and recrystallization have a significant influence on the tensile strength and a reduction of the yield strength with an increase of the elongation in alloy 5454. The tensile strength of this alloy is relatively high even in the annealed condition because of the alloy strengthening due to magnesium atoms in solution.

---

[2] "Recovery" is a broader term that includes changes in various properties of the deformed structure as a function of temperature and time. It is the first process that causes marked changes in structure when a deformed metal is heated, by rearrangement of the dislocations into a distribution of lower energy without a very marked decrease in the total number of dislocations.

[3] Temperatures of 100–200°C are used for "stabilizing" or "stress relieving" the Al-Mg alloys, which gradually "age soften" at room temperature. Partial-annealed tempers are identified by an "H2x" designation and stabilized tempers by "H3x" in the Aluminum Association system. Stabilized tempers have only a slightly decreased tensile strength compared to as-worked properties.

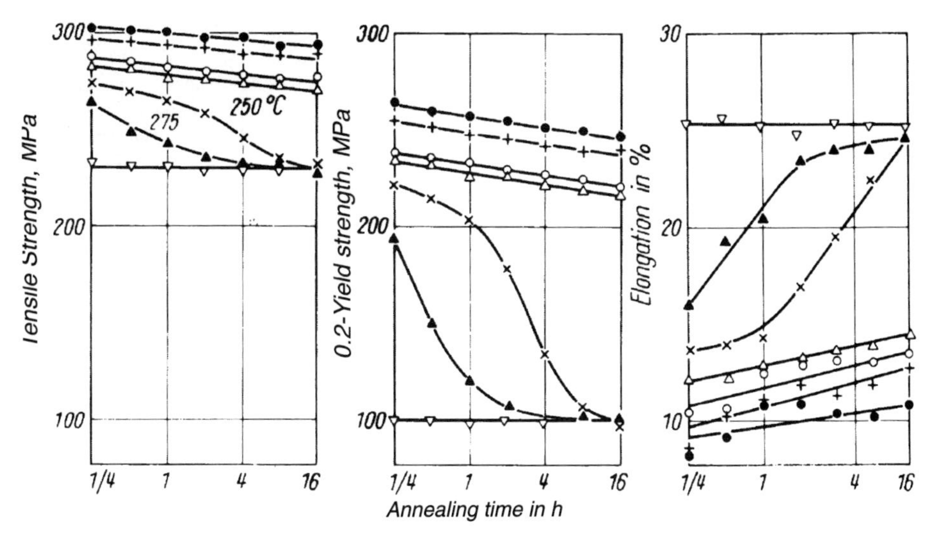

Fig. 10.9: Influence of annealing time and temperature on the mechanical properties of 5454 sheet (Y. Bresson and M. Renouard). Initial values: UTS = 335 MPa, TYS = 292 MPa, El = 10%. Composition: 2.8% Mg, 0.5% Mn.

Commercially, partial annealing is carried out at follows: The onset of recrystallization for a hard-rolled 5454 sheet, 1 mm thick, is 250°C. This means that recrystallization takes place above this temperature after annealing several hours. In the present example, recrystallization begins after about an hour at 250°C and is completed after about 16 hours. At 275°C, the recrystallization is completed after approximately four hours and in less than 15 minutes at 300°C. If a "half-hard" sheet is required for a special end use, a partial anneal 10–20°C below the recrystallization temperature removes approximately half of the lattice stresses. This causes a significant increase in elongation while the tensile and yield strength is not reduced so very much.

It should be pointed out that strain-hardened aluminum-magnesium alloys are a special problem because they tend to age soften at room temperature. Therefore, to eliminate the problem, it is industry practice to speed-up the age softening and increase ductility of these alloys through a thermal treatment very similar to partial annealing, which is called stabilizing. These tempers are identified in Aluminum Association standards with H3x designations in contrast to H2x for partial-annealed tempers.

The beginning of recrystallization, schematically shown in the right part of Fig. 10.10, is dependent on a series of factors, especially the combination of annealing time and temperature. The illustrated case is indicative for relatively short annealing times.

10.7 Two Methods for Producing Medium-Strength Material

Medium strength sheet is intended for certain forming processes in which a part of the sheet undergoes severe deformation. Soft sheet is not suited for such purposes since the

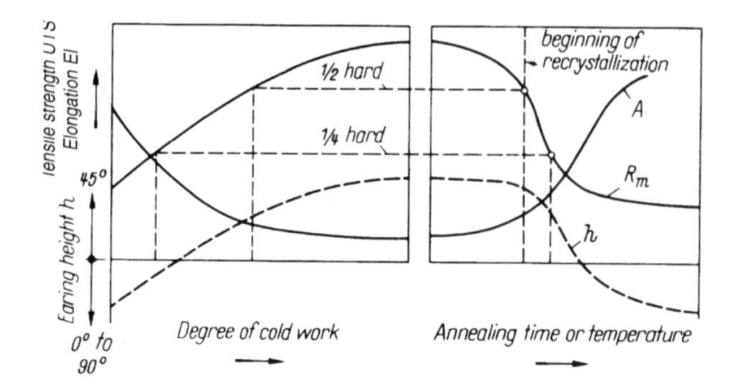

Fig. 10.10: Schematic representation of the behavior of tensile strength, elongation, and earing height as a function of cold work, softening, and recrystallization for 99.5% Al.

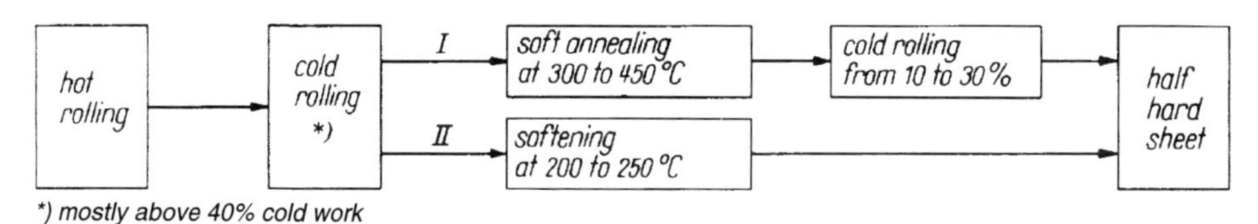

Fig. 10.11: Two ways (I and II) to produce medium-strength sheet (unalloyed or non-heat-treatable aluminum alloys). I: half hard, H14 temper; II: half hard, H24 temper.

sections of the material that are not formed in the process will be too soft. On the other hand, hard sheet is not suitable, since it would not have sufficient formability to make the part.

Fig. 10.11 and Fig. 10.12 present schematic summaries on the influence of cold rolling and subsequent anneal on tensile strength, elongation, and earing. It can be seen that the desired condition can be achieved by two methods. Fig. 10.13 shows a more detailed explanation for the production of sheet by the two methods. These are designated H14 and H24.

The first method to achieve a medium-strength temper (H14) consists of annealing (recrystallizing) the sheet, then giving the material 10–30% cold work, according to alloy (pure aluminum requires about 30%).

The second method (H24) is through partially annealing the sheet after cold deformation. One gets the impression from Fig. 10.9 and Fig. 10.10 that a definite range of mechanical properties may be attained either by cold rolling after an anneal or with a heat treatment after cold rolling. However, the relationship of tensile strength to elongation is not the same for the two methods (see Fig. 10.12 and Fig. 10.13). Partial annealing can give comparatively higher earing values, which is disadvantageous for deep-drawing, although elongation values are higher in the tensile test.

Fig. 10.12: Schematic representation of the relation between elongation and tensile strength for progressive strengthening after full annealing of a strengthened metal. (D. Whitwham and J. Hérenguel)

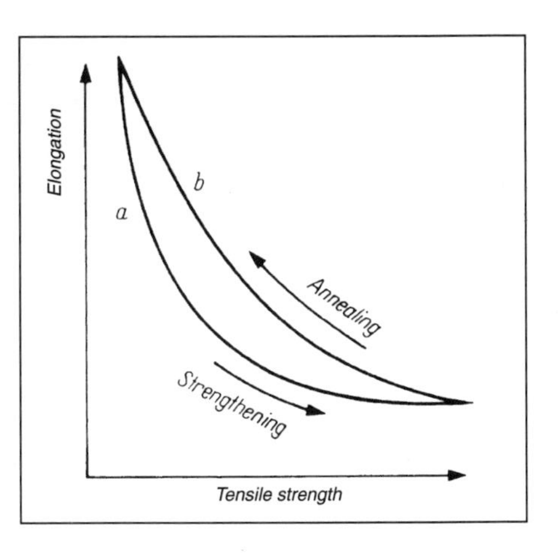

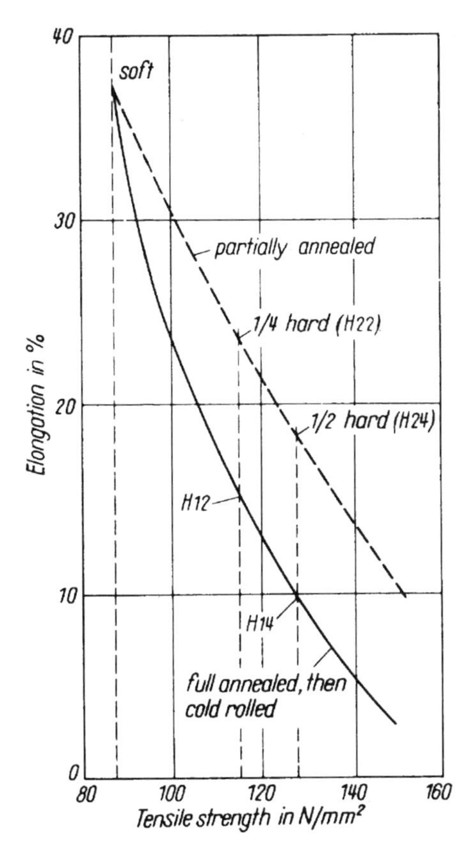

Fig. 10.13: Relationship between elongation and tensile strength for 1100 sheet, either partial-annealed or cold-rolled after a full anneal. In the H22 and H24 partially annealed condition, besides softening, some recrystallization has taken place. (J. Hérenguel)

Chapter 11. Alloy Hardening

As is the case with other currently used metals, the strength of aluminum may be increased by two methods, namely "work hardening" and "alloy hardening."[1] The two methods may be employed simultaneously.

Work hardening, as pointed out in the last chapter, depends on the mutual hindrance and "piling-up" of dislocations with increasing cold work.

Alloy hardening is based on the reaction of dislocation lines with foreign atoms. The foreign atoms have a different atomic diameter and electron structure than aluminum atoms, and for this reason, the addition of such atoms creates a disturbance in the aluminum lattice (for example, see Fig. 8.1, lattice imperfection b). Different types of foreign atoms affect the lattice to differing degrees. Moreover, the hardening depends on whether the foreign elements are in solution or precipitate in a more or less fine distribution within the aluminum lattice. Depending on their distribution, the foreign atoms impede the movement of the dislocations and, thereby, the progress of plastic deformation to widely differing degrees. For this reason we must divide alloy hardening into two kinds: first, solid-solution hardening, which is used in non-heat-treatable alloys; second, precipitation hardening or age hardening, produced by the controlled precipitation of alloying elements that were previously in solution in heat-treatable alloys.

11.1 Non-Heat-Treatable Alloys ("Work-Hardening Alloys")

Commercially pure aluminum (99.5%) contains about 0.4% iron plus silicon and, therefore, could be regarded theoretically as an alloy based on super-purity aluminum. The tensile strength in the annealed state is almost double that of super-purity aluminum because of the presence of iron and silicon atoms. Normally, commercially pure aluminum is not considered as an alloy, since the iron and silicon content is an inherent part of the electrolytic production of aluminum. The term "alloy" is generally used to designate an intentional addition of some other element to aluminum.

The non-heat-treatable, wrought alloys contain magnesium additions of 0.5–5.5% and/or manganese up to 1.5%. In these alloys, the increase in strength is mainly due to lattice distortion by the atoms in solution. Considerable strengthening is produced with magnesium atoms in solution (Table 11.1). Manganese atoms also cause strengthening, but not to the same degree as the same weight percent magnesium addition. Constituents often known as heterogeneities, which are clearly visible through an optical microscope and which occur with higher iron or manganese contents, have only a slight strengthening effect. It is understandable that coarse precipitates in a wrought structure do not greatly inhibit the movement of the dislocations along the slip planes. In the aluminum matrix around such coarse precipitates, the lattice planes are relatively undistorted, and disloca-

[1] There is also a third method involving the dispersion of insoluble constituents. For example, the dispersion of extremely fine oxide inclusions in aluminum increases the strength. Such a high temperature sintered material produced by powder metallurgy techniques is known as SAP (Sintered Aluminum Powder).

Table 11.1: Tensile properties of 1 mm thick non-heat-treatable aluminum alloy sheet

Material	Condition	Ultimate Tensile Strength, UTS min.–max. (MPa)	Tensile Yield Strength, $TYS_{0.2}$ min.–max. (MPa)	Minimum Elongation in 50 mm (%)
1100-O	soft	75–105	25	22
1100-H14	1/2 hard	110–145	95	3
1100-H18	hard	150	—	2
3003-O	soft	95–130	35	22
3003-H14	1/2 hard	140–180	115	3
3003-H18	hard	185	165	2
5005-O	soft	105–145	35	19
5005-H14	1/2 hard	145–185	115	2
5005-H18	hard	185	—	2
5454-O	soft	215–285	85	13
5454-H34	1/2 hard	270–325	200	5
5083-O	soft	275–350	125–200	16*
5083-H323	1/4 hard	305	215	10*

* for 1.6 mm thickness; for compositions, see Appendix D, Table D2.

tions can move freely on them between the precipitates. Atoms in solution or precipitated in very fine dispersions, on the other hand, create distortions in the entire lattice, and thus, inhibit the mobility of all dislocations.

The non-heat-treatable alloys have a wide range of application. They are used instead of commercial-purity aluminum when increased strength is required.

Sheet products from non-heat-treatable alloys are usually ordered in a temper, which means that, in addition to the alloying hardening, there is also additional strength imparted by cold working (see Table 11.1).

11.2 Heat-Treatable Alloys

The first commercially heat-treatable alloy was discovered by Alfred Wilm in 1906, when he produced the "Duralumin" alloy type. Duralumin has a nominal composition of 4.5% Cu, 0.5–1.0% Mg, and 0.5% Mn. The alloy was later identified as 2017.

It could almost be said that the discovery of Duralumin was accidental. Wilm had given a laboratory technician the assignment of testing the physical properties of various Al-Cu-Mg alloys. Since the weekend was approaching, the laboratory technician left the samples unattended until the following Monday. On Monday, he found unusually high strength in the samples, which was the result of the Duralumin sample being stored for two days at room temperature. This process is known as "natural aging."

11.2.1 Which alloys are heat-treatable?

As has already been pointed out, certain alloying elements, such as copper, are highly soluble in aluminum in the solid state, and solid solutions arise. The resulting structure is called homogeneous if the alloying element is completely dissolved and heterogeneous if some is precipitated as a separate crystal phase.

Very similar relationships exist for the solution of a salt (e.g., borax) in water. As can be seen from Fig. 11.1, borax is soluble in water in varying amounts at different temperatures: 1.3% at 0°C, 15% at 60°C, and 34.3% at 100°C. By comparison, Fig. 11.2 shows the solubility of several metals in solid aluminum, where the percentage of the alloying element in solution increases with increasing temperature, just as in the case of borax in water.

The addition of two alloying elements in appropriate combination can greatly alter the solubility curve, when compared to the addition of either of the elements separately. If for example, magnesium and silicon are added to aluminum in an atomic ratio of 2:1, the intermetallic compound Mg_2Si is formed, which greatly reduces the apparent solubility of magnesium, since the magnesium atoms tend to combine in the form of Mg_2Si. The same is true for the solubility of zinc, if zinc and magnesium are added corresponding to the intermetallic phase $MgZn_2$. All the alloying elements or combination of elements listed in Fig. 11.2 can be used for heat-treatable aluminum alloys. However, the age-hardening effect of the commercial Al-Si, Al-Mg, and Al-Zn alloys is relatively weak.

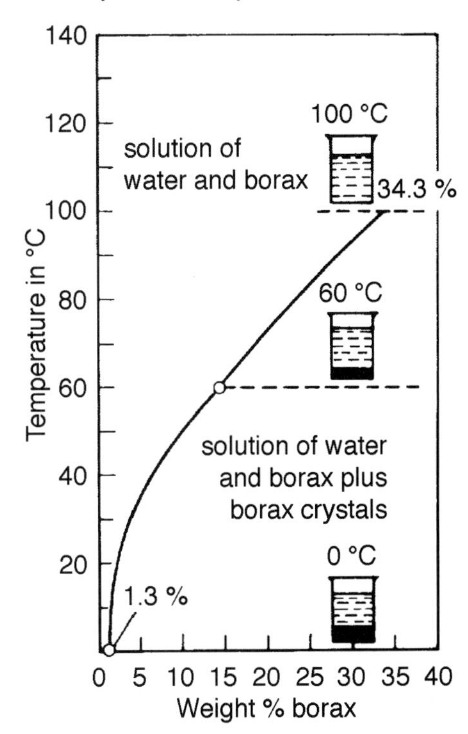

Fig. 11.1: Solubility of borax in water.

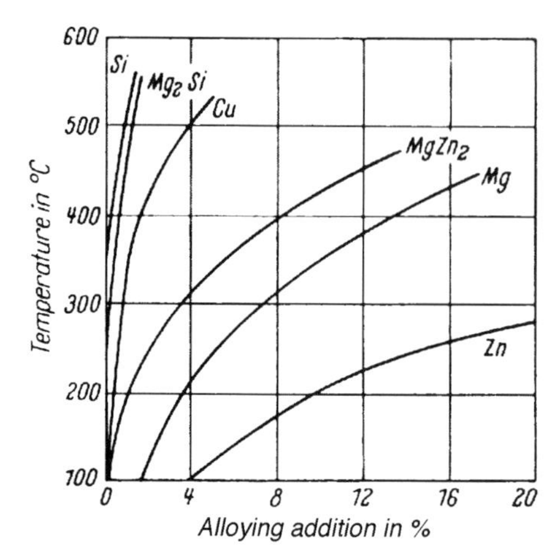

Fig. 11.2: Solubility of alloying additions in aluminum as a function of temperature. To the right of each curve lies the heterogeneous region, while to the left of each curve lies the homogeneous region, within which the alloying addition is completely soluble in aluminum.

For wrought alloys, it is not desirable to exceed 8% of the major alloying element because this tends to make the alloy too brittle for forming and, for many of such alloys, results in reduced corrosion resistance. Wrought alloys suited for heat treatment are usually those in which the solubility of the alloying constituent decreases sharply with decreasing temperature (e.g., copper and the combination of Mg_2Si or $MgZn_2$).

11.2.2 Some heat-treatable wrought alloys

For the reasons discussed above, heat-treatable alloys usually contain at least two alloying elements. Manganese or chromium is often added to inhibit recrystallization. Heat-treatable alloying groups of special importance are listed in Table 11.2.

Alloys of the Al-Cu series (2xxx) contain 1–7% copper as the prime hardening agent. A magnesium addition of 1.2–1.8% will increase the "natural aging"[2] of the alloy and also

Table 11.2: Minimum tensile properties of extruded heat-treatable aluminum alloys (compositions in Table D2, Appendix D)

Material	Condition	Ultimate Tensile Strength UTS (MPa)	Tensile Yield Strength $TYS_{0.2}$ (MPa)	Elongation in 5D El_5 (%)
2014-T4	Naturally aged	345	240	10
2014-T6	Artificially aged	470	400	5
6063-T4	Naturally aged	125	60	12
6063-T6	Artificially aged	205	170	9
6061-T4	Naturally aged	180	110	14
6061-T6	Artificially aged	260	240	9
7075-T6	Artificially aged	560	495	6

[2] "Natural aging" refers to precipitation hardening, which takes place during storage at room temperature, whereas "artificial aging" implies an elevated temperature treatment to accelerate precipitation.

Fig. 11.3: The influence of precipitation temperature on natural aging of 2017 wire (Zeerleder).

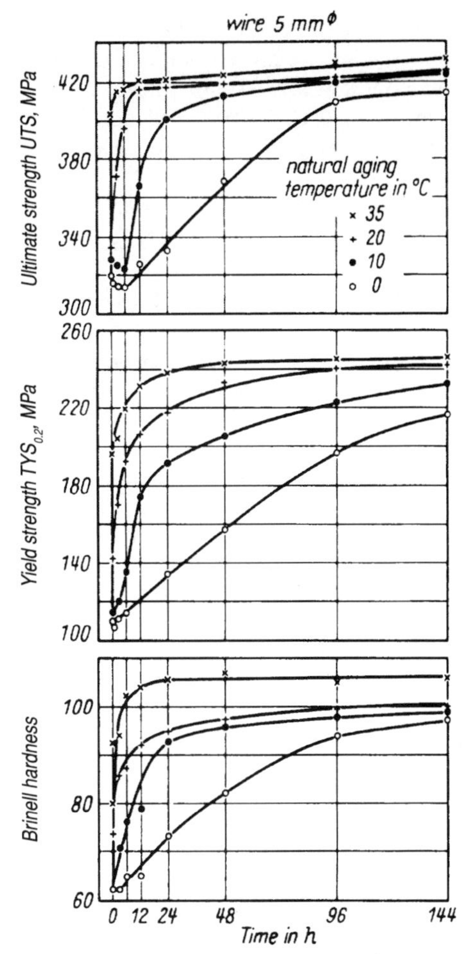

the maximum strength. After storage for several days at room temperature, the alloy attains the strength of normal structural steel (Fig. 11.3). In Fig. 11.4, the age hardening of an Al-Cu-Mg alloy at different temperatures is shown for comparison.

Alloys of the Al-Mg-Si series (6xxx) contain magnesium and silicon additions of 0.4–1.2% each and sometimes smaller amounts of copper, chromium, or manganese. The alloys are usually artificially aged at temperatures of 140–185°C. Al-Mg-Si alloys attain medium- to high-strength values combined with good corrosion resistance. In alloys of the Al-Mg-Si series, the age hardening takes place in a ternary "solubility field" (i.e., the magnesium and silicon atoms combine to form Mg_2Si, which precipitates causing the hardening). For this reason, Al-Mg-Si alloys usually have an atomic ratio of Mg:Si near 2:1.

Alloys of the Al-Zn-Mg and Al-Zn-Mg-Cu series (7xxx) are high-strength materials similar to Al-Cu-Mg, except for a zinc addition instead of copper or in combination with copper. The Al-Zn-Mg alloys are used in varying compositions. The most widely used at present are the alloys with approximately 4.5% Zn and 1.3% Mg, which combine good formability during extrusion or rolling with satisfactory corrosion resistance. Al-Zn-Mg alloys attain medium to high strength after aging for a few weeks at room temperature or for a few months at lower temperatures. Maximum properties are attained by artificially aging at 130–170°C. The Al-Zn-Mg-Cu alloy group attains the highest strength of all aluminum alloys, exceeding normal structural steel. Artificial aging is carried out at 120–160°C.

Wrought alloys were chosen for illustrating the principles of heat-treatable alloys; similar principles apply to casting alloys.

11.2.3 Factors in the selection of a heat-treatable alloy

An important advantage of semi-finished products from heat-treatable alloys lies in the attainment of high yield strength combined with high elongation (i.e., good formability). An example is shown in Fig. 11.4.

It can be seen that after heat treatment, the elongation of 2024 sheet is 10–20%, while the yield strength is approximately 350 MPa. Herein lies the basic difference between heat-treatable and non-heat-treatable wrought alloys. In non-heat-treatable alloys, a yield strength of 250–300 MPa could be obtained through severe cold work; however, the elongation would be greatly reduced. Fig. 11.4 also shows that during the process of artificial aging for maximum strength, the elongation is reduced. This is due to the development of relatively large precipitates that form small internal notch effects in the ductile alumi-

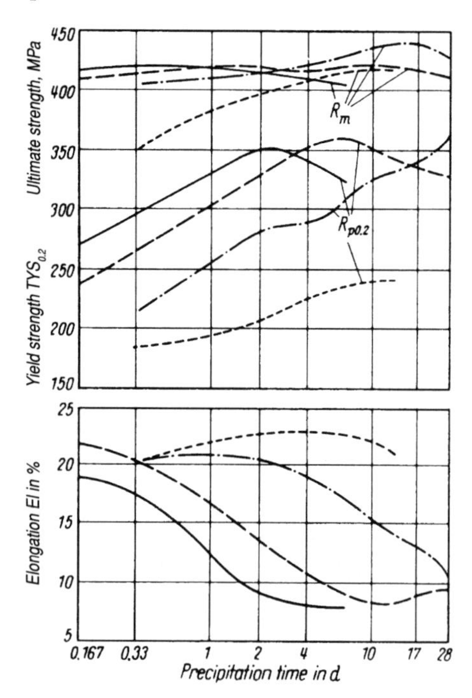

Fig. 11.4: Variations in mechanical properties of 2024 sheet by aging at different temperatures (P. Brenner). ————: 175°C; — — —: 160°C; — · — · —: 140°C; -------: 20°C.

num matrix, causing a reduction in the elongation. This phenomenon is especially noticeable after "overaging" the heat-treatable alloys.

The high mechanical properties of the age-hardening alloys come with some disadvantages. A heat-treatable alloy is selected only if higher yield strengths are required. If low or medium strength is sufficient, a non-heat-treatable alloy or commercial-purity aluminum is selected, which is easier to fabricate and not as sensitive to thermal treatment.

If the decision is made to use a heat-treatable alloy, the next question is whether maximum strength is required. In order to obtain the highest strengths, some corrosion resistance must be sacrificed by the selection of a copper-containing alloy. However, the sensitivity to corrosion can be substantially reduced by cladding (e.g., with commercially pure aluminum) or other protective coatings. If an intermediate strength is required, either Al-Zn-Mg or Al-Mg-Si alloys, which are also heat-treatable, may be considered. Al-Mg-Si alloys have much better corrosion resistance, but their use (e.g., in vehicles) is limited by softening that occurs at welds and hot bends. To recover their strength, they require new solution and precipitation treatments, which may be impractical after assembly. In this connection, the Al-Zn-Mg alloys offer interesting advantages. After welding and hot-deformation, they re-harden naturally with time; for this reason, they have been used recently, especially in welded construction. These alloys should not be cold formed after heat treatment, since this promotes stress corrosion.

Almost every alloying addition has other effects besides the intended ones, and sometimes these side effects are undesirable. For example, certain additions used for grain refinement in the as-cast structure (Fe or Ti) or to inhibit recrystallization and coarse grain during solution heat treatment (Mn, Zr, or Cr) will degrade the appearance of anodic films.

11.3 Metallurgical Principles of Precipitation Hardening

11.3.1 Heat-treating operations

For age hardening, heat treatment is carried out in three successive steps.

1. Solution heat treatment at a temperature above the solubility curve. For extrusions, a separate solution heat treatment is often unnecessary, since the freshly extruded shape already has a temper and condition that corresponds to that of a solution-heat-treated piece.
2. Quenching, usually in water, sometimes in air or water spray. Castings are also quenched in heated oil to reduce quench stresses.
3. Aging (precipitation) at room temperature (natural aging) or at elevated temperatures up to 200°C (artificial aging).

The three steps as a function of time are represented in Fig. 11.5 for alloy 2024. For a given alloy, the age hardenability, as well as the temperature range for the solution heat treatment, can be determined from the lower part of the equilibrium diagram, which shows the solid state.

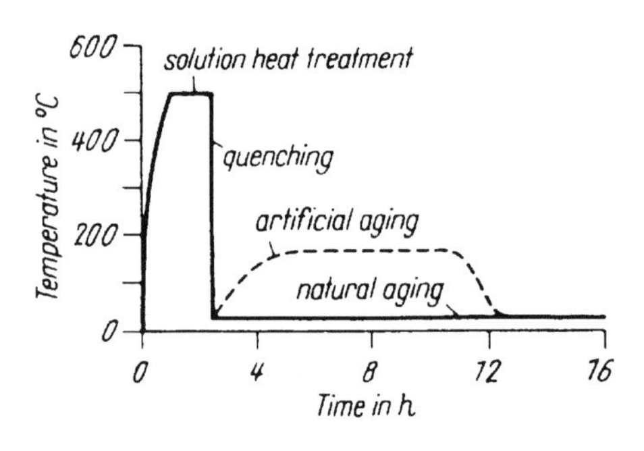

Fig. 11.5: Schematic showing the age-hardening of 2024.

In the following section, the steps taken during heat treatment are considered along with the changes observed in the structure.

11.3.2 Survey of structural conditions

The states through which the microstructure of a heat-treatable alloy passes from casting to heat treatment are presented schematically in Fig. 11.6 for the Al-Cu system. The sequence is as follows:

1. All metal is liquid.
2. Liquid is located between the primary crystals. This structure also results from overheating during solution-heat-treatment, if the temperature of the treatment accidentally exceeds the solidus temperature.

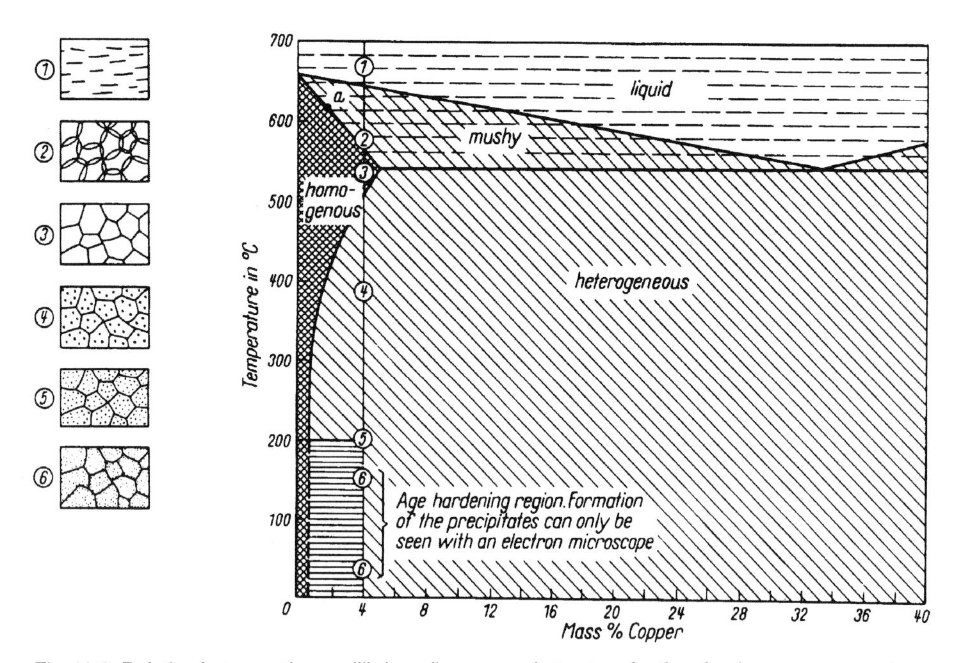

Fig. 11.6: Relation between the equilibrium diagram and structure for the aluminum-copper system. The schematic representation of the structure is for a 4% copper content.

3. Homogeneous structure is obtained by solution heat treatment at 520°C, for example, and subsequent rapid quenching.
4. This structure forms from the homogeneous structure by annealing at 400°C. A relatively coarse precipitate of a copper-rich phase ($CuAl_2$) forms. The fully annealed condition is obtained by this "heterogenization."
5. Structure after precipitation at approximately 200°C (heterogenization). This structure is obtained if the aging temperature is too high so that copper precipitates of medium size are formed. Such a structure is undesirable, because it reduces the corrosion resistance and mechanical properties as compared to Structure 6. Nevertheless, it is much harder than Structure 4.
6. Structure of the fully aged alloy. This structure is obtained as follows. First, a homogeneous structure is obtained by solution heat treatment. The material is then quenched as rapidly as possible in water (Structure 3). If the quench rate is too slow, visible precipitates will begin to form (i.e., the undesirable Structure 5). The rapidly quenched structure shows no precipitation, even after aging, under the microscope. (Precipitates containing iron and manganese are an exception, but they have little effect in the heat treatment.)

11.4 Atomic-Scale Processes

11.4.1 Solubility and diffusion of foreign atoms in the solid solution lattice

The atoms in the metal vibrate constantly around their equilibrium position. As can be seen from Fig. 11.7, these vibrations become larger with increasing temperature and facilitate the migration of atoms within the lattice ("diffusion").

At the same time, diffusion in the metal lattice is enhanced by the simultaneous increase of vacancies in the lattice with increasing temperature.

Thus, oscillation of the atoms increases with temperature, causing the atoms to change their places in the lattice more often by movement through the vacancies.

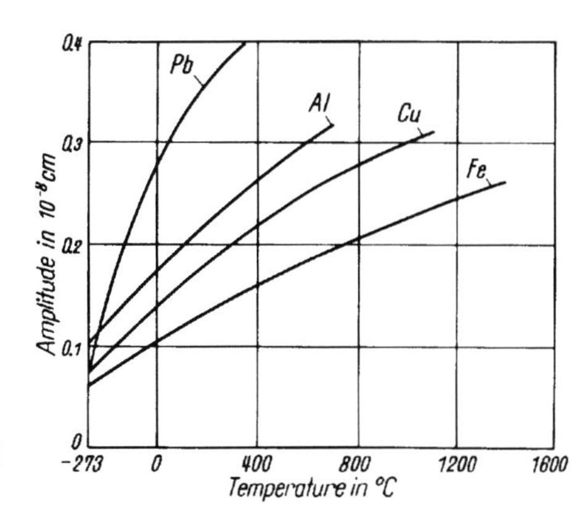

Fig. 11.7: Thermal vibration of atoms in metal crystals as a function of temperature.

11.4.2 Precipitation in the solid state

Fig. 11.8 illustrates the different states of precipitation on an atomic scale and presents associated terminology. Fig. 11.8.a shows the random distribution of the foreign atoms.

According to the phase diagram, the quenched solid solution should decompose at room temperature, leaving a purer aluminum lattice plus particles of a different phase. How-

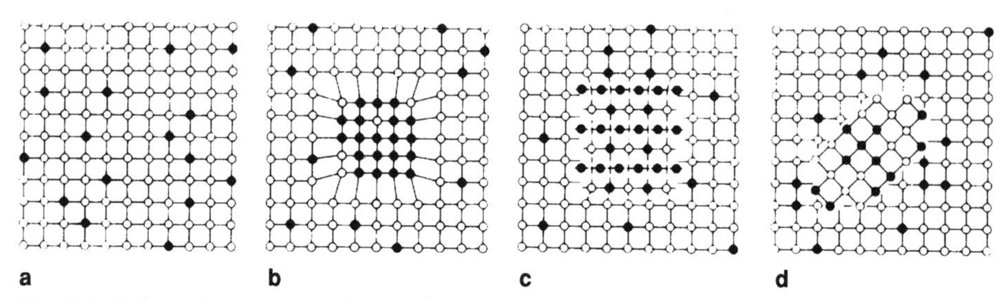

Fig 11.8: Schematic representations of (a) a solid solution with a random distribution of foreign atoms; (b) coherent precipitation; (c) a partially coherent intermediate phase in a solid solution. The vertical crystal planes are coherent, the horizontal planes incoherent; and (d) an incoherent precipitate. Open circles represent aluminum atoms; solid ones denote foreign atoms (e.g., copper).

Table 11.3: Age hardening behavior of an Al–4% Cu alloy

Stage	Heat treatment	Disposition of copper atoms	Size of precipitates	Attainable hardness (HV)	Figure
I	None, apart from previous processing	Mostly precipitated as equilibrium $CuAl_2$ phase	D: 1–10 μm	30–70	11.5
II	Solution treatment + rapid cooling	all Cu in solution	—	60	—
III	Natural aging at room temperature	Segregation into GP I zones, coherent	D: up to 100 Å S: approx. 2 Å	100	—
IVa	Brief heating to ca. 150–200°C	Resolution of GP I zones (retrogression)	—	40–70	—
IVb	Age hardening at150°C	Segregation into GP II zones, coherent	D: 100–700 Å S: 10–50 Å	120	11.10
	Further aging at 150°C	Increased diffusion into GP II zones and precipitation as Θ′ theta prime (as in Fig. 11.11) phase, partially coherent	D: 0.5–1 μm S: 30–100 Å	130	11.11
V	Overaging resulting from temperature too high or treatment time too long	Precipitated as Θ phase (equilibrium phase), incoherent	D: ca. 0.1–3 μm S: ca. same as D	ca. 30	11.3

HV = Vickers hardness; D = diameter; S = thickness.

ever, at temperatures below about 150°C, the foreign atoms have very little thermal energy and cannot diffuse far enough to form particles visible under an optical microscope, even after several weeks time. Nevertheless, they do begin to coalesce.

11.4.3 Formation of "zones" or coherent particles

The precipitates represented in Figures 11.8.b and 11.8.c are no more than clusters of foreign atoms called "zones," which have come together within the solid solution. Although locally somewhat distorted, the aluminum lattice structure remains more-or-less continuous with that of the precipitate. Therefore, they are called coherent or semi-coherent precipitates according to the degree of continuity or "coherence" they have with the parent lattice; they are typically formed during aging at room temperature or up to about 150°C. The zones may be in the form of platelets, needles, or clusters and usually measure ten to a few hundred inter-atomic distances, so that they cannot be seen under an optical microscope. (These zones are not always detectable with an electron microscope either, especially if there is not a great deal of contrast between the lattice and the foreign atoms.)

The maximum hardness during age hardening is often obtained by partly incoherent precipitation (or intermediate phase; see Table 11.3).

Fig. 11.12 shows a section through an incoherent particle magnified 71,000 times. At the coherent boundary surface, the stress regions can be recognized in the aluminum lattice as an indication of strong internal stresses.

11.4.4 Precipitation of incoherent particles

Larger particles tend to break away from the aluminum lattice, forming small crystals, sometimes with aluminum atoms (e.g., $CuAl_2$) or intermetallic compounds of the foreign atoms such as $MgZn_2$ or Mg_2Si. In this case, the lattice structure of this precipitate is completely divergent from that of the aluminum matrix (Fig. 11.8.d). The diameter of these precipitates (also called heterogeneities) measures between 0.02 μm and 5 μm. Only those particles larger than approximately 1 μm in diameter are visible under an optical microscope.

11.5 Precipitation Processes

It was previously believed that natural aging occurred in heat-treatable aluminum alloys together with zone formation, while artificial aging brought about incoherent precipitation.

The structural changes during the age-hardening process, in reality, are not so easy to distinguish. With increasing aging temperatures, different decomposition and precipitation processes take place in sequence in some cases and simultaneously in others. This can be shown in detail by considering an alloy of aluminum and 4% copper. In such an alloy, five stages of structural changes can be determined during the heat-treating operations, which are summarized in Fig. 11.9 and Table 11.3. Stage I represents the structure

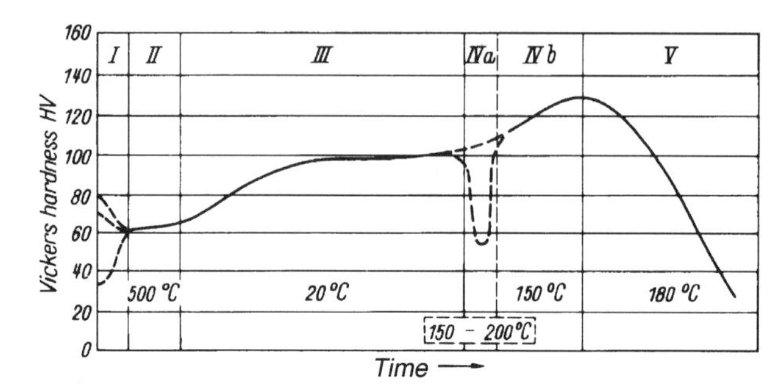

Fig. 11.9: Age-hardening behavior of an Al-4Cu alloy. Step I: previous thermal history; Step II: solution-heat-treatment and quenching; Step III: natural aging (GP I zones); Step IVa: retrogression at 150–200°C (resolution of GP I zones); Step IVb: artificial aging (GP II zones, plus Θ′ theta prime phase); Step V: overaging (Θ Phase). See structural features in Table 11.3.

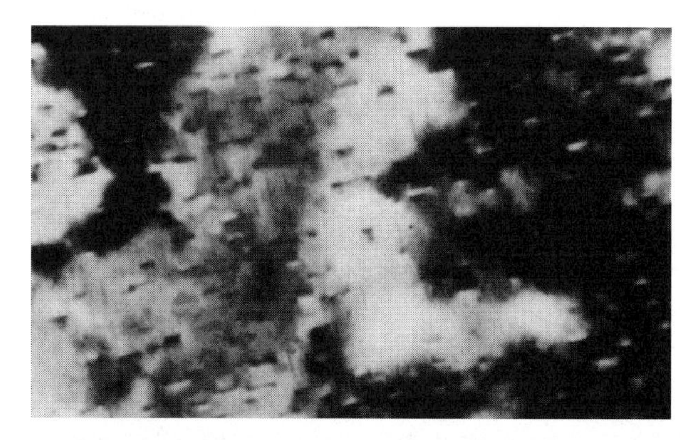

Fig. 11.10: Coherent GP II zones in an Al-5% Cu alloy (M. v. Heimendahl). 77000×

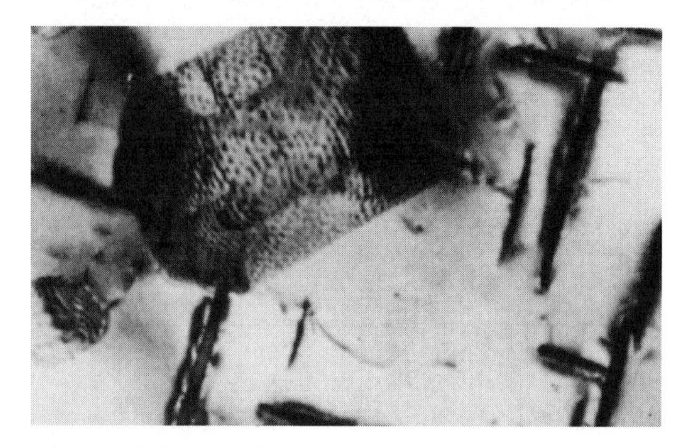

Fig. 11.11: Semi-coherent Θ′ plates on all three cube directions in the Al-Cu system. A large particle with surface dislocations possesses a coherent interface parallel to the sample surface. The dislocations bridge small differences in lattice constants between the aluminum matrix and the Θ′-particles. This is somewhat analogous to subgrain boundaries where dislocations permit compensation for small differences in orientation. (U. Koster) 20,000×

prior to solution heat treating. The precipitate may originate either in the cast structure or any previous thermal treatment. The hardness values, therefore, may vary widely. In Stage II, the copper atoms that had precipitated previously are taken into solution at 500°C. This solution-heat-treatment also removes any previous cold work or precipitation hardening. The material is then quenched from the solution-treatment temperature to room temperature.

During natural aging (Stage III) the dissociation of the supersaturated solid solution begins by coherent precipitation of copper-rich zones only a few atoms thick. These zones are called Guinier-Preston Zones in honor of their discoverers (abbreviated GP zones). They stress the aluminum lattice and, because of their great number, form a dense network that impedes the movement of dislocations during deformation. The extremely small zones that originate at room temperature are called GP I zones and do not grow any more during subsequent artificial aging, but actually go back into solution in a short time at 150–200°C. This leads to a temporary reduction in hardness (Fig. 11.9, Step IVa). This "retrogression" suggests that the arrangement of the strengthening atoms, after artificial aging, is basically different from that after natural aging. Artificial aging at elevated temperatures, according to Step IVb, causes new enrichment of copper atoms in GP II zones (Fig. 11.10) soon after the end of the reversion.

As the artificial aging continues at approximately 150°C, the formation of a metastable phase Θ′ takes place, which corresponds to the equilibrium phase $CuAl_2$ in its composition but has a different crystal lattice (Figures 11.11–11.13). Maximum hardness is obtained during the appearance of the GP II zones and sometimes for a mixture of GP II and Θ′.

As soon as only Θ particles are present, the stage of overaging is attained, which is characterized by a reduction in hardness, tensile strength, and elongation.

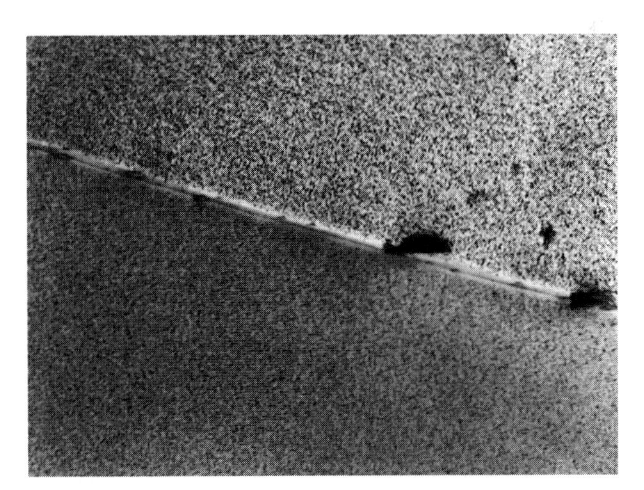

Fig. 11.12: Θ′-particle in the Al-Cu system with strong stress fields along the coherent interface. (U. Köster) 71,000×

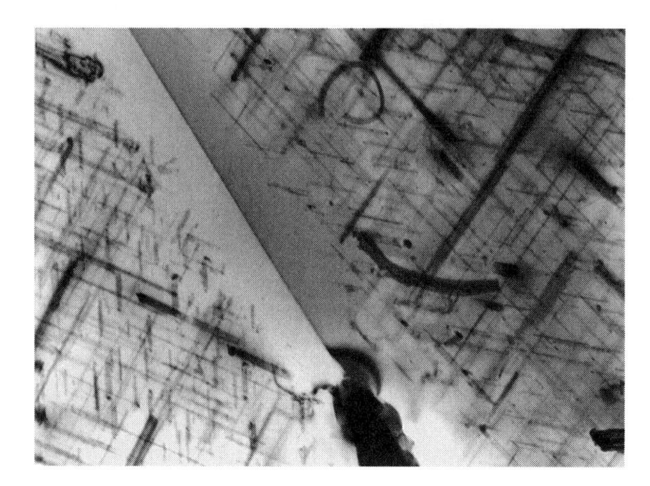

Fig. 11.13: The start of overaging in an Al-Cu system through growth of the Θ' particles within the grains. Relatively coarse Θ-precipitates (incoherent) lie on the grain boundaries. (U. Köster) 15400×

If the aging is carried out at temperatures between 170°C and 300°C, the equilibrium phase $CuAl_2$ forms a relative coarse precipitate, which is visible under an optical microscope. The aluminum is now softer than in the quenched state, since the solid solution strengthening by copper atoms in solution is lost.

11.5.1 Strengthening through precipitation of foreign atoms

The hindrance of dislocation movement through reaction with foreign atoms may be explained as follows (Fig. 11.14).

Foreign atoms in solution, or present in zones, create stress areas in the aluminum matrix, which resist the approach of dislocations. After overcoming this barrier, the migrating dislocations traverse very fine precipitates up to 150 Å diameter with the expenditure of the appropriate energy. Together, these two cause a significant increase in the stress necessary to move dislocations through the lattice (recognized as an increase in yield strength).

Precipitates with a diameter of more than around 150 Å are usually not traversed by the dislocations. More often, the dislocations are locked in place by these precipitates but have a certain mobility in the free space between the precipitates.

As long as the precipitation particles are traversed by the dislocations, the yield strength increases generally with the increasing diameter of the precipitate. The transition in microstructures from Fig. 11.14.a to Fig. 11.14.b thus results in an increase in yield strength. But if the bypass mechanism described in Fig. 11.14.c occurs, then the yield strength drops with increasing distance between the precipitates.

This brief description of the reaction between precipitates and dislocations does not take into account the relationship between separation and diameter of the particles. The main purpose was to show the connection between the various steps taken and the resulting phenomena in the structure and to explain, from an atomic standpoint, how heat treatment affects the mechanical properties of the final product.

11.5.2 Influence of vacancies in the precipitation process

The influence of vacancies in the heat-treating process has not been discussed in detail up to this point. It has been stated that the movement of atoms in the solid state is dependent on the presence of vacancies. It was further noted that the number of vacancies in equilibrium in the aluminum lattice increases with temperature.

An excess number of vacancies can be frozen into the structure by rapid quenching from approximately 500°C. However, these vacancies are highly mobile at room and slightly higher temperature, making possible the precipitation processes in heat-treatable alloys during aging.

Two examples may be used to clarify this phenomenon:

1. If an Al-4% Cu alloy, which has been solution heat treated and quenched, is heated for approximately one minute at 200°C, the supersaturated vacancies disappear into "sinks" (e.g., grain boundaries); upon subsequent aging, all aging processes by diffusion of copper atoms are drastically slowed. In this case, the vacancies at 200°C have much greater mobility than the copper atoms.
2. An Al-Cu-Mg alloy exhibits exceptional hardening by aging at room temperature, which is not the case in the Al-4% Cu alloy. The basis for this difference may be due to the fact that the relatively large magnesium atoms hold the vacancies trapped in their vicinity. These vacancies then diffuse slowly to the sinks at room temperature, making possible the movement of the copper atoms into the GP zones.

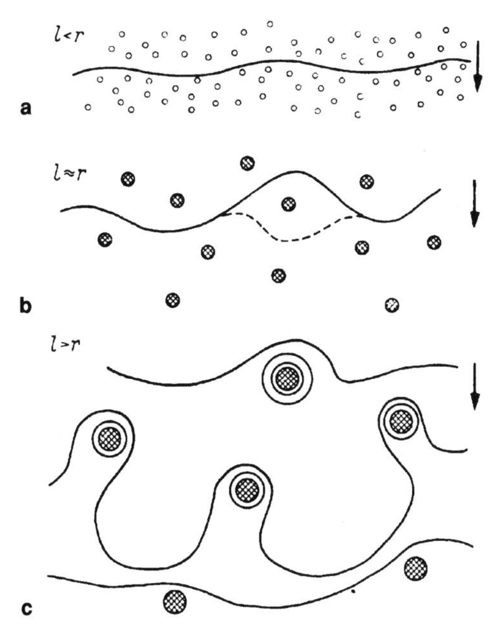

Fig. 11.14: Movement of a dislocation line through a crystal containing differently dispersed precipitates (schematic); l = average distance between precipitates, r = radius of curvature of the dislocation line under the influence of the stress present. Arrow = indicates direction of dislocation movement.

11.6 Possible Errors in the Heat Treatment of Age-Hardening Alloys

Structure 2 in Fig. 11.6 shows what happens if the solution-heat-treatment temperature is too high. Melting takes place at the grain boundaries, making the material unusable. It must then be remelted and recast.

With typical Al-Cu-Mg alloys, it is important that quenching takes place without delay and at a sufficiently high rate. Even with a delayed quench of 10–30 seconds with cooling to 300–450°C, for example, coarse precipitates start to form (Structure 4, Fig. 11.6), which result in a loss of properties after quenching and aging. This is also very detrimental to the corrosion resistance. The same thing occurs in heat-treatable alloys during welding. In the region of the weld bead, strength is lost due to the heterogenization, if the whole assembly is not solution treated and aged again.

Some other heat-treatable alloys may tolerate a much slower quench rate without detrimental effects (e.g., forced air-cooling after solution heat treatment). Alloys that belong to this group include 7005, Al-Cu-Mg alloys with up to two percent copper content, and Al-Mg-Si alloys with a low manganese content. It is generally true that the cooling rate after the solution heat treatment may be slower for a less concentrated alloy composition. Thus, the lower the supersaturation at 300°C, for example, the slower the precipitation rate at that temperature. One caution, however: while strength may not be affected by quench rate, other properties such as toughness may be and should be checked if they are important in the intended applications.

In several alloy types, including Al-Mg-Si alloys, the precipitation rate is greatly increased by the addition of the recrystallization inhibitors manganese and chromium.

In order to minimize internal quenching stresses, the cooling rate is preferably maintained as slow as possible without the loss of strength or corrosion resistance (e.g., Al-Cu-Mg alloys). Such stresses can cause distortion, calling for straightening operations at con-

Table 11.4: General guidelines for quenching heat-treatable alloys after solution-heat treatment

Alloy	Solution temperature	Quenching to <200°C	Typical quench medium
2017 (with about 4% Cu)	500°C	<10 sec	Water
Low-alloyed AlCuMg (with 2% Cu)	475–530°C	<40 sec	Water, with sheet under 1.5 mm, also forced air
6061 (with 1% Si 1% Mg + Cu and Cr)	540 °C	<30 sec	Water with >3 mm thickness, and forced air with <3 mm
6063 (with 0.5% Mg and 0.5% Si)	530 °C	<5 min	Water with >5 mm thickness, forced air with <5 mm
7008 (with 4–5% Zn)	450 °C	5–20 min	Forced air
7075 (with 6% Zn, 2% Mg 1.5% Cu)	475 °C	<30 sec	Water or spray mist

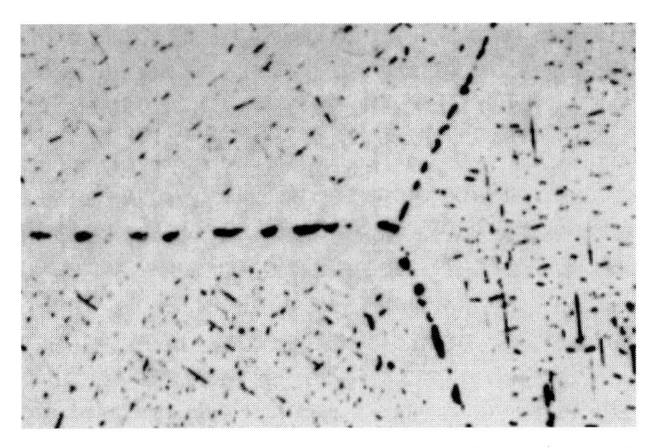

Fig. 11.15: Micrograph of Al-Cu-Mg sheet heterogenized at 400°C (Hanemann and Schrader). Copper containing grain boundary precipitate with depleted boundary region. 600×

siderable expense. It is for this reason that air-quenching is used with extruded thin-walled shapes of 6063, while the more highly supersaturated 6061 generally needs to be quenched in water spray because of the significantly higher precipitation rate of the Mg_2Si (see Chapter 13).

The method of quenching also depends on the wall thickness. Table 11.4 lists some general guidelines for quenching.

It is characteristic of age-hardenable alloys that they are affected by temperatures above 120°C, if held at such temperatures for longer periods of time. A coarse precipitate gradually develops, as pointed out in Fig. 11.13. This means that it serves no purpose to attempt to thermally "stress relieve" heat-treatable alloys, since the properties of the structure will be damaged, not improved. This is also true for Al-Mg alloys with more than four percent magnesium, which become sensitive to corrosion at between 100°C and 200°C.

Stress relieving the heat-treatable alloys at temperatures up to 250°C does cause a reduction in the tensile strength due to the precipitation that occurs, but the material becomes very sensitive to corrosion, which primarily attacks the grain boundaries. Precipitation processes take place especially rapidly at the grain boundaries. This is clearly seen in Fig. 11.15. It has been pointed out that the lattice structure is distorted in the area of the grain boundary so that there is more room between the individual aluminum atoms, which allows increased mobility ("diffusion rate") among the foreign atoms. The region denuded of precipitates in the neighborhood of grain boundaries originates because the supersaturated vacancies rapidly vanish into the grain boundaries. Therefore, close to the grain boundaries, the precipitation processes are slowed. If a heat-treatable alloy is heated to the stress-relieving temperature range, the precipitate at grain boundaries creates a locally strong difference in electrochemical potential with the precipitate-free region. This makes such a structure susceptible to "grain boundary corrosion."

11.7 Annealing Age-Hardenable Alloys

As stated in the previous chapter, annealing at a sufficiently high temperature removes the hardening effect of a previous heat treatment. For example, in an Al-Cu alloy, the equilibrium phase $CuAl_2$ forms and grows into large precipitates that are easily visible under the optical microscope (Fig. 11.15). These large particles have only a minimal hardening effect. Heat-treatable alloys are usually annealed at temperatures between 350°C and 400°C.

After this high temperature anneal, slow cooling (preferably less than 30°C/hour) is necessary to reach the fully soft condition. Fig. 11.2 shows that at 350–400°C all of the alloying elements that make up the heat-treatable alloys have a higher solubility than at room temperature. If the annealed material is rapidly cooled, a certain amount of age hardening can take place during storage at room temperature, which is clearly undesirable for annealed material. By slow cooling to around 200°C, the atoms, which were in solution at the higher annealing temperature, produce coarse precipitates and cannot form "hardening zones" later. This is the way to obtain an annealed condition in a heat treatable alloy; in this state, the alloy is suitable for forming, especially if recrystallization took place during the high-temperature stage. Annealed and subsequently deformed material may be restored to the heat-treated state, if desired, by solution heat treatment, which takes all of the precipitates back into solution, followed by quenching and aging. This is the basic difference between heat-treatable and non-heat-treatable alloys, including pure aluminum. Once the non-age-hardening alloys are annealed, softening is permanent, unless the material is further cold worked.

11.8 Deformation of Age-Hardenable Alloys

Hot deformation requires a structure with no coarse heterogeneities or pores, because these act as stress risers, causing crack propagation, at least in the high-strength alloys that are notch sensitive.

Shaping of sheet by cold forming (bending, stretching, deep-drawing, etc.) is best carried out in the solution-heat-treated state or, in some cases, after full annealing.

In a few very special cases, the sheet may be deformed after a short "retrogression" anneal (e.g., a few minutes at 180–200°C). This corresponds to Fig. 11.9 and Table 11.3, Step IVa. Which heat treatment is most suitable prior to cold deformation depends on a number of factors. Especially important are how much age hardening occurs between solution heat treatment and cold working, the allowable deformation stress, the degree of deformation, and whether it is possible to solution treat after the deformation. However, solution treatment and quenching often create problems with grain growth and quench distortion.

In general, it is preferable to carry out forming in the solution-heat-treated condition.

Chapter 12. Corrosion Resistance and Protection[1]

All metals attempt to return to the form in which they occur in nature. The expression corrosion comes from the Latin word *corrodere,* which means to nibble away or gnaw at. The term corrosion is used to describe a great number of different aggressive and oxidation processes that a metal component may undergo during its useful service life. The visible and measurable changes are called corrosion symptoms; the changes that impair the component's function are called corrosion damage. The term corrosion protection covers all measures designed to reduce corrosion damage. General terminology for the corrosion of metals is defined in ASTM Standard G15 and ISO 8044.

Aluminum must be considered a corrosion-resistant metal since it can often be used without any special protective coating or surface treatment. Unprotected aluminum is resistant to weathering, fresh and salt water, and many foodstuffs and chemicals. It is important, however, to choose the correct alloy, thermal treatment, and joining technique. There are many examples of the long life of aluminum, including the roof of St James's church in Rome, which has been in place without any protection since 1897. It is still in good condition, with only small surface pits less than 0.1 mm deep. The material is unalloyed aluminum with 0.5% Fe and 0.9% Si as impurities.

Because of the frequent use of unprotected aluminum in service, it is important to understand the causes of corrosion on aluminum surfaces. A basic understanding of the corrosion process is only possible if aluminum is "viewed from within."

12.1 Basic Relationships

Corrosion processes are mainly oxidation processes, and metal oxides—compounds of the metal with oxygen—are formed as the corrosion products. In the case of iron, the product is rust (i.e., iron oxide); in the case of aluminum, the corrosion product is a white hydrated aluminum oxide.

The commonly used metals are not usually found in their metallic state in nature. They usually occur in combination with oxygen. Chemically, this is the most stable state since the chemical bond between the less noble metal and oxygen is very strong. Aluminum occurs in the earth in the form of a hydrated oxide known as bauxite. A similar type of oxide forms in corrosion products. If one applies enough energy under the proper conditions, for example, in a blast furnace in the case of iron or in an electrolytic cell in the case of aluminum, metal can be separated from oxygen. This brings the metal atoms to an artificially higher energy level.

The relationship is similar to pumping water into an elevated reservoir. Special precautions must be taken to prevent leaks, since the water will use every opportunity to back down to a lower energy level. The same principle applies to metals. They are stored in a shell that offers as much protection as possible against corrosion processes that tend to-

[1] By W. Huppatz.

ward the lower energy level of the oxide. There are different types of protective shells to prevent the atoms that have been forced into the metal lattice structure from reverting to their natural state. These can be protective coatings made of a nobler metal, for example chromium in the case of iron. Alternatively, the coating may be enamel or a protective paint, which must be renewed periodically. The larger the number of weak points in such a protective coating, the larger the number of metal atoms that return to the energy level of the oxide. In the case of iron, these would be recognizable as rust spots.

Fortunately, nature has provided aluminum with a highly protective shell in the form of a natural oxide skin. If the oxide skin is scraped away, the exposed aluminum immediately forms another skin with oxygen from the air. The oxygen blocks its own way in a very short time since the aluminum oxide, as it forms, produces a barrier film on the surface. This hinders deeper intrusion of oxygen or other gases or corrosive liquids to the subsurface aluminum atoms. This situation does not occur with iron since the reaction of iron with oxygen from the atmosphere does not lead to the formation of a protective oxide skin. In fact, the oxide layer usually flakes away. Rust offers almost no protection to the commonly used steels.

With iron, the corrosion problem is relatively simple. It is accepted that unprotected iron (or steel) will rust. The corrosive attack on an iron component can be considered as being layer by layer. Therefore, it is possible to estimate that after a certain number of years under specific conditions, unprotected iron will have reverted back to iron oxide.

The situation is different for aluminum, in that the oxide layer provides a protective barrier that does not allow the whole component to be reduced to aluminum oxide. It is widely believed that aluminum will be completely immune to corrosive attack. The situation is not quite as simple as that, however, sometimes to the surprise of users of aluminum.

12.2 Corrosion Processes

The processes of metal corrosion may be divided into two groups.

- Attack without the presence of water.
- Decomposition in the presence of water.

The first group is very easy to handle, and normally has only minor significance in practice for aluminum. At room temperature, aluminum reacts with oxygen from the air; however, the oxidation process stops after the formation of a very thin oxide layer, only about 0.01 μm thick. If the surface is abraded and the oxide layer continually disturbed, then oxidation will continue. This happens in so-called "fretting corrosion."

At elevated temperatures above 500°C, aluminum reacts with oxygen to form thicker oxide layers. The aluminum oxide formed on the surface under these conditions also protects the underlying metal from further attack by oxygen. At approximately 600°C the oxide layer attains a thickness of around 0.1–1 μm.

The second and more important group of corrosion phenomena, namely attack in the presence of water, causes more problems. Even on aluminum that appears dry, there is often a thin layer of moisture present as a result of adsorption and condensation. If an aluminum sheet is exposed to an aggressive industrial atmosphere, the thin moisture layer on the surface takes part in the corrosive reactions. One should not conclude from this that water always aggravates corrosion. Frequent washing by rain in regions with aggressive atmospheres helps to rinse away the corrosive deposits.

12.2.1 Surface corrosion

Aluminum suffers an overall surface attack upon exposure to a marine or industrial atmosphere that usually comes to a stop, or is at least greatly retarded, because of the formation of a heavy oxide layer (Fig. 12.1). The metallic luster of the aluminum surface is lost with the growth of the corrosion product layer. Also, there are usually a large number of small shallow pits that are closed up by further oxidation. Humidity in combination with traces of aggressive agents in the air, especially sulfur dioxide and salt, play an important role in atmospheric corrosion. Depending on the amount of impurities in the air and the temperature cycle to which the aluminum surface is exposed, atmospheric corrosion increases with relative humidity above 60%.

The formation of condensation in aggressive industrial atmospheres is often the trigger for corrosion, for example, on the surface of extrusions or sheet products. On extrusions, a thin oxide film, which offers little protection, forms immediately as the hot metal emerges from the die. Later, depending on the alloy and atmospheric conditions during storage, a more or less protective hydrated layer develops. Extrusions should not be stored in conditions where there are large variations in temperature and, hence, in dew point. Because moisture is attracted to hygroscopic dust particles on the surface, condensation can form even above the dew point. This condensation can start corrosion. Since any such corroded extrusions require considerable polishing before they can be used, it is very important to store extrusions in a dry atmosphere.

Strip in coil is also liable to damage from condensed moisture, since the water will penetrate several centimeters between the wraps due to capillary action. Damage to the strip surface results by staining or through crevice corrosion (see Section 12.2.8).

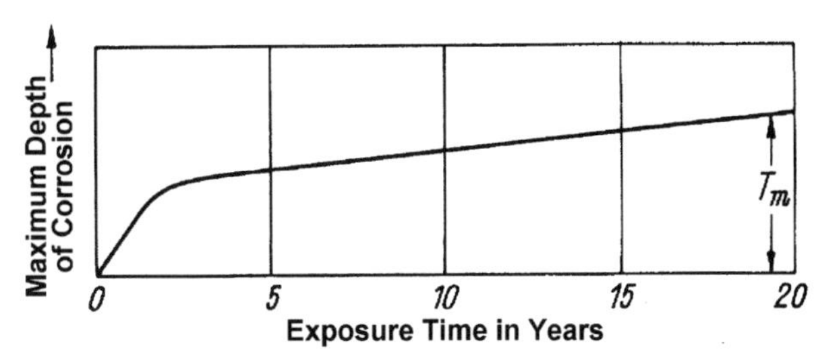

Fig. 12.1: Typical atmospheric corrosion-rate curve. T_m = about 0.01–0.1 mm, depending on the aggressiveness of the atmosphere.

The corrosion behavior of aluminum in water depends much on whether oxidants such as atmospheric oxygen are present. In neutral, oxidant-free water, for example in modern hot water heating systems, aluminum is not normally attacked, even when in metallic contact with brass fittings.

In oxygenated water, the nature of aluminum's corrosion is different. Drinking water, swimming pool water, sea and river waters naturally contain oxygen from the air or from treatment plants, in the case of drinking water, up to a saturation level. In these more oxidizing waters, pitting corrosion must, therefore, be expected. Nevertheless, pitting does not occur until the galvanic potential exceeds a critical level, the pitting threshold potential.

12.2.2 Corrosion resistance under chemical attack

In general the oxide layer on aluminum is resistant to aqueous solutions with pH values between 5 and 8.[2] Aluminum tends to be unsuitable for contact with water-based solutions below pH 4.5 (acids and acid solutions) and above 8.5 (alkaline solutions), but there are notable exceptions in the presence of certain ions. For instance, concentrated nitric acid at pH = 1, acetic acid at pH = 3, and ammonia at pH = 13 have very little effect on aluminum. However, if aluminum comes in contact with certain reagents, such as sodium hydroxide, potassium hydroxide, hydrofluoric acid, hydrochloric acid, sulfuric acid, or sulfurous acid, it forms soluble compounds. Uniform surface corrosion occurs. Metal dissolves uniformly from the total contact surface and, depending on the attacking solution, the rate of attack may be linear or may increase or decrease with time.

12.2.3 Localized corrosion (pitting)

This type of corrosion is characterized by the formation of pits (Fig. 12.2). The pits are usually small and may heal themselves. The aluminum oxides that form as corrosion products close the small pits like a plug so that any water present is not able to gain access to the base of the pit. This stops the pitting attack. If the aluminum is exposed to an aggressive solution, a large number of small pits that have been closed and healed over by the corrosion products can often be seen on the surface. Under unfavorable conditions, pitting may be so severe as to perforate the sheet.

Deposits of carbon (graphite) or copper on the surface favor pitting corrosion through the formation of galvanic cells.

Corrosive media in the near neutral pH range can cause pitting corrosion on aluminum. Since aluminum oxide has very low solubility in this range, general surface attack remains negligible. The aluminum surface maintains its passive oxide protection as long as oxidizing substances in the corrosive environment prevent the reaction from reaching the pitting threshold potential.[3]

---

[2] pH value is a number from 1 to 14, which represents how acidic or alkaline a solution in water is. pH 7 is neutral. Acidity decreases from pH 1 to pH 7. Alkalinity increases from pH 7 to pH 14.

[3] The pitting threshold potential is an electrochemical potential beyond which passive behavior gives way to pitting corrosion.

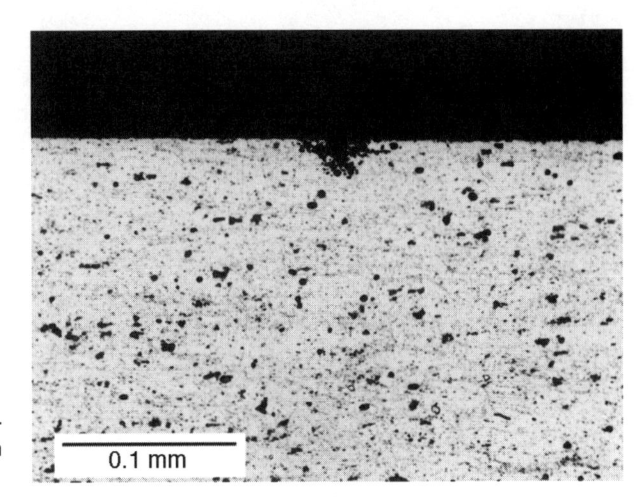

Fig. 12.2: Corrosion site on aluminum with localized corrosion or pitting.

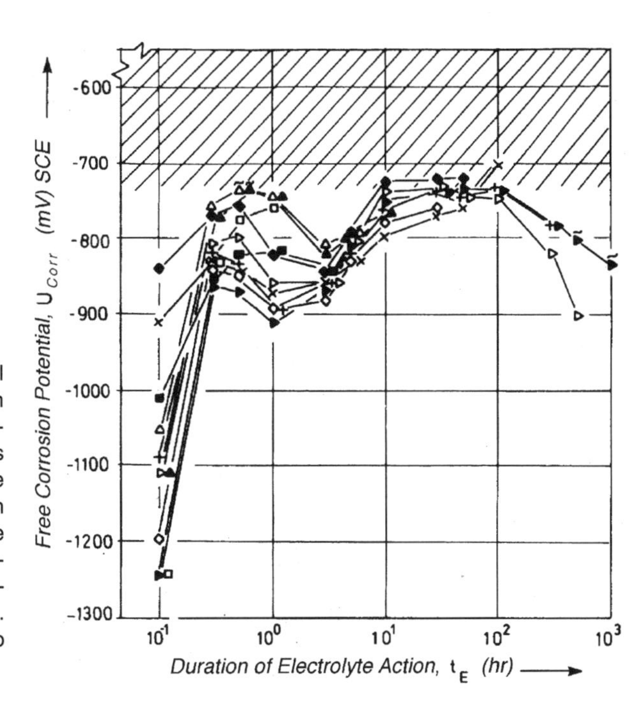

Fig. 12.3: Electrode potential change over time for ten samples of alloy 5049 with different (undefined) prehistories in aerated sea water; the hatched area is the corrosion domain, the white area is the passive domain. Voltage is plotted relative to a reference electrode of saturated calomel. Most samples were found to remain passive to corrosion.

The structure and thickness of the oxide film affect the incubation time before pitting can start. Laboratory measurements using reference electrodes quickly show whether aluminum components in a corrosive environment behave passively or suffer pitting attack. Changes in corrosion potential indicate increasing corrosion risk with more electropositive potential and decreasing risk with more negative potential. Thus, these measurements can provide a rough estimate of the alloy's probable corrosion behavior (Fig. 12.3).

12.2.4 Intergranular corrosion

Intergranular corrosion starts along the grain boundaries and can cause an extensive loss of strength. This type of corrosion can occur in wrought alloys with more than 3.5% Mg as well as Al-Cu-Mg (2xxx) or Al-Zn-Mg-Cu (7xxx) alloys if incorrectly heat treated. If Al-Cu-Mg alloys are heated to temperatures between 200°C and 400°C, small copper-containing precipitates form at the grain boundaries. Just inside the grains, the copper content is lower. A similar structure is obtained if this alloy or an Al-Zn-Mg-Cu alloy is not quenched rapidly enough after solution heat treatment. If this material is then placed in an aggressive environment, corrosion will rapidly take place along the grain boundaries. This can cause extensive disintegration of the structure (Fig. 12.4). A similar situation may be observed in materials with higher percentages of magnesium. For example, if a material containing supersaturated magnesium in the cold-worked state is heated to 100–200°C, Al_8Mg_5 (β phase) precipitates at the grain boundaries relatively rapidly. Since this intermetallic compound is less noble than the solid solution matrix, it is preferentially attacked in any corrosion reaction.

Fig. 12.5 shows a 5083-type Al-Mg alloy in three different structural states that differ extensively with regard to intergranular corrosion. As soon as electronegative precipitation forms in the grain boundaries, intergranular corrosion can occur rapidly, particularly if the surface of the material is stressed in tension. An alloy with 4% or 5% magnesium is thus susceptible to intergranular corrosion and stress corrosion if not properly heat treated. This phenomenon can be controlled by an intermediate precipitation heat treatment of the material, which reduces the level of supersaturated magnesium. An anneal at the final thickness can also be used to modify precipitation seams at the grain boundaries. After a heat treatment at 300°C, for example, one obtains the structure shown in Fig. 12.5.c, in which the β phase has coarsened into a spherical shape. This greatly reduces the susceptibility to grain boundary corrosion. The corrosive environment can only dissolve the precipitates located at the surface, which means that the grain boundary corrosion cannot penetrate deeper. Fig 12.6 illustrates the way corrosion depends on the electrode potential.

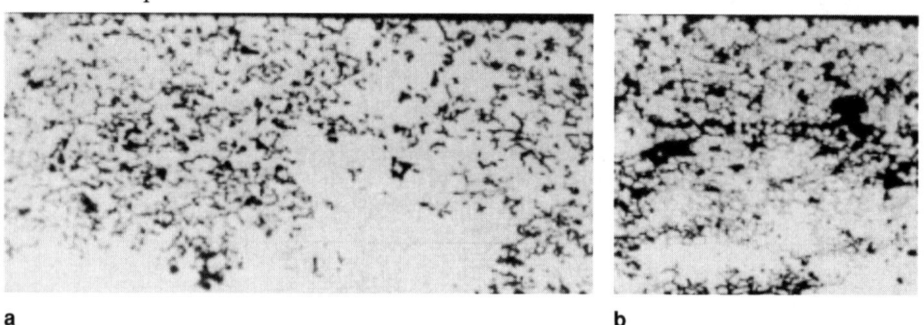

Fig. 12.4: The influence of delay in quenching on the intergranular corrosion of Al-Cu-Mg (2xxx) alloys. (a) A 20 second delay in quenching; (b) a 60 second delay in quenching. The sample was exposed for three days in a 3% NaCl solution with an addition of 1% HCl. With too long a delay before quenching, the temperature falls slowly from 500°C and allows fine copper-containing precipitates to form on the grain boundaries. This condition favors grain boundary corrosion. Magnification: a = 110×, b = 130×.

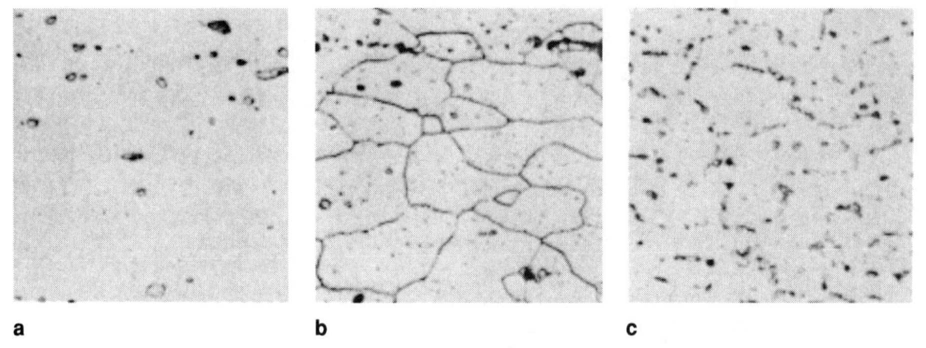

Fig. 12.5: Micrographs of 5083-type Al-Mg alloy sheet, first solution-heat-treated for six hours at 400°C, then quenched. Three different thermal treatments are represented: (a) only solution-heat-treated and quenched, (b) solution-heat-treated, quenched, aged four days at 100°C, and sensitized, (c) solution-heat-treated, quenched, aged ten days at 300°C, "string of pearls structure." Structure b is sensitive to intergranular corrosion, c is completely immune. Magnification 130×.

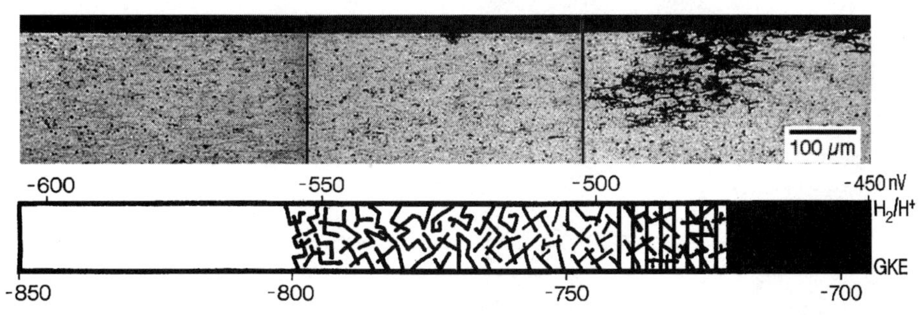

Fig. 12.6: The influence of electrochemical potential (galvanic voltage) on the type of corrosion on the critical boundary potential of aluminum alloy 5083 (AlMg4.5Mn) with sensitized structure in synthetic sea water. White potential field: passive domain; crackled potential field: intercrystalline corrosion domain, coming before pitting; cross-hatched potential field: transition domain; black potential field: domain of pitting and intercrystalline corrosion. Electrochemical potentials are given relative to the hydrogen and the saturated calomel (SCE) standard reference electrodes.

12.2.5 Exfoliation

Corrosion of this type proceeds in layers parallel to the surface, see Fig. 12.7. This is a special case of corrosion that is particularly evident in fibrous structures; however, it may also occur in recrystallized structures. It takes place when segregation from the cast structure has been elongated parallel to the surface. The resulting corrosion products cause the undermined layers near the surface to peel away.

The other, transcrystalline type of exfoliation corrosion tends to occur in Al-Zn-Mg-(Cu) alloys in the naturally aged (T4) temper. Again, it follows segregation bands from the cast structure, which hot working has stretched parallel to the surface. The tendency to exfoliation corrosion disappears after a sufficiently long precipitation heat treatment at temperatures between 120°C and 140°C. Therefore, rolled and extruded products of these alloys should only be used in service in precipitation treated tempers. After welding, a

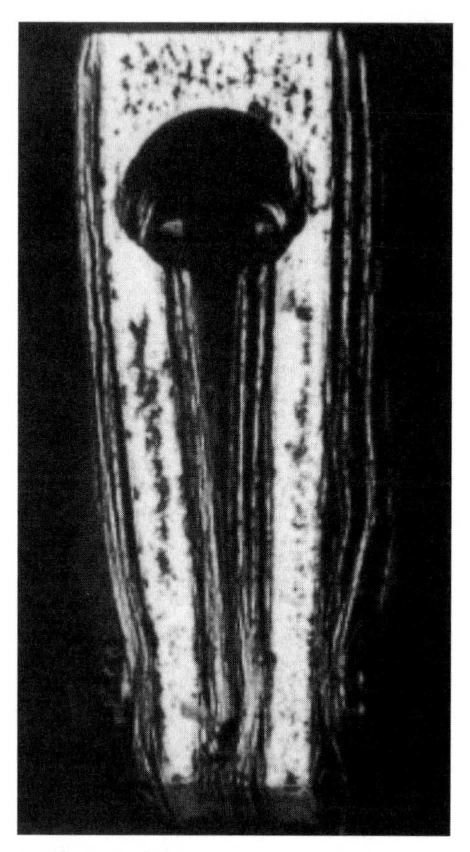

Fig. 12.7: Extreme case of exfoliation in a section through a naturally aged, forked test piece of 7020 (AlZnMg1) after corrosion by an aggressive environment. The sample had a fibrous structure, and the attack proceeded along grain segregation from the as-cast structure. About twice the actual size.

further precipitation treatment is recommended to protect the heat-affected zone adjacent to the weld from exfoliation corrosion.

12.2.6 Stress corrosion cracking

During stress corrosion, cracks can grow very rapidly in sections where the surface is under tension. Even high strength alloys of the types Al-Zn-Mg-(Cu) and Al-Li-(Cu) can show this type of corrosion if their heat treatment is unfavorable. In the Al-Zn-Mg-(Cu) alloys, one suggested mechanism is that hydrogen released by corrosion reactions diffuses under stress along the grain boundaries, where it makes the grain boundaries brittle (Fig. 12.8).

This is rarely accompanied by a noticeable loss in weight. Alloys containing more than 3.5% Mg as well as Al-Zn-Mg and Al-Zn-Mg-Cu alloys are to a certain extent sensitive to stress corrosion.

Stress corrosion in aluminum is always intergranular (Fig. 12.8). There are alloys and structural conditions that tend toward intergranular corrosion, but not to stress corrosion, such as Al-Mg-Si under unfavorable corrosion conditions. In view of this, it is necessary to discuss these two corrosion phenomena separately.

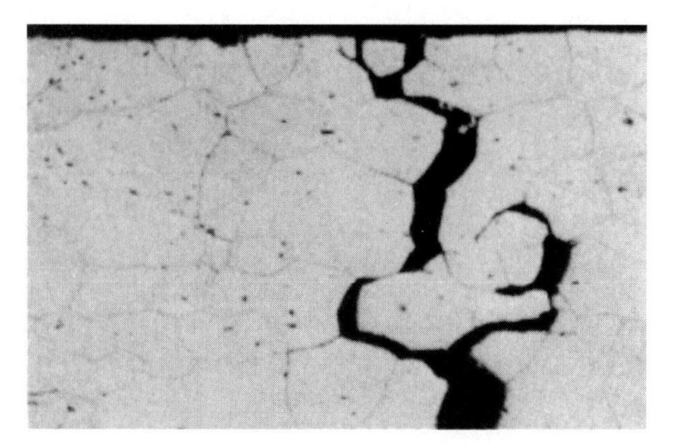

Fig. 12.8: Intergranular stress corrosion crack on the surface of an extrusion of 7020 (AlZnMg1) alloy. The structure is recrystallized, favoring development of stress-corrosion cracking.

Standard test methods serve to estimate resistance to stress corrosion. For instance, ASTM G44, 47 (ISO 11845, ISO 11539) describe various shapes of test pieces and various methods of applying stress. Stress corrosion is always due to the combined action of four factors:

1. Alloy composition.
2. Presence and type of corrosive environment.
3. Microstructure.
4. External or internal tensile stresses.

Factors 3 and 4 interact; a structure with planar longitudinally stretched grains will show the most harmful cracking if it is stressed in the short-transverse direction, that is, with stress normal to the plane of the product or through the thickness.

Because certain alloys such as the Al-Zn-Mg (7xxx) series have an innate tendency to stress corrosion due to their composition and since sufficiently aggressive corrosion media are usually available, factors 3 and 4 are of special importance when dealing with these alloys. Specific thermal treatments like T7XX tempers have been developed to enhance the stress corrosion resistance of susceptible alloys such as the Al-Zn-Mg-Cu (7000 series) alloys.

Holding the alloys for a longer time at the precipitation treatment temperature leads to over-aging, which is desirable for some applications. Overaging generally improves corrosion resistance, since precipitation of alloying atoms reduces their free energy, and fewer remain in super-saturated solid solution. A progression of tempers and properties to improve resistance to various types of corrosion, including stress-corrosion cracking, has been developed in the USA and Europe and are defined in the European Standard EN 515, as Table 12.1 shows. One can see that with increasing over-aging, corrosion resistance improves at the cost of some loss in strength. In the range of 100–150°C, most wrought alloys reach their load-bearing limit and begin to creep under load due to softening through recovery or over-aging.

Table 12.1: The effect of increasing overaging on heat-treatable aluminum alloys

(extract from European Standard EN 515 - 1992)

Property/Temper	T79	T76	T74	T73
Tensile strength				
Stress-corrosion resistance				
Fracture toughness				
Exfoliation corrosion resistance				

Explanation of the tempers:
T7X: Solution heat-treated and artificially overaged ("stabilized")
T79: Very slight overaging
T76: Limited overaging in order to achieve maximum tensile strength with good exfoliation corrosion resistance
T74: Longer overaging in order to achieve a good balance between tensile strength and corrosion resistance
T73: Full overaging to guarantee maximum stress-corrosion resistance

There are several measures that will inhibit stress corrosion, the most effective of which is the correct heat treatment, as noted above. Refinement of the alloy composition, structural condition, and shape of the grains also play important roles. A fibrous structure, for example, is generally more resistant to stress corrosion than a recrystallized structure. In the latter case, there will always be a certain number of grain boundaries that run perpendicular to the surface, favoring the propagation of a crack generated by stress corrosion. Additions of chromium, zirconium, and manganese increase the temperature of recrystallization during hot working and can be effective in retaining a fibrous structure in extrusions and hot-rolled sheets.

Also, it is important that no severe internal stresses are introduced after age hardening during further fabrication of this material or, if so, that appropriate stress-relievable practices are utilized. With a given alloy and heat treatment, the degree of the elastic tensile stress near the surface is a factor in controlling the life-span of a material.

12.2.7 Galvanic corrosion

If two dissimilar metals are immersed in an electrolyte, a potential difference between the two metals results. This corresponds to the potential difference between the two metals in the electrochemical series. If an electric current flows and thereby concentrates electrochemical attack on the less noble (relatively electronegative) of the metals, this is called galvanic, or bimetallic, corrosion. Conversely, the more noble (relatively electropositive) metal benefits from much reduced corrosion. This can be used to provide cathodic protection.

The classic situation for galvanic corrosion occurs if aluminum is joined in direct metallic contact with an electrochemically nobler metal, for example copper or iron, in an oxidizing corrosive environment. Direct metallic contact may occur in riveted or bolted joints. In a corrosive environment, and even in simply moist conditions, the less noble metal is electrochemically attacked, and a measurable electric current flows.

Table 12.2: Electrochemical potential of various metals and alloys relative to pure aluminum

(nominal values in millivolts [mV] against Al 99.5% in a 2% NaCl solution saturated with air)

Metal or alloy	mV	Metal or alloy	mV
Gold	+1000	Cast AlSi12	+30 to +60
Stainless steel 18/8	+850	Cadmium	0 to +20
Mercury	+750	Al-Mn (3003)	+10 to +20
Silver	+700 to +800	Al-Mg-Si 1 (6082)	0 to 10
Copper	+550	Aluminum 99.5%	0
Brass	+500	Al-Mg-Mn (3004)	–10 to 0
Nickel (7075-T6)	+480	Al-Zn-Mg-Cu	–20 to –10
Tin	+300	Al-Mg (5754)	–30 to –20
Lead	+150 to +180	Al-Zn (7072)	–150
Al-Cu-Mg (2024-T4)	+100	Zinc	–300
Iron	+100	Magnesium	–850

Aluminum ions become positively charged when they go into solution and surrender three electrons into the metal surface. The potential of a metal in a solution of its own ions can be precisely measured in volts. These values determine the metal's position on the scale of "electrochemical potential," as in Table 12.2. More noble metals have a higher positive potential, whereas less noble metals have a lower positive potential, the lowest even registering negative values. The lower a metal stands in the series, the greater its tendency to pass into solution (i.e., to corrode). The standard potential is defined as the potential of the metal in a molar solution of its ions, with reference to the standard hydrogen electrode.

Aluminum's oxide film has a major effect in hindering the reactions and electric currents that occur during galvanic corrosion. It is, therefore, not nearly so strongly attacked as its low position in the electrochemical series would suggest.

12.2.7.1 Deposition of a nobler metal on an aluminum surface

This is a special case of galvanic corrosion. If an electrolyte containing salts of nobler metals such as copper, tin, or mercury in solution comes into contact with aluminum, the nobler metals precipitate out in small areas on the surface of the aluminum, and localized corrosion can begin. The precipitation of copper from copper-containing tap water is an important practical example of this phenomenon.

Mercury and its salts in aqueous solution are particularly dangerous. Not only does it form galvanic cells, but in addition, mercury destroys the protective oxide layer and inhibits the formation of a new oxide layer. Even relatively small traces of mercury can quickly lead to massive pits and subsequent embrittlement.

12.2.7.2 Conditions for galvanic corrosion

As can be seen from Fig. 12.9, the attack takes place preferentially at the points where the three components (i.e., the two metals and the electrolyte) come together. The electrical corrosion current is most intense at this point. Attack can continue until the aluminum is completely consumed in the vicinity of the nobler metal. Although the physical arrangement is different from that of a classical galvanic cell, the same processes take place. In this case, one can speak of a contact element. Under these circumstances, the two metals of differing potential are in direct contact with each other in the corrosive environment and, therefore, meet the essential requirements for a galvanic circuit:

- Current-conducting connection between two metals of differing electrochemical potential.
- Immersion of the two metals in the same aggressive, current-conducting electrolyte.

If the neutral electrolyte contains no oxidizing agent, such as dissolved atmospheric oxygen, then contact corrosion is impossible, even if the two conditions above are fulfilled.

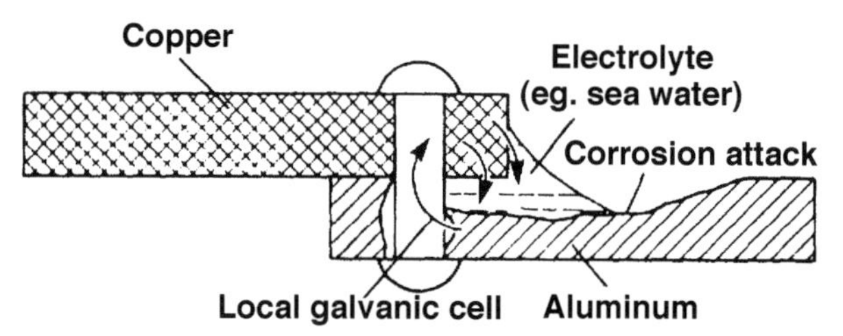

Fig. 12.9: Galvanic corrosion in an Al-Cu cell improperly assembled (riveted with direct metallic contact).

12.2.7.3 Measures against galvanic corrosion

The electrochemical series shows clearly which elements are nobler than aluminum (meaning that they can cause galvanic corrosion of aluminum when in direct contact). This is notably the case of copper, brass, tin, lead, nickel, and iron. If aluminum is in intimate contact with these metals (for example, by screw fasteners or riveting), there is a definite risk of galvanic corrosion in the presence of moisture, which can act as an electrolyte. In spite of this, it is possible to prevent corrosion resulting when two metals of different potentials are joined, for example aluminum and iron. This requires eliminating at least one of the two basic requirements for galvanic corrosion mentioned above:

- *Prevent electrical contact* between the two metals by inserting an insulating layer such as a thick paint film between them and around the bolts used to make the joint. This prevents electrons from flowing from the less noble to the more noble metal and reduces the galvanic current to a negligible level. This prevents galvanic attack of the less noble metal.
- *Prevent the intrusion of moisture,* for example, by painting. As mentioned above, it is almost impossible to prevent the occurrence of thin layers of moisture, so that in practice any joint between aluminum and a nobler metal should also be electrically insulated.

Cathodic protection offers another solution for immersed structures in cases where the above measures are insufficiently effective or are not technically or economically feasible. Cathodic protection involves deliberately creating a galvanic cell by connecting a sacrificial anode made of less noble magnesium, zinc, or a specially designed aluminum alloy. Galvanic corrosion attacks these sacrificial anodes, releasing electrons that raise the galvanic potential of the rest of the structure. Thus, the otherwise less noble aluminum components recover a galvanic potential, which is again in the passive domain. Cathodic protection has been successfully used to protect the aluminum used in the hulls of ships, which are in direct metallic contact with copper alloys and stainless steel.

12.2.8 Crevice corrosion

Crevice corrosion can occur in narrow gaps between different metals (and sometimes between sheets of the same metal) where there is usually an oxygen gradient between the outside and the inside region of the gap. The oxygen concentration is highest at the outside edge, where the oxygen supply is renewed, and lowest at the inner end, where oxygen has been consumed by corrosion reactions. Due to this oxygen gradient, polarization occurs, causing an attack that is very similar to galvanic corrosion, but usually less severe. Crevice corrosion usually involves contact with the air. The most effective prevention method is to seal the crevice. Crevice corrosion is a danger when transporting or storing coils of rolled product. If cold coils are allowed to contact relatively warm air even for a short time, water can condense on the flat face of the coil and then be drawn in between the wraps by capillary action. For this reason, coils of sheet or foil are often hermetically sealed in plastic for storage or shipping.

12.3 Corrosion Testing

The purpose of corrosion tests is:

- To enable confident prediction of a material's service behavior.
- To determine the material's properties such as resistance to intercrystalline and stress corrosion.
- To make comparisons of corrosion resistance.
- To clarify the origins of corrosion reactions.
- Frequently, to accelerate corrosion so as to obtain results in a fraction of a normal service lifetime.

A variety of methods are commonly used to examine and test a material's corrosion behavior. These methods differ essentially in the corrosive environment they employ. Corrosive media that dissolve the oxide skin, like strong acids and alkalis, allow only very limited conclusions. In any case, the degree of correspondence between test results and evidence from use in service has to be verified. More suitable corrosion test media tend to be solutions with near neutral pH (between 4.5 and 8.5). Such solutions hardly dissolve aluminum oxide so that pitting and other localized forms of corrosion can develop.

In these less aggressive media, the effect of the oxide film dominates. What matters initially is whether pores in the oxide film heal themselves or not (see Fig. 12.13). If not, then the metal can often be protected by artificially strengthening the oxide layer.

The different categories of corrosion testing are:

- Corrosion tests under near-service conditions.
- Corrosion tests under intensified service conditions.
- Accelerated corrosion tests.
- Laboratory tests.
- Service tests.
- Natural exposure tests.

Standard ASTM methods are available for testing aluminum alloys under various conditions in ASTM Standards Volume 03.02.

12.4 Key Elements of Corrosion Resistance

12.4.1 Corrosion resistance of various alloys

Knowledge of bimetallic corrosion provides the key to understanding the differing corrosion resistance of individual aluminum alloys. Foreign atoms, especially in the form of second phase particles, embedded in aluminum can generate local galvanic cells that accelerate the corrosion. Super purity aluminum is not attacked by dilute acid to any appreciable extent, commercial purity aluminum is to a significantly larger extent, and copper-containing alloys are severely attacked. Super purity aluminum (Al 99.99%) has a rather homogeneous structure (i.e., it is generally free of embedded second phase particles). Unless a foreign sliver is pressed into the surface (e.g., by rolling), super purity aluminum represents a metal that is largely free of embedded local cells. As a result, it can be immersed in very aggressive electrolytes for long periods. This is because the electrons that are liberated as the aluminum goes into solution cannot easily find sites to reduce hydrogen ions to gaseous hydrogen. On the other hand, commercial purity aluminum corrodes noticeably faster in an aggressive liquid. The second phases containing iron and silicon are nobler than the aluminum matrix and, thus, set up small local cells (Fig. 12.10).

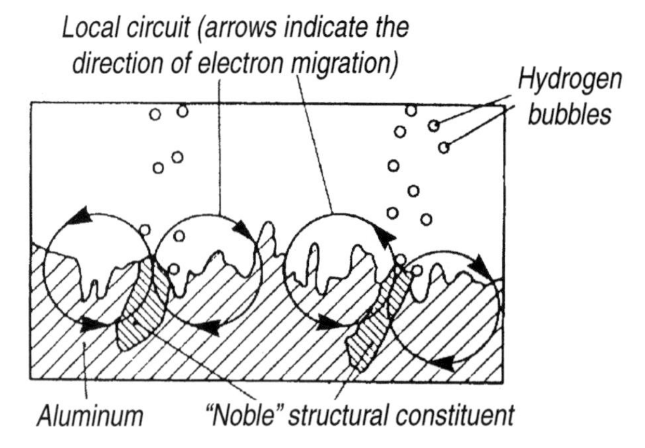

Fig. 12.10: Minute local galvanic cell caused by the presence of noble secondary phases (heterogeneities).

Fig. 12.11: Deposit or sliver of a nobler metal forming a local galvanic cell.

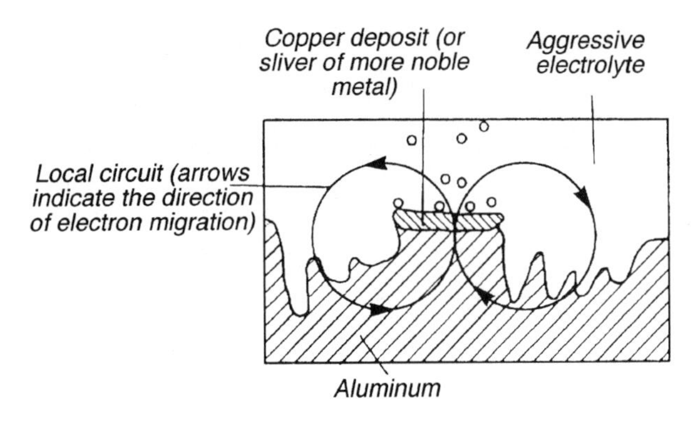

Commercial purity aluminum of at least 99.5% Al with an iron content of less than 0.40% can be regarded as corrosion-resistant for most purposes, thanks to the protection the oxide layer provides. It should be kept in mind that the schematic example described in Figs. 12.10 and 12.11 and the observation that super purity aluminum resists corrosion better than commercial purity both relate to their reactions with a relatively aggressive corrosive environment. Nevertheless, in commercial purity aluminum as well as in most copper-free wrought alloys, corrosion resistance decreases significantly with increasing iron level (e.g., Fe from 0.2% to 0.7%), even in quite weakly aggressive corrosion media. The same applies to the silicon content, especially in Al-Mg alloys, and to some extent in commercial purity metal.

The presence of copper accelerates the corrosion of aluminum. Fig. 12.11 shows how such local cells work in copper-containing alloys or through the introduction of slivers of more noble metals. The copper in the copper-containing alloy works in an indirect way. First, the copper goes into solution, then reprecipitates to complete the conditions for a local galvanic cell. On the other hand, a very small amount of copper can have a beneficial effect, as long as it is homogeneously dispersed in the crystal structure. The attack then results in a large number of uniformly distributed tiny pits, which is far less harmful than fewer and bigger pits.

The corrosion-resistant aluminum alloys contain magnesium and manganese as their chief alloying elements. These alloys and alloys with silicon all have about the same electrochemical potential in liquids, including sea water, so that contact between them does not aggravate the risk of corrosion. If zinc-free and zinc-bearing alloys are in contact in corrosive conditions, however, the zinc-bearing alloy, having a more negative potential, will be preferentially attacked.

12.4.2 The influence of heat treatment on corrosion

This has been the subject of intensive studies over many decades, resulting in procedures to avoid allowing the temperature of sensitive alloys to reach the range where their corrosion resistance is impaired. This range may reach as low as 70°C for alloys with a high magnesium content. The subject has already been dealt with in Chapter 11 and earlier in this chapter. It offers many opportunities for applying the A-B-C sequence of measures taken, the corresponding structural changes, and the resulting chemical and physical properties.

12.4.3 The natural oxide skin

Aluminum is permanently covered with a thin oxide layer. This layer is transparent and maintains aluminum's metallic luster, in contrast to iron's reddish oxide. The oxide layer on aluminum might best be compared to a thin glaze that conducts electricity only slightly or not at all, thereby largely inhibiting galvanic corrosion.

Fig. 12.12 shows that the natural oxide layer of aluminum consists of an extremely thin and compact base layer called the barrier layer and a hydrated upper layer whose thickness ranges from 0.005 µm to 0.01 µm. Such a thin skin has a number of pores and weak points where the metallic aluminum is able to go into solution. However, if aluminum goes into solution at such a pore, it usually precipitates as corrosion products that fill up the pore, so bringing the attack to a stop (Fig. 12.13, top).

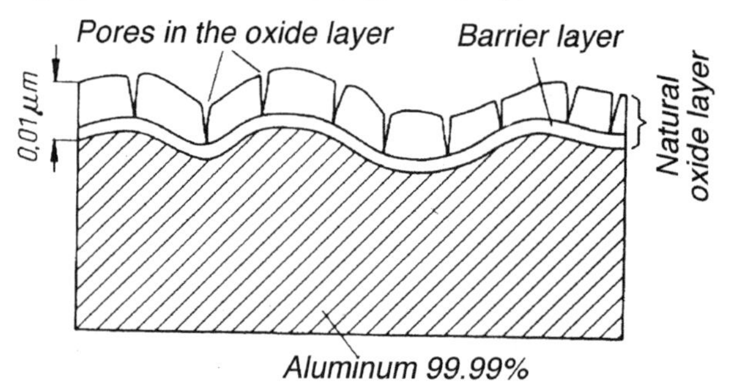

Fig. 12.12: Natural oxide layer on superpurity aluminum. The layer covers the aluminum like a glaze.

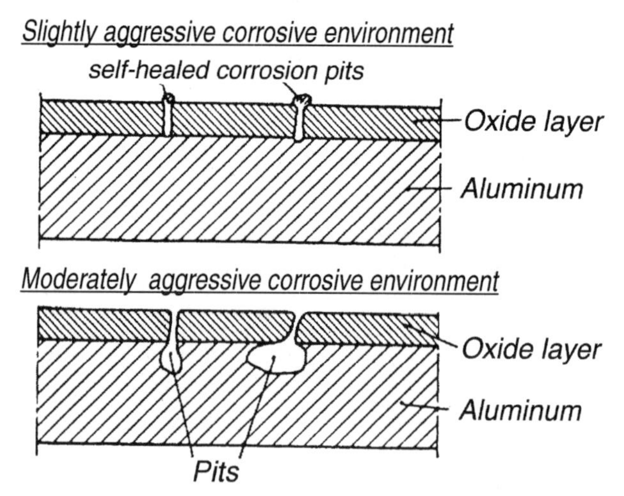

Fig. 12.13: Behavior of aluminum in weaker and stronger corrosive media. The pores and pits in the upper part of the figure are to a large extent plugged with hydrated oxides, the corrosion products.

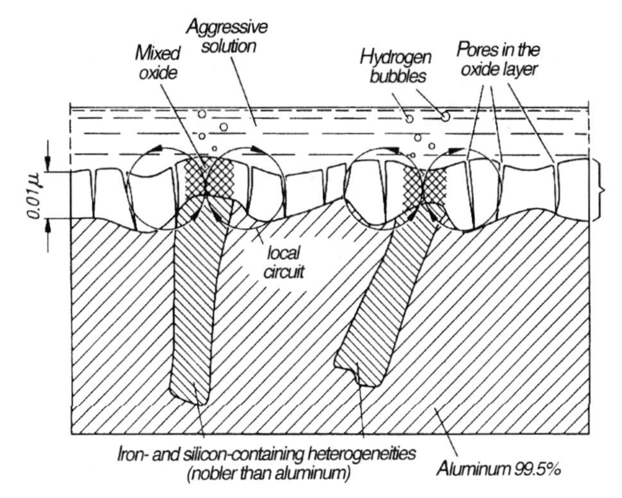

Fig. 12.14: Corrosion of 99.5% aluminum covered with a natural oxide layer in an aggressive solution. The figure illustrates that the precipitates in the aluminum are considerably bigger than the thickness of the natural oxide layer. Since the natural oxide layer grows into the metal, an anomalous composition results in places where a precipitate particle lies in the structure. If the surface comes in contact with an aggressive environment, small local cells form, despite the presence of the oxide layer. (The barrier layer formed at the base of the natural oxide skin is not shown.)

If the aluminum contains alloying elements in solid solution, they also appear in the oxide layer. In Al-Mg alloys, the concentration of magnesium in the oxide layer is higher than in the metal itself, especially after annealing. Hence, the surface of Al-Mg semi-fabricated products can take on a gray cast after a full anneal, which is why such alloys are often annealed in an inert gas atmosphere rather than in air. Relative to the original surface, the oxide layer on aluminum grows partly outward, because it occupies more volume than the metal from which it formed, and partly inward by consuming the metal substrate.

Since the natural oxide layer grows into the metal, it exhibits a different chemical composition at points containing inclusions in the aluminum structure (Fig. 12.14). The aluminum oxide itself is a good insulator, but an aluminum oxide containing iron, for example, must be considered a semiconductor that allows the electrons to pass to a certain degree, making galvanic corrosion possible. This still holds true even if the oxide layer is artificially thickened (e.g., by anodizing). Thus, the oxide layer is not a totally impermeable barrier, concealing every aspect of the underlying aluminum; the structural constituents of the aluminum will influence the composition of the oxide layer as well as those of the parent metal.

12.5 Corrosion Protection[4]

The thicker the oxide layer is and the fewer pores or semiconducting regions it contains, the greater is the protection it offers. A number of different processes can increase the

[4] By A. Blecher.

thickness of the oxide layer, resulting in different structures or chemical compositions of the layer. The thicker the layer, the more it resists the passage of electrons, and, thus, the more it hinders dissolution of the underlying metal. In this context, the oxide layer should itself be considered a product of corrosion. By far the most frequent corrosion product is hydrated aluminum oxide. This may form as a tightly adhering protective layer, as a porous layer, or it may go into solution in the corrosive environment, depending on the conditions. An oxide layer that has been thickened for protection is known as an "artificially thickened" oxide layer, and this technique has been successfully used for years to protect aluminum against corrosion. There are two oxide-thickening basic techniques: heat and / or chemical means and electrolytic oxidation (anodizing). Whichever technique is used, it is essential to prepare the surface properly in order to ensure a uniform thickening of the layer.

12.5.1 Obtaining a defined surface condition in preparation for thickening the oxide layer

Before a thickened oxide layer can be produced, the surface must be thoroughly cleaned. Often, the natural oxide layer is contaminated from previous processing, for example by rolling or drawing lubricants or by oil films used to protect the surfaces during transport. The pre-treatment processes used in practice range from simple wiping or mechanical roughening, through dipping and vapor degreasing with organic solvents, to degreasing and etching with subsequent conversion coating. With wiping, and even dipping, the contaminating oils or greases are usually simply spread out evenly as a thin film over the entire surface. Vapor degreasing is better, as the freshly condensed distillate removes the grease without disturbing the oxide layer.

Organic solvents are being replaced to an ever-increasing extent by neutral water-based cleaners, which also degrease without affecting the oxide layer. Chemical pre-treatment with a medium that has some etching action proves far more effective. For this purpose, acidic water-based media or, more especially, alkaline ones based on polyphosphates or borates have been used. With this kind of degreasing, the slight etching removes the oxide layer and with it all contamination that had been chemically combined in or on the oxide film. When more aggressive alkaline etching is used, the deposits ("smut") remaining on the surface afterwards should be removed by washing with acid. Etching residues on alloys with a high silicon content can only be thoroughly removed by washing with added hydrofluoric acid or fluorides. The resulting degreased and precleaned surface is then ideal for subsequent thickening of the oxide layer by chemical or by electrochemical means.

12.5.2 Reinforcing the natural oxide film by chemical reaction

The reaction between boiling water (or steam) and aluminum can increase the thickness of the natural oxide layer ten to a hundredfold. The resulting hydrated oxide layer ("boehmite") offers a certain amount of protection against corrosion and provides a good base for paint as well as preventing "tap water blackening." Depending on the required thickness of the oxide layer, the duration of the treatment can extend to several hours, which accounts for the diminishing economic importance of "boehmiting." To ensure a corrosion-resistant oxide layer, it is absolutely essential to use distilled water or at least water that is entirely free of chlorides and alkalis.

If the water contains chemical substances such as alkalis, acids, or added fluorides, chromates, and/or phosphates, the oxide layer is at first dissolved. The resulting pH change in the boundary layer leads to the reduction of the chromates and the formation of a hydrated oxide layer composed of aluminum oxides and chromium oxide or chromium phosphate. Chemically produced protective layers of this kind are so tightly bound to the aluminum surface that they will not flake off even under bending or impact stress. Oxide layers produced from alkaline solutions, which contain chromium compounds, are amorphous, porous, and hence, absorbent, so that they can be colored. Commercial processes of this sort include MBV (Modified Bauer-Vogel process), EW (Erftwerk process), and the Alrok and Pylumin processes. The alkaline processes are now only rarely used and are of little economic importance. Oxide layers produced in the acidic pH range are clearly superior to those obtained in the alkaline range in respect of uniformity and durability of the film and cost less to make. Today, the main uses for these chemically thickened oxide layers are for inhibiting corrosion and improving the adhesion of organic coatings on aluminum. The production and properties of these films will be considered in more detail in the section on organic coatings.

12.5.3 Reinforcing the natural oxide film by anodic oxidation ("Anodizing")

Anodizing is an electrolytic process for producing artificial oxide layers 200 to 2500 times as thick as the natural oxide layer. The anodic film thickness ranges from 2 μm (e.g., for anchoring organic coatings) to 25 μm (e.g., for maximum corrosion protection). It provides effective surface protection and can often have a decorative appearance. Transparent anodic films can preserve the decorative appearance of a mechanical or chemical pretreatment of the surface. The mechanical treatments cover, according to the General Specification of ISO 7599, grinding, brushing, and polishing. For rolled products, there are also the various surface finishes obtained by using polished or patterned workrolls. Among the chemical pretreatments, matte etching is of particular importance for use in buildings, especially for facades. Partly through a slight leveling of the surface and partly by causing random reflection of incident light, matte etching makes even a mechanically untreated "mill finish" surface appear smooth and uniform. Chemical and electrolytic brightening before anodizing require special "bright alloys" based on higher purity aluminum and tend to be used for inside architectural applications as well as for automotive bright trim and lamp housings of all kinds.

All anodizing processes have one thing in common: the aluminum workpiece, with a well-defined surface state, is immersed in a conductive liquid and is connected to the positive pole of a source of direct current. The electric field attracts oxygen-bearing anions toward the metal surface, where they react to form aluminum oxides. In this way a very thin, non-porous, electrically insulating base or blocking film is formed at first. Since the electrical resistance of the film increases rapidly with increasing thickness, it might be expected that with a fixed anodizing voltage the current would decrease correspondingly quickly, halting the growth in the surface film. Indeed, this happens when using electrolytes that do not redissolve the oxide layer, for example dilute boric acid.

However, the most common electrolyte for anodizing, dilute sulfuric acid, tends to partly redissolve the film, forming a porous covering layer. After a short time, an equilibrium is established between the newly forming base layer and its transformation into the porous covering layer. Hence, the covering layer grows ever thicker while the thickness of the base layer remains unchanged. If, as is usual, anodizing takes place at constant current density, the anodizing voltage increases only slightly as the film thickness increases. The rate of increases in the thickness of the oxide layer is proportional to the current density. Since the oxide layer has a greater volume than the metal layer from which it is formed, it appears to grow to the extent of one-third of its thickness above the original level of the metal surface and to two-thirds below it. At the cathode, which is usually of lead or of aluminum, hydrogen is formed and escapes as gas.

Uniform, perfectly color-fast oxide films can be produced in this way, which may be transparent, gray, light to dark bronze, or black, depending on the alloying elements, the microstructure, the composition of the electrolyte, and the process parameters. Freshly formed anodic films are extremely absorbent due to their finely porous structure and their large active area, and at this stage they are easily dyed. They are not scratch-resistant in this state and are prone to staining and weathering. It is, therefore, essential to seal any anodic film by some form of finishing treatment. After rinsing, the film can be sealed in simmering distilled water (> 96°C) or in saturated steam (> 98°C) to transform part of the oxide layer into hydrated aluminum oxide. The increase in volume resulting from this reaction seals the pores in the oxide layer. Such sealed anodic oxide films about 20–25 μm thick provide excellent corrosion resistance. For example, they enable aluminum to withstand aggressive industrial atmospheres in which the natural oxide layer is no longer able to maintain aluminum's original metallic sheen. Alternatively, the pores can be sealed by depositing nickel ions in them, the so-called cold impregnation process, which is made permanent by aging in warm water (> 60°C). It is also worth mentioning that the pores can be sealed with a non-yellowing clear lacquer. Once the oxide layer has been sealed, it is no longer possible to impregnate it with any subsequent coloring, and it also loses its ability to anchor organic coatings.

As for other properties of anodic films, it is worth noting that anodized aluminum can withstand a pH range of 3.5 to 8.5. It is also many times harder and more abrasion resistant than the base metal. Depending on the anodizing process used, the film reaches Vickers hardness values of 200 to 500.

12.5.4 Influence of alloy composition and microstructure on the appearance of the anodic film

The structure and alloy composition of aluminum strongly influences the appearance of the anodic film. This is not surprising, since the oxide layer forms from the parent metal (Fig. 12.15). The anodic film on high-purity aluminum containing more than 99.8% aluminum is generally clear, so that light is mainly reflected at the boundary between the metal and the oxide layer. This still holds true if magnesium has been added to increase strength. As the level of this and other alloying elements increases, the anodized layer

Fig. 12.15: The difference in relative position between an applied coating and an oxide film formed from the aluminum itself (anodic film, natural oxide, etc.).

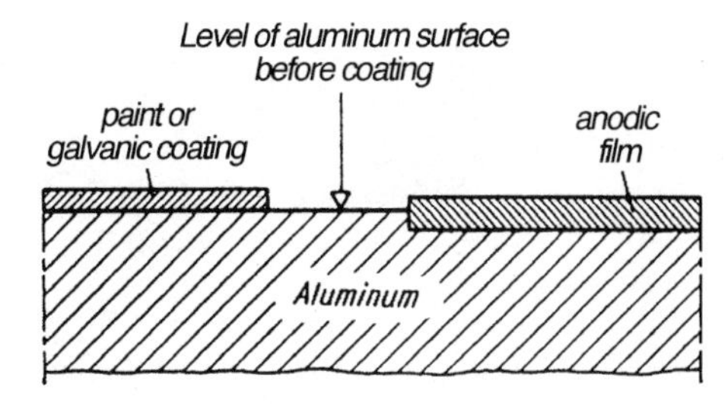

becomes milky, often darker in color, and rather less resistant to corrosion. With aluminum contents of 99.3–99.5% and in the homogeneous alloys based on this purity, the layer is only slightly discolored. Thus, almost transparent anodic films providing good protection against corrosion are commercially available on Al-Mg-Si and Al-Mg alloys.

12.5.5 Processes for producing colored anodic oxide films

The most important processes for producing transparent oxide layers are the DC sulfuric acid process and the DC sulfuric acid + oxalic acid process. Both are economic because the electrolyte is cheap and they do not use much energy. The surface hardness can be increased within 250–350 Vickers by reducing the electrolyte concentration and temperature. Alloys with 3–5% silicon have a gray oxide layer. For producing colored oxide layers there are essentially three types of process available:

- Integral color
- Electrolytic coloring
- Pigment impregnation

In the integral color process, light to dark bronze and light gray to black tones can be obtained by appropriately selecting the alloy and by using special organic acids as the electrolyte. Table 12.3 summarizes the most common integral color processes. These sunlight- and weather-resistant oxide films are also important for their exceptional hardness (350–700 Vickers) and abrasion resistance. These electrolytes have less capacity to redissolve the base oxide layer than conventional sulfuric acid, and consequently the covering layer grows thicker and less porous. Because the electrical resistance of the oxide layer increases with decreasing pore density, these organic acid processes need a significantly higher voltage compared with that typical of the usual sulfuric acid anodizing. The source of the coloring is the inclusion in the oxide layer of very fine, incompletely oxidized aluminum particles, which scatter the incident light. The depth of the color depends on

Table 12.3: Some examples of integral color anodizing processes

Process name	Company	Main component of the electrolyte
Colodur	Fried. Blasberg GmbH	Maleic acid
Duranodic	Alcoa	Sulphophtalic acid
Kalcolor	Kaiser Aluminum & Chemical	Sulfosalicylic acid
Permalux	Alusuisse	Maleic acid
Veroxal	VAW	Maleic acid

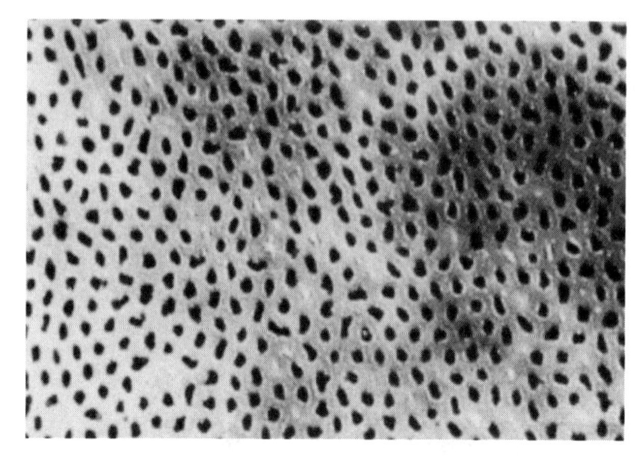

Fig. 12.16: A sample electrolytically colored, with the pores filled with electrolytically precipitated nickel. Transmission electron micrograph, ×130,000.

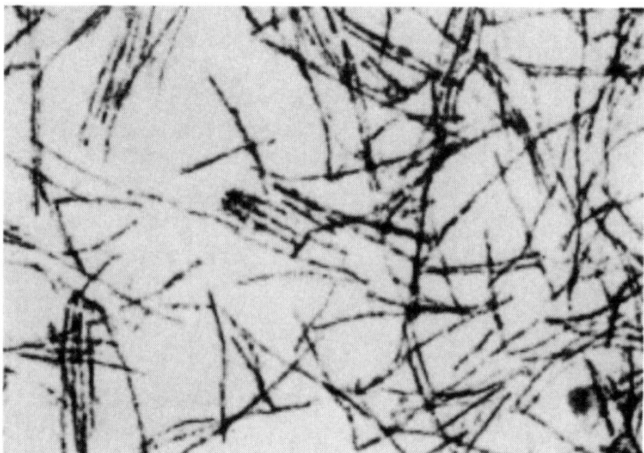

Fig. 12.17: Metal precipitated into the pores during electrolytic coloring remains behind in the form of fibers on the surface after the oxide layer has been dissolved. Transmission electron micrograph, ×32,600.

the thickness of the layer, which in turn depends on the current density. The film retains its color even at temperatures up to and above 400°C.

Ensuring uniform and reproducible integral colors requires great care in manufacturing the semi-fabricated product, including the thermal treatments and finishing procedures. For this reason, the coloring processes currently preferred are those that first create a transparent film and only then impregnate it with a color. Such processes are also cheaper.

Electrolytic coloring produces color-fast and weather-resistant anodic films in colors ranging from light bronze through medium and dark bronze to black. The procedure consists of coloring the oxide layer produced by the sulfuric acid process in a second stage of electrolysis, in which an AC current passing through an electrolyte containing salts of tin, nickel, or cobalt causes fine metallic particles to precipitate irreversibly in the pores of the oxide layer. Fig. 12.16 shows an oxide layer whose pores have been filled in this way. In the laboratory, it is possible to dissolve the film and to examine the metallic fibers that had filled its pores (Fig. 12.17). A section analyzed under the electron microprobe shows the composition of the different layers. The color is to a large extent independent of the alloy, being determined by the electrical conditions during coloring and the amount of

Table 12.4: Some industrial electrolytic coloring processes

Process	Developer	Main coloring metal employed
Almecolor	Henkel (Germany)	Sn
Anolok	Alcan (Canada)	Ni, Co
Carmiol	I.S.M.L. (Italy)	Ni, Co
Colinal 2000	Alusuisse (Switzerland)	Ni+Co or Sn
Colorox	Josef Gartner (Germany)	Sn
Elektrocolor	Langbein-Pfanhauser Werke AG (Germany)	Ni or Sn, Ni
Endacolor	Endasa (Spain)	Sn-Ni
Eurocolor 800 and 1000	Pechiney (France)	Sn+Ni or Ni
Korundalor	Korundalwerk (Germany)	Sn
Metoxal	VAW (Germany)	Sn
Metacolor	Metachemie (Germany)	Sn
Oxicolor	Riedel & Co. (Germany)	Sn-Ni or Sn
Rocolor	Rodriguez (Spain)	Sn or Ni
Sallox	Italecno (Italy)	Sn
Summaldic	Sumitomo (Japan)	Ni (d.c.)
Trucolor	Reynolds (U.S.A.)	Sn
Unicol	N.K.K. (Japan)	Ni

metallic particles precipitated. After coloring, the oxide layer is sealed in the usual way. The hardness of the finished film corresponds to that of the film resulting from the first stage. Color-fastness at high temperature compares well with that of the integral color process. The many commercial variants of this process, of which a few are mentioned in Table 12.4, are much used in exterior architectural applications.

In pigment impregnation, colorless oxide films, formed by conventional anodizing in sulfuric acid, are dyed by dipping in solutions carrying inorganic or organic pigments, and then sealed to fix the pigment. Only a limited number of pigments are light- and weather-resistant. The range of light- and weather-resistant colors can be extended by a combination of electrolytic coloring and pigment impregnation. Two different coloring media, precipitated metal and organic dyestuff, overlie each other in the pores of the oxide layer, with the two base colors combining to give a different color. The hardness of such films corresponds to that of the anodic film.

In all the anodizing processes described so far, whether in natural or in artificially colored films, the aluminum surface retains its metallic luster. The range of colors allows full scope for architectural use in combination with traditional construction materials such as stone, glass, and wood. Anodizing has proved itself for many years as an excellent means of protecting aluminum components against atmospheric corrosion. Furthermore, regular cleaning guarantees that the decorative appearance stays unchanged over a long lifetime. Being integrally bound into the underlying metal from which it formed, the anodic film will not peel or flake off. There have never been any problems with inadequate covering or protection of the edges. Surface finish is preserved by a layer of exceptional hardness and abrasion resistance. Aluminum's metallic character remains unaffected by the light-resistant and weather-resistant coloring process, and its loss of surface luster is negligible.

For completeness, we may here mention functional oxide layers (as opposed to decorative ones). For construction applications the standard processes described above are sufficient. Where especially good abrasion resistance is required, hard anodizing is used, in which special process conditions with suitable electrolytes yield oxide layers with hardnesses up to 700 Vickers. The electrolyte is often sulfuric acid at lower temperature (< 5°C) and a lower concentration than usual, which means a very low rate of redissolution of the oxide layer, higher voltage, and a much longer anodizing cycle. All these lead to the formation of dark-colored, very thick anodic films (50–170 μm), which are very hard and abrasion-resistant. Usually, such films cannot be conventionally sealed because this would reduce abrasion resistance. To further improve the abrasion resistance by incorporating permanent lubricity into the surface, polymers such as PTFE are applied to hard anodized oxide layers and sealed in by heating.

12.5.6 Inorganic protective coatings

Enamel is an inorganic glass-like coating mainly composed of oxides. Enamels with relatively low melting points, requiring furnace temperatures of approximately 500°C, have been developed for aluminum.

The solidus temperature of alloys to be enameled must not be less than 600°C. Best results are to be had with pure aluminum and AlMn alloys, in which the copper content is held below 0.3% and that of magnesium below 0.01%. Recommended pretreatment is by alkaline etching with acid rinse and a final rinse in demineralized water. Enameling produces colorful, matte, or shiny protective layers 60–80 μm thick. Such coatings are colorfast and have good weather- and corrosion-resistance. They are mainly used for facades, signs, and household appliances.

12.6 Chemical Conversion Coatings and Primers

The resistance of coated aluminum components is judged according to a wide variety of criteria that may be divided into two main categories:

- Changes to the appearance of the coating itself, such as reduced gloss, chalking, fading, and so on.
- Defects in the underlying surface, such as failures of adhesion, blistering, or corrosion under the paint film.

Changes to the appearance depend entirely on the properties of the coating system, such as the binder medium, the pigment, and so on. On the other hand, the adhesion of the coating and the other criteria of the second category depend not only upon the coating system and its tendency to penetration by gases and humidity but also, and vitally, on the pretreatment of the aluminum surface itself. A coating is only as good as the preparation of the ground on which it is laid. Not even the best coating can compensate for deficiencies or compromises in prewriting the substrate.

12.6.1 Barrier films

All organic materials, whether plastic or other coatings, take up more or less water, which diffuses to the surface of the metal with which the polymer is in contact. Activated aluminum, etched or mechanically roughened, reacts very readily with humidity to form hydroxides, which weaken the adhesion of the coating because they are not strongly bound to the metal itself. It is essential to coat the metal surface with a passive film, insoluble in water, which serves both to protect the metal against corrosive influences and as a binding layer for the coating. Such passive films may be either chemically or electrolytically thickened oxide layers. For years, the standard has been to use chromate films.

12.6.2 Chromate films

There are two chief forms of chromate film, yellow and green, distinguished from each other by the conditions in the bath itself and by the composition of the resulting film.

The yellow chromate film is formed in acid solutions at room temperature by the reaction of valency 6 chromium with the aluminum surface in the presence of active fluorides and metal complexes as catalysts. During the reaction, the valency 6 chromium is partly reduced to valency 3, with a simultaneous increase in pH and the formation on the aluminum surface of a complex mixture of basic chromate and hydroxides of chromium and aluminum. The iridescent yellowish to gold-colored film still contains Cr^{6+} ions, which are unwanted because they are poisonous.

The green chromate film is formed using a mixture of chromic and phosphoric acids with active fluorides at 40°C to 50°C. The film consists essentially of $Cr^{3+} PO_4 \cdot 4H_2O$ and contains no detectable Cr^{6+} ions. Depending on its thickness, the film is colorless to a matte light green. The usual treatment times today range from a few seconds to several minutes. The film thickness depends on the treatment time and temperature, but is chiefly determined by the ratio of F^-/CrO_3. Cr^{6+}-free green chromating is accepted for use in pretreating aluminum for food packaging, being in accordance with the legal requirements.

Both these chromium-containing conversion coatings, whether applied by dipping or spraying, offer optimum corrosion inhibition and adhesion properties. It is, however, necessary to prepare the surface carefully before chromating in order to assure a uniform conversion coating. Between the separate pretreatment steps as well as after chromating, the surface must be thoroughly rinsed. The final rinsing should be done with deionized water or other suitable rinsing solution.

12.6.3 The "no-rinse" process

This process, which generates no waste water, was developed in order to avoid the problem of disposing of the chromate-bearing rinsing water resulting from traditional chromating. The chrome-bearing pretreatment solution is applied to the moving metal surface by a contra-rotating coating roll, and reacts completely, so that no water-soluble salts remain behind on the surface; the remaining wet deposits are dried off. Since no final rinsing is needed, there is no waste water to dispose of. The film thickness depends on the chemical concentration and is regulated by adjusting the speed of the coating roll and the pressure with which it is applied. Test results show that this process, until now used only in coil-coating, ranks equally with conventional chromating in respect to corrosion inhibition and enhancement of adhesion.

12.6.4 Chromium-free conversion coatings

Despite these techniques, the development of chrome-free conversion coatings is being vigorously pursued in order to avoid the environmental problems caused by using Cr^{6+} compounds. Chrome-free conversion coatings are being produced using active fluorides and zirconium or titanium compounds or complex salts. Thin, colorless films with good adhesion characteristics and acceptable corrosion inhibition are applied at room temperature by dipping, spraying, or roll-coating. The ability of these films to withstand corrosion over the long term still appears to leave room for improvement. The processes commercially available at present have not yet found wide application except in individual cases of sheet and coil pretreatment or in special applications such as the pretreatment of aluminum beverage cans.

Film-forming zinc phosphating belongs also to the class of chrome-free conversion coatings. It also provides a good base for organic coatings. Zinc phosphate films are precipitated from an acidic solution on to a clean metal surface, which thereby loses its metallic appearance. A phosphate film can only be formed on aluminum if the Al^{3+} ions liberated by the acid etching are bound in complexes with the fluorides. All the component parts of the bath that influence the formation of the film (metal cations Mn^{2+} or Ni^{2+}, H^{+}, fluorides, and oxidizing catalysts) must be rigorously controlled in order to assure uniform quality. The considerable effort in control and regulation that this implies is a significant disadvantage of the process. There is a body of practical experience, notably in the automobile industry, with film-forming phosphating of aluminum as well as steel and galvanized materials. There is little practical experience of the suitability of the process in connection with powder and liquid paint coatings, but cathodic electrodip coating has yielded very good results.

A thin anodic film has long been used as an alternative for pretreating extruded profiles and constructional components. Liquid and powder coatings adhere well to an unsealed anodic film.

12.6.5 Primer coatings

A primer is a paint undercoat used to improve the performance or properties of the final coat. Here, a brief mention must be made of the pretreatment of workpieces that are too big to fit in any available chromating tank. They may be given a primary coat of two-part epoxy resin containing corrosion-inhibiting pigments.[5] Again, it is important that the surface be thoroughly degreased with solvent and mechanically roughened by grinding or shotblasting.

[5] The term pigment is used not only for coloring agents, but also for corrosion-inhibiting agents, whether colored or not.

12.7 *Organic Protective Coatings*

Organic coatings have been used since the very beginning to protect aluminum surfaces. Especially in more recent years, when architects have become increasingly color-conscious, coated aluminum has been used ever more extensively in the facades of buildings. Integral color and pigmented anodic films provide a somewhat limited range of colors, whereas organic coatings can satisfy almost any requirement as regards color, gloss, and texture. Along with these purely visual aspects there are, however, certain more technical requirements for coatings, for example the need to improve long-term corrosion resistance. Thus, today's high-grade coatings possess outstanding protective qualities, for example, against alkaline substances like mortar splashes, against sooty acid atmospheres with sulfurous and sulfuric acids such as occur in industrial areas, and against chlorides such as are encountered at sea or at the coast.

12.7.1 The coating processes

The numerous possible ways of applying organic coatings are the same for aluminum as for other metals. They can all be used, as the choice of aluminum alloy imposes hardly any restriction. For each coating process there is a wide range of coating materials from which to choose a coating system suiting the requirements of the final product. As regards the coating process, it is convenient to distinguish two main classes: those for coating individual workpieces and those for coating strip in coil form.

12.7.1.1 Piece-by-piece coating

For piece-by-piece coating of aluminum components, building facade parts, and profiles there are two chief methods:

- Spray application of liquid paints containing solvents.
- Electrostatic powder coating.

The usual liquid paints consist of 50% to 70% solvents together with pigments and binders. The main task of the solvent is to thin the viscous binders so as to obtain an optimal working viscosity, while not affecting their chemical properties. During application and drying, the solvents evaporate. Depending on the kind of paint, the formation of the solid film can take place by physical drying or by chemical reaction.

The advantages of liquid coating are:

- Flexibility in application.
- A very wide choice of colors.
- Easy adjustment of the color.
- Convenience in touching up.

The disadvantage is:

- Depending on the shape of the workpiece, the type of spraying system, and the boundary conditions, 10–80% of the paint is wasted in overspraying.

The 2K PUR system has proved itself for coating aluminum. It is chemically very stable and has excellent weather resistance and good chalking resistance. Reaction-hardening 2K acrylic systems are also used, yielding high-class coatings having properties comparable with those of the 2K PUR paints. In the last few years, systems have been available that are based on the fluorocarbons (e.g., PVDF). They are both more colorfast and corrosion-resistant than the usual paint systems, but they are not entirely free of problems in the application process.

In powder coating, very fine thermosetting or thermoplastic particles are applied to the surface of the metal. With heating, the particles melt to form a sealed paint film, polymerized in the case of the thermosetting compounds. In electrostatic powder application, the coating powder with particle sizes 20–60 µm is sprayed onto the cold, electrically grounded component using a special pistol, forcing the electrically charged powder to stick to the surfaces. The air that carries the powder is ionized by a high-voltage electrode, carrying the powder particles along the electrical field lines between the pistol and the grounded component to be painted. Because the field extends also around the back of the component, hidden surfaces are also well coated, except for the inside of so-called "Faraday cages," where the coating is much thinner. To ensure effective coating of these hollow spaces, extra air-flow is directed into them. The powder particles already sticking to the surface repel new particles approaching them, so that the resulting layer is very uniform although of limited thickness (25–125 µm). Powder that by-passes the component ("overspray") is recovered and fed back into the powder supply.

The tribo-charging process makes use of an ionization channel, in which the particles acquire their charge by friction. A cloud of such tribo-charged particles penetrates more readily into Faraday cages and hollow spaces. Special pistols and powder are required, and there are problems with recovering overspray; for these reasons, the process is not widely used. Stoving at 120°C to 250°C first melts the powder and simultaneously starts the polymerization reaction.

The advantages of powder coating are:

- No solvents, so it is environmentally friendly.
- Efficient use of the coating medium by the recovery of the overspray, reducing costs.
- Ease of operation, making it easy to automate.
- Good quality, including good edge coverage and coating layers of more than 40 µm in one operation.

Its disadvantages are:

- Long downtime for color changing.
- Relatively high furnace temperatures.
- The Faraday effect and uneven coating thickness in components with complex shape, except for tribo-charging.

Aluminum for building construction has been successfully coated for years using polyester-based powders with triglycidylisocyanate (TGIC) as a binder. Powder coatings using this binder possess good chemical stability, weather resistance, and chalking resistance, especially in acidic atmospheres. Trials are being made of TGIC-free systems. Mixed polyester powders and epoxy powders have no significant outdoor uses, but they have good to very good chemical stability. It should be mentioned that these are thermosetting systems, that is, a polymerization reaction takes place during furnace treating. The coating allows only limited further forming of the workpiece.

Thermoplastic powders (e.g. polyamide, polyethylene, and ethylene vinyl acetate copolymer) can also be applied electrostatically. Alternatively, they can be applied by instant sintering in a fluidized bed of powder, using particle sizes up to 200 μm. The aluminum part to be coated is heated above the melting point of the plastic and then plunged into the fluidized bed of powdered plastic. Subsequent heat treatment improves the properties of the coating. With a modified fluidizing process even pipes can be coated on the inside and / or outside.

12.7.1.2 Coil coating

Material in coil form can be coated very economically. Here, the strip is first pretreated, rinsed, and dried, and then the paint is applied and baked in one continuous process. Strip speeds are usually in the range of 60–150 m / min. As a rule a two-coat system is used, consisting of the primer (about 5 μm) and finishing (about 20 μm) coats. The back side of the strip is usually given a protective coating. Often, however, both sides are simultaneously given the same coating treatment. In the same installation it is possible to apply a laminating glue and to laminate a plastic film onto the strip surface. The coated strip can be decoratively embossed and shaped, for example roll-formed into corrugated sheet. For this reason, not only must the coating be able to withstand such forming, but the aluminum itself must have very good formability. In coil coating, the paint system is selected according to the requirements of the product. Polyester systems are widely used, not only for building facades but in a great many other application areas. They are relatively hard and easily formed, and possess good weather resistance and chemical stability. Their gloss-maintenance and chalking resistance can be further improved by combining them with silicone resins.

Very high-grade systems are those based on polymers containing fluorocarbons (PVDF, PVF). A stoved PVDF coating shows good formability with outstandingly good surface properties; it withstands weathering and strong UV radiation as well as aggressive industrial and marine atmospheres and can, therefore, be recommended in the most critical application areas.

Coil-coated aluminum sheet has an enormous variety of applications. Besides the architectural uses already mentioned, it is used among other things for clock dials, scale-faces, and signs of all kinds as well as for cladding household appliances.

When the surface of aluminum sheet is pretreated and then coated by one or another of the systems described above, the result is a high-grade composite material that combines the superior properties of the core material, aluminum, with the protective and decorative functions of the organic coatings.

Chapter 13. Aluminum's Edifice of Knowledge, and Its Main Impacts

Technological developments, both major and minor, have already been mentioned in earlier chapters, for example, in casting and heat treatment. Developing new manufacturing processes and improving existing ones, finding new end-uses for aluminum: these activities have founded whole new branches of study, and in the course of time a vast body of knowledge has been built up. To outline this, we have chosen a metaphor—*aluminum's edifice of knowledge.*

13.1 The Edifice

The basic idea of this book was stated at the outset in Chapter 3: to describe the sequence of operations (A) used in fabricating a product and define at each stage the structural changes (B) the metal undergoes in relation to the resulting properties (C). At first, the cases studied were those where a simple cause-and-effect could be demonstrated. For example, it was shown how the solidification rate affects cell size in the cast structure, and then how the cell size affects the evening out of grain segregation during annealing in order to obtain a streak-free surface for decorative anodizing. Later in the book more complicated cases are dealt with. A sequence of fabrication steps is observed that are interrelated by the A-B-C chain of process-structure-properties to enable understanding the whole thermal cycle before and during extrusion of sections (see Chapter 9.2.3.4).

Fig. 13.1 shows how the influence of the process conditions on the microstructure and, finally, on the properties operates at three different levels from the semi-fabricated to the end product. It is important to realize that the microstructure resulting from each earlier step carries over into the succeeding steps, forming the link between them. Thus, each stage provides building blocks for the whole "edifice" of understanding about processing aluminum.

We see in Fig. 13.2 that the edifice has four stories following each other in time from the top to the bottom of the structure. It is important to understand that in each of the four stories a given relationship exists between the process steps taken, the resulting microstructural changes, and the way in which the latter affect the properties of the material. The figure also makes graphically clear that process steps that are taken, for example in casting or in hot working, at the very start of the life-cycle often influence the microstructure and, hence, the properties at intermediate fabrication stages right through to those of the final product.

There is a *vitally important idea* that forms the basis of a fundamental understanding of how the properties of aluminum materials develop and change. It is necessary, of course, but not sufficient to understand the interactions between process steps, the corresponding microstructural changes, and the properties that take place, for example within a rolling mill or an extrusion plant. However, we must go further and grasp the fact that there is a kind of "inheritance" that is carried forward in the microstructure from one stage to all the following stages and has both positive and negative effects on properties.

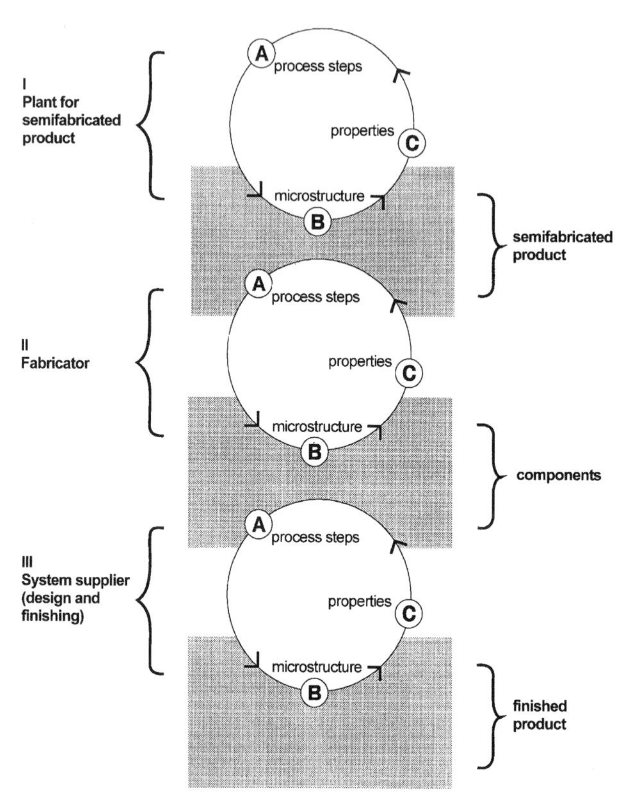

Fig. 13.1: How process conditions (A) affect the resulting microstructure (B) and properties (C) at three different levels in the chain of manufacture. (schematic)

In the metaphor of the edifice, this is shown as a spiral staircase running from top to bottom of the building and leading from one floor to all the others; it is the microstructure that forms this spiral staircase.[1]

There are two further points to be made concerning the organization of the edifice. First, the process steps named in the left-hand side of Fig. 13.2 (A) concern not only the semifabrication plant, but also the fabrication of individual components. Similarly, steps that are shown as belonging to the stage of component fabrication can also be taken in the semifabrication plant. Second, forged and mold-cast parts as well as extruded shapes are included in Group II (components) not only because these processes result in a shape very close to that of the finished part (near net shape), but also to give emphasis to factors having a decisive influence in producing high-grade components.

The number of firms, enterprises, and plants involved increases enormously from the top to the bottom of the edifice, as does the number of products. Here are a few figures in this regard: there are about 150 aluminum smelters in the world, more than 1000 casthouses, and hundreds of extrusion plants. The numbers in the second floor from the bottom are

[1] There is a parallel here with the helical structure of DNA, which carries the genetic inheritance from one generation down to those which follow.

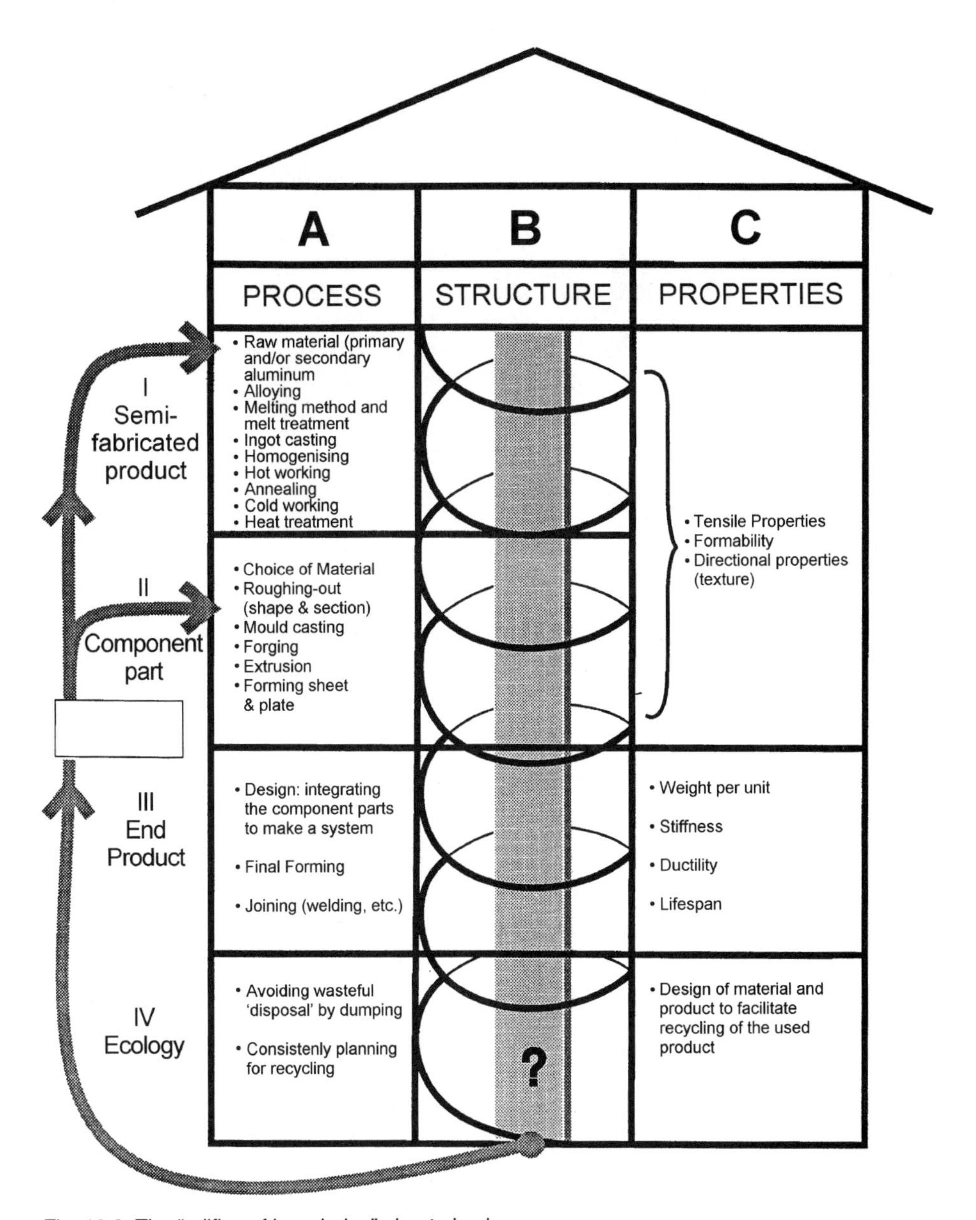

Fig. 13.2: The "edifice of knowledge" about aluminum.

particularly impressive: aluminum can rightfully be called the metal of a thousand and one products! These products range from vehicles of all kinds to household utensils and appliances and from building facades to packaging. One could go on for a long time citing product groups that fan out at the bottom of the edifice, owing their existence not only to the wide choice of materials and processes, but also to the range of surface finishes available.

As for the very bottom floor of the building, while it is true that from the beginning of the century it has been customary to remelt used aluminum products and that the proportion of secondary aluminum in the raw material supply has increased steadily to reach nearly 40% today, it is only recently that the need to recycle has become an important consideration at every stage of the way from raw material to finished product.

In this great edifice of knowledge about aluminum, the experience of many decades provides the knowledge base for changing the fabrication methods and extending the range of application of aluminum and aluminum-based materials, now and in the future.

To illustrate the power of the edifice, three examples, each important building blocks of technology representing alloy, process, and system development, will be outlined briefly below:

- Alloy development (Section 13.2)
- Process development (Section 13.3)
- Systems development (Section 13.4)

13.2 Alloy Development

Alloy development runs like a connecting thread from top to bottom of the edifice, creating interactions with the main applications of aluminum in the marketplace. Based on a few examples, the following text on alloy-related process development will demonstrate that the edifice of knowledge contains the seeds of further innovation.

13.2.1 Conventional alloys

The main aluminum alloys were developed during the 50 years from 1910 to 1960. The beginning was marked by the development of Duralumin (2017) even before the war of 1914–1818. Most of the current standard alloys were, however, developed between 1930 and 1960. Since then, these alloys have been tailored for production use and their properties refined. Even today, only a few standard alloys have found a really wide range of application, although fundamental research into new alloys continues intensively.

There are three main reasons why the industry is so cautious in adopting new alloys:

- The currently available alloy systems, ranging from soft, unalloyed aluminum to high-strength alloys, cover a very wide range of properties.
- The aspects to be considered when choosing an alloy comprise its cost, availability, and properties, such as formability, strength (especially in fatigue), and corrosion resistance. When all of these are taken into account, it often turns out that only a well-established alloy with proven properties is worth considering.
- The internal dynamics of fundamental research generate an enormous number of new publications, to be sure; but it takes a long time before a new "exotic" alloy can be economically produced and applied, and appropriate standards and quality assurance procedures adopted.

Developing and marketing a new alloy or a new composite material represents a risk for original producer, component manufacturer, and final user alike. A thorough cost-benefit risk analysis often reveals three objections to an improved design with a new material:

- The designer needs a large amount of data about the material, among which information in the form of long-term test results is obviously unavailable.
- There are no generally accepted design guidelines for new materials.
- Specifications already exist concerning applicable design standards, and they refer to the existing materials.

When we consider these objections in the context of the automobile industry, we find, in addition, that a component made of a new material or even a known one with an untested microstructure is hardly ever used in production before it has undergone several years of testing and prototyping under conditions similar to those that it will encounter in service. Even outside the transport sector, it is generally true that because of the dominance of the internationally accepted alloy standards, it is both costly and time-consuming to introduce a fundamentally new material, and often the sponsor has no guarantee that his or her competitors will not reap the most benefit. For all these reasons, in the foreseeable future no marked change in the range of available and regularly used standard alloys is to be expected.

Meanwhile, aluminum materials continue to be successful in the never-ending struggle to substitute them for other, more traditional materials. With a planning horizon of about ten years, however, this struggle is not to be won with exotic new materials. On the contrary, success is to be found by offering specialty products made of well-proven alloys, tailored to the specific requirements of the end-product. By doing this, a supplier can carve out a market niche by offering a product with an attractive combination of properties, better functionality, and high added value. This continued tailoring of semi-fabricated and finished products made from well-proven alloys will be illustrated using as examples the contemporary further development of magnesium-silicon alloys for extrusion and for casting.

13.2.1.1 Al-Mg-Si extrusion alloys

Fig. 13.3 gives a brief survey of the compositions of four dominant alloys. Market segmentation and the development of extrusion process technology have resulted in three main alloy subgroups:

- Low-strength alloys such as 6063 with about 0.4–0.6% each of Mg and Si, mainly used in buildings for windows and doors. The profiles are air-quenched after extrusion.
- Medium-strength alloys such as 6005, which reach the required strength levels without rapid quenching after extrusion, with the advantage of avoiding the extra costs of the straightening operation before artificial aging. Large hollow extrusions are made from these alloys today for general use in the transport sector of the market.
- High-strength alloys such as 6061 and 6082, which require rapid quenching after extrusion and a subsequent straightening operation.

The need for tailored quenching of a specific alloy, temper, and extrusion profile has driven the development of sophisticated press-quenching of the Al-Mg-Si alloys using water-air mixtures with computer-controlled sprays to achieve optimal cooling of the

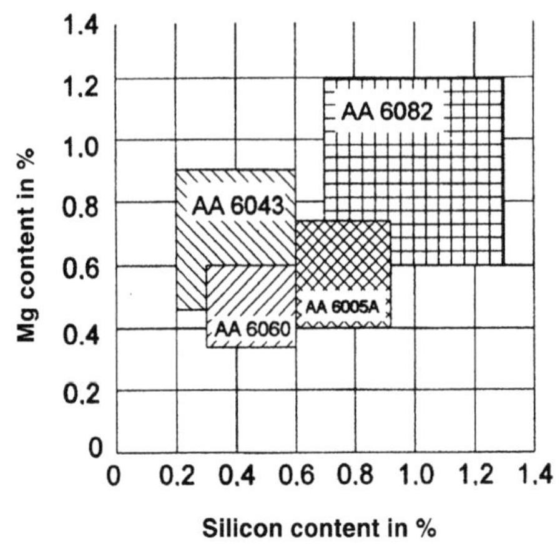

Fig. 13.3: Composition survey for Mg and Si in four Al-Mg-Si alloys. (Aluminium Taschenbuch)

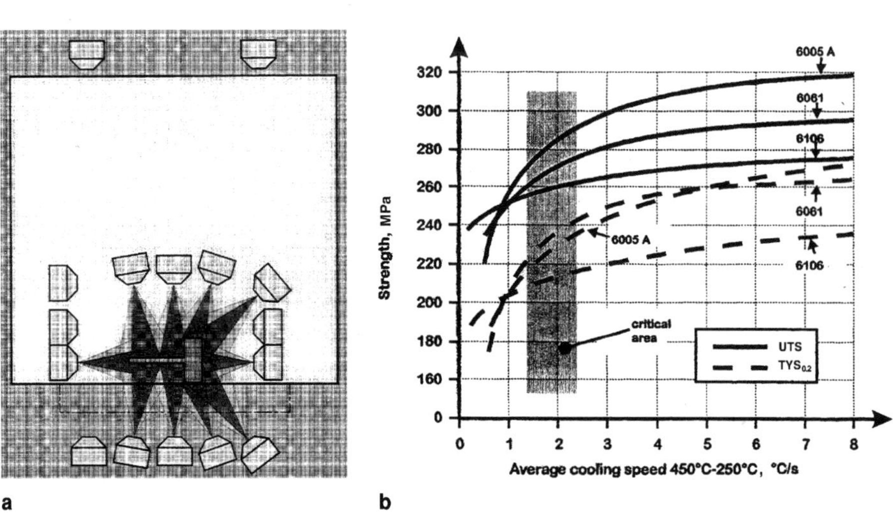

Fig. 13.4: (a) A schematic of computer-controlled spray-quench facility and (b) the effect of cooling rate during quenching on tensile yield strength of Al-Mg-Si alloys. (Alusuisse)

different wall thicknesses, as illustrated in Fig. 13.4.a. Fig. 13.4.b shows the critical quench-sensitive cooling speeds of medium-strength alloys. This outline demonstrates the way in which alloy development leads, in turn, to the development of process technologies to satisfy the highly demanding new markets for aluminum extrusions.

13.2.1.2 Al-Si9-Mg-Mn-Sr casting alloys

We saw in Chapter 6 the profound influence of a wide variety of process steps, before and during casting, on the success of the evolution from a simple product to a systemic component. Producing car spaceframe nodes by vacuum pressure diecasting requires an alloy that combines high ductility with medium-to-high strength for crashworthiness. Fig. 13.5 shows such a node for the spaceframe structure of the Audi 8.

Table 13.1: Chemical composition of Silafont-36 alloy in weight %

(Aluminium Rheinfelden)

	Si	Fe	Cu	Mn	Mg	Zn	Ti	Other
Minimum	9.5	—	—	0.5	0.1	—	—	—
Maximum	11.5	0.15	0.03	0.8	0.5	0.1	0.15	Sr

Table 13.2: Variation in tensile properties of Silafont-36 with varying Mg content

Mg content (wt. %)	UTS (MPa)	TYS (MPa)	El (%)
0.15	250	117	11.2
0.28	264	121	10.2
0.30	280	133	8.1
0.33	261	141	6.3
0.42	286	146	5.8

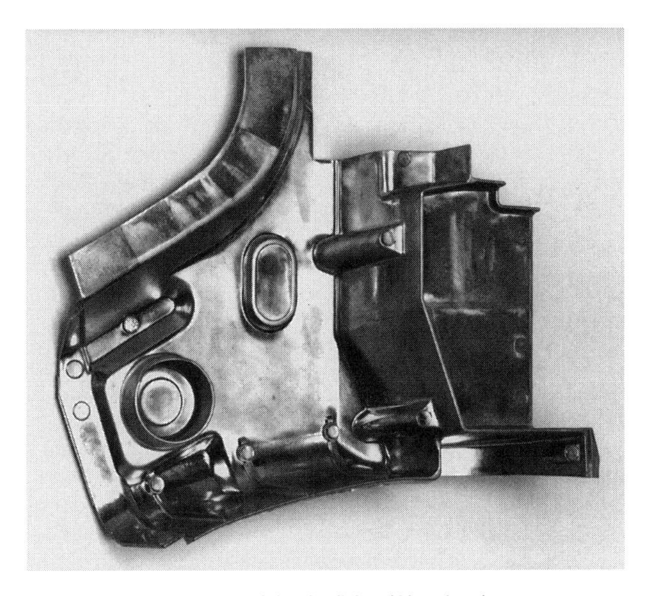

Fig. 13.5: Cast node for the space frame of the Audi A8. (Alusuisse)

Table 13.1 shows the composition of an alloy used at present for this purpose. Table 13.2 shows the enormous influence of the magnesium content on strength and ductility in F temper, and Fig. 13.6 illustrates its effect on yield stress and elongation for various tempers. A very low iron content is also required to ensure ductility for safety-critical parts made from castings.

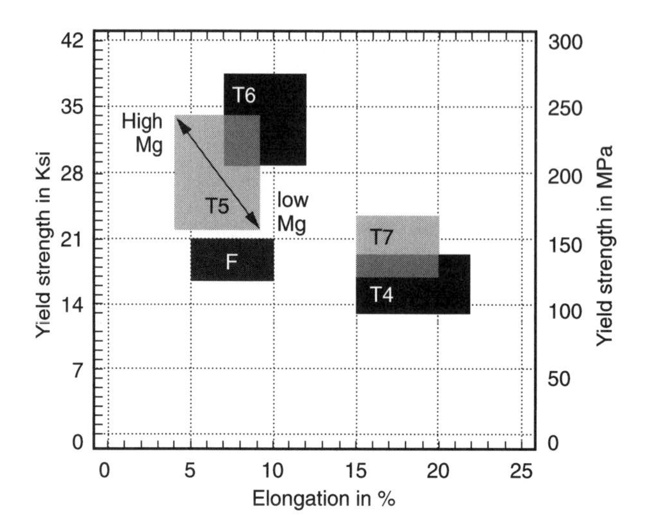

Fig. 13.6: Ranges of tensile properties for Silafont 36, an Al-Si-Mg-Mn-Sr alloy.

13.2.2 New alloys[2]

13.2.2.1 High-strength lithium-bearing alloys

Adding 2% to 3% of lithium to an alloy brings a double advantage in the search for stiffer, lighter components. The alloy's density is decreased by several percent, and at the same time its modulus of elasticity is increased to about the same extent. For the same stiffness, components made of high-strength alloys can be made up to 10% lighter if they contain lithium. This is particularly important for aerospace applications (see Chapter 15).

13.2.2.2 Superplastic alloys, processes, and applications[3]

Superplasticity means the capacity to withstand very large deformation under certain conditions. Superplastic forming imposes some specific requirements both on the alloy and on the method of working it. Certain aluminum alloys exhibit superplastic behavior at temperatures above 300–400°C. The following microstructural characteristics are important:

- The grain structure must be homogeneous, that is globular and with no preferred orientation.
- The grain size must be very fine, normally not more than ten microns.
- The grain structure must be stable under high temperature deformation, with no significant grain growth.

These microstructural properties can be produced in a number of aluminum alloys by appropriately choosing the composition and the fabrication conditions. The first practical applications involved alloys that had been specially developed for superplastic work-

[2] Selected examples.

[3] By P. Furrer.

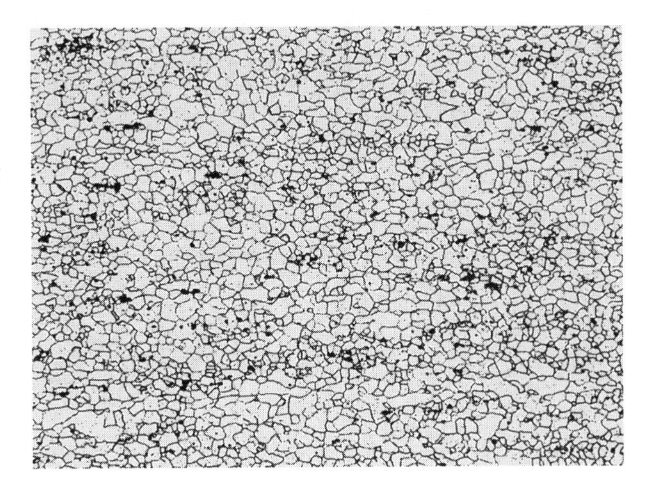

Fig. 13.7: Microstructure of sheet made of a superplastically formable alloy of the Al-Mg type (5083, Formall-545) with about 20,000 grains/mm^2 and an average grain diameter of 5.4 μm.

ing. Superplastically formed parts in the alloy group AlCu6Zn have been used for more than 20 years, chiefly in the aircraft industry. Today, conventional alloys are being used that have been thermomechanically worked under particular conditions. The high-strength Al-Zn-Mg-Cu alloy 7475 is an important example, as well as the Al-Li alloy 8090 and, in particular, Al-Mg alloys similar to 5083.

The recrystallization, under suitable conditions, of a very severely deformed rolled structure produces the necessary extremely fine grain in these alloys, as shown in Fig. 13.7. It is of prime importance to heat up the material very quickly to recrystallization temperature. The resulting fine grain size is stabilized by small particles of a finely precipitated secondary phase.

Deformation processes

Forming at a high enough temperature and at a relatively low strain rate is crucial to successful superplastic deformation. This requires close control of the temperature and the speed of deformation from beginning to end of the deformation. The ideal temperature range for 5083 is 495–535°C, according to circumstances, at a strain rate of about 10^{-3} s^{-1}. Under these conditions, elongations of 200–400% can be attained without difficulty, which are amply sufficient for fabricating most shapes required in practice. At the microstructural level, slipping along the grain boundaries forms a significant part of the deformation mechanism.

Because of the very low strain rates required, the time to form a part superplastically is relatively long, ranging from ten minutes to an hour according to the size of the part. The necessary deformation forces are, however , quite low, that is, the sheet can be "blown" into the mold by a slight overpressure (Fig. 13.8). Various tooling designs are used, depending on the geometry of the part to be formed.

So-called "membrane forming" is a special case, in which a sheet of ordinary (i.e., not superplastic) material is inserted between two sheets of superplastic alloy. Supported in this way, the ordinary material can also successfully undergo extreme plastic deforma-

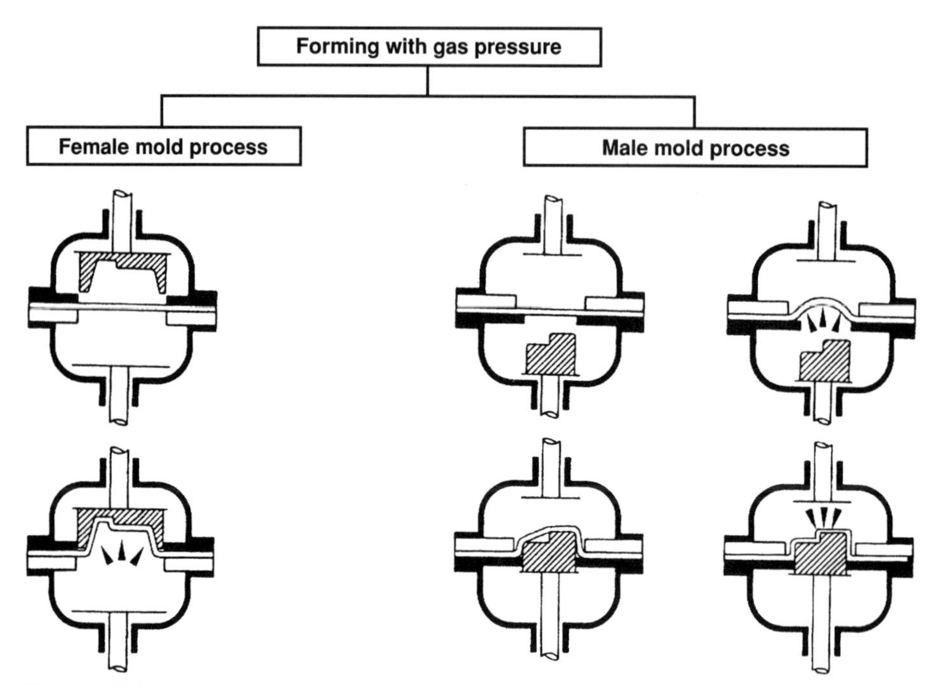

Fig. 13.8: Schematic representation of the conventional superplastic blowforming process: (left) with female mold, (right) with male mold, in two stages.

tion. Because its edges are firmly clamped between the two halves of the tool, the sandwiched sheet cannot be drawn in (i.e., the increase in its surface area takes place entirely through reduction in its thickness).

Thus, an essential problem in superplastic forming is to ensure the most uniform possible wall thickness over the entire surface of the formed part. During very large deformations there is a tendency to form cavities wherever three grains come together. This represents a further difficulty in superplastic forming, because this cavitation adversely affects the strength and elongation of the formed material, thus, indirectly limiting the degree of deformation that can be used in practice. A significant reduction in the volume of such cavities results from applying back pressure to the convex side of the part during the forming process. This means increasing absolute pressure while maintaining the differential pressure necessary for forming.

Applications

Advantages of superplastic forming are:

- Each shape can be made with a single, one-piece tool. This allows great versatility in design.
- The formed part is true to the designed shape and has close dimensional tolerances with no springback.
- Tooling costs little, as relatively inexpensive ceramic tools are often used.

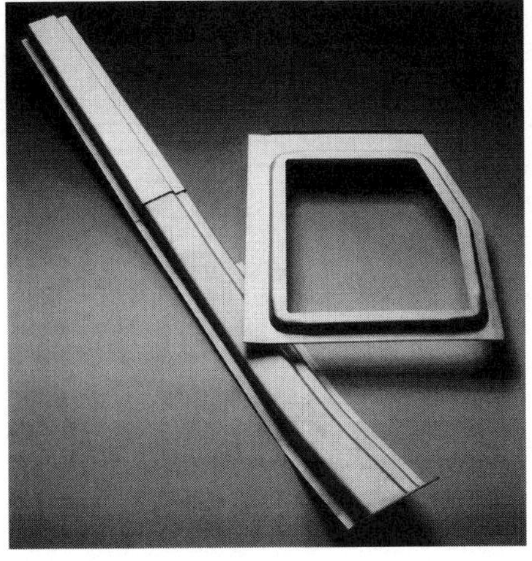

a

b

Fig. 13.9: (a) Superplastically formed parts for cladding pillars (joined by welding) and door frames for railway cars (Alusuisse) and (b) a superplastically formed part for the wheel of a solar-powered vehicle. (Alusuisse)

Its disadvantages are:

- A relatively long time is required to form a part.
- Non-heat-treatable alloys emerge in the soft annealed condition.

Thus, the process represents a technically versatile method for medium-length production runs from about 50–10,000 units. Given good process control, the properties of the resulting formed parts correspond largely to those of the base material (i.e., to those in the fully annealed state). Parts formed from a heat-treatable superplastic alloy can be heat-treated to obtain higher strength in a subsequent operation, or in certain cases, quenched straight from the press.

Whereas the high-strength superplastic alloy systems Al-Cu, Al-Zn-Mg-Cu, and Al-Li are still used mainly in aircraft construction, the Al-Mg alloys could perhaps find a wider range of application (e.g., in vehicles for road and rail and for household appliances). Decisive advantages in these applications are the good corrosion resistance and weldability of the Al-Mg alloys, as well as the wide range of surface finishes available with them. Fig. 13.9 shows some typical applications.

Further development is not primarily aimed at expanding the range of superplastic elongation, as the already available 200–400% is adequate for any conceivable practical application. It is much more important to reduce the cost of producing superplastically formable sheet as well as that of the forming process itself. Optimization of the process and the tooling design by means of computer simulation that is already available could open up more fields of application for superplastically formable aluminum alloys.

13.2.3 Modeling microstructure and properties[4]

The changeover from producing products to producing systems requires the optimization of materials. Materials play a significant role in the value of parts, component assemblies, and systems. Materials and their associated production techniques are the basis for achieving success and ensuring that the end product meets the required quality and cost expectations. In many instances, new or improved products can only be created because of the existence of new and innovative materials; that is to say, material development improves chances in the marketplace.

For each application there is a basically optimal material. This depends on the requirements of the product and other physical properties that can be obtained from the material. It is understood that the word material can imply either a homogeneous material or a combination of materials (i.e., a composite).

Material development makes use of the well-known relationships illustrated in Fig. 13.10. The problems facing modern material scientists are, however, more complex than those dealt with in the disciplines of conventional metallurgical or materials science. Often the knowledge obtained from established data only makes possible very general statements. A semi-empirical approach is often the only way in which the exact and precise requirements of a particular material development can be met.

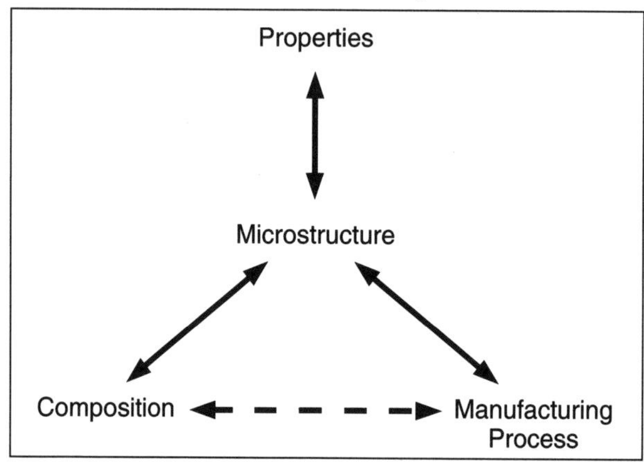

Fig. 13.10: The relationships in material development.

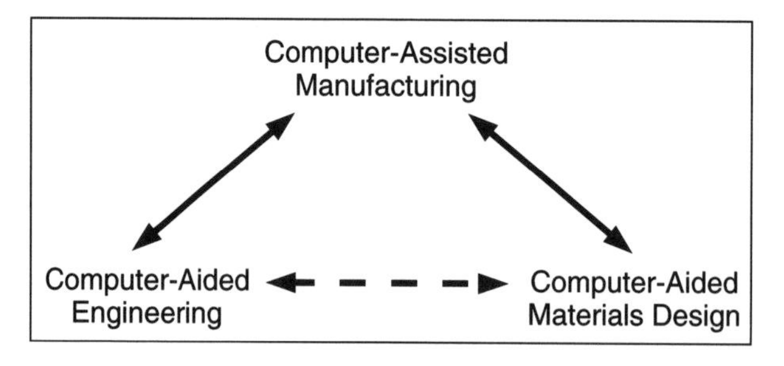

Fig. 13.11: Schematic of a material science expert system, combining computer-aided materials design (CAMD), computer-assisted manufacturing (CAM), and computer-aided engineering (CAE).

[4] By P. Furrer.

Unfortunately, a material science expert system (Fig. 13.11) that would serve as an aid to computer-aided engineering (CAE) and computer-assisted manufacturing (CAM) to make possible computer-aided materials design (CAMD) are not yet available. The use of tailor-made materials will, however, soon be possible because of continuous improvements in theoretical knowledge, more precise methods of characterizing structures, and more exact methods for determining properties. The modeling of materials, product processing, and operating conditions in service is becoming more and more important.

Statistical process control (SPC) is the usual starting point for developing a model for a particular finishing process. This is the first step toward understanding and controlling complex processing procedures. However, SPC can also provide a bridge between scientific fundamentals and a general engineering overview of a process. There are many macroscopic models based on existing data that can be used, for example, to predict forces or temperatures resulting from the various deformation processes. By linking the process parameters established by using such a model with experimentally determined material properties, for example grain size, texture, or mechanical strength, it is possible to generate empirical data on the behavior of a given alloy. In many instances this makes it possible to predict, depending on the precise processing parameters, changes in the structure of the material during a process step. This type of model, created by using qualitative physical observations, can be used to predict behavior in many aluminum manufacturing processes, such as the production of canstock or foilstock. These models make use of the very large amount of established data to provide a statistical representation of the likely end product.

The aim is to provide the basic knowledge that will make it possible to link manufacturing process steps to the microstructure and the mechanical properties that result. Modeling the relationships between processing parameters, microstructure, and properties for the promotion of a specific product is a prerequisite for developing a system.

13.3 Process development: joining technology as a key element for structural applications of aluminum

The development of methods for welding aluminum materials has opened new market segments in stationary uses such as bridges and buildings as well as in transportation uses such as ships, trains, and automobiles. This will be demonstrated in Chapter 15. In what follows, we will first discuss welding as a typical case for the A-B-C sequence of studies where process development is a cornerstone to arrive at new original systems for the applications of aluminum.

Welding and other joining technologies should be understood on several levels:

1. The welding process in its contemporary versions and developmental aspects.
2. Quality criteria of the weld seam and its surrounding microstructure.
3. Comparison with other joining methods such as mechanical joining, with regard to the properties of the resulting structure, especially the fatigue strength.

The quality aspects will be dealt with first, following the basic concept of this book. Next we shall examine a recent process development, that of a welded structure for an automotive application. Finally, we will briefly outline the advantages of the mechanical joining processes and make a comparison of properties for automotive structures joined mechanically with those for a welded structure.

13.3.1 Fusion welding processes, a general description

Fusion welding involves melting together the edges to be joined, often adding molten metal to fill a V-shaped channel. The joint, therefore, consists partly or entirely of resolidified parent metal with a cast structure. Traditional oxyacetylene flame welding uses a flux of molten salt to dissolve the aluminum oxide and cover the liquid metal. Most modern methods that expose liquid aluminum shield it with an inert gas, either argon or helium. The two principal processes are widely known by the following names / abbreviations:

- With a consumable electrode: gas metal arc welding, GMAW (or metal inert gas welding, MIG).
- With a non-consumable electrode: gas tungsten arc welding, GTAW (or tungsten inert gas welding, TIG).

Brazing, although it involves melting only added metal of a lower melting point, has metallurgical effects that can be understood in the light of this section.

Several specialized welding (fusion) processes make joints almost devoid of cast structure in the weld seam. They do this by rapidly melting the parent metal surfaces and then forcing them together, thus squeezing out most of the liquid metal. Such processes include flash-butt and stud welding (variants of electric arc welding) and magnetic induction welding. Resistance spot welding melts only a very small volume of metal for an extremely brief period. It is beyond the scope of this book to address each of these specialized processes in detail.

For some applications, several joining methods are used together. Weld bonding, for example, is a combination of resistance spot welding and adhesive bonding.

13.3.1.1 Solid-state welding methods

For the sake of overview, solid state welding methods should be mentioned here. Several methods have been applied to aluminum materials for a long time. One is pressure welding, which usually involves a hot working process (e.g., cladding, roll bonding, or porthole die extrusion). Another method is explosive welding, which propels one component against the other at supersonic speed, so as to intimately mix the surfaces.

Fig. 13.12 illustrates cold pressure welding for extrusions. This well-proven technique combines two or more small or medium sized extrusions into a large cross-section. This offers two important advantages over the use of a larger single profile:

- The investment for a larger extrusion press is avoided. Instead, standard presses such as those used for window sections may be satisfactory for the production of large profiles.
- Complex sections that may be difficult or impossible to extrude can be produced.

Fig. 13.12: An electronic heatsink with housing parts for frequency converters for continuous speed control of three-phase current motors. Individual shapes are cold pressure welded, providing large surfaces with technically simple profiles. (Honsel Profilprodukte, GmbH)

For the welding, a modified sheet forming press may be used for lengths at least up to 8 m.

13.3.2 Effect of fusion welding on structure and properties[5]

13.3.2.1 Temperature in the heat-affected zone near a fusion weld

Whether or not filler metal is added to help join the parent metal edges, fusion welding needs enough heat input to partially melt these edges. Due to aluminum's high thermal conductivity (about twice that of steel), much of the added heat escapes into the surrounding parent metal. The shorter the time taken to complete a weld, the less heat diffuses into the parent metal. Thus, paradoxically, a more intense and concentrated energy input allows less overall heat diffusion into the surrounding metal, because welding can be completed more quickly. Deep, narrow fusion welds are only possible with a highly concentrated energy input.

Welding produces a joint with a characteristic structure. From the metallurgical standpoint the weld consists of two distinct regions.

First, there is the weld seam with a cast structure where the molten weld zone solidified. Due to strongly directional heat flow into the parent metal, the grains are typically dendrites directed toward the weld seam center line. Depending on the solidification condi-

[5] By W. Kuhlein with comments of A. Bushnell.

tions, there may be defects such as piping voids, segregation, and gas porosity, which can significantly affect weld strength.

The second region is equally important for weld strength: this is the heat-affected zone, or HAZ, which immediately borders the weld seam. The thermal history (temperature and time) depends on the quantity of heat per unit length of weld. The heat-affected zone may be considered to end wherever the temperature has not exceeded 100°C during welding. In TIG welding, a region about 30 mm from the weld seam would typically reach about 100°C. Fig. 13.13 shows a typical plot of the thermal history at a point in a heat-affected zone. Fig. 13.14 illustrates the much deeper and narrower weld seam that a very concentrated energy beam can achieve.

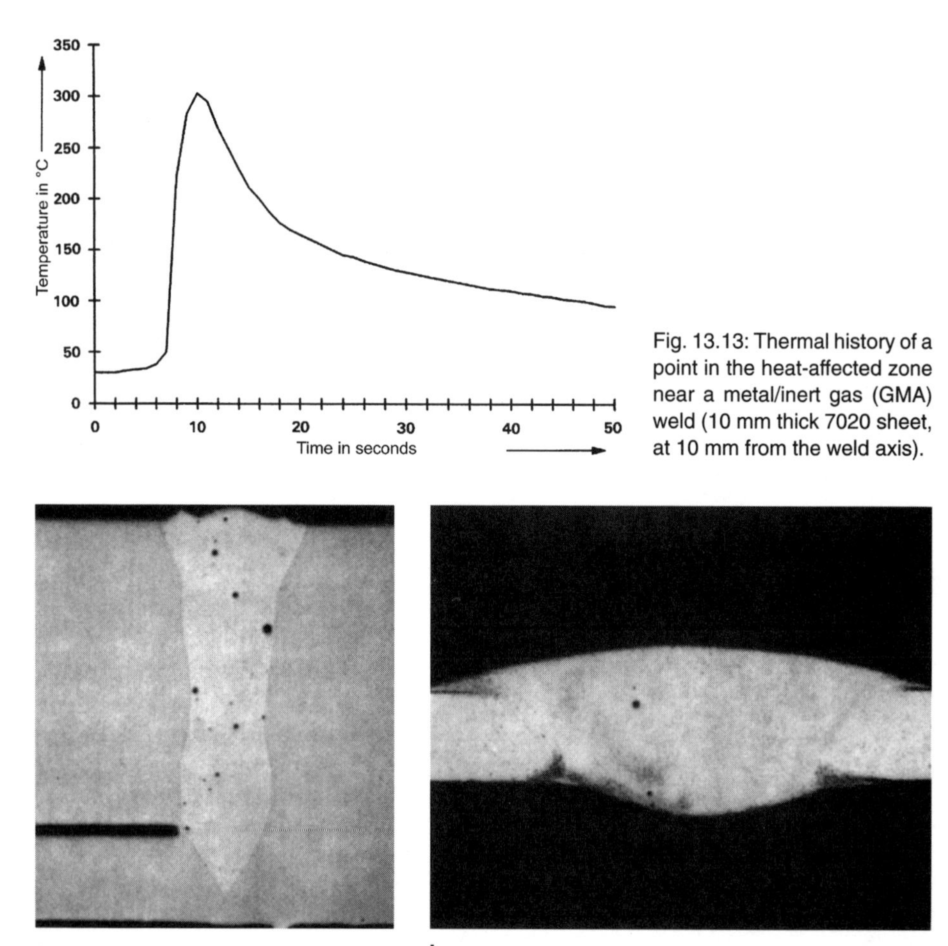

Fig. 13.13: Thermal history of a point in the heat-affected zone near a metal/inert gas (GMA) weld (10 mm thick 7020 sheet, at 10 mm from the weld axis).

Fig. 13.14: Weld depth to width ratio: the effect of highly concentrated versus less concentrated energy input. (a) Laser weld seam, edge-to-edge (butt) weld, alloy 6082, sheets 6 mm/4.5 mm, seam width 2.5 mm, seam depth 5.7 mm. (b) Metal/inert gas (GMA) weld seam, butt weld, 1.2 mm thick 5019, seam width 5 mm (courtesy AMAG).

13.3.2.2 Behavior of different alloy types during welding

For non-heat-treatable alloys in the annealed condition, any strength change in the heat affected zone will be insignificant. In the work hardened (H) conditions, by contrast, several processes occur in the heat affected zone, depending on the temperature and the degree of cold work:

- Grain growth at temperatures well above the recrystallization threshold.
- Recrystallization of the deformed structure at temperatures just above the recrystallization threshold.
- Recovery and stress relief in less intensely heated regions further from the fusion zone.

Heat-treatable alloys are generally more sensitive to the temperatures reached in the heat affected zone. The structure in this zone is, therefore, always likely to show metallurgical changes. In the very hot region near the molten metal of the weld pool, precipitates can largely dissolve as in solution treatment, but for lack of rapid quenching, this solid solution will not be frozen as it cools to room temperature. Indeed, the resulting condition is typically closer to the soft annealed condition than to the quenched condition.

There is an exception that allows a satisfactory quench effect. Alloy 7020 not only readily achieves full solution treatment through the heat of welding, but is sufficiently tolerant of a slow quench rate so as to maintain this solution treated condition down to room temperature. This results in subsequent age hardening. Within 30 days, for instance, natural age hardening raises the strength to nearly that of the fully heat treated (T6) parent metal. Post-weld precipitation treatment for artificial aging provides maximum strength levels practically indistinguishable from those of the parent metal (Fig. 13.15) and is also recommended to improve corrosion resistance at the weld.

Fig. 13.16 summarizes how the heat of welding affects the structure beside the weld seam, according to the type of alloy. It is important to note that these structural changes affect not only strength but also chemical behavior, such as corrosion resistance and anodizing quality.

If weld filler metal is used, it must be borne in mind that these are generally of non-heat-treatable alloys. Moreover, the strength of these alloys as-cast will be similar to that in their annealed condition. Therefore, in work hardened or heat treated parent metal, the weld tends to be the weakest part of the joint.

13.3.2.3 Causes of pores in weld metal

Welds in aluminum are particularly susceptible to shrinkage pores and gas bubbles (Fig. 13.17). Weld porosity can be tolerated if it is finely distributed in the weld metal. This means that pores must not exceed a certain size, and they must not form rows or "piping" lines. If the pores do cluster, then the welds are unsuitable for critical mechanical loads, because such porosity significantly weakens the metals, especially in fatigue and in impact loading.

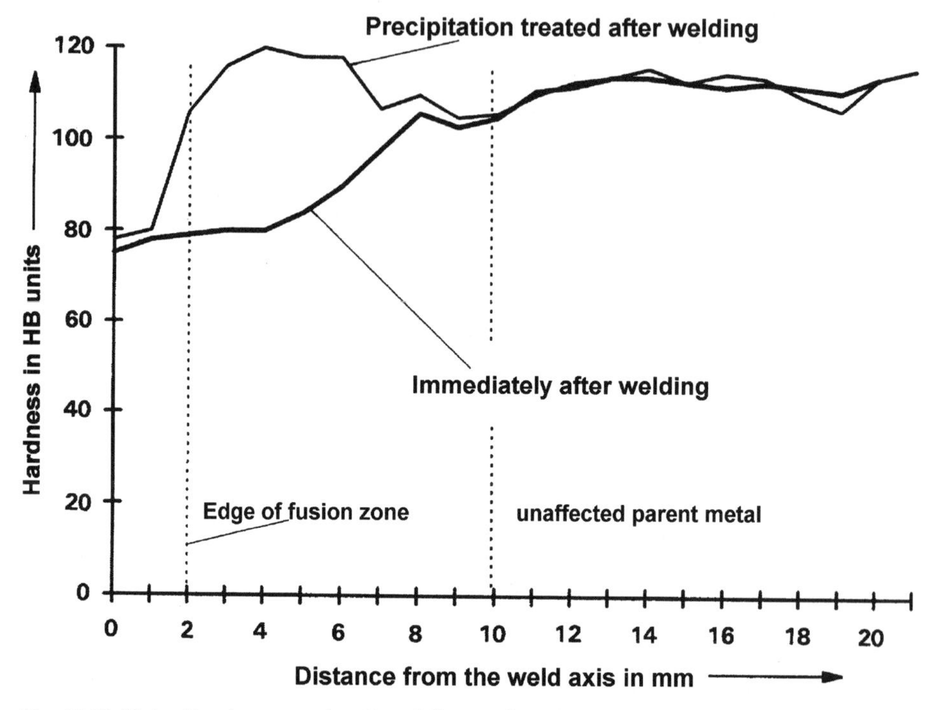

Fig. 13.15: Plots of hardness as a function of distance from the weld axis in a sheet of the alloy 7020, welded with a non-heat-treatable filler metal.

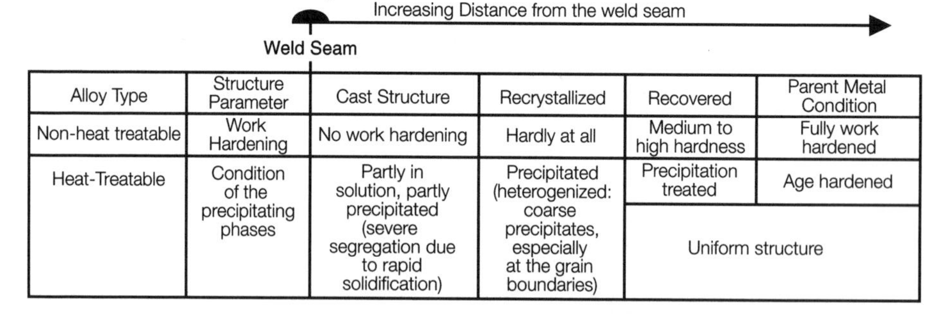

Alloy Type	Structure Parameter	Cast Structure	Recrystallized	Recovered	Parent Metal Condition
Non-heat treatable	Work Hardening	No work hardening	Hardly at all	Medium to high hardness	Fully work hardened
Heat-Treatable	Condition of the precipitating phases	Partly in solution, partly precipitated (severe segregation due to rapid solidification)	Precipitated (heterogenized: coarse precipitates, especially at the grain boundaries)	Precipitation treated	Age hardened
				Uniform structure	

Fig. 13.16: Summary of how welding affects the structure in and beside the weld seam.

Solidification shrinkage can lead to porosity within the weld seam when the local solidification conditions hinder the feeding of liquid metal into the shrinkage cavities that are forming. However, weld seam porosity is predominantly due to gas inclusions.

The weld pool solidifies quickly, therefore, there is a risk of gas being trapped in the metal and forming pores in the weld bead. The main danger here is from hydrogen, which is formed by the reaction between the metal and water vapor and is readily taken up by the weld pool. The moisture may come from humid surroundings, impure inert

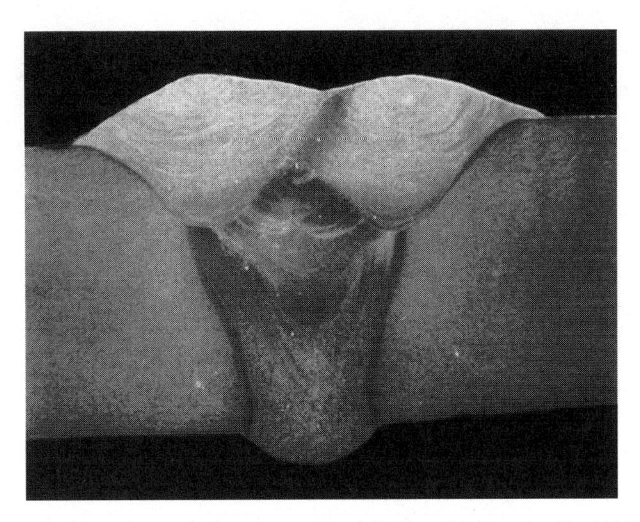

Fig. 13.17: Finely distributed porosity in an automated GMA weld seam in alloy 7020. (AMAG)

gas, or from paint and grease contaminating the welding rod or the weld region. The solubility of hydrogen in the liquid metal is up to a hundred times greater than in the solid metal, therefore, large amounts of the gas are released as the weld pool solidifies.

The faster the weld solidifies, the more it tends to trap hydrogen bubbles, possibly together with stirred-in inert gases and nitrogen, because there is less time for the bubbles to escape. Therefore, oxyacetylene and TIG welding are less at risk from porosity, thanks to their high energy input per unit length and resulting slow solidification rate. By contrast, MIG and especially laser welding nearly always show finely distributed gas porosity, because the liquid metal freezes too quickly for the gas to escape.

Besides the welding process itself, certain parent material characteristics cause gas porosity. Cast components are particularly at risk because of their casting conditions (e.g., hydrogen pick-up from a sand mold or pressure die casting trapping compressed air bubbles). Primary and secondary porosity were already described in Chapter 6. The secondary pores are less important for welding. Primary porosity in the parent metal, as well as a high overall gas content, is a major cause of weld porosity and, hence, of mechanical weakness in the weld seam. As mentioned in the description and discussion of mold casting processes in Chapter 6, there are processes and treatments that permit subsequent welding and heat treatment. These may involve changing the component design, metal treatment, or casting process.[6]

Furthermore, if one chooses a welding process with a narrow heat affected zone, then higher temperatures will reach only zones of the casting that have few pores (i.e., the rapidly solidified, superficial zones). Electron beam welding and laser welding permit such narrow heat affected zones.

[6] To locally suppress primary porosity in sand casting and pressure die casting, it is the state of the art to arrange for faster heat extraction and, hence, faster solidification in zones that are subsequently to be fusion welded.

13.3.2.4 Tendency to weld cracking

A comparison of different aluminum alloys under similar welding conditions reveals that certain alloys tend to develop characteristic cracks during or after welding. One type of crack occurs along the weld center line following the axis and is particularly pronounced in the end crater where welding stops. These cracks often appear immediately, only a few millimeters behind the weld pool, and cause immediate rupture of the weld.

A different type of crack occurs at the weld edge just beside the weld seam, in the hottest part of the heat-affected zone. These are typically microcracks and can only be detected with considerable technical effort. Nevertheless, with the stress of a further fabrication step, or under applied load, these microcracks can suddenly grow to macrocracks and cause the joint to fail.

Both types of crack result from welding stresses in the momentarily weakest part of the structure, but their formation mechanisms are different. Weld center line cracks are always associated with marked dendrite growth. These dendrites grow nearly at right angles to the edges and meet at the axis. Thus, they concentrate impurities and gas bubbles in the middle of the weld, reducing the load-bearing section and creating a notch. The same mechanism gives weld end craters their typical cracks: alloy elements pushed ahead of the solidification front are most concentrated there. Weld edge microcracks can occur when the heat of welding results in the melting of grain boundaries in the hottest part of the heat-affected zone. Low-melting-point phases in the grain boundaries are typical of medium-strength heat-treatable alloys. These grain boundaries can resolidify with microcracks, which when under load can quickly grow and cause failure. These are also known as hot cracks (Fig. 13.18).

Added weld metal alloys provide a means to prevent these cracking phenomena. As a rule, the added weld metal, usually called filler metal, will have a lower solidus temperature than the parts being welded. Such weld metal alloys, like 4043, form liquids with low viscosity (i.e., are not tough flowing); they readily feed the shrinkage spaces to heal

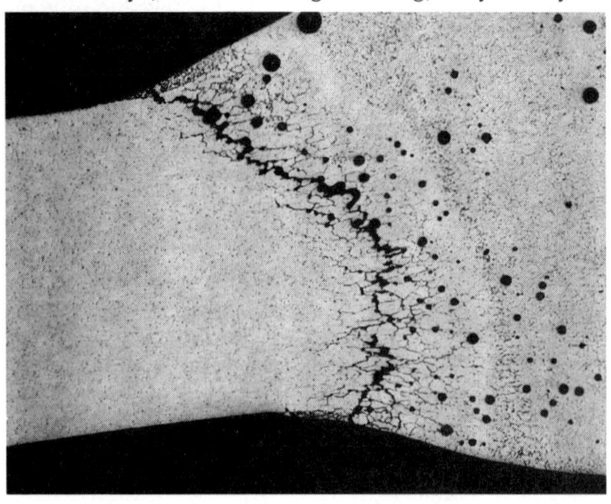

Fig. 13.18: Hot cracks and pores in the heat-affected zone of a GTA weld on a tube of 7020. (AMAG)

cracks. Indeed, without suitable weld metal it is practically impossible to make conventional fusion weld seams in the alloys most subject to hot cracking, that is the heat-treatable Al-Cu (2xxx) and Al-Zn-Mg-(Cu) (7xxx) series. A further means to alleviate the tendency to hot cracking is to prevent or reduce stresses, notably those due to thermal contraction during cooling. Thus, the method of clamping the parts and the form and sequence of the welds is very important.

13.3.3 Welding process development for automotive structures[7]

13.3.3.1 Recent events

New concepts in automotive frame structures have led to significant welding process developments. The Audi/Alcoa space frame described in Chapter 15 is one example. Another is the all-aluminum front and rear axle structures of the BMW 500 series cars.

The rear BMW axle alone is composed of some 25 components welded together into a rather complex structure. As illustrated in Figs. 13.19–13.21, 6063 extrusions and 357.0 castings are welded to the main tubular axle frame with tubes about 90 mm in diameter.

This is the first well-known case where large tubes of seam welded 5083 sheet are hydroformed to rather exotic shapes to provide the highest stiffness and fatigue strength combined with a weight reduction of 40% in comparison with the original steel structure.

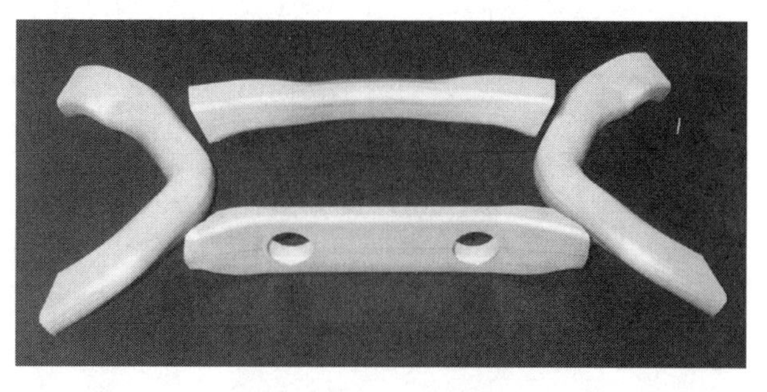

a

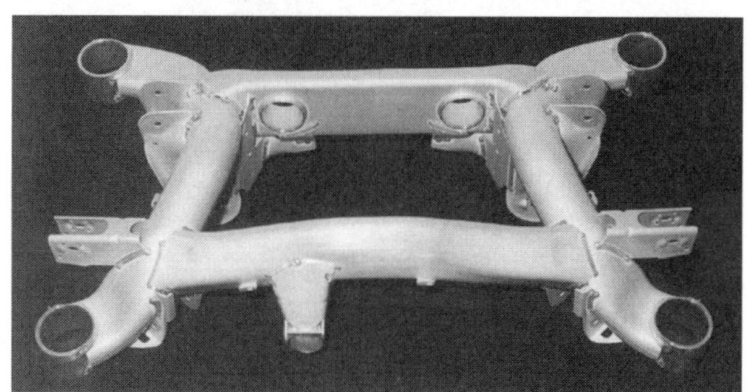

b

Fig. 13.19: The component assembly for the BMW Model 5 series: (a) Four 5083 aluminum alloy tubes (82 mm OD, 4 mm wall and 89 mm wall diameter, 4.95 mm wall), which were prebent, hydroformed to final shape, and then cut to length prior to GMA welding to the axle body and (b) rear axle body assembly.

[7] By M. Weitzer.

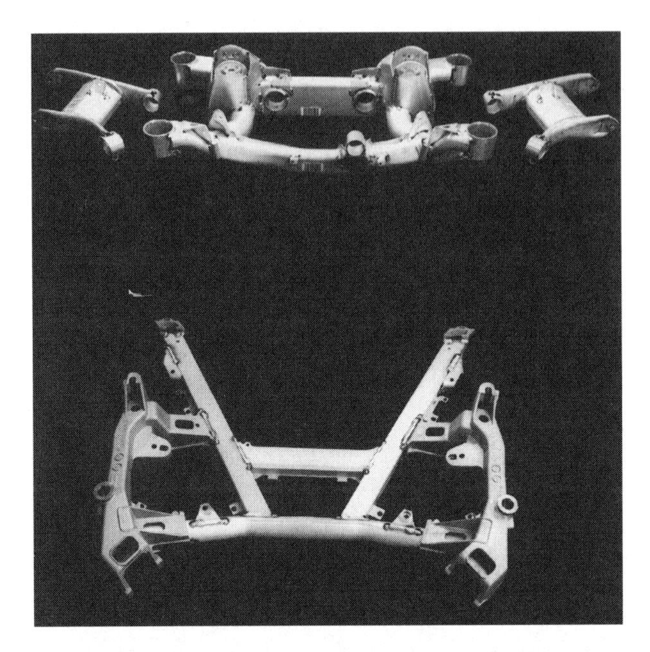

Fig. 13.20: Front and rear axle bodies for the BMW Model 5, employing 5083 tube, 6063 extrusions, and AlSi7Mg castings.

Fig. 13.21: Illustration of a complete axle assembly for the BMW model, including the all-aluminum axle structure.

Further developments of vehicle components employing formed sheet welded into car components are foreseen based upon this experience.

13.3.3.2 Current process development

Some of the most recent advances in welding process development can be illustrated with the use of aluminum in vehicle suspension elements. Suspension components are subjected to both high static and dynamic loading. In crash situations, adequate and predictable deformation is needed to protect the vehicle's occupants.

There are cost advantages if the suspension element can be installed without the need to paint. Corrosion protection can be provided by aluminum's natural oxide skin. The use of aluminum with steel bolts, however, has to be treated with caution to avoid possible bimetallic corrosion.

Economic production can only be ensured if the appropriate aluminum alloy is used with the correct welding filler wire and with optimized welding parameters. Structures manufactured in this way will be consistent in quality and meet the high quality standards required for safety parts. Stable welding process conditions provide the basis for process consistency. These stable conditions can readily be achieved using conventional Al-Mg-Si and Al-Mg-Mn alloys.

The service life of suspension components is determined to a very large extent by the notch sensitivity of the base material. It is, therefore, essential to optimize both the design and component finishing practice to avoid the creation of sensitive notch features. Aluminum has to be used intelligently in these applications. This involves optimized detail design supported by static and dynamic test results. In order to make the correct choice of material properties for the special requirements of the application, the strength and load performance of parts have to be tested in conditions as close as possible to those encountered in practice. In order to make use of welded aluminum components in automobile construction, it is essential that design features are optimized through close cooperation between production and component development specialists.

13.3.3.3 Automated welding of aluminum in the car industry

Fully automated production requires precise material processing practices combined with careful selection of the welding parameters. In order to produce high quality welds in aluminum and its alloys, process equipment capable of providing a high energy input is essential. Because of the high thermal conductivity of the base material, heat readily flows away into the component during welding. The welding equipment must, therefore, continuously replace this lost energy. A concentrated thermal energy input makes possible rapid heating and keeps the heat affected zone small.

The oxide skin has a considerable influence on the welding process and quality. An optimized thin oxide layer can be obtained on the surface of the newly formed weld bead if surfaces of the aluminum parts are adequately prepared by degreasing, neutralizing, and drying. The prepared surfaces allow the formation of stable welding arcs and give reproducable weld penetration depth with minimum oxide and gas inclusions.

Finishing tolerances of the individual parts are affected by the size of the weld gap. Large weld gaps tend to cause weld bead irregularities, cracks, failure to effectively bond, etc. Large weld gaps also affect economics and cause high scrap rates. Aluminum's high welding requirements are met by the following techniques.

13.3.3.4 GMA and GMA impulse welding

The GMA (also known as MIG) and the GMA-impulse processes arc function with a positive electrode. In GMA welding, the torch is fed with a central aluminum alloy wire electrode filler rod. This is the most favored technique for fully automated welding. The choice of alloy for the welding electrode can have a major affect upon the GMA welding process. Alloy 4043 is a proven filler rod for welding safety sensitive suspension structures. The GMA-impulse arc technique makes possible a controlled droplet formation. During the high power phase, the droplet is formed and the weld depth is established. In the low power phase, the welding wire electrode is melted, and the depth of weld penetration is limited. Switching between the two phases occurs at frequencies of 50–250 Hz.

The advantage of GMA and GMA-impulse welding compared to other techniques is their ability to accommodate a wide range of finishing tolerances.

13.3.3.5 Double welding wire techniques

With double wire welding, two welding arcs are struck from two parallel wire electrodes. This type of welding process is only stable with impulsed power input (i.e., it has to have a controlled droplet formation). The heat energy input from double wire electrodes via the two welding arcs, which generally operate one behind the other, has the advantage that the component to be welded is subjected to lower thermal energy concentrations. This limits the weld depth penetration and makes possible optimized high performance arcs that function well in the thickness range 2.5–8 mm. The technique offers a distinct improvement over single electrode arcs, and in the automobile industry it is being increasingly adopted because of the need for higher welding speeds. The welding speed is two to three times faster than conventional GMA welding but can only be used for long weld beads. The method makes possible a clear reduction in the input energy and reduces the size of the heat affected zone.

13.3.3.6 GTA Welding

During GTA welding (also known as TIG welding), an arc is formed between a nonmelting tungsten electrode and the work piece. GTA welding is primarily carried out using an alternating current power source. Modern power generators supply alternating power with a rectangular wave form. Alternating current power is used because it is necessary to give the tungsten electrode time to recover thermally during the negative phase of the cycle. During the positive half cycle, the oxide skin is destroyed, and this causes a high thermal loading on the electrode. Modern power sources, therefore, distort the cycle so that the positive phase is reduced to a minimum. Alloy tungsten electrodes with differing included oxide levels or with rare earth additions and the use of a reduced positive half cycle make possible optimized service life of the electrodes and, thus, have a positive effect on the economics of the aluminum GTA welding process. The GTA welding process is most suitable for butt-joints with the highest quality requirements. Because welding torch and welding wire supply are separate it is possible to make better welds.

13.3.4 Prospects for new welding methods

New developments in welding aluminum generally aim to reduce the heat input into the seam and, hence, the width of the heat affected zone or to develop new applications. Two

Fig. 13.22: Laser welded heater component.

recent developments that are beginning to find commercial application are laser welding and friction stir welding.

13.3.4.1 Laser welding

Welding with lasers is an example; heat input and the width of the heat-affected zone are much reduced. This largely eliminates the loss of strength in the heat-affected zone, so that it is possible to design for higher working stresses at the welds. Modern laser welding installations that are equipped with numerical control or computer aided manufacturing systems can usefully combine cutting and welding functions. For instance, the two edges can be overlapped so as to be laser cut to precisely matching contours in three dimensions and then placed edge to edge and laser welded. The need for highly precise positioning makes handling by robots essential.

Such laser welding equipment is also able to weld edge-to-edge two sheets of different thickness after prior contour cutting. The intermediate products so made are known as tailor-welded blanks. This method, already used for steel blanks in the automobile industry, leads to high-strength components that also save weight.

Laser welding has made an enormous leap in development in the last few years. For a long time, steel was successfully laser welded, but the process seemed to be unsuitable for aluminum. These problems have now mostly been overcome. The automobile industry and other branches recognize the advantages of laser welding and will promote it to better exploit the advantages of aluminum, such as weight saving and recyclability. An early and successful application of this technology is illustrated in Fig. 13.22.

13.3.4.2 Friction stir welding

Friction stir welding is a very recent development, the parameters of which are still being studied. A high-speed rotary rod, much like a drill, advances along the abutting edges of pieces to be welded. Heat input and, hence, the width of the heat-affected zone, is relatively small. For the highest quality (e.g., regarding fatigue strength), melting of the stir weld zone is strictly avoided. Instead, a quasifluid structure for welding in the solid state is the goal. However, local melting at grain boundaries may occur during stir welding of alloys with wide solidification ranges.

A first commercial application of this process has been in the deck panels of a Norwegian bridge. Experiments are continuing to better define the effects of variables such as forward speed and rotation speed of the rod and preheating and post-heat-treating the components.

13.3.5 Mechanical joining: a brief comparison with welding

13.3.5.1 Riveted joints

Riveted joints are generally used where welding is difficult or uneconomic or where heating would cause unacceptable loss of strength, distortion, or other problems. Aluminum structures are usually riveted cold. With a cold-upset rivet, the forces acting upon the joint are mainly taken up by the compressive friction on the walls of the hole and the shear and bending resistance of the rivet shank. The main subdivision is between solid and blind riveting. The use of solid rivets requires that both sides of the parts to be joined are accessible so that rivets can be closed with the appropriate tools. Blind rivets are driven from one side only and thus are used when the joints are accessible from just one side.

To prevent corrosion, the materials used for the rivets and the parts to be joined must be compatible (i.e., the contacting parts should have similar electrochemical potentials). Where riveted joints are exposed to weathering or chemical attack, insulation may be needed between the different alloys. Also, in composite structures, for example steel and aluminum members in bridges, materials with greatly differing electrochemical potentials may need to be separated from each other as well as from the rivet material.

13.3.5.2 Special bolted fixture joints

Often the most convenient and technically optimum means of joining two aluminum extrusions is to use a specially designed mechanical fixturing arrangement. The combination of relatively few welds with a high proportion of mechanical joints has become the standard for helidecks, as illustrated in Fig. 13.23. For well over a decade, specially designed mechanical joints such as those in Fig. 15.47 have been used extensively in bus skeleton structures. These joints have exceptional fatigue properties when compared to welded joints (Fig. 13.24).

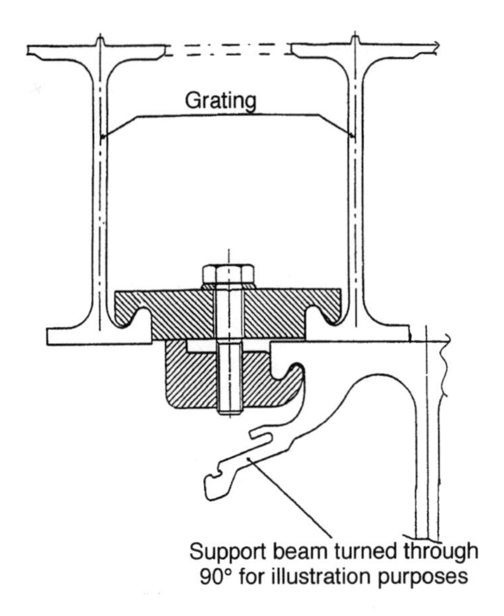

Fig. 13.23: Bolted grating and support beam structure for a helideck assembly.

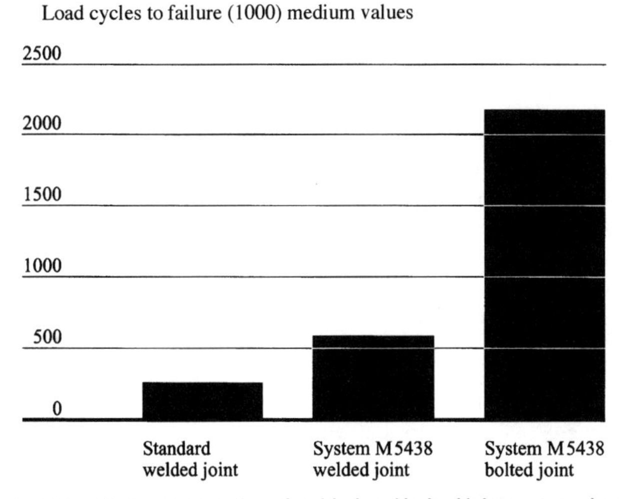

Fig. 13.24: Comparison of fatigue properties of welded and bolted joint systems in aluminum for bus skeleton structures. (Alusuisse)

13.4 Systems Development: the Aluminum Can as a Classic Example[8]

13.4.1 Introduction

The edifice of knowledge has provided the basis for the steady evolution of a wide variety of processes and products, but its crowning achievement is systems development. The forces driving systems development derive both from inside the edifice and from outside (the marketplace).

In 1963, the Reynolds Metals Company pioneered the first commercially viable method of making aluminum cans, producing the 335 ml (12 oz.) can. This process, which came to be known by the name of "draw and iron" (D&I), involved reducing the wall thickness of a cup to form a full can. In 1965, Reynolds introduced the first necked cans, which helped to reduce the amount of metal used in the ends. The successful evolution of lightweight cans has been in part due to a synergistic endeavor between the canstock suppliers and the canmakers. Productivity in canmaking plants is increased if the canstock is customized to suit the canmakers' needs. A clear understanding of all the process variables and their interactions both in making the sheet and in the can/end plant is essential to the consistent production of high-quality sheet.

13.4.2 Canstock

Aluminum cans are made from 3004/3104 alloy in H19 temper. Typical sheet production starts with ingot casting and continues through scalping, homogenizing, hot-rolling, and annealing to cold-rolling.

Before casting, the molten metal is cleaned by fluxing with gas and filtering through ceramic foam. The sheet ingot is produced by conventional DC casting or by electromagnetic casting (EMC). EMC casting results in a better surface quality than that obtained with conventional DC casting. In consequence, EMC ingot does not require scalping, resulting in higher productivity and less scrap.

Ingot homogenization is a crucial processing step. Its purposes are to:

- Remove microsegregation.
- Reduce manganese in solid solution.
- Provide a coarse size distribution of the submicron size dispersoids (to reduce earing).
- Transform the $Al_6(FeMn)$ constituents to a more desirable α-$Al_6(FeMn)_3Si$ phase that is relatively harder than the regular $Al_6(FeMn)$.
- Improve the resistance to galling during D&I.

The hot-rolling and annealing steps are critical for developing cube texture. This texture is needed to balance the 45° texture developed during cold-rolling such that the finished cansheet has low earing. Optimal microstructure in terms of intermetallic type and size distribution in the sheet at final gauge is important for controlling earing. In Fig. 13.25, a

[8] By R.G. Kamat.

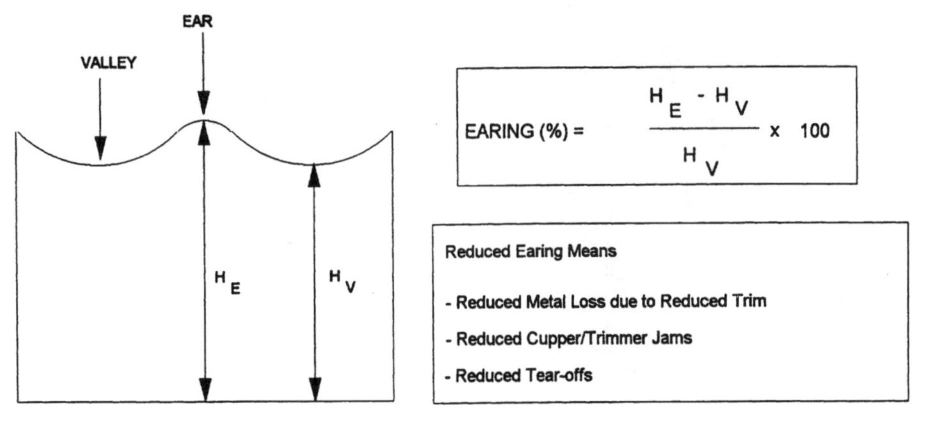

Fig. 13.25: Schematic illustration of earing in a cup/can, and the definition of % earing.

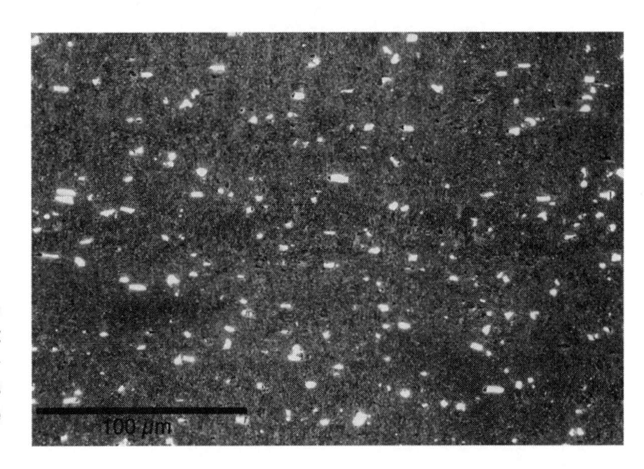

Fig. 13.26: Scanning electron micrograph of can body sheet showing $Al_6(FeMn)$ and α-$Al_6(FeMn)_3Si$ constituents (large bright particles) in the aluminum matrix.

cup with ears is shown schematically and the definition of % earing is given. This is measured for each coil produced. The ears formed on the cups result in uneven can height and leads to loss of metal as the can is trimmed to the required height. If the ears are too high, the cans may jam in the conveyor or clipped ears may lead to tearoffs. When this happens, the canmaking line is stopped, and the machine is cleared of torn cans.

Constituent size control is critical also, because of the continuing trend toward thinner can walls. Fig. 13.26 shows a scanning electron microscope image of cansheet at final gauge, with typical microstructure. The bright $Al_6(FeMn)$ and α-$Al_6(FeMn)_3Si$ intermetallics help to clean the ironing dies during the D&I operation. The evolution of the structural properties of canstock and can end stock has ensured success in downgauging.

13.4.3 D&I process

The D&I process consists of two operations: cupping, in which the sheet is blanked and drawn into cups, and body making. To form the can body, the cup is first redrawn to the

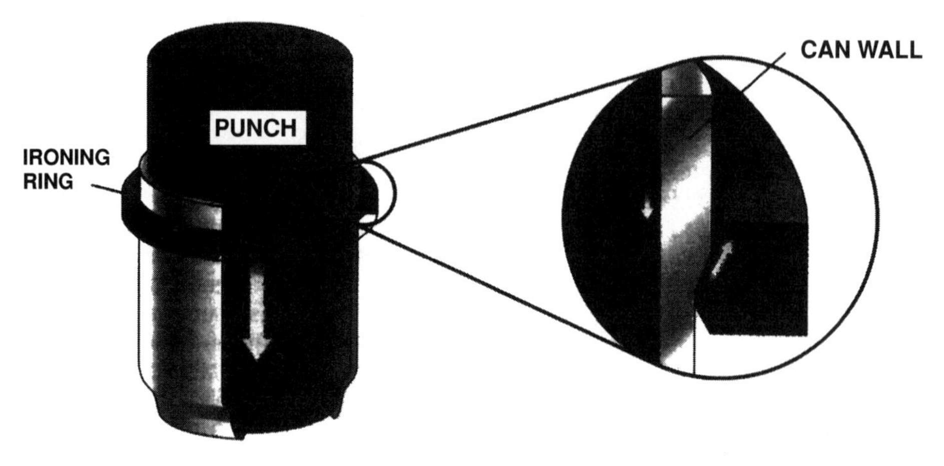

Fig. 13.27: Ironing the can walls in a can body maker.

can diameter; next, the bottom is formed to a dome, and the walls are ironed to the full height of the can. Fig. 13.27 illustrates schematically the process of ironing the walls. Ironing, using three dies reduces the sheet thickness by almost 65% and is performed at high speed (more than 250 cans per minute).

13.4.4 Can end stock

Like body stock, can end stock is made from non heat-treatable aluminum alloy and derives its strength from work-hardening during cold-rolling. Alloy 5182 (Table 13.3), with 4.5% magnesium, is currently the alloy most commonly used for endstock.

13.4.5 Can end making

The high-magnesium alloys (5xxx) used for can ends are more costly than the 3xxx series used for can bodies. In recent years, therefore, the trend to lighter gauge in can end stock has been more pronounced than that in can body stock. It has been possible to reduce gauge mainly because of the progressive reduction in the can end diameter. In the early 1980s, the stay-on tab (SOT) was introduced, replacing the pull-out ring tab. The SOT makes less litter and is safer.

Table 13.3: Nominal compositions (% by weight) of can end stock alloys

Alloy	Si	Fe	Cu	Mn	Mg
5017	0.30	0.35	0.25	0.75	1.85
5349	0.30	0.40	0.24	1.10	1.90
5182	0.10	0.24	0.03	0.35	4.50
5082	0.10	0.17	0.03	0.05	4.50
5042	0.10	0.24	0.03	0.30	3.50

13.4.6 Developments

Numerous technological developments have led to a steady improvement in the quality of canmaking sheet as well as in the processes of making the can body and end. This has made the aluminum can economical and able to compete with other packaging materials such as glass and plastic. One of the main reasons for the success of the aluminum can has been the fact that it can be recycled with no degradation of its properties. Any further increase in the recycling rate from the current U.S. national average of 65% can only improve the economic advantage of the aluminum can compared with its competitors. Higher recycling rates of 70% and 90% have already been achieved in Brazil and Sweden, respectively. Technological development continues and will further evolve, especially in view of its synergistic influence on quality and productivity.

13.5 Conclusion—Simultaneous Development of Microstructure, Process, and Product as a Prerequisite for Innovation

Many case histories show that a holistic approach integrating elements and influences from both inside and outside the edifice are needed to create promising new alloys, processes, or systems.

As an illustration, the success of the D&I can owes itself not only to the controlled fine precipitates, but also to another important structural criterion: the H19 temper, first introduced almost 30 years ago. Conventional experience and even physical metallurgy would have predicted that extra full hard cold rolled alloy sheet would have been unsuitable for precision forming to a new shape with a considerable reduction in gauge as in the D&I process. However, in high speed cold-rolling to the H19 temper, an unexpected phenomenon occurs. Within a minute fraction of a second, the increase in temperature together with a huge surplus of vacancies creates a "quasi-fluid" lattice that results in a kind of work-softening during the formation of the H19 temper.

Successful creation of new alloys, processes, and systems depends on understanding the microstructural changes which result from each process step.

Chapter 14. Aluminum in the Materials Competition

14.1 Positioning of Aluminum and its Alloys

14.1.1 The role of established alloys

The important properties of aluminum have been described in Chapters 1, 7, and 12 and its versatility in production in Chapters 8–11. Table 14.1 summarizes how its useful properties and convenient fabrication processes have led to the chief applications of aluminum materials.

Thanks to its superior properties, aluminum can (a) be the unique choice, (b) make a good substitute for another material, or (c) serve as an ideal partner in combination with another material. This versatility in application is unsurpassed by any other material.

Table 14.1: Important properties of aluminum materials and the main fields of applications resulting from them (after Murakami)[1]

Properties	Easily applied fabrication processes	Principal uses
Lightness (low density)	Many forming methods, with or without work-hardening	Transport (road and rail)
Low melting point	Casting, in standard forms or in molding	Aerospace
Medium to high strength	All machining processes	Building and architectural
Resistance to weathering	Many joining methods	
Good formability	Many processes for surface finishing	
High electrical and thermal conductivity		Electrical conductors and heat exchangers
Nonmagnetic, not poisonous		Packaging, especially food,drinks, pharmaceuticals

14.1.2 A summary of principal characteristics and applications of wrought aluminum alloys[2]

The competitiveness of aluminum alloys is based upon the principal characteristics of its main alloys. A summary of the major classes of wrought aluminum alloys, their characteristics, and major fields of applications is presented below as a bridge between the discussion on alloy development in Chapter 13 and the illustrations of many individual applications in Chapter 15. Refer to Chapter 6 for comparable background on aluminum casting alloys.

[1] Items in the three vertical columns are not related in horizontal positioning.

[2] By J.G. Kaufman.

1xxx—Pure Al; representative examples, 1100, 1350: The 1xxx series represent commercially pure aluminum, ranging from the baseline 1100 (99.00% min. Al) to relatively purer 1050/1350 (99.50% min. Al) and 1175 (99.75% min. Al). Some, like 1350, which is used especially for electrical applications, have relatively tight controls on those impurities that might lower electrical conductivity. The 1xxx series are strain-hardenable, but would not be used where strength is a prime consideration. Rather the emphasis would be on those applications where extremely high corrosion resistance, formability, and/or electrical conductivity are required (e.g., foil and strip for packaging, chemical equipment, tank car or truck bodies, spun hollowware, and elaborate sheet metal work).

2xxx—Al-Cu alloys; representative examples, 2014, 2024, 2219: The 2xxx series are heat-treatable and possess in individual alloys good combinations of high strength (especially at elevated temperatures), toughness, and, in specific cases, weldability; they are not resistant to atmospheric corrosion and so are usually painted or clad in such exposures. The higher strength 2xxx alloys are primarily used for aircraft (2024) and truck body (2014) applications; these are usually used in bolted or riveted construction. Specific members of the series (e.g., 2219 and 2048) are readily welded, and, hence, are used for aerospace applications where welding is the preferred joining method. Alloy 2195 is a new Li-bearing alloy for space applications providing a very high modulus of elasticity along with high strength and weldability. There are also high-toughness versions of several of the alloys (e.g., 2124, 2124, and 2419) that have tighter control on the impurities that may diminish resistance to unstable fracture, all developed specifically for the aircraft industry. Alloys 2011, 2017, and 2117 are widely used for fasteners and screw-machine stock.

3xxx—Al-Mn alloys; representative examples, 3003, 3004, 3105: The 3xxx series are strain-hardenable, have excellent corrosion resistance, and are readily welded, brazed, and soldered. Alloy 3004 and its modification 3104 are among the most widely used aluminum alloys because they are drawn and ironed into the bodies of beverage cans. Alloy 3003 is widely used in cooking utensils, chemical equipment, and builders' hardware. Alloy 3105 is a principal for roofing and siding. Variations of the 3xxx series are used in sheet and tubular form for heat exchangers in vehicles and power plants.

4xxx—Al-Si alloys; representative examples, 4032, 4043: Of the two most widely used 4xxx alloys, 4032 is a medium high-strength, heat-treatable alloy used principally for forgings in applications such as aircraft pistons. Alloy 4043, on the other hand, is one of the most widely used filler alloys for gas-metal arc (GMA) welding 6xxx alloys for structural and automotive applications. The same characteristic leads to both applications: a good flow characteristic provided by the high silicon content, which in the case of forgings ensures the filling of complex dies and in the case of welding ensures complete filling of crevices and grooves in the members to be joined. For the same reason, other variations of the 4xxx alloys are used for the cladding on brazing sheet, the component that flows to complete the bond.

5xxx—Al-Mg alloys: representative examples, 5052, 5083, 5086, 5183, 5754: Al-Mg alloys of the 5xxx series are strain hardenable, have moderately high strength, excellent corrosion resistance even in salt water, and very high toughness even at cryogenic temperatures to near absolute zero. They are readily welded by a variety of techniques at thicknesses up to 20 cm. As a result, 5xxx alloys find wide application in building and construction;

highway structures such as bridges, storage tanks, and pressure vessels; cryogenic tankage and systems for temperatures as low as –270°C (near absolute zero); and marine applications. Alloys 5052, 5086, and 5083 are the work horses from the structural standpoint, with increasingly higher strength associated with the increasingly higher Mg content. Specialty alloys in the group include 5182, the beverage can end alloy and among the largest in tonnage; 5754 for automotive body panel and frame applications; and 5252, 5457, and 5657 for bright trim applications, including automotive trim.

6xxx—Al-Mg-Si alloys; representative examples, 6061, 6063, 6111: The 6xxx alloys are heat treatable and have moderately high strength coupled with excellent corrosion resistance. They are readily welded. A unique feature is their extrudability, making them the first choice for architectural and structural members where unusual or, particularly, strength- or stiffness-criticality is important. Alloy 6063 is perhaps the most widely used because of its extrudability; it was a key in the recent all-aluminum bridge structure erected in only a few days in Foresmo, Norway, and is a choice for automotive space frame members. Higher strength 6061 alloy finds broad use in welded structural members such as truck and marine frames, railroad cars, and pipelines. Among the specialty alloys in the series are 6066-T6, with high strength for forgings; 6111 for automotive body panels with high dent resistance; and 6201 for high-strength conductor wire.

7xxx—Al-Zn alloys; representative examples, 7005, 7050, 7075, 7475: The 7xxx alloys are heat treatable and among the Al-Zn-Mg-Cu versions provide the highest strengths of all aluminum alloys. There are several alloys in the series that are produced especially for their high toughness, notably 7150 and 7475, both with controlled impurity levels to maximize the combination of strength and fracture toughness. The widest application of the 7xxx alloys has historically been in the aircraft industry, where fracture-critical design concepts have provided the impetus for the high-toughness alloy development. These alloys are not considered weldable by routine commercial processes and are regularly used in riveted construction. Their atmospheric corrosion resistance is not as high as the 5xxx and 6xxx alloys, so in such service they are usually coated or, for sheet and plate, used in an alclad version. The use of special tempers such as the T73-type are required in place of T6-type tempers whenever stress corrosion cracking may be a problem.

8xxx—Al + other elements; representative examples, 8017, 8090: The 8xxx series is used for those alloys with lesser used alloying elements such as Fe, Ni, and Li. Each is used for the particular characteristics it provides the alloys; Fe and Ni provide strength with little loss in electrical conductivity and are used in a series of alloys represented by 8017 for conductors. Lithium provides an exceptionally high modulus and is used for aerospace applications where increases in stiffness combined with high strength reduce component weight.

14.1.3 Aluminum in comparison with its main competing materials

This leads naturally to the question of how aluminum lines up in competition with other materials. These competing materials include magnesium, titanium, and, above all plastics, but also steel in lightweight construction.

Table 14.2: Properties of some selected competing materials (after Murakami)

Property	Aluminum	Magnesium	Titanium	Stainless steel
Freezing range (°C)	638–474	610–455	1660–1604	1427–1400
Density ($g \cdot cm^{-3}$)	2.80	1.84	4.43	7.76
UTS (MPa)	300–525	280	1205	510–860
Modulus of elasticity (GPa)	72	45.5	119	203
Coefficient of thermal expansion ($10^{-6} \cdot K^{-1}$)	23.6	26.0	8.0	16.5
Thermal conductivity ($W \cdot m^{-1} \cdot K^{-1}$)	130	84	5.6	18.0

The way in which goods are produced in industrial society depends upon their functionality. This requires that the materials used to make those goods possess certain definite properties. Table 14.2 contrasts some basic properties of a few competing materials. This juxtaposition shows clearly those individual properties that mark out a material as dominant for particular functional uses and those that make it unsuitable for others. Hence, the optimal solution to the problem of choosing a material today is often only to be resolved with a composite of two or more different materials whose properties complement each other.

14.2 Magnesium

Like aluminum, magnesium can be obtained by electrolysis. Another, less commonly used process is silicothermic reduction of magnesium oxide. The raw material for the electrolytic process is magnesium chloride from seawater, mainly in large dried-up deposits, so there is no limit to its availability. The leading producers are in Norway and North America. With a density of only 1.8 g/cm^3, magnesium alloys count among the lightest known construction materials. This is their chief advantage when compared with aluminum and titanium. However, a low yield stress and modulus of elasticity together with poor thermal and electrical conductivity limit their range of application. Nevertheless, they have a whole range of useful properties for particular applications. As Fig. 14.1 shows, for example, sandcast parts made of magnesium and aluminum alloys are very similar in strength.

Accordingly, the automobile industry has made its decision; for example, two major car firms in the USA use cast magnesium parts in the drive trains of mass-production models. The situation is different as regards components that have to be formed in the solid state, because magnesium has a hexagonal crystal lattice structure, which means that it cannot easily be plastically deformed to make sheet, plate, and extrusions. This is in marked contrast to the ductile metals such as iron, aluminum, and other non-ferrous metals that, having a cubic crystal lattice, can easily be plastically deformed.

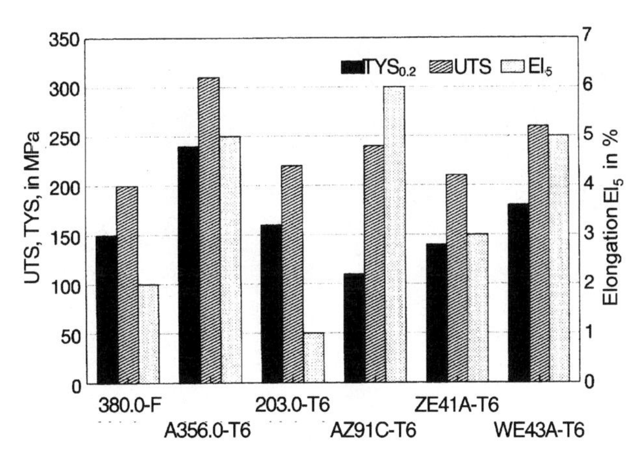

Fig. 14.1: A comparison of the mechanical properties of sandcast aluminum and magnesium at room temperature. (Honsel)

14.2.1 Applications

As long as 50 years ago, magnesium was being used in vehicle and aircraft construction; the Volkswagen Beetle had an engine block and a clutch housing of magnesium alloy. At that time, cast magnesium had the disadvantages of poor corrosion resistance and low mechanical strength, which led to its being replaced by aluminum. However, the development of alloys based on high-purity magnesium has in the meantime led to significant improvements in strength and corrosion resistance.

Magnesium has several advantages over steel and aluminum when it comes to machining. They are low specific cutting force, natural breaking of the cuttings into short pieces, less tool wear, better surface finish, higher cutting speeds, and, for the same reason, higher feed speeds.

The long-awaited breakthrough of magnesium for large-scale use in the car industry has not happened so far. This will likely remain as long as magnesium components stay appreciably more expensive than aluminum ones having similar properties, despite the greater weight of the latter. Fig. 14.2 shows typical applications of cast magnesium parts in motor vehicles.

Table 14.3 shows several industry sectors that use magnesium alloys in relation to some typical properties. The use of magnesium has increased markedly in aircraft construction and in the armaments industry because of its favorable combination of high stiffness with low weight and good machinability. The right-hand side of the table shows products in magnesium alloys, which are coming increasingly into prominence. The worldwide consumption of primary magnesium has now reached about 300,000 tonnes per year. If we consider the comparison in terms of physical volume rather than mass, then the present production of magnesium is 2% that of aluminum. However, the lion's share goes for alloying and other metallurgical uses; only one-quarter goes to structural applications (pressure die-casting). In the present European motor industry, magnesium accounts for less than 2% of vehicle weight, ranking as little more than an "also ran" among the competing materials. The magnesium industry, however, predicts vigorous growth in the world market for magnesium castings.

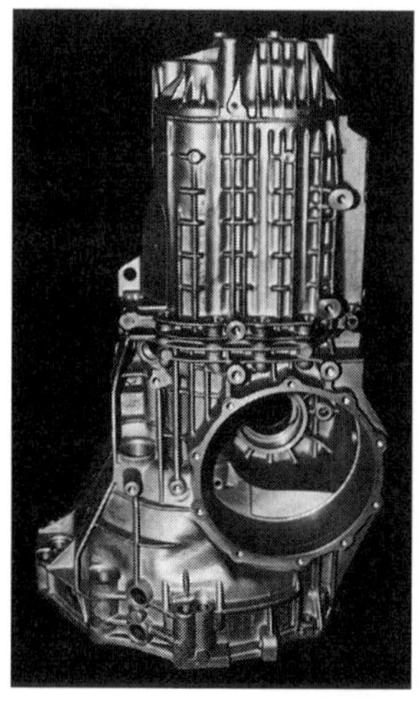

a

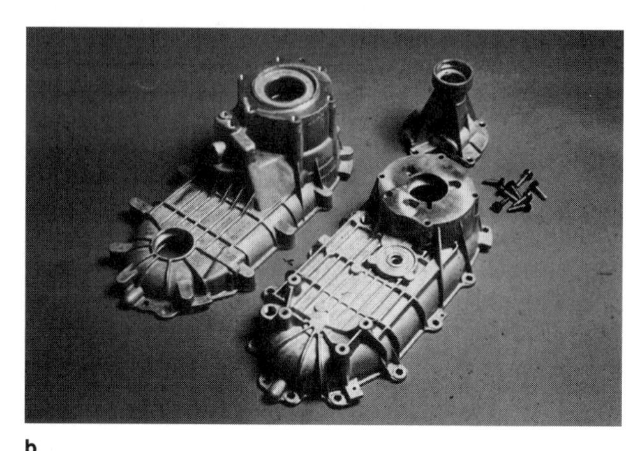

b

c

Fig. 14.2: Examples of components cast in magnesium. (a) Gear box of a 1996 VW Passat consisting of three AZ91 HP castings with a total weight of 13.49 kg compared with the original 17.6 kg aluminum gearbox (Volkswagen AG, courtesy of IWW TU Clausthal). (b) Differential gearcase, pressure die-cast, alloy AZ91HP, weight 7.7 kg (Hydro Magnesium). (c) Truck instrument panel back, pressure die-cast, alloy AM50HP, weight 4.2 kg. (Hydro Magnesium)

Table 14.3: Important properties of magnesium alloys and their resulting fields of use (modified, after Murukami)

Low density, high stiffness per unit weight	Aerospace and aircraft construction	Parts for rockets, aeroengines, helicopters, aircraft (not safety-critical)
Good corrosion resistance	Automobile industry	Wheels, engine components, dash board
Good noise damping	Computer industry	Parts for computer peripherals, enclosures for electronic assemblies
Easy machining	Machine tools	Frames for portable saws, drills Machine housings

Table 14.4: Typical applications of high-purity magnesium alloys in the car industry

Alloy designation	Mechanical properties UTS (MPa)	TYS (MPa)	El (%)	Applications
Mg-9Al-1Zn (AZ91HP) casting alloy	200–250	150–170	0.5–3.0	Easily castable universal alloy for cam covers, housings (among others)
Mg-4Al-.5Si-1Mn (A741HP) casting alloy	200–250	120–150	3–6	Creep-resistant alloy for use at high working temperatures (e.g., for gear cases, oil pumps, and engine parts.)
Mg-5Al-Mn (AM50HP)	180–220	110–140	5–9	Ductile alloy for safety-critical parts such as steering wheels, car seats and instrument panels

High-purity magnesium alloys will play an important role (Table 14.4), and today the high purity specification "H.P." is defacto the industry standard for die cast alloys.

Thus, the proportion of magnesium in vehicles could grow significantly, depending on the thrust toward reducing vehicle weight. The Japanese car makers in particular have set the goal of reducing vehicle weight considerably by the year 2000. The first step in this direction will be an increased use of aluminum parts; second, some of these will be replaced by magnesium castings, and, according to industry forecasts, even by magnesium forgings. Substitution is believed to be economic when the unit price of the magnesium part is less than 30% higher than that of the aluminum one.

14.2.2 Casting methods

The usual methods of casting magnesium are sandcasting, permanent mold casting, and pressure die-casting, the latter being the far most important with 80–90% of the total. The alloys most often used in sandcasting and permanent mold casting are AZ81 and AZ91. Other alloy groups contain no aluminum, containing instead elements such as zinc, zirconium, rare earths, and silver. Such alloys, requiring special melting and casting techniques, are mostly used for making aircraft parts. Practice in sandcasting follows broadly the same lines as with aluminum, but it must be kept in mind that magnesium reacts vigorously with the oxygen in the air and with humidity from the molding sand. Protective gases and other protective materials are used in order to prevent or slow down these reactions.

By far, the most economical way of making magnesium castings is pressure die-casting. In many cases, the somewhat higher raw material cost is offset by the higher productivity resulting from the faster casting rate, the doubled tool life, the closer dimensional tolerances of the thinnest sections, and the ease of working relative to those used with aluminum. Of all the casting processes, pressure die-casting yields the best surface finish and the closest dimensional tolerances. The alloy AZ91 is the one most often used for pressure die-casting. Both cold-chamber and hot-chamber casting machines can be used. The metal has a lower thermal capacity per unit volume than aluminum, so the time it takes to solidify in permanent mold casting is commensurately shorter. The structure is fine-grained, and the cast part is mechanically stronger than with sandcastings of the same wall thickness. Thixocasting is a new casting process being tried, but it is not yet in wide use with magnesium.

The fabrication of cast products from magnesium is burdened with concerns for environmental protection, which are common to all non-ferrous foundries. Modern melting technology uses gases instead of salt fluxes to prevent the liquid magnesium from oxidizing; however, extra care is generally taken to minimize the use of protective gases such as SF_6 and SO_2. Environmentally hostile melt treatments are not used in pressure die-casting magnesium. Magnesium castings are very easily recycled. High-purity scrap can be recycled straight back into high-purity alloys; even mixed scrap or metal from shredders can be remelted or otherwise used as alloying material.

14.3 *Titanium*[3]

Titanium is the youngest of the light metals, having been used to a significant extent only since the beginning of the 1950s. Its combination of low density (4.43 g/cm^3) with outstandingly good corrosion resistance and high strength at elevated temperatures have assured the metal a certain share of the market. Table 14.5 shows its typical properties and the end uses for which they are suitable. Because of its high price, titanium is not really in the race to supplant other metals, despite its very superior properties. It has its own niche markets in aircraft construction, the chemical industry, and armaments.

Table 14.5: Important properties of titanium alloys and their corresponding chief uses

Properties	Main area of application	Products
Low density, high strength	Automobile industry, aerospace, weapons, electricity generation	Aeroengines, airframes and skins, cardan shafts, valves and springs
Very good corrosion resistance	Chemical industry (withstands boiling seawater corrosion)	Condensers, turbine blading, heat exchangers, reactors and pumps, desalification equipment
Non-poisonous, well-tolerated by the human body, low modulus of elasticity	Implants	Prostheses for knee, hip, and tooth implants
Good combination of properties	Superconductivity, cryotechnology, shape memory alloys	Strong and lightweight magnets, sensors

[3] Source: H. Sibum and G. Stein, DTG Essen in *Ergebnisse der Werkstofforschung*, Bd. 5, Swiss Academy for Material Science, as well as G. Lütgering, BDU Werkstofftag 1991.

The reason why it was so late coming into regular use is the great expense of reducing it from its oxide. High-technology processes (the Kroll and the Hunter processes) have made possible the economic extraction of the metal only since the end of the 1940s. The intermediate step, in which titanium tetrafluoride is formed from titanium dioxide, is common to both processes. The Hunter process uses magnesium, and the Kroll process uses sodium to bring about the final reduction from which the metal emerges as sponge. There are large-scale producers in Russia, Japan, and the USA, which together cover the entire present world demand.

14.3.1 Titanium alloys

The titanium alloys are divided into groups that are clearly distinguished according to their structure and/or their field of application (Fig. 14.3):

- $\alpha + \beta$ titanium alloys to be used in the medium temperature range up to a maximum of 300°C. The microstructure consists of about 85% hexagonal α-phase and about 15% of the body-centered cubic β-phase. The most used alloy is Ti-6Al-4V.
- β titanium alloys, also for the medium temperature range. These alloys are also basically $\alpha + \beta$ alloys, but they consist preponderantly of β matrix with α precipitates. Examples are Ti-023 and "Beta-C" alloys.
- High temperature alloys to be used up to a maximum of 600°C. These alloys contain little or no β-phase and, hence, are referred to as "near α." Well-known examples are Ti-6242 and the newer alloys IMI-834 and Ti-1100.
- New alloys, still in the development phase, are those based on the intermetallic compound $Ti_3Al(\alpha 2)$. These alloys are designed for use at temperatures above 600°C. The alloy "Super Alpha 2" (Ti-14Al-20Nb-3.2V-22Mo) is already on the market; it contains the well-ordered hexagonal Ti_3Al phase.

14.3.2 Fabrication methods

The raw material for making semi-fabricated products is titanium sponge, which has to be remelted because of its unsuitable physical form and the impurities it contains. This refining operation gets rid of the easily vaporized residues (e.g., sodium) and, depending

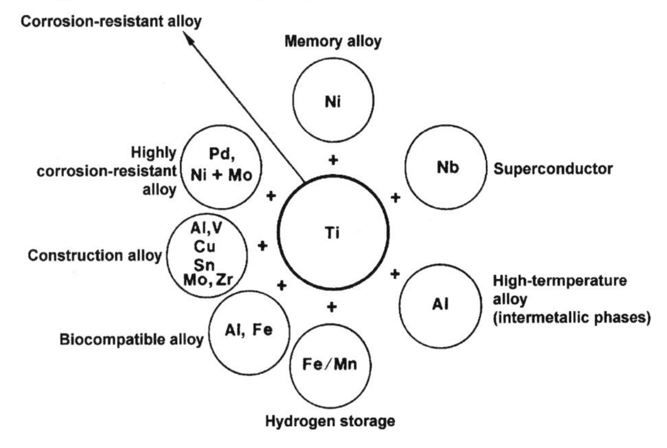

Fig. 14.3: The application range of titanium alloys.

on the remelting process used, results in a product in the form of a molded casting, an ingot, or a slab. Since pure titanium has no direct practical use because its mechanical strength is too low, the melt is modified at this time so as to reach the required composition. With pure titanium, this is done by ensuring a prescribed level of oxygen (embedded foreign atoms). Generally, however, it is done by additions of alloying elements such as aluminum, vanadium, tin, molybdenum, or zirconium. Alloying elements stabilize the various crystallographic phases of the titanium in the desired direction. Fig. 14.3 shows the very wide spectrum of properties that can be obtained in this way. The production methods are closely related to those used in making specialty steels. Remelting units such as vacuum electric arc furnaces, electron beam, and plasma furnaces have proved their worth in the production of slabs and ingots in large sizes. Similarly, investment casting and the main specialty steel casting methods are used, as well as the common forming processes like forging, extruding, rolling, and drawing.

14.3.3 Applications

14.3.3.1 Aerospace

The use of titanium alloys to reduce weight and allow higher working temperatures in the compressor and high-pressure stages of jet engines has long been state of the art. The working temperatures of the near-α titanium alloys have been steadily increased through alloy development, so that components that formerly were usually made of superalloys can now be made of titanium. On the other hand, similar developments have led to the replacement of titanium by aluminum in the lower-pressure, cooler turbine stages of jet engines. In the future, flying at several times the speed of sound will require skin materials that maintain their high strength at substantially higher temperatures than those used at present. For the hypersonic speed range, and especially in the contingency planning for the development of the SST, titanium alloys are in the forefront.

14.3.3.2 Automotive

Titanium has demonstrated its reliability in motor-racing for many years. As with aircraft, reducing weight may be a good reason for using titanium. Here, it is important to consider the movements a component must make and, in particular, the accelerations to which it is subjected. Parts subject to cyclical or reciprocating motion, especially in the engine, offer the greatest potential for fuel saving by reducing weight. Titanium is technically completely satisfactory as a car making material, having all the necessary characteristics, including:

- Load-bearing capacity determined by its chemical, mechanical, and other technical properties.
- Fitness for the function (e.g., in valves, connecting rods, springs, and spring backing-plates).
- Safety, long lifespan determined by the reliability of the material.
- Recyclability

It may, however, be hard to justify its use economically.

14.3.3.3 *Energy production and storage*

In the future, special alloys based on titanium could play a useful role. Following the use of niobium-titanium as a superconducting material in medical technology (nuclear magnetic resonance tomography), a substantial research and development effort is under way to put superconductors to the test in electricity generators and in the mass storage of electrical energy. The Japanese government has declared high priority for a project to develop electrical energy storage in enormous superconducting storage loops. The SMES system (superconducting magnetic energy storage) can supply its stored energy on demand with virtually instantaneous response and no losses.

Niobium-titanium will remain, at least for the next one or two decades, the only practically usable superconducting material despite the very low operating temperature required. Its superiority has been proven in the Japanese Levytrain.

The hydrogen-absorbing alloy based on titanium is worth mentioning for its potential use as an energy storage element in motor vehicles, because it offers a way to store the hydrogen cleanly, safely, and not under pressure.

14.3.3.4 *Chemical industry*

Titanium has a high resistance to corrosion because its passive surface layer is extraordinarily resistant to chemical attack. The layer consists of a non-porous oxide layer that forms spontaneously and clings tightly to the underlying metal. Titanium's corrosion resistance can be improved still further by additions of palladium or small amounts of nickel and molybdenum. These low-alloyed titanium materials are the preferred choice of the chemical industry. High resistance to both corrosion and erosion even at elevated temperatures makes these materials the chemical engineer's automatic choice for building equipment such as tanks and plate- or tube-type heat exchangers.

14.3.3.5 *Medical*

For years, titanium alloys have proved themselves for use in implants because of their outstanding compatibility with body tissues. Hip joint replacement imposes the most severe requirements for load-bearing capacity and long life; false teeth, tooth implants, bone replacements, and operating instruments must also possess this virtue of being tolerated by the human body. Today, when many people complain of allergies (among others, of being allergic to nickel, an important component of stainless steel) there is a change in attitude as regards any metallic materials in constant contact with the skin. It is for this reason that eyeglass frames and wristwatch bands are, to an increasing extent, being made of titanium alloys.

14.4 Ferrous Materials[4]

There is a vast edifice of knowledge about steel and other ferrous materials, which we cannot hope to cover herein. Instead, in what follows, we just outline some differences in their properties as compared with aluminum materials. These are illustrated with a few selected examples of steel applications and recent developments.

In Table 14.2, 18-8 stainless steel was quoted as a competing material for several applications in the context of the keen competition between stainless steel and aluminum. Examples of this are rail vehicles, cladding for building facades, tanks, and other large containers, to name but a few. In most countries where the two materials are in competition for these applications, aluminum is used more than stainless steel in terms of area covered. The picture is quite different, however, when other grades of iron and steel are included, and the worldwide production of ferrous metals is at least twenty times that of aluminum in terms of physical volume. In spite of this, the areas where aluminum is being substituted for steel are not negligible—almost exclusively in packaging and transport, but to some extent also in machine tools.

Apart from lightness, it is nearly always the same properties that have a bearing when we consider substituting aluminum for ferrous materials: corrosion resistance, decorative surface appearance, and versatility in forming by casting, extruding, and rolling (right down to thin foil). This is covered in Chapter 15 for a large number of end uses. The mechanical properties of steel and aluminum are compared in Table 14.2. An essential point to grasp is that although steel is three times as stiff as aluminum due to its higher modulus of elasticity, it is also three times as dense. This means that the stiffness of sections with the same weight is about the same in either material. However, the stiffness of ribs or flanges running parallel to the loading direction increases with the third power of their height, so that by using such stiffening elements it is possible to replace steel structures with aluminum ones roughly half their weight, albeit with a deeper cross-section.

The weight of structural components in substitute materials depends not only on the cross-section and the joining methods used, but also on the nature of the applied loads. This is especially clear in the case of torsional loading or in sharp bending. From Fig. 14.4, we see that 6063 has lower tensile strength than 5083, but an extruded section (with stiffening members) made of the former alloy is clearly superior to a plate made of the latter, because it is lighter. On the other hand, the weight of a part with the same buckling strength is the same for both alloys, that is, almost exactly half that of an equally stiff steel structure. For structural applications, therefore, the general rule applies that for equal stiffness, 2 kg of steel can be replaced with 1–1.4 kg of aluminum. Today, the competition between steel and aluminum in vehicle construction has become so severe both on the technical and the economic level that frequent reference is made to this fact in the course of this book. In fact, steel, aluminum, and plastics are today engaged in a kind of marathon race in the American, Japanese, and European car industries, a race brought about by the legal requirement to reduce average car weight by 150–200 kg.

---

[4] By M. Speidel.

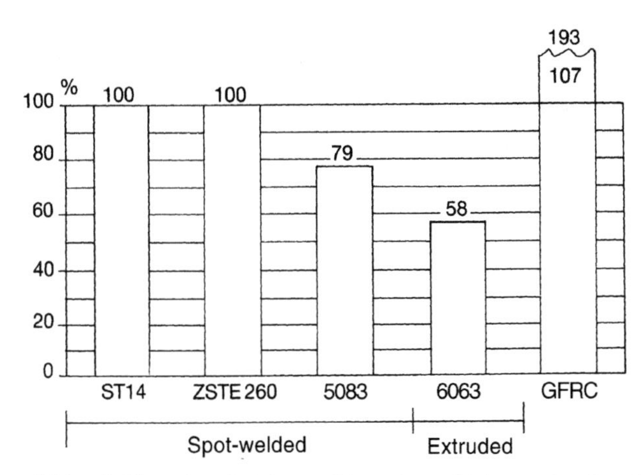

Fig. 14.4: The weights of different materials used in motor vehicle construction for a component having the same torsional stiffness. (after Ostermann)

Lightweight construction materials are not the only winners in this marathon, however, as improved grades of steel and especially new methods of making steel components are also resulting in weight savings. Steel's share in car making has fluctuated for many years around 65%. The proportion for 1993 was 60%, and it is likely to remain at this level for many years to come. How can this be, when the producers of aluminum, magnesium, plastics, ceramics, and composite materials spare no effort to chase steel out of the market, with the argument above all of lower density and, hence, lighter components? There are several solid reasons for the continuing competitiveness of steel. In the first place, the material itself as well as its fabrication methods are being improved and made cheaper by steady, ongoing research and development. Second, steel has been recycled since its very beginnings by remelting scrap. Third, the very wide distribution of iron ore in the world, together with the ease with which it can be reduced to iron, provides a guarantee of future supplies of this cheap material.

Steel and its alloys have improved steadily and in small steps for decades. Starting with the long-proven grades like A7/St37 and A36/St52, which have been industry's workhorses for many years, today's steels offer a wide range of improved properties to suit many applications. High-strength low alloy steels (HSLA), for example, are very fine-grained and contain very small and accurately controlled amounts of alloying elements such as niobium, vanadium, titanium, aluminum, boron, and nitrogen in order to achieve the very uniform mechanical properties that are so important for hot- and cold-rolled strip as well as for the subsequent annealing operation. After the introduction of the deep-drawing grades St14 and St15, steady persistent development over 20 years (notably by using modern metallurgical processes) has led to a substantial reduction in the levels of carbon, nitrogen, sulfur, and phosphorus. This, in turn, has improved their deep-drawing qualities, especially the elastic limit, and made them more uniform.

The most modern fabrication plants, using high performance processes and "smart" systems, justify perhaps even more than does the material itself the assertion that steel will long remain a basic material for car making. Tailored blanks, compound steel sheets made to measure for specific types of application, offer an example of what such installations can do. Tailored blanks use laser beam welding to make compound sheets or plates out of individual sheets of different steel grades, surface coatings, or thicknesses. They offer some totally new opportunities to the designer. For example, a 1.5 mm sheet of hot-dipped soft St05Z and a sheet of 1 mm thick cold-rolled high-strength ZStE300 can be butt-joined by laser beam welding with a 0.8 mm sheet of hot-dipped high-strength FHZ220. Such a compound sheet can then be used to make car parts such as support brackets, doors with integral strengthening in the hinge area, wheel arches, integrated McPherson struts, floor plates reinforced in the area of the transmission tunnel, collision-resistant sidewalls, and so on. Besides laser beam welding, pinched-seam welding can also be used.

The advantages of tailored blanks can be summarized as follows: lower weight; reduction in the number of parts and the consequent simplification in production methods; higher productivity; lower costs of investment, assembly, material, and logistics; closer dimensional tolerances; improved creep and local corrosion resistance; better performance under crash conditions; easier repair; and recycling. These all enhance steel's attractiveness as a material for making cars. The disappearance of overlapped spotwelded joints when tailored blanks are used means that the costs of sealing and sealing compounds vanish, and corrosion resistance improves. It is especially important that the introduction of tailored blanks is directly in line with the trend of the automobile industry away from backward integration and toward buying finished subassemblies from subcontractors. This may give fresh impetus to the competition between steel and aluminum in car making, benefiting the trend toward lighter vehicles and easier disposal at the end of their lives. This competition is, by the way, not restricted to bodywork alone; it extends to cast parts for the engine (connecting rods and gearcasings) and to hang-on parts as well as to the wheels, where the two metals compete directly.

14.5 Plastic Materials[5]

Historically, plastics were seen as substitutes for metals. Since then, the synthetic polymers have become a group of materials in their own right, comparable with metals as much in quantity as in variety, application range, and price. The worldwide production of plastics in 1991 amounted to some 80 million tonnes or cubic meters. Synthetic plastics have only been known for about 70 years (celluloid, phenol resins) and mass produced plastics (polyethylene, polystyrene) since the 1940s.

All plastics are chemical compounds (giant molecules called polymers) based on carbon, with hydrogen, oxygen, and sometimes other elements. Most industrially produced plastics do not have a crystal lattice. This results in a characteristic that makes these polymers still appear as a sort of substitute material—they are affected by heat, chemicals and micro-organisms, and may absorb water; in short, they degrade. Their properties change during their service life, and change not only on the surface, as with metals, but in their whole cross-section. For this reason, there are hardly any large, industrial, or architectural objects that are made entirely from plastics.

[5] By J. Schrade.

14.5.1 Properties and fabrication

From a manufacturing point of view there are two main groups of plastics:

- Thermoplastics. These soften or melt at higher temperatures and harden on cooling. Thus, they can be hot formed, remelted, and recycled without chemical change.
- Thermosetting resins. These undergo an irreversible chemical reaction that hardens them. In general they can only be recycled as powdered fillers.

Table 14.6 compares the properties of aluminum with those of selected thermosetting and thermoplastic polymers. Conspicuous differences are the thermoplastics' lower values for elastic modulus, strength in bending and tension, and working temperature. On the other hand, some of the properties of some thermosetting resins are similar to those of aluminum. This imposes a major distinction between the applications of the two groups of plastics.

Thermosetting resins cost more to make than thermoplastics. Thermoplastics can be very cheaply formed to profiles and plates continuously (by extrusion) or made into molded shapes discontinuously (by injection molding or blowing). By contrast, thermosetting resins are still fabricated almost exclusively by discontinuous processes. The injection molding process for plastics closely resembles pressure die casting of aluminum. However, it costs less to mold plastics at about 250°C than aluminum at about 700°C, because the lower temperature allows higher productivity with shorter cycle times and cheaper mold materials.

The thermoplastic polymer with the highest rigidity is POM, a partially crystalline polymer based on formaldehyde, which is itself made from methane/methanol, a readily available and cheap raw material. Curiously, it has so far not proved possible to rationalize the polymerization process so as to bring the price level of POM down to that of the common polyethylene and polypropylene. Polyethylene and polypropylene are cheaply produced on a large scale all over the world and are, therefore, used in many mass produced components, for instance quite large parts in vehicles. Because they are cheap, the required stiffness can be achieved simply by increasing the wall thickness. Since plastics have no continuous crystal lattice (even the so-called crystallized polymers like PA and POM have only limited crystallized zones), they tend to creep under load. As a rule of thumb, the allowable creep-free permanent load at 20°C, expressed as a fraction of its short-term tensile and bending strength, is between 30% and 40% for thermosetting resins, and between 20% and 25% for thermoplastics. It is a common experience that designers ignore these limits, so that, for instance, tightly screwed plastic parts deform because the creep limit was exceeded.

14.5.2 Plastics in vehicle construction

It is a general rule that plastics dominate for inside fittings, whereas aluminum plays the main role in the structural and outer skin elements (aircraft, railway, and commercial vehicles). Automobile construction is a special case, in which aluminum and plastics often replace each other. Fig. 14.5 shows two cases where plastic was substituted for a typically aluminum component.

Table 14.6: Comparison of the typical properties of aluminum with those of selected plastics

		Thermosetting				Thermoplastic					
	Al	EP (glass fiber reinforced)	UP	EP (with carbon fibers)	PF (pressed solid)	PVC (hard)	PE (low density)	PP (high density)	PA	POM	PUR (easily cross-linked)
Maximum continuous service temperature (°C)	150	130	120	130	130	70	60	80	90	100	100
Elastic modulus (GPa)	70	20	7	40	10	2.5	0.3	0.9	1.2	2.5	3.5
Bending (tensile) strength (MPa)	100–400	400	150	500	50	60	25	35	65	80	70
Coefficient of friction between like materials	1.9	0.5	0.5	.04	0.5	0.3	0.2	0.2	0.2	0.3	0.2

EP: epoxy resin; PVC: polyvinyl chloride; PA: polyamide;UP: unsaturated polyester; PE: polyethylene; POM: polyoxymethylene; PF: phenoplast PP: polypropylene; PUR:polyurethane.

Fig. 14.5:(a) Air inlet manifold made of glass fiber reinforced 66-Polyamide, stabilized to resist high temperatures (BASF). (b) Car spoiler made of resin-transfer-molding (RTM) of epoxy-glass fiber prepreg. (Alusuisse Automotive).

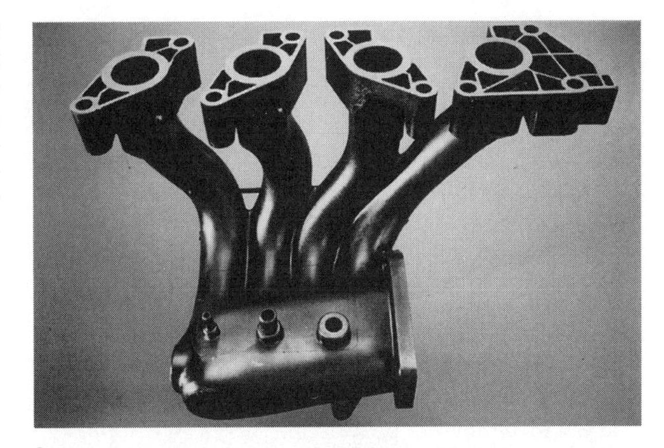

a

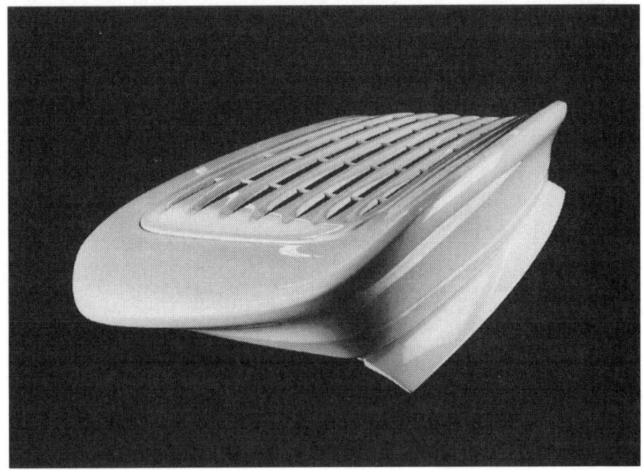

b

The central theme here is not that of plastics replacing metals; rather, the question is to what extent have plastics become indispensable to automobile construction, and what general direction of change does this impose? Components now being made of plastic, such as spoilers, bumper bars, door components, motor hood, and trunk lid often replace steel sheet, but not so often aluminum. However, there are cases where plastic can be economically molded into shapes not practical for either steel or aluminum. Most analysts agree that with the need to further reduce vehicle weight, the proportion of plastic will significantly increase.

In the domain of mechanical components, aluminum pressure die cast components have in recent years been replaced by cheaper injection molded plastic parts. The plastics used resist the temperatures in the engine compartment, and when sited in places exposed to splashes of sometimes salty melt water from the road, they are more corrosion resistant than metals. Examples of these plastic components are carburetor and fuel pump housings, fuel injection accessories, air inlet pipes, and brake cylinders.

14.5.3 Joining aluminum and plastic

While it is true that in many applications the two materials compete directly as alternatives, they are more often used together in composite materials (e.g., packaging). Careful preparation of the surfaces is a decisive factor in producing a durable bond. A glued joint offers a good example of this.

Aluminum is normally bonded to plastics by gluing. Many different types of glue are used, depending particularly on the type of plastic. The advantage of gluing instead of spot welding, screwing, or riveting is that gluing transfers the force uniformly, without local stress peaks at attachment points or edges of holes and so improves fatigue life. Gluing is performed at temperatures below 200°C, which for a short duration has no significant effect on aluminum. The basis for industrial gluing of aluminum was laid with the invention of epoxy resin. These can be applied in liquid form without added solvents and can then be reacted in various ways with so-called hardeners, without releasing troublesome by-products. The result is an insoluble, non-melting thermosetting polymer.

The further development of epoxy resins lead to glues that are called single component or two-component glues, which may harden at room temperature or by heating and mechanically may be made in varieties varying from rigid-brittle to soft-elastic combinations of properties.

Besides epoxy resins, there are other polymers like polyurethane (soft-elastic) and polyimide (resistant to high temperatures) that have earned an important place as glues for aluminum. In addition, a whole series of elastomers are used for flexible, elastic connections to aluminum. If moisture produced during the glue-hardening reaction can escape (as in gluing to honeycomb structures or to balsa wood), then phenolic resin, usually modified, is also very suitable for gluing aluminum. Special glues are needed to attach aluminum to polyolefins (polyethylene, polypropylene), since the usual glues do not stick to these non-polar polymers.

14.5.4 Disposal and recycling

The state of the art at present only permits the re-use of clean thermoplastic scrap that is separated by type. This means process scrap and trimmings from manufacture by injection and blow moldings, extrusions, foil, and plates. Scrap may also come from further fabrication steps (such as the assembly of windows and furniture, deep drawing of components, and packaging) and from packing materials. It would also be possible to recycle trimmings when installing building fittings (floor coverings, etc.), but this has hitherto been too awkward for the building workers, so that such trimmings usually end up as general building site rubbish. The only type of recovery so far available to treat contaminated plastic scrap invariably containing mixed types of plastic is pyrolysis (breakdown at high temperatures). Pyrolysis does not yield complete polymers, but only small, basic molecules. So far there are no large-scale pyrolysis facilities because it is not profitable enough.

Composites of plastics with aluminum are particularly difficult to recycle. Laminated or printed aluminum foil less than about 0.1 mm thick cannot be recovered by melting, because the large organic fraction reacts chemically with the aluminum, producing more dross than metal. Aluminum parts thicker than about 0.2 mm can certainly be melted out of composites, but intensive cleaning of the melt is needed to produce good quality. It is therefore necessary to separate intimate mixtures (e.g., by grinding). From the ground materials of uniform particle size, aluminum can be separated by density, and both components can be recycled.

Thermosetting polymers cannot be recovered by melting. They are generally finely ground and used as fillers with fresh thermosetting materials. An interesting alternative way to use such scrap is to grind it to fine particles for "sand blasting" to remove paint. The skin panels of aircraft must be periodically cleaned of paint, so that they can be checked for hairline cracks and repainted without accumulating additional weight. Plastic grit blasting removes paint very well and saves the purchase and disposal of strippers and solvents. The mixture of plastic grit and old paint can be burnt to recover its heat energy.

14.6 Composite Materials[6]

There is available a very wide range of well-proven wrought and casting aluminum alloys with outstanding chemical and physical properties from which to make finished products. Despite this, these properties do not always manage to satisfy the customer's requirements. This has led to the consideration of novel combinations of different materials with correspondingly useful combinations of properties. Designers have accordingly gone over to using hybrid materials where occasion demands, which opens the way to increasing the strength of a component in whole or in part. Of course, wherever composite materials are considered, the need for recyclability must always be kept in mind.

14.6.1 Composites of different aluminum alloys

For many years the aluminum industry has made composite products, chiefly by rolling, by a process known as "cladding." It consists of bonding a layer of a different alloy to the surface of the core material during hot rolling. The cladding material, in the form of a plate that has already been hot-rolled, is temporarily attached to one or both faces of the sheet ingot. This assembly is preheated in the usual way and immediately hot-rolled using a special, predetermined pass schedule (Fig. 14.6). Under the high temperature and pressure in the rollbite, the cladding plate becomes metallurgically bonded to the core ingot. Once this bond has been firmly established, the ingot is rolled down to the required hot-rolled gauge as usual, resulting in a semi-fabricated composite product combining the properties of the core and cladding alloys.

Sheet clad to obtain high corrosion resistance for use as skin sheet in aircraft construction is the earliest known use of composite materials in the aluminum industry. The core material consists of a heat-treated high-strength alloy, usually 2024/2124 or 7075/7475. The corrosion resistance of these high-strength alloys does not meet the extreme requirements

[6] By G. Scharf.

Fig. 14.6: Aluminum sheet ingots with cladding plates temporarily attached, before hot rolling. (VAW)

Fig. 14.7: Prestressed plate heat exchanger ready for the furnace brazing operation. (VAW)

of the aircraft industry, and for this reason they are clad, typically to 3–5% of the total thickness, with pure aluminum or 7072. In this way, the surface of the core material is shielded against corrosion by the cladding. Cut edges and bored holes benefit from cathodic protection by sacrificial corrosion of the cladding layer. Therefore, it is most important to take into account the corrosion potential of the cladding alloy. If the cladding has a more negative electrochemical potential than the core, then the core will receive effective cathodic protection.

Car radiators and heat exchangers are examples of the commercial application of brazing sheet, which is a composite material clad on one or both faces. The cladding, suitable for brazing, consists of an aluminum alloy whose melting point is 30–50°C lower than that of the core alloy. This is an essential requirement for brazing (i.e., binding together metallic materials with a layer of molten metal [the braze]). The condition is satisfied by a eutectic or hypoeutectic Al-Si alloy. The core is Al-Mn or Al-Mg-Si alloy. Core and cladding are bonded by hot-rolling, as already explained. The thickness of the cladding is usually 5–10% of the total thickness.

Taking the fabrication of a multiplate heat exchanger as an example, the separator plates and the main plates are prefabricated from clad sheet cut to size, then cleaned and dried. They are then assembled in a stack and squeezed together (Fig. 14.7). The brazing process itself takes place at about 600°C with or without flux. If without flux, then brazing must be done under high vacuum or in an inert gas atmosphere. At every place where the individual clad elements are in contact with each other, a brazed joint is formed, binding the elements firmly together.

A great many components require not only good mechanical properties, but also a decorative, glossy surface. On both technical and economic grounds, these requirements can seldom be simultaneously satisfied with a single material. The solution is to use a composite material consisting of a core of high-strength alloy that is clad on one or both sides with a high-purity bright alloy. A typical composite semi-fabricated product for making parts with high surface reflectivity has a core of a suitable Al-Mg alloy clad with a 5–10% layer of high purity aluminum (e.g., 1160). Such a material is very suitable, for example, for making headlamp reflectors. The core material is easily formed and has satisfactory strength and stiffness after forming. At the same time, after polishing and chemical or electrolytic brightening with subsequent protective anodizing, the clad layer provides an outstandingly good reflective surface.

14.6.2 Composites of aluminum and other metals

Composite materials consisting of aluminum together with other metals have also been developed. However, this means that because the two metals differ greatly both in their resistance to deformation and in the nature of their surfaces, procedures must be used to fabricate the components that differ from those used with aluminum materials alone. On top of this, the method of making the semi-fabricated product also differs from that described. Both may often lead to hot-rolling at much reduced temperature or even to cladding at room temperature.

At the beginning of the 1960s, a clad material was introduced for making evaporator plates for refrigerators, deep-freezers, and industrial cold rooms. It consisted of a sheet of Al-4Zn alloy (e.g., 7005)—an almost eutectic zinc alloy—sandwiched between two aluminum sheets. The special feature of this composite material is that the zinc alloy has a much lower melting point (400°C) than the aluminum, as well as an extremely short freezing range. Hence, when this sandwich is heated, the zinc layer melts as soon as the temperature reaches 400–420°C, whereas the aluminum remains solid. Before heating, the sandwich is clamped between the two halves of a tool in which semicylindrical

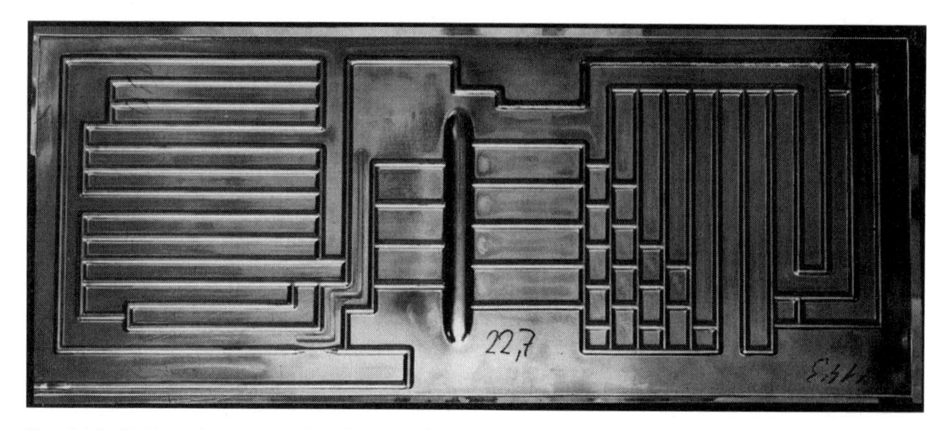

Fig. 14.8: Refrigerator evaporator plate made from a composite consisting of a sheet of zinc sandwiched between two of aluminum. (VAW)

grooves have been worked in each face. After the temperature of the assembly of sandwich sheet and tool has reached the melting point of the zinc phase, compressed air is blown into the hollows of the tool, forcing the aluminum to conform to the shape of the hollows and forming a network of tubes (Fig. 14.8). During subsequent cooling, the Al-4Zn layer resolidifies, so that all the areas that had not been separated by the action of the compressed air are brazed together.

There are many fields of application for composites of steel and aluminum. Typical examples are exhaust systems and slide bearings for motor vehicles, heat-conducting bottoms for pots and pans, housings for household appliances, and bicycle mudguards. Cladding is carried out not by hot-rolling as with composites of two aluminum alloys, but by cold-rolling. The steel is rolled typically to 3–5 mm thickness and the aluminum to 0.2 mm in separate operations. The surfaces are then prepared by etching and brushing immediately before the cladding operation, which consists of rolling the assembled sheets with a 40–60% reduction. Under such a heavy reduction, the aluminum reaches 150–250°C and bonds to the steel by cold pressure welding. The coiled composite then undergoes immediate interannealing and further cold-rolling. It is important to set the annealing conditions so as to avoid forming brittle intermetallic phases of Al-Fe-Si. The process is suitable for one- or two-sided cladding. Either pure aluminum with 0.7–1.5% silicon or an Al-Mn alloy is used as the cladding material. Cladding thickness is usually 5–10% of the total. In most applications, the aluminum cladding serves to improve the corrosion resistance of the steel, give it a decorative appearance, or even out temperature differences. Slide bearings are a special case, in which the steel is first clad on one side with pure aluminum and in a subsequent operation clad with the Al-20Sn-Cu bearing alloy 8081(Fig. 14.9).

Copper-clad aluminum plates are used mainly in electrical engineering applications. They are used for busbars and can be fastened to copper busbars by screws without the risk of corrosion. An interesting new product is a weld insert that is used where copper and aluminum busbars have to be butt-welded. Apart from these main uses, copper-clad aluminum sheet is used for all kinds of metalware for its decorative surface appearance.

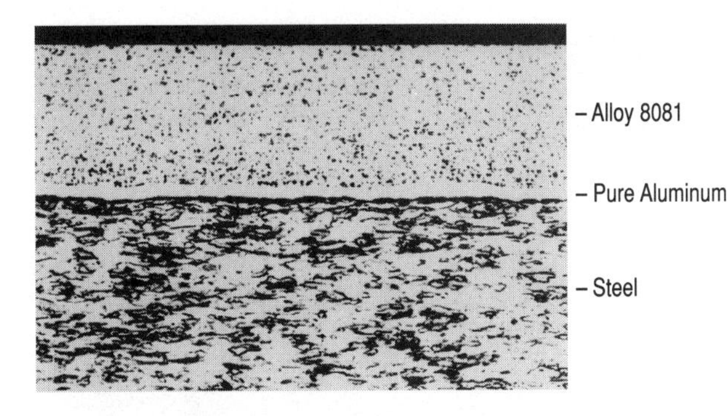

Fig. 14.9: Microstructure of a slide bearing made of a steel/ aluminum composite.

Copper-cladding is done in the same way as aluminum/steel composites—by etching and brushing the surfaces of the pre-rolled sheets and then forming the bond by cold-rolling with a heavy reduction. The technique will not work for relatively thick plates, however, because it is not practicable to reach the required temperature to form the bond; in such cases, explosive cladding is used.

14.6.3 Composites of aluminum and non-metallic materials

Composites of aluminum with plastics and ceramics have been growing in importance recently, because they possess properties that are unattainable with metals alone. These properties are high resistance to bending and cracking, lower density, low heat conductivity, and the capacity for noise- and vibration-damping as well as a higher modulus of elasticity and improved abrasion resistance. The bond is made using high-strength, stable glues that adhere well to aluminum. Such a composite sheet, Alucobond®, a composite consisting of two sheets of aluminum enclosing a polyethylene core (Figure 14.10.a) is used in large quantities for architectural curtain facades (Fig. 14.10.b). The aluminum sheet, 5050 alloy, can be delivered with mill finish or with the surface anodized or painted. The separate layers in coil form are glued together in a continuous process. Alucobond® panels possess outstandingly good flatness and are very stiff with low weight per unit area. Reynobond®, introduced later by Reynolds Metals Company, is somewhat similar to Alucobond®.

Besides these flat panels, other sandwich panels are made for undemanding applications using corrugated and/or embossed sheet on one side and a plain sheet on the other, the space between them being filled with heat-insulating plastic foam. Such sandwich panels are often used for ceiling panels and for cladding the outside of commercial and industrial buildings.

A layered composite material made of aluminum and fiber-reinforced plastic in coil form is sold under the name "Arall" for use in the aerospace industry. The outer skins are of 2024 or 7075 glued to a sandwich filler that is made of plastic reinforced with aramide fibers. The particular advantages of this material are reducing density by 20% and increasing the fatigue strength in the direction parallel to the fibers (i.e., the mechanical properties are anisotropic). This composite material is remarkable for its outstanding noise-damping capacity in comparison with plate made entirely of aluminum alloy.

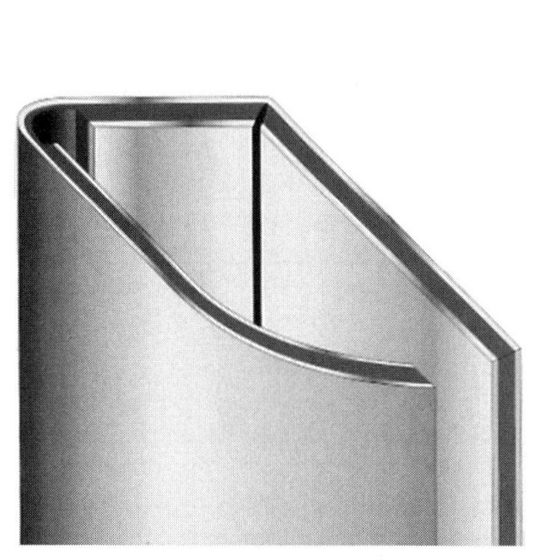

a

b

Fig. 14.10: (a) Alucobond® panels and (b) their application in building construction. (Alusuisse)

Fig. 14.11: Insulating double-glazed aluminum window frame making use of plastic extrusions to block the flow of heat from the inner to the outer aluminum profiles of the window frame and window casing, respectively. (Alusuisse)

Various heat insulating systems have been developed both to save energy and to avoid water condensation in window frames made of aluminum extrusions. They all have in common a barrier element made of heat-insulating plastic, sometimes fiber-reinforced, to block the flow of heat through the high-conductivity aluminum (Fig. 14.11).

14.6.4 Composites with aluminum

The so-called metal matrix composites (MMC) are novel composite materials. They consist of conventional aluminum alloys with embedded reinforcing elements, which may be in the form of fibers, whiskers, or particles of ceramic material, most commonly aluminum oxide (Al_2O_3) or silicon carbide (SiC) (Fig. 14.12).

Ceramic fibers have a very high modulus of elasticity in the direction parallel to the fibers, up to four or five times that of aluminum alloys. The mechanical properties of each of the two components separately and the degree of interaction between the metal matrix and the fibers determine the properties of the fiber-matrix composite. The properties of the composite material are the weighted averages of its components' properties. Different degrees of anisotropy can, however, occur in the composite, depending on its composition and structure. The addition of long fibers results in a particularly marked improvement in strength in the fiber direction. Nevertheless, it has to be mentioned that little plastic deformation is possible once the long fibers are embedded, otherwise the ceramic fibers would break. Hence, reinforcement with long fibers is only suitable during forming processes that result in a shape that is close to that of the finished component (near net shape).

There are several casting processes that lend themselves to the fabrication of such composite materials. Their common characteristic is that the melt infiltrates the fiber-containing structure under pressure. Particular reference may be made here to the squeeze-casting and vacuum die-casting processes. Even today, however, the very high cost of producing long fibers is an obstacle to introducing these materials. Composites with fine ceramic particles are much more promising. They possess the further advantage of having no marked anisotropy. The classical fabrication methods can be used for producing particle-reinforced semi-fabricated products (Fig. 14.13). Conventionally produced castings can be replaced by particle-reinforced composites produced by Squeeze-compo casting or by the spray deposition process. The particular advantages of particle-reinforced composites as compared with unreinforced aluminum alloys are their high abrasion resistance, higher strength at elevated temperature, and their capacity to withstand thermal shock.

14.7 Summing it up: Aluminum in the Materials Competition

In the competition between light metals, the properties of magnesium make it a serious, high quality alternative to aluminum. The automotive industry, as well as producers of molded castings, are increasingly gaining experience in working with magnesium. In the near future, they will more often be looking at cost/benefit analyses to decide whether to

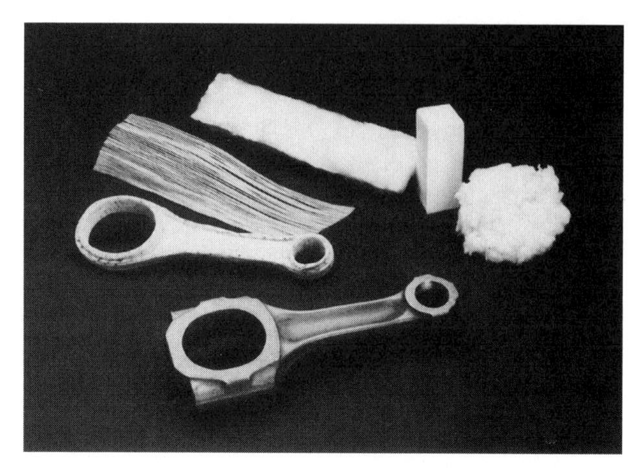

Fig. 14.12: Connecting rod as an example of strengthening a component by means of embedded fibers of aluminum oxide. (VAW)

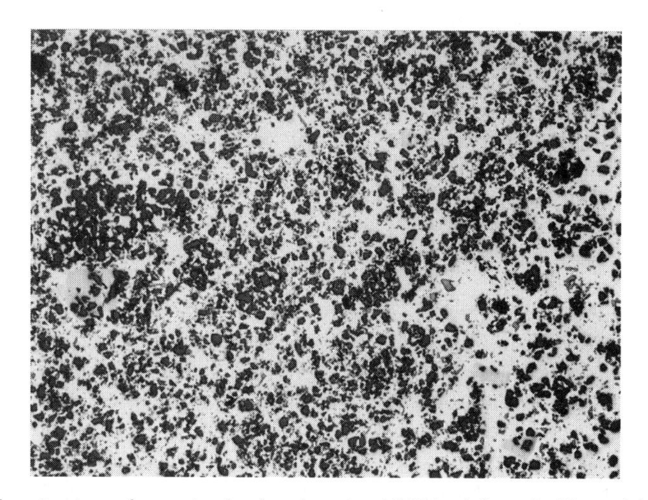

Fig. 14.13: Microstructure of an extrusion ingot made of 6061 reinforced with particles of SiC; polished section, unetched, ×100.

replace steel by one or another light metal. Magnesium has so far taken an insignificant share of the market for structural parts of vehicles, but it is to be expected that there will be substantially more use of magnesium castings in the future.

Is there any likelihood of substitution competition between titanium and aluminum in the foreseeable future? Surely not, because it is so expensive, and because its properties make it a more natural competitor for steel than for aluminum. Perhaps progress toward more competitively priced titanium fabrication will in the medium term enable titanium to replace some steel components in the automotive industry.

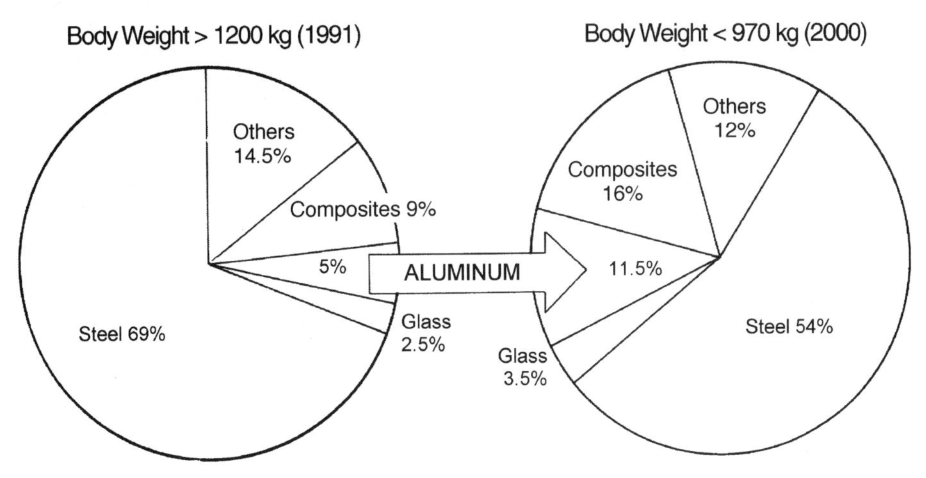

Fig. 14.14: Forecast growth in use of materials in the European automotive industry from 1991 to 2000. (U.B.M.)

The competition between aluminum and steel in automobile construction will enter a new phase in the next 5–10 years. Gains and losses can be expected on both sides. Aluminum still has the better cards to play, especially as regards energy economy, recyclability, and life-cycle cost. But the steel industry continues to innovate, and one must beware of wishful thinking about lightweight construction in the coming automobile generation.

Fig. 14.14 shows the forecasted growth in the use of both aluminum and plastics in automotive applications. Plastics are, however, more often partner materials than substitution competitors for aluminum. In both roles, plastics suffer from problems with the ever increasing demands for recyclability and the ever more stringent pollution regulations. In both respects, composite materials are increasingly at a disadvantage, and their success may depend on developments in recycling technology. Therefore, a shift is to be expected in the balance between plastics and aluminum in the marketplace. The details of this new equilibrium, for instance in the packaging sector, are still uncertain.

At present, design solutions using composites of two (or more) materials are coming to the foreground. Even here a substitution competition is taking place, this time between groups of materials rather than between single materials. However, every leading materials industry is doing its utmost to innovate, so that advances in a particular group of materials are soon answered by innovation in the competing industry. All these considerations lead inevitably to the conclusion that a total ecological audit comparing various materials for an identical component will hardly be possible in the near future. Efforts to present a total life cycle analysis for a single material such as aluminum are gaining momentum (see LCA in Chapter 16).

Chapter 15. Important Aluminum Applications

15.1 Aluminum as a Useful Material[1]

Aluminum alloys are being used successfully in a wide range of applications, such as building, aerospace, transportation, ship construction, machine tools, conveyer systems, domestic appliances, electrical conductors, packaging, etc. Designers have a choice of many heat-treatable and non-heat-treatable alloy variants in the form of sheet, plate, formed sheet, extruded profiles, forgings, and castings. Joining methods include the use of screwed connections, rivets, crimping, brazing, welding, and adhesives. Because aluminum has only been available as an industrial material for the last 100 years, its use has very often involved a change away from one of the traditional materials, wood or steel. The most vigorous impulse to substitute aluminum for a traditional material has occurred where aluminum has several advantages when compared to the material that it has displaced. In the following, some examples of the use of aluminum have been selected to show the particular advantages gained from substitution.

The machine tool industry was a classic domain for ferrous materials. However, aluminum, because of specific properties, has made serious inroads. Its high thermal conductivity, good machinability, and the fact that sparks are not produced from its surface at higher temperatures makes aluminum an ideal mold material for the plastics industry. In all machines in which components are subjected to high accelerations, the combination of stiffness, low density, and strength give aluminum a very great advantage. Because of the possibilities of the extrusion process, the number of joints that have to be made in the manufacture of a particular component can be dramatically reduced. This aspect will be discussed in more detail later.

Wood was the traditional material for window and door frames. In this application, the arguments in favor of the use of extruded aluminum hinge on its exceptional two-dimensional form flexibility. Notches, slots, protrusions, and grooves to be used for the fitting of all types of fixtures, glass, and sealing elements can be integrated into an extruded aluminum profile design. Further advantages of architectural profiles are their stiffness and their extremely long life. This characteristic is an important factor in their high performance as sealing elements. Last but not least is the maintenance-free decorative surface of the metal. A negative aspect for aluminum window profiles is the high thermal conductivity of the base metal. This disadvantage can be overcome by building in an insulating "thermal barrier" element to separate inside surfaces from outside surfaces of the profile.

Engineering uses of aluminum include the roofs of industrial buildings, large green houses, so called "winter gardens," and, recently, bridges. These applications make full use of the particular advantages of aluminum as a material that acts as the prime motivator for change.

---

[1] By J. Zehnder.

An especially successful area of application is in conveyor technology. Rails for conveyor equipment can take advantage of the wide range of shapes available from the extrusion process.

Containers of all types are stable, durable, pose no contamination problems for food stuffs, and are light. Their stability in terms of shape make aluminum pallets especially suitable for automatically operated high-bay storage systems where deformation could cause a malfunction.

The high corrosion resistance of selected alloys in maritime atmospheres, good weldability, and high strength after welding make aluminum an interesting material for shipbuilders. In special cases, its magnetic properties are an advantage. The advantage of lightweight construction is used in cruise ships to reduce draft and so allow vessels to enter shallow harbors. The use of aluminum in the superstructure provides additional volume for passenger comfort and/or lowers the ship's center of gravity. Fast ships such as hydrofoils, wave piercing catamarans, and fast monohulls together with hovercraft are of increasing economic interest for high speed ferries. Their speeds are directly related to their ability to achieve surface skimming effects because of their lightweight construction.

For some decades, aluminum has been used in increasing quantities instead of steel for railway vehicles. The first substitutions were made of movable parts on freight cars. Use was then made of aluminum in freight trains for the transport of bulk commodities such as ores or coal over long distances. In the last decade, the use of aluminum in the form of large extruded profiles has been the most important application for aluminum in the railway sector. Large extruded profiles are now standard for modern designs of high speed inter-city, local, and regional passenger trains.

The automotive sector has for decades awaited an increase in the use of aluminum. Aluminum first began to be increasingly used in trucks. In automobiles, traditionally up to 90% of the aluminum used took the form of castings. In the 1990s, however, the advent of the aluminum frame design provides opportunities for more rapid growth of rolled, extruded, and cast components in automobiles.

For many decades aluminum has been the main material used by aircraft builders. High strength aluminum alloy sheet, often combined with extrusions, proved ideal for the application. More recently, heavy gauge plates have been increasingly used; these can be processed to complex components because of good machining characteristics.

In the transport sector, a significant number of other products are worthy of mention, products such as bicycles, motorcycles, cable cars, and air freight containers, to name but a few examples. This short list shows that a careful analysis of a structure and a thorough evaluation of required material properties produce a number of interesting applications for aluminum. The possibilities are virtually unlimited if the designer is prepared to leave behind traditional ways of thinking. In what follows, certain important areas of use will be considered in more detail.

15.2 Structural Applications[2]

15.2.1 Advantages and range of structural applications

Applications that are primarily determined by engineering or structural considerations are covered by the term "structural applications." Such applications include hangars and other large functional buildings, but also include a very large number of fringe areas associated with buildings, machine tools, and general purpose equipment construction. As far as aluminum is concerned, this general construction market is of great interest even if it is not, in contrast to steel, dominated by spectacular structures such as large bridges and skyscrapers.

Engineering applications usually take advantage of the two classic properties of aluminum: low mass and high corrosion resistance. A considerable number of aluminum bridges and other functional structures are already in use, but they are typically applications in which a specific property of the material is of importance (Fig. 15.1). The fact that not many such aluminum structures exist is also due, to a large degree, to the fact that many engineers at the conceptual stage are not aware of the interesting possibilities offered by aluminum. An exception has occurred in the field of mobile military bridges, where aluminum is widely used.

In recent years, a number of large extruded aluminum sections have been used in offshore applications such as helicopter landing decks (Fig. 15.2), accommodation modules,

Fig. 15.1: Aluminum foot bridge that can be raised. Span 16 m, Width 1.20 m. (Alusingen)

[2] By R. Gitter.

stair towers, and personnel bridges. These special applications, however, are concentrated geographically on the few manufacturing companies that possess the necessary specialist knowledge and personnel.

Major technological advances in extrusion practice in the last 20 years have led to significant cost savings in construction. Cleverly designed profiles can be multi-functional, which can provide significant cost savings during assembly. It is possible in many cases to more than compensate for the higher price of the semi-fabricated aluminum product when compared to steel. Table 15.1 gives an overview of current applications in engineering construction. A number of important applications are considered here in more detail.

An example of a use of aluminum that has been furthered because of its good corrosion resistance is its use for construction of structures for the mounting of road signs. The atmosphere close to main roads is generally quite aggressive. Most materials would be seriously affected in this polluted environment. Continuous maintenance work in the presence of free flowing traffic is not only expensive, but constitutes an accident risk. Aluminum highway signs and poles have been used very successfully to provide a solution to this problem.

Fig. 15.2: Helicopter landing deck on a small North Sea offshore platform with a 240 bed aluminum accommodation module on the Snorre Platform in the Norwegian sector of the North Sea. This module is five stories high and has a helideck on top capable of taking the largest helicopters. The accommodation module and helideck are aluminum structures. The platform is a tension leg platform (TLP), moored in 340 m deep waters; weight aspects were, therefore, critical—hence, the aluminum hotel. The welded design needed some 780 tons of large extrusions, and the total weight of the accommodation module when finished was 2100 tons. (Alu–singen; platform builders: Lervik Svies)

For the same reason, aluminum has been used in support structures for water and sewage treatment installations. In this type of environment, aggressive gases develop, especially mixtures containing ammonia, which are extremely corrosive to steels. Galvanizing protection for steel is only effective for a relatively short time. Ammonia also permeates through concrete and destroys the steel reinforcing bars.

Telescopic inspection platforms are a classic example of the choice of aluminum because of its low mass. The low mass makes possible the improved reach of such equipment because of the reduced effect on the center of gravity as the telescopic arm is extended (Fig. 15.3).

Table 15.1: Aluminum applications in civil engineering and related areas

Divided by function into the following main groups: C—corrosion resistance, L—light (low mass), E—extrusion process offers design flexibility.

C	C + L	L
Containers		
Flood lighting masts	Flag poles	Mobile cranes
Roofing	Aircraft access bridges	Mining aids
Overhead conductor masts	High voltage masts	
Covered sewage processing	Inspection stairs	Bridge cranes and inspection platforms
Sound protection	Offshore accommodation, bridges, and stairtowers	
Crash barriers		Scaffolding
Silos	Tank decks	Ladders
Road sign supports		Telescopic lifting gear
		Tent spars
Transformer boxes		

C + E + L
Covers
Helidecks
Ship structures

C + E	E	E + L
	Complete balconies	Ramps
Boat ladders	Suspended transport rails	
	Automatic doors	Lifts
Railings	Tunnel cladding	
Greenhouses		Scaffolding planks
Domes		Loading ramps and bridges
Avalanche barriers		
	Pavilion structures	Tent frames
Swimming pool roofs		Jig supports
		Transmitter masts
Winter gardens		Stages

Fig. 15.3: Aluminum telescopic raised platform. (Alusingen)

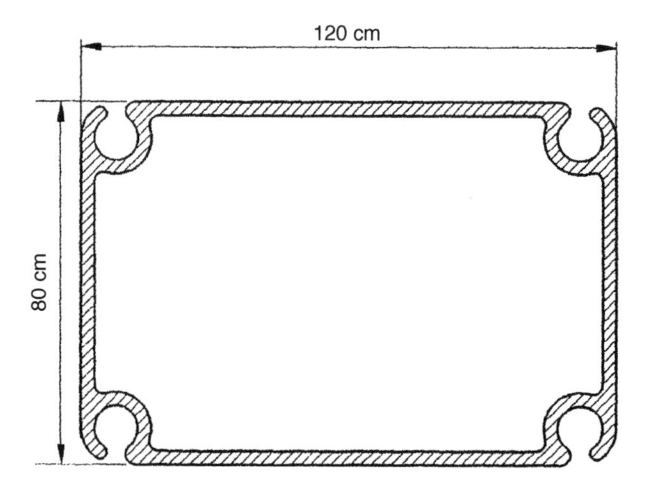

Fig. 15.4: Cross section of an extrusion for transportation and the use of a tent. (Alusingen)

The use of aluminum profiles in large tent / marquee structures provides a good example of the exploitation of the design possibilities offered by the extrusion process. Fig. 15.4 shows the cross section of a hollow rectangular profile with an inbuilt channel / slot at each corner. The purpose of these slots is to accept the rope sewn into the edge of the tent canvas. This system has completely replaced wooden spars in modern large tents. The use of these specially designed profiles makes it possible to erect these temporary structures very efficiently in an entirely new way. Before the tent frame is set up, with the aluminum spars all at ground level, ropes are threaded through the channels so that after the frame has been raised, the canvas can be pulled into position from the ground without the need to climb aloft onto the frame. This new method makes the erection of these large tents much safer.

Like large tent structures, mobile bridges are applications where solutions in aluminum can be both elegant and the only technically feasible design (Fig. 15.1).

15.2.2 Aluminum bridges and bridge decks[3]

The momentum in the use of aluminum as a major structural material for bridges and bridge decks is increasing. This is due to a number of driving forces in society, the fact that there have been many successful aluminum bridge applications, and the new technological advances recently made in this field.

Worldwide, it is estimated that there are more than two million highway bridges, many of which are in poor condition, are too narrow, offer too little vertical clearance, or are underdesigned to meet current and projected traffic levels. In addition, in developing countries there is a need for new durable bridges that is being driven by the demands of both population and economic growth.

In the USA alone there are some 590,000 highway bridges, of which 32% are categorized as structurally deficient or obsolete. On the North American continent, the bridge problem is worsening due to the effect of de-icing salts on both steel and concrete bridges, higher than originally calculated traffic volumes, and inadequate bridge and highway maintenance affected by costly environmental regulations requiring repainting. In the future, with the advent of the North American Free Trade rules, larger truck loads will be required. Improved seismic-resistant designs are being specified because of recent experience gained worldwide from bridge failures due to earthquakes.

Aluminum offers distinct advantages when attempting to solve these problems. It can be very effective in many applications, particularly on a life-cycle-cost basis. The main properties of aluminum that are useful to a bridge builder are its light weight, high specific strength and fatigue performance, and excellent corrosion resistance.

Aluminum bridge decks, including the wearing surface weighing approximately 93 kg/m^2, have recently been developed and meet USA federal HS-20 highway standards. Concrete decks that meet these standards weigh typically 465 kg/m^2 or more. Steel decks equipped with a wearing surface usually weigh between 210–375 kg/m^2. Aluminum girder or stringer designs for the underlying structure can be 50% of the weight of steel and 30% of the weight of concrete structures. This applies as long as the additional vertical web height is not constrained by the specific requirements of the application. Aluminum typically requires 30–40% more height to reduce its deflection to the same level as would be obtained from steel.

The much lighter weight of an aluminum deck or girder reduces the size and cost of the bridge footings or, alternatively, allows an increase in the live traffic loads and a removal of weight restrictions. It is often possible to substantially widen a bridge without strengthening the existing foundations while maintaining or increasing traffic loads. The much lighter deck weight will also decrease the risk of earthquake damage.

Aluminum's low weight makes possible the construction of large high quality prefabricated deck sections or even complete bridges, including girders or box beams and wearing surfaces, which can be moved from the place of assembly to the bridge site. This can

[3] By R. Hanneman.

Fig. 15.5: Installation of the 38 m prefabricated all-aluminum bridge in Forsmo, Norway, in 1995.

dramatically reduce installation time and cost and avoid traffic delays. Fig. 15.5 shows a 38 m prefabricated aluminum bridge being installed in Forsmo, Norway, in 1995.

Aluminum has a long history of satisfactory performance in bridges. Ten aluminum highway bridges and many pedestrian bridges have been constructed in North America and have been in service for more than 60 years. At least one of these bridges has handled automobile and truck volumes of over 100,000 vehicles per day. In Scandinavia, more than 40 bridges with the Svenssen-designed 6063-T6 multi-hollow extruded aluminum decks have been built in the last decade; some have been in regular use for more than 10 years At least 25 bridges with aluminum decks or sub-structures have been built in other parts of Europe since World War II. This depth of experience has demonstrated the excellent corrosion resistance and overall durability of unpainted aluminum structures, even in high salt environments. Fig. 15.6 shows the arch of the 88 m aluminum span bridge that was constructed in 1950 over the Saguenay River near Arvida, Canada. This bridge remains in very good condition after more than 45 years of service.

The Smithfield bridge at Pittsburgh, Pennsylvania, had more than 60 years of service even though nonoptimal 2xxx series aluminum alloys were used for the floor beams. It has had nearly 28 years of service with a welded orthotropic aluminum deck system fabricated from 6061-T6 extrusions and 5083-H113 plate. By using aluminum, the rebuilt Smithfield bridge superstructure was reduced by 751 tons, and this made possible an increase in the bridge's live load capacity by an equivalent amount. Apart from some spalling problems of the wearing surface, some localized galvanic corrosion, and some fatigue cracks that are now understood to be due to design shortcomings, the bridge performed remarkably well and stood the effects of many years of service. This evaluation was supported by a recent detailed structural study of a number of the bridge's component elements.

Reynolds Metals is currently pioneering the development of a new generation of isotropic aluminum bridge deck panels that are based on 6063-T6 alloy multivoid extrusions. These specially designed 1.86 m^2 deck system modules can be factory fabricated with full

Fig. 15.6: The all-aluminum Arvida bridge constructed over the Saguenay River in Canada in 1950.

penetration high speed welds in more than 30 m × 4.25 m sizes. A cross section of an extruded element making up one of these modules is shown in Fig. 15.7. Once the module is welded together, a proprietary pretreatment is followed by bonding a high performance epoxy resin that incorporates a high wear-resistant abrasive filler onto the top surface. The deck is then shipped out for use in the field.

The installed isotropic deck module design incorporates galvanic protection. Welding in the field is avoided by use of a proprietary bolting system together with a new method of grouting. The system is designed to achieve a composite action when used with either aluminum structures or underlying steel girders, such as in bridge deck replacement applications (Fig. 15.8). Extensive government verification testing of these bridge modules is currently being conducted at the Oak Ridge National Laboratories.

An all-aluminum 98 m suspension bridge deck was installed in Pennsylvania in 1996 that had its maximum allowable live load increased by a factor of three through use of aluminum with a hard surface, versus the previous asphalt coated, corrugated steel deck. Several other new aluminum bridges are already in the design and planning stages. In the future, very long span suspension bridges should show aluminum's combined cost and performance superiority over steel or any other material.

Movable highway bridges over waterways, under-designed historic bridges, and temporary reusable bridge panels for construction projects can all benefit from aluminum's light weight and corrosion resistance.

One of the most significant short term problems that prevents a more general use of aluminum is the lack of experience of many federal and state highway engineers, contractors, fabricators, and erectors. A second problem is related to the difficulties in persuading the authorities that they should consider a life-cycle-cost concept and not just initial cost. Further education and changes in legislation are needed to overcome this problem. Other difficulties range from the availability of qualified aluminum welders to a lack of general knowledge of damage repair procedures.

Fig. 15.7: Cross-section of the Reynolds isotropic aluminum bridge deck panels based upon 6063-T6 multivoid extruded shapes.

Fig. 15.8: A schematic of an isotropic aluminum bridge deck with longitudinal girders.

There are a number of spin-off applications for the currently emerging aluminum bridge deck technology. These include: structural decking in high speed oceangoing ships; new cost-effective mobile military bridges; bridges between clustered offshore oil platforms; railroad bridges; pedestrian bridges or tunnels, including those between high-rise buildings; and upgrading of railroad bridges to increase live loads.

15.2.3 Future prospects

Apart from major transportation applications, building products, and packaging and containers, aluminum structures and especially bridges and replacement bridge decks present one of the largest new growth opportunities for the aluminum industry in the next few decades.

15.3 Buildings and Architecture[4]

15.3.1 Advantages for a variety of applications

The increase in the use of aluminum as a building material began in the years after World War II. Today, aluminum is one of the most important materials in high-rise construction and is especially used for the cladding of buildings. Here, its good decorative appearance and high corrosion resistance are key factors. Since the energy crisis of the mid 1970s, there was a move first in Europe and soon after in Japan toward the use of thinner aluminum extrusions and, in the window sector, the introduction of insulating inserts. Particularly interesting in relation to windows for dwelling houses is the continuing competition between aluminum and plastic profiles. Development progresses cyclically, with victories and defeats on both sides. About two decades ago in Europe, then in Japan, and more recently in America, aluminum plastic sandwich materials with cores approximately

[4] In collaboration with H.E. Hartmann and R. Voegtin.

Fig. 15.9: Building facade in aluminum clad sandwich material Alucobond®. (Alusingen)

3–6 mm thick have been increasingly used as elements in facades. Well known trade names are, for example, Alucobond, Alpolit, and Reynobond. Their processing is based on simple application techniques that require less investment than is needed for wood or components made out of other metals. Fig. 15.9 shows a typical application (see also Chapter 14.5).

Aluminum has certain unchangeable properties that might be considered as disadvantages when compared to steel. It has a modulus of elasticity that is three times lower than steel, and this makes larger cross-sections necessary for beams and supports. Its higher linear coefficient of expansion means that higher allowances for dilatometric effects have to be calculated into designs. Greater protection may be needed for certain fire resistant elements because of aluminum's low melting point.

The problem of recycling building material from a demolished building at the end of its life has to be taken into consideration in the initial design. This poses interesting new challenges in terms of the choice of materials.

The following list provides an overview of the wide range of applications for aluminum in today's building sector:

- Windows and doors and most other types of opening generally made of extruded profiles with integral thermal barrier elements; combinations of wood and aluminum windows are favored by many architects.
- Facades with integrated self-supporting elements or skeleton structures with sills, lintels, and window elements.
- Curtain walls and rear ventilating facades made from aluminum-clad sandwich materials.
- Industrial facades assembled from profile-stiffened sheet and coil.
- Sun shading elements of various forms such as horizontal venetian blinds and either horizontal or vertically mounted large curtain wall elements.
- Balcony balustrading, applications in roofing, profiled roof cladding, corrugated sheeting, drain piping, spacing elements, and water down pipes.
- Interiors, ceilings, lighting surrounds, reception areas, stair balustrading, separating walls, floor channeling for electrical installations, use in heating and air conditioning, in lifts, and as wall cladding.

In the following sections some of the more important uses of aluminum in buildings will be discussed.

15.3.2 Roofing

Roofs are very much affected by weathering and by the influence of pollutants in the atmosphere. Because of aluminum's natural corrosion resistance it stands alongside copper as the most suitable metallic roofing material. It is generally used in the form of flat or profiled sheeting and is easy to erect because of its formability and because of the absence of any difficulty in forming it to shape on site. As a roofing material, aluminum is very much in competition with a number of other traditional materials. Aluminum roof cladding has become established in Europe as the material for a number of specific market sectors, particularly for commercial and agricultural buildings.

15.3.3 Winter gardens

Winter gardens are very much favored by today's architects. Their function is to increase space and to improve the atmosphere of the inhabited area by the passive use of sunlight. The use of aluminum extrusions as a structural material has played an important role in furthering this trend. Aluminum winter gardens provide high stability with low weight and offer opportunities for a freedom of form in architectural design. This has promoted their use in many modern public building projects (Fig. 15.10).

15.3.4 Window designs

Windows have always played an important role in architectural design, a situation that is not likely to change in the foreseeable future. As an important part of the outer shell of a building, windows have a major affect on possible use, the quality of any living space, and the energy losses from the structure. In the past, windows were usually installed with an eye to architectural appearance or with cost considerations as a major factor. Since the energy crisis of the 1970s, however, the requirements placed on windows, particu-

Fig. 15.10: Door and entrance to a town hall with lofty glass and aluminum vestibule/winter garden. (Wicona)

Fig. 15.11: Example of an architecturally appealing aluminum facade. (Geilinger)

larly in terms of their ability to conserve heat, have increased dramatically. Sound insulation has also become an important consideration in recent years.

Aluminum plays a very important role in modern high-rise construction. Hollow aluminum extruded profiles in aluminum magnesium silicide 6000 series alloys are generally used for this application. In order to improve the thermal insulating characteristics of the hollow profiles, the inner and outer surfaces are separated by use of dimensionally stable, extruded plastic inserts that have low thermal conductivity. The weathering performance of the aluminum surface is improved by use of surface treatment techniques, such as anodizing, painting, or (molten) resin powder coating (Fig. 15.11).

A further variation in the use of aluminum for windows is in combination with wood. In wood / aluminum systems, wood has the function of providing the thermal insulation and at the same time creating the impression of a warm and comfortable interior. The purpose of the aluminum is to provide protection against the weather. A special type of window design that is becoming increasingly important includes integrated fiberglass. In this system, the aluminum substructure that, for the most part, is invisible is an important and highly developed component of the design.

15.3.5 Facades

In North America, aluminum facades have been popular for residential buildings for some decades. The material has shown itself to be very suitable for retrofitting onto existing buildings as siding. This offers many homeowners the chance of providing for themselves an attractive but inexpensive facelift for the outer surface of the house. For this application, prepainted coil, roll formed for stiffness and optical acceptance, offers the advantage of watertight protection.

Aluminum sheet materials are used worldwide for commercial buildings where the weathering resistance of their anodized or stove lacquered surfaces is accepted as being a major advantage.

Since the energy crisis of 1973–1974 in Europe, the importance of high thermal insulation values for facades has become a factor of great importance. This is making itself felt in the new building standards that are coming into use. This aspect will be discussed in the following section.

15.3.5.1 Energy saving facades

High thermal insulation, air tight building shells with windows providing k-values of kF = 0.80 W / m^2K, make possible structures with very low energy requirements. Such buildings can provide excellent levels of comfort. Even though the outside temperatures may be extremely low there are no cold window surfaces because of the high levels of insulation provided by modern designs. It is not necessary to compensate for heat losses with warm currents of convected air. The use of highly insulating facades has also had a positive influence on such structures as indoor swimming pools, providing cost savings for heating and air conditioning.

It has been proved that the requirements of high insulation construction can be very elegantly met with a suitable window design. High thermal and sound insulation (40–50 dB), low ultraviolet transmission levels (which prevent fading of interior fittings), and excellent storm rain insulation can be provided by the latest windows. Their self-supporting frames made from extruded aluminum with integrated insulating elements provide a long service life (Fig. 15.12).

15.3.5.2 Energy-active facades

Aluminum profiles and sheet play a significant role in combination with insulating materials and special glass in so called energy active facades. The word "active" means that the energy flow into the building through the facade is regulated or controlled through a sub-system. Depending on the orientation of the building, specially installed systems can provide controlled shading from the sun's energy. Rooms can be protected by means of a movable array of aluminum elements installed on the outside of the structure (Fig. 15.13).

In a further development, controlled energy facades can be operated with a photoelectric cell regulated sun blind system. This allows controlled quantities of sunlight into the interior of the building. Complex energy facades are in the pilot stage of development. These systems make use of thermal reservoirs and heat pumps. A further type of energy saving facade (Fig. 15.14.a) works as follows: a black coated aluminum profile siding element is heated by solar energy. Even on cloudy days diffuse sunlight is sufficient to provide a usable heating effect. Outside air flows through specially dimensioned tiny perforations in the facade and absorbs some of the solar energy. In a specially constructed

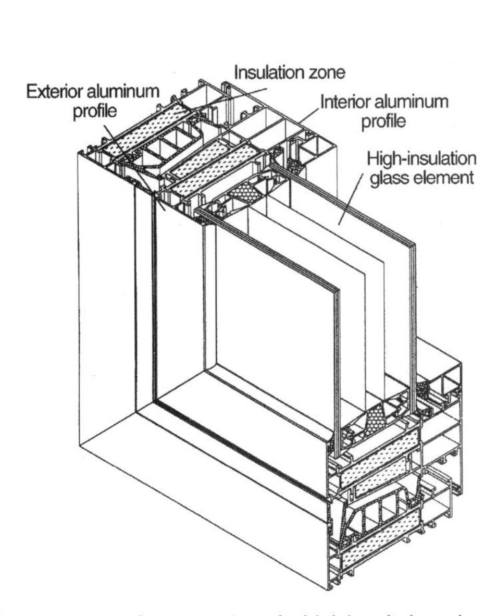

Fig. 15.12: Cross section of a high insulation window system. (Geilinger)

Fig. 15.13: Facade with movable sun shade elements made from sheet and profiles.

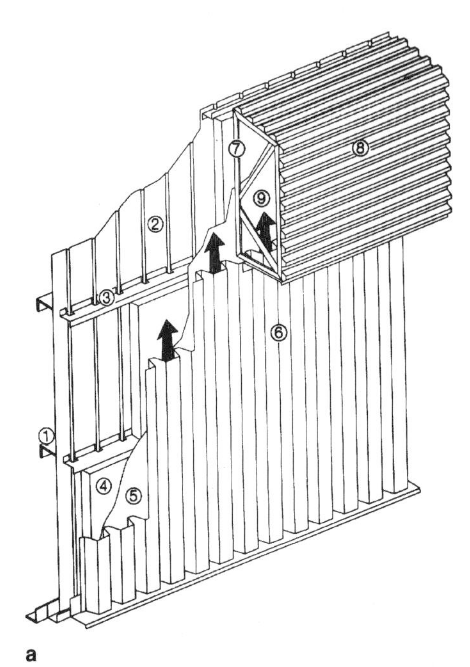

a

Fig. 15.14: (a) The working principle of a new type of energy saving facade. 1—cladding element supports mounted on the building; 2—inner wall; 3—Z-shaped extrusion; 4—insulation; 5—surface cladding sheet; 6—extruded aluminum plate with special perforations; 7—air chamber; 8—air chamber cladding; 9—warmed air to the air vent. (Alcan). (b) A modern aluminum facade with photo-voltaic collectors. (Gartner)

b

chamber just below the roof level, this warm air is collected and then blown into the building.

The energy saving facade shown in Fig. 15.14.b operates on the same principle as a photoelectric cell. This type of facade can provide energy from all shadow free vertical or suitably angled surfaces that face from southeast to southwest. The total installation consists of 145 m^2 amorphous optical facade and 435 m^2 of solar modules that are arranged as shading elements. The total installation provides approximately 52% of the estimated annual requirements of 50,000 KWh. The power is conducted via a transformer into the building's electrical system for general use. Surplus energy is made available to the public grid.

Today, these photoelectric modules are architecturally and structurally suited to meet the requirements of civil engineers and architects. Such facades can provide weather protection, meet thermal, sound, or fire standards and, at the same time, act as sources for electrical energy. With photoelectric facades, there is a degree of architectural acceptance that is far above that accorded to other solutions. No additional secondary surface or extra substructure behind the facade is needed. Since electrical circuits are required throughout the building, no high extra costs have to be added to link the solar system to the building's infrastructure. Five percent of the solar energy reaching the solar panels is converted into electrical energy. The deciding factors are, therefore, the amount of sun light available, the area of the solar cells, and the efficiency of the system.

15.3.6 Reassessment of the ecological balance

More and more aluminum scrap from abandoned buildings and renovation projects is becoming available for possible reuse. Aluminum is an ideal material for recycling. Old window and door frames, balustrading, facade profiles, roofing sheet, etc., can be processed into secondary aluminum and subsequently be used to remake aluminum components for the building industry.

In order to calculate the ecological balance of complete building shells or facades, it is necessary to consider not only the aluminum in the structure but also assess the value and "hidden energy" of glass, insulating material, and the supporting substructure. Such a total "ecological" balance is also influenced by such factors as the make-up of the interior of a building and the prevailing climatic conditions. Such calculations are both time consuming and complex. The aluminum industry is actively engaged in studying further the economic/ecological balance of aluminum used in buildings (see LCA, Chapter 16).

15.3.7 Future prospects

Aluminum will continue to play an important role for such products as window systems and building facades, and it is anticipated that, because of the wide range of products and shapes, this role will become even more important. Aluminum will be increasingly used in combination with other materials (especially insulating materials) that are suitable for separation prior to recycling. Aluminum will be used in greater amounts in high-rise buildings, systems to collect solar energy, and passive thermal insulations.

15.4 *Aircraft Construction*[5]

Aluminum continues to be the dominant material used in the construction of civil aircraft structures. The airframe is about 80% aluminum by weight. Air travel has continued to grow, and this trend is expected to continue. The growth of the industry, however, has not been directly translated into increased use of aluminum. This is because the air transport sector has been strongly affected by changes following the worldwide deregulation of air travel, which began in the USA. In a perverse way, these industry-wide changes may, in fact, benefit aluminum as a structural material in its continuing competition with alternatives such as composites.

In the years 1990–1993, the major world airlines lost more money (approximately $9 billion) than they had made since World War II. This was due to the intense competition following deregulation, the general economic recession, and the emergence of "no frills" airlines. In Europe the industry has some way to go with its restructuring, but in the USA, the airlines have emerged from this period as more pragmatic and competitive organizations. This, is turn, has directly influenced the manufacturers Boeing and McDonnell-Douglas in terms of their choice of structural materials for their aircraft.

Cost reduction in aircraft construction has become paramount and many airlines (e.g., United Airlines and Lufthansa) are evaluating materials based on a life cycle approach. The introduction of new materials has not been so rapid. Changes have been gradual and evolutionary rather than revolutionary. In the continuing competition between materials and especially between aluminum and composites, this more gradual approach is of benefit to the traditional airframe material aluminum. Composites are generally considered to have higher initial cost since they require more manual labor in their production, and they are also more expensive to maintain.

Against this background, there has been considerable progress in the development of aluminum alloys with improved strength, toughness, and corrosion resistance that are easier to manufacture. These new alloys are generally variants of the existing Al-Cu-Mg (2xxx) and Al-Zn-Mg-Cu (7xxx) series of alloys. Typically, the new alloy variants have lower levels of iron and silicon and are produced using closer processing parameters to provide better mechanical performance.

15.4.1 Al-Cu-Mg alloys (2xxx series)

The Al-Cu (2xxx series) alloy 2024, which was originally introduced in the 1930s as an aircraft material, remains an alloy of major importance. Alloy 2024 shows extremely good damage tolerance with high fracture toughness and high resistance to fatigue crack propagation in its mid-strength, room temperature aged (T3) condition.

It is important when selecting materials for their damage tolerance that any cracks that may be formed do not grow rapidly. Slow crack growth allows time for cracks to be detected at an early stage during the regular structural inspections, before they approach

[5] By J.A.S. Green and G. Tempus.

critical proportions (the "fail safe" principle). In today's aircraft construction, the aim is to ensure that the primary structure has at least twice the design life of the airplane (approximately 200,000 hours) without the development of critical cracks (the "safe life" principle).

Alloy 2024 is still used in many civil aircraft applications for fatigue critical components such as the highly sensitive fuselage skin (Fig. 15.15) and the lower wing surfaces. However, the thin 2024 sheet used in the fuselage skin has to be protected with a cladding layer of pure aluminum to provide protection against corrosion. This is particularly important since the resistance of this alloy to intercrystalline corrosion is not good enough for this application.

The stressed part of the structure, particularly the lower wing surfaces where improved durability and damage tolerance are required, are produced in 2024 alloy in the form of thick plates or as thick walled extruded profiles. In the case of the new Airbus A330/340, these are up to 18 meters long.

By reducing the content of the impurity elements Fe and Si and optimizing the processing method, alloy 2024 has been further developed into the variants 2124, 2224, and 2324 (Table 15.2). These alloys provide, in comparison to 2024, higher mechanical properties and also considerably higher fracture toughness because of their lower volume fraction of coarse Fe- and Si-containing intermetallic phases. These purer variants were especially developed for use in the latest aircraft types by the Boeing company. One of these variants has a 17% improvement in toughness and a 60% slower crack growth rate than 2024-T3, and this is being used for the fuselage of the Boeing 777. Since most of these alloys are modifications and improvements of existing alloys, they minimize the risks the aircraft manufacturer must assume for design and in-service performance.

The alloy 2219 is especially important in aerospace applications. It combines high strength with acceptable weldability and has been widely used for the fabrication of cryogenic tanks for aerospace applications (e.g., the external tank of the Space Shuttle); this application is described in greater detail in the section on Al-Li alloys.

Fig. 15.15: Thin walled stringer profiles used to stiffen the sheet skin and floor transverse support profiles in the Airbus body cross-section. (VAW)

Table 15.2: Composition of conventional aircraft alloys together with the more recently developed Al–Li variants

(Aluminum Association International Alloy Register. Values in weight %; maximum, unless a range, is shown)

Alloy	Si	Fe	Cu	Mn	Mg	Cr	Ni	Li	Zn	Zr
2024	0.5	0.5	3.8–4.9	0.3–0.9	1.2–1.8	0.10	—	—	0.25	—
2224	0.12	0.15	3.8–4.4	0.3–0.9	1.2–1.8	0.10	—	—	0.25	—
2324	0.10	0.12	3.8–4.4	0.3–0.9	1.2–1.8	0.10	—	—	0.25	—
2618	0.10–0.25	0.9–1.3	1.9–2.7	—	1.3–1.8	—	0.9–1.2	—	0.10	—
7075	0.4	0.50	1.2–2.0	0.3	0.18–0.28	0.18–0.28	—	—	5.1–6.1	—
7475	0.1	0.12	1.2–1.9	0.06	1.9–2.6	0.18–0.25	—	—	5.2–6.2	—
7150	0.12	0.15	1.9–2.5	0.10	2.0–2.7	0.04	—	—	5.9–6.9	0.08–0.15
8090	0.2	0.30	1.0–1.6	0.10	0.6–1.3	0.10	—	2.2–2.7	0.25	0.04–0.16
2091	0.2	0.30	1.8–2.5	0.10	1.1–1.9	0.10	—	1.7–2.3	0.25	0.04–0.16
2090	0.1	0.12	2.4–3.0	0.05	0.25	0.05	—	1.9–2.6	0.10	0.08–0.15

15.4.2 Al-Zn-Mg-Cu alloys (7xxx series)

The original alloy of the high-strength 7xxx series was 7075. This alloy has been used since the mid-1940s in various forms (as plate, sheet, extruded profiles, and forged parts) for highly stressed components (Table 15.2). The minor alloying elements Mn, Cr, and Zr, which are included to form finely dispersed intermetallic phases, act as strong recrystallization inhibitors that cause the fibrous crystal structure to be retained during processing to the semi-fabricated product stage.

In order to obtain high tensile and yield strengths, the semi-fabricated material is artificially aged to maximum properties by a thermal treatment of 12–24 hours at approximately 120°C (Table 15.3). In this temper (T6 type), the stress corrosion resistance, particularly in heavy gage material in the short transverse direction, has been shown to be inadequate. The stress corrosion characteristics are lowered by the action of forces at right angles to the grain boundaries. This problem led in the 1970s to the introduction of a two-step artificial aging process that improved stress corrosion behavior in the 7075 alloy even in the critical short transverse direction. The modified thermal treatment also improved resistance to exfoliation corrosion. This temper (T73 type) is achieved by an artificial aging treatment to the T6 temper (12–24 hours at 120°C), followed by an overaging treatment at higher temperature (8–12 hours at 170°C), which causes the tensile properties to fall by 11–14 percent from the T6 condition (Table 15.3).

Table 15.3: Tensile properties of semifabricated products from various aircraft alloys

(minimum properties for 6.30 mm thickness)

Alloy	Product form	Temper designation	Ultimate tensile strength (UTS) (MPa)	Tensile yield strength ($TYS_{0.2}$) (MPa)	Elongation in 50 mm El_{50} (%)
2024	Sheet	T3	435	290	15
7075	Extrusion	T6	540	485	7
7075	Extrusion	T73	470	400	7
7150	Extrusion	T76	615	580	7

In contrast, thin semi-fabricated products can be used in the T6 condition because their loading in service in the critical short transverse direction is too low to cause stress corrosion. The higher sensitivity to exfoliation corrosion of thin profiles in the T6 condition can be considerably improved without a loss in properties by a relatively mild overaging treatment T79 temper (3–5 hours at 170°C). Transverse profiles in the floor region of an aircraft passenger cell can be used in this T79 condition, whereas in those areas with severe corrosion resistance requirements, such as in the lower tail stringer profiles, only material in the T73 condition can be used (Fig. 15.15).

In the same way that purer variants of the 2024 alloy were developed by lowering the Fe and Si contents, purer versions of 7075 have been developed, notably 7175 and 7475. Alloy 7475 (Table 15.2) has up to 75% higher fracture toughness (K_{Ic}) in the critical short transverse direction, allowing a three times longer critical crack length to develop without the danger of failure.

One of the most recent alloys introduced into aerospace applications is 7150, which shows both higher strength and considerably reduced quench sensitivity when compared with the original 7075 alloy. The lower quench sensitivity was achieved by replacing the Cr addition with Zr, and higher strength was obtained by increasing the Cu and Zn content (Tables 15.2 and 15.3). This alloy is particularly well suited for extremely thick semi-fabricated products (from 150–170 mm) because its low quench sensitivity makes possible high mechanical properties even in the center of regions with large cross sections. The alloy is principally used in the highly stressed upper wing surfaces in the form of thick plate and thick walled extruded profiles. Because the upper side of the wings in flight are mainly subjected to compressive stresses, it is possible to use the material in a less than optimum condition in terms of its stress corrosion resistance by making use of a slightly overaged T76 temper.

Alloy 7055 also relies on strict compositional control of alloying additions and thermomechanical processing to produce a material that has a higher strength than 7178-T6, together with improvements in its susceptibility to stress corrosion cracking and exfoliation corrosion, fracture toughness, and fatigue resistance.

15.4.3 Al-Li alloys (in 2xxx and 8xxx series)[6]

Additions of lithium have a strong effect on both the modulus of elasticity and on the density of aluminum alloys. For every weight percent of lithium added, the elastic modulus is increased by 6% and the density is lowered by 3%. These effects have prompted extensive research for several decades into the development of alloys for aircraft and aerospace applications. Three generations of Al-Li alloy development can be distinguished.

The Al-Li alloys developed in 1950s, 1960s, and 1970s suffered from relatively low ductility and fracture toughness. Alloy 2020, which was adopted for the compression wing skins of a fighter aircraft, is the only one that achieved commercial application.

Alloys that emerged from research conducted in the 1980s contained relatively high levels of lithium (2.0–2.7 wt.%) in order to maximize potential property advantages. While these alloys had significantly improved elastic modulus and low density, they also developed highly unfavorable anisotropy. This reduced the benefits gained in mechanical properties. The alloys 2090 and 8090 are compositions from this period.

It was not until the most recent high strength Weldalite-type alloys were developed at Martin Marietta Laboratories that it was clear that the Al-Li alloys would eventually see widespread use as aerospace materials. As the name implies, these Weldalite alloys (2195 is an important example) were designed to be both weldable and lightweight. They were based of the Al-Cu-Li system with small additions of both magnesium and silver. These additions encourage the formation of an extremely fine platelet microstructure similar to the Al2CuLi or T1 phase, which provides considerable strength to the alloy. It was found that in an alloy of Al-6.3Cu-0.4Ag-0.4Mg+Li the maximum tensile and yield strength values were obtained at lithium levels of between 1–1.4%, a much lower lithium level than had previously been explored. At lithium levels higher than 1.4%, the strength was found to decrease rapidly. This is believed to be associated with the delta phase precipitation.

The good weldability of these alloys, together with their extremely high strength and lower density, suggested that they would be ideal for use in the cryogenic tanks of space launch systems. This is a significant niche application that justifies the development costs and the higher production costs associated with the more complex alloying processes involved when manufacturing lithium-containing alloys. In view of the reactivity of lithium, alloying additions to the liquid metal must be made under an atmosphere of inert gas.

Many space launch vehicles are designed with a large number of welds, particularly in the oxidizer and fuel propellant tanks of launch systems. With the cost of launching a payload into low earth orbit running at approximately $8000 per kg, it is logical to target large space launch systems as the ideal application of these alloys.

The use of these alloys for cryogenic tanks has been shown in practice. The welding parameters and the manufacturing methods needed to fabricate a subscale cryogenic tank

[6] By J.A.S. Green.

have been demonstrated. It has also been proven that these alloys can be safely used with liquid oxygen.

Martin Marietta Corporation (now Lockheed Martin Corporation) is the prime contractor and major fabricator of the two largest launch systems, the Titan family of launch vehicles and the external tank of the Space Shuttle. After the successful fabrication of a subscale tank, it was decided to specify the use of alloy 2195 for the fabrication of the Super Lightweight Tank program for the Space Shuttle. This tank is intended to be 3500 kg lighter than the current external tank. This development will increase the payload of the Space Shuttle by the same amount. The Al-Li alloy has also been specified for the future use in the Titan launch system following the next major design change.

Fig. 15.16 shows the Space Shuttle during lift-off from the Cape Kennedy launch site. The large orange-brown external tank acts as the structural backbone of the launch system and accordingly must be sufficiently rigid to hold all the components together during the launch. An exploded view of the external tank is shown on the right hand side of Fig. 15.16. The bulk of the volume is taken up by the liquid hydrogen tank. The smaller liquid oxygen tank is mounted on top of an intertank assembly that separates the two. The combination of the hydrogen and oxygen propellants provides the thrust for the Space Shuttle to reach earth orbit.

In the Super Lightweight Tank, the alloy 2195 has been substituted for 2219 in both the liquid hydrogen and liquid oxygen tanks. This has enabled some components to be reduced in size. In the innertank assembly, the Al-Li alloy 2090 has replaced 2024. As a result of these changes, a weight saving of about 600 kg was achieved in the oxygen tank, about 350 kg in the intertank, and about 2000 kg in the hydrogen tank. With additional savings of about 400 kg from the thermal protection system, a total weight reduction of about 3500 kg was obtained from the new tank design.

Fig. 15.17 shows the prototype Al-Li tank undergoing final assembly and testing at the NASA Assembly Facility at Michoud, Louisiana. The complex welded structure of the dome section of the tank is clearly evident in this picture.

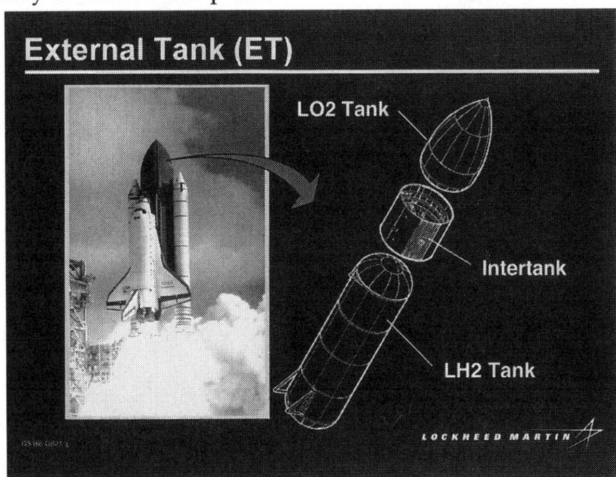

Fig. 15.16: Space Shuttle with aluminum tanks during lift-off from Cape Kennedy.

Fig. 15.17: Prototype Al-Li alloy tank undergoing final assembly and testing at the NASA facility at Michoud, LA.

Given their early success in the US Space Program, the most recent family of Al-Li alloys are planned for increased use in military aircraft. Alloy 2197 is being used to refurbish the F-16 jet fighter. Alloy 2197 is being used for a significant bulkhead application and its substitution and consequent weight saving provides improved performance and range for the aircraft. The Al-Li alloys are proceeding along the traditional development path of advanced alloys in space and military vehicles. This will eventually lead to widespread use in large commercial transport aircraft.

In summary, the load-carrying capability of the Space Shuttle is being extended through the use of Al-Li alloys. Alloy 2195 offers a 40% increase in strength, a 5% increase in elastic modulus, and a 5% reduction in density compared to alloy 2219, which it will replace in the external tank. It is considered most likely that, following this successful application, Al-Li alloys will see increasing use in advanced aerospace and aircraft structures.

15.4.4 Competition with other materials

The availability of Al-Li alloys has considerably improved the competitive position of aluminum as a material compared to carbon fiber reinforced plastics (CFRP). However, as studies for the Airbus aircraft have shown (Fig. 15.18), the proportion of CFRP materials in aircraft structures will increase up to and beyond the year 2000 largely as a result of a reduction in the proportion of aluminum materials in the structure. The latest Airbus type, the A340, shows an increase in CFRP components of 14%; nevertheless, aluminum remains the dominant material (70%). The most important mechanical properties of the CFRP materials in comparison to conventional aluminum alloys are their up to three times higher strength and stiffness together with their exceptional fatigue characteristics. Limitations are their less-predictable failure modes and their high maintenance costs.

Use of the property improvements provided by CFRP materials can, for example, achieve an almost 20% weight saving in the fin and tail unit of an Airbus A340 if these components are made completely out of CFRP material. Although CFRP materials offer a potential of up to 30%, they have been used relatively sparingly to date in civil aircraft con-

Fig. 15.18: The relative proportion of materials in the Airbus series from the A300 up to the 600–800 seat Airbus 2000, which is planned for the year 2000. (VAW)

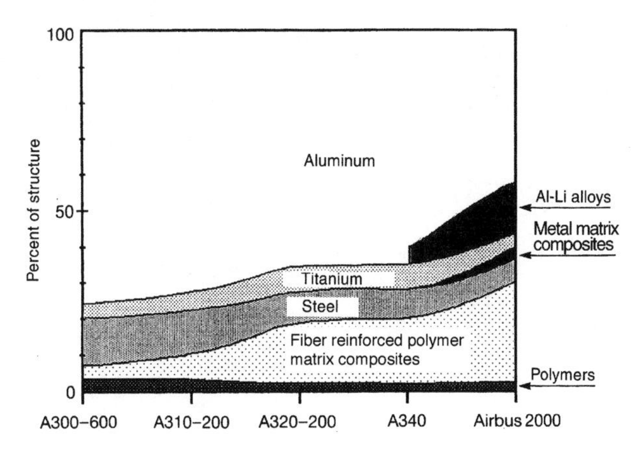

struction. The reason for this, apart from their very high price, is unsolved material problems such as low ductility and toughness as well as high damage sensitivity. In addition, the new CFRP materials have complicated manufacturing and quality control procedures involving entirely new inspection and repair methods. In contrast, the Al-Li alloys, with the single exception of problems in casting caused by lithium's high reactivity with oxygen, can be used as conventional aluminum materials during processing and in service.

Airbus-Industrie has recently started development of a new supersonic passenger aircraft planned as a replacement for the Concorde. Because of the aerodynamic skin friction heating effects that warm the outer skin of this proposed new aircraft up to a maximum temperature of 190°C, use of improved heat resistant alloys will be necessary. For this application, the high Fe- and Ni-containing 2618 aluminum alloy is being considered as a possible starting point. This alloy is used as the outer skin of the present Concorde without problems (Table 15.2). The 130°C limitation on the use of 2618 for surface cladding means that new alloy variants need to be developed to meet the 190°C requirement. The 2618 alloy composition will require further optimization. Apart from the need for higher elevated temperature tensile properties, it will be necessary to ensure that the toughness and fatigue behavior of the material are satisfactory after exposure to long times at the high operating temperatures.

15.4.5 Future prospects

The extreme material property requirements in aerospace applications and the increased competition with other materials, such as CFRPs, has provided a great challenge for aluminum manufacturers. The few examples here show that the aluminum manufacturers are meeting this challenge by a continuous optimization and further development of the conventional aluminum alloys as well as by the development of completely new materials, such as the AlLi family of alloys. It has been demonstrated that the full potential of aluminum based alloys is far from being exhausted. In the future, aluminum alloys will continue to play the dominant role as the material for aircraft construction.

15.5 Aluminum in Ships[7]

15.5.1 Merchant ships

In conventional merchant ships, there is little difficulty in designing steel vessels with a center of buoyancy at an optimum height above the center of gravity to provide the ideal metacentric height for the vessel's stability. There is no pressing need to make use of the weight advantages of aluminum. Its use is, therefore, generally limited to ancillary low maintenance fittings.

A significant specialized application makes use of the cryogenic properties of aluminum for liquefied natural gas (LNG) tanker vessels. The large spherical tanks that are the distinctive feature of these ships are welded from formed 5083 plate.

15.5.2 Passenger vessels

In passenger ships of between 30,000–70,000 gross tonnes, approximately half of the empty weight is due to the vessel's structure. Therefore, optimization of weight in the design is of major importance. The main constructional material in ship building is steel, however, in certain classes of vessel, the advantages offered to ship's designers by the low weight and excellent corrosion resistance of aluminum alloys are very evident. Typical alloys for marine applications are 5052, 5454, 5083, 5086, 6061, 6063, and 6082.

Aluminum is frequently used in the superstructure of passenger vessels. The position of the ship's center of buoyancy above its center of mass is the factor that determines a vessel's stability. Use of aluminum makes possible an increase in the volume and height of the superstructure without loss of stability. This allows the space in the part of the vessel most favored by passengers to be increased in size. The designer of ships combining steel and aluminum in structural functions has to solve very specific problems:

- Because of the significant difference in modulus of elasticity between steel and aluminum it is necessary to carry out a full three-dimensional finite element analysis of the superstructure in order to ensure that the elastic deformation and resulting local stresses remain within acceptable limits. Use of computer aided design (CAD) techniques are unavoidable in order to optimize these more complex structures in terms of both weight and functionality.
- In order to avoid problems from bimetallic corrosion caused by differences in electrical potential in steel-aluminum joints in the structure, the use of permanently bonded bimetallic elements is generally necessary. These elements are frequently manufactured by use of explosive bonding techniques.

15.5.3 Cruise ships

A major application area for aluminum is in modern cruise ships. These vessels are becoming more and more like floating hotels. They are generally designed as shallow draft vessels so that they can take advantage of the many beautiful harbors unable to accept

[7] In collaboration with R.J. Dean.

Table 15.4: Use of aluminum in cruise ships

	Raffello	Crown Princess	Eugenio Costa
Length (Meters)	244	204	189
Speed (Knots)	29	22	2
Displacement (BRT)	46,000	70,000	30,560
Aluminum weight (Tonnes)	500	310	150

deep keel ships. Passengers like to have cabins above the water line. More passengers can be accommodated on a vessel without loss of stability if the upper decks and superstructure are built in aluminum. As an additional benefit, maintenance on these appearance-sensitive upper deck areas is reduced when aluminum is used. Table 15.4 gives details of three Italian cruise ships with aluminum superstructures.

15.5.4 Fast and unconventional craft[8]

In the past two decades, there has been an accelerating demand for higher speed ships for passenger and vehicular ferries, cargo container vessels, and commercial, military, and patrol applications around the world. Aluminum is playing an increasing role for many of these applications, partly because of its combination of favorable properties, including its light weight and excellent corrosion and fatigue resistances, as well as relative ease of fabrication. Fig. 15.19 shows a fast all-aluminum motor yacht. The structure (Fig. 15.20) was welded primarily of 5083-H112 plate.

A more recent and increasingly important marine application is in the field of "dynamic lift" or "surface effect" ships. The lighter these vessels are, the greater the payload that they can carry at speed for a given engine power. The weight of the hull is inversely related to the vessel's earning power. The lighter the vessel, the greater the "deadweight" (e.g., passengers, fuel, cars, lorries, freight, etc.) that it can carry. The choice of construc-

Fig 15.19: The Italian all aluminum yacht *Destriero*; it won the "Blue Riband" in 1992 when it beat the previous record by almost two days.

[8] By R. Hanneman.

Fig. 15.20: The rib structure of the all-aluminum yacht shown in Fig. 15.20.

tional materials for this class of vessel is, therefore, restricted in practical terms to aluminum, graphite-reinforced polymers, or special high strength steels.

Studies have shown that ferry users are prepared to pay more for faster, more frequent services. The market for this class of vessel is growing worldwide. Since the end of the 1980s, fast ferries capable of transporting cars and lorries have been available from a number of shipyards. These 60 meter plus ships were first introduced by the Australia yard, Incat, based at Hobart on the island of Tasmania. This yard, with its Phillip Hercus designed 74 m and 79 m wave piercing catamarans, was the first to begin building the new road-vehicle-carrying generation of larger fast ships. They now offer an all aluminum "fast vehicle ferry" designed for inter-island transport in the Pacific Basin. Other Australian all aluminum fast catamaran builders are based in Perth and Cairns.

Global competitive forces are also increasing the need for shorter product cycle times from their manufacturing locations to their customers, including shorter shipment and load / unload times. This is particularly true for transoceanic high unit value cargo shipments that cannot justify higher air freight costs. Development work is underway on aluminum-intensive cargo ships with load capabilities up to 3000 tonnes that could eventually operate at up to 60 knots and handle Class 6 seas for regional or transoceanic operation. Such vessels will be capable of crossing the Atlantic in under 60 hours. An example of a particularly fuel efficient hull design is the Quadrimaran tapered four-hull vessel shown in Fig. 15.21.

Future military requirements are moving toward much higher speed, somewhat smaller, and more agile lower profile vessels, often with nonconventional hull designs and sophisticated missile weaponry capabilities. Speeds of 60–80 knots or more by some of these ships are anticipated, along with lower draft and wakes. This will require rugged lightweight materials such as aluminum and unconventional hull designs involving surface effect air lift principles.

Fig. 15.21: A schematic of a Quadrimaran high speed surface effect ship under development.

Fig. 15.22: The Stena HSS-1500 fast ferry operating in Europe.

Several non-monohull aluminum-intensive civilian fast ferries have been designed and built in recent years, capable of handling up to 1500 passengers and 375 cars at speeds in the range of 30–50 knots. The market for such vessels is continuing to grow. The important fast ferry trends include increased speed and size requirements, enhanced fuel economy, sea worthiness, passenger comfort and safety, high reliability, and low maintenance. Many of these ships have used twin catamaran type hulls, such as the HSS 1500 vessel shown in Fig. 15.22. Most of the technology advances and commercial shipbuilding success of such aluminum-intensive high speed ferries in recent years has occurred in Australia, Europe, and the Far East.

Key technology factors contributing to such successes in fast ships have included sophisticated computer-aided design; innovation in hull designs; enhanced propulsion and control systems; and improved aluminum processing, fabrication, and welding capabilities.

Fig. 15.23: A partially assembled hull frame of a 79 meter 40 knot "Super Sea Cat" catamaran vessel.

The scale of building the new large high-speed aluminum multihull ships and the welding operations involved in their fabrication are impressive, as can be seen in Figure 15.18. Several hundred tonnes or more of aluminum sheet, plate, and extrusions with many thousand welds can go into building a single ship, as illustrated by Fig. 15.23 for a high-speed vessel.

The most common plate and sheet alloys currently used for high speed aluminum-intensive ships include 5083 or its recently developed cousin, 5383. Also, alloys 5086 and 5454 have been used in some sheet and plate applications. The stronger alloy 5456, which was popular for many years in marine applications, is disappearing due to its slight susceptibility to intergranular and exfoliation corrosion and difficulties with edge cracking during production, which affects yields.

For 5083 and 5383, partially work hardened tempers such as H111, H116, or H32 can give a definite advantage for achieving better toughness, buckling resistance, and stiffness in the region of the weld heat affected zone. The intergranular corrosion and exfoliation resistances of these alloys are excellent in marine environments. Some 6xxx sheet alloys, such as 6061, have been successfully used in above deck applications. To avoid or minimize excessive hydrogen and porosity or other weld defects, high quality TIG or MIG welding techniques are generally used with 5356 or 5183 alloy filler wire.

Further advances in nonconventional welding methods and use of structural adhesive bonding of certain aluminum components are anticipated in the near future. In the meanwhile, aluminum will continue to be used in a wide variety of smaller pleasure boats and yachts as well as the applications cited above.

15.5.5 Future prospects

Highly stressed aluminum ship structures have only become possible through careful application of computer aided design techniques using finite element methods. Aluminum will continue to be used for both passenger vessel superstructures and fast surface skimming craft. The fast ship market is growing rapidly.

15.6 Rail Transport Vehicles[9]

In the rail transport sector, aluminum has become a major structural material for a number of distinctly differing applications.

15.6.1 Movable side walls on freight cars

Low mass, ease of manufacture, and low maintenance—these are the principal reasons for the choice of aluminum as the material for relatively simple rail car components. A rail car's side walls should hold its load in place on the flat floor area, should be capable of being used during loading as a bridge between loading ramp and rail car, and should, where possible, be capable of being put into the required position by one person. The classic steel design for this type of side wall part consists of a flat sheet with upper and lower longitudinal U-shaped profiles. The steel side wall has vertical U-profiles and is welded together. Frequently, the upper side wall edge has additional closing elements welded into place. The hinges are made from forged parts and are also welded. A typical steel rail car side wall of this type is made from 25 individual parts and weighs 106 kilos. Two people are needed to move it. Because of wear and tear in service, such side walls soon become scratched and damaged. Rusting of the areas devoid of paint rapidly follows. Within a few years, the side wall of a steel car usually needs to be repainted.

[9] By J. Zehnder.

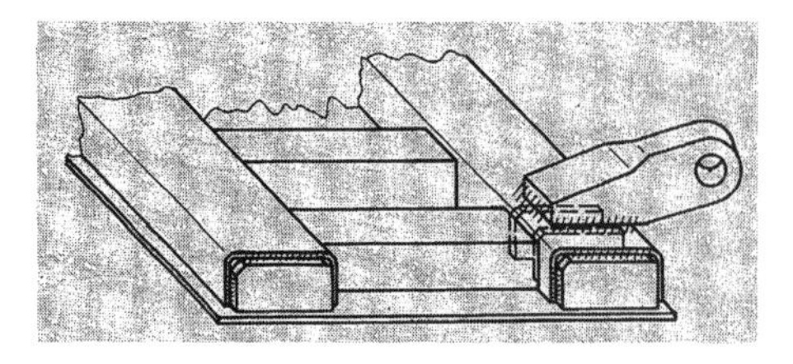

a

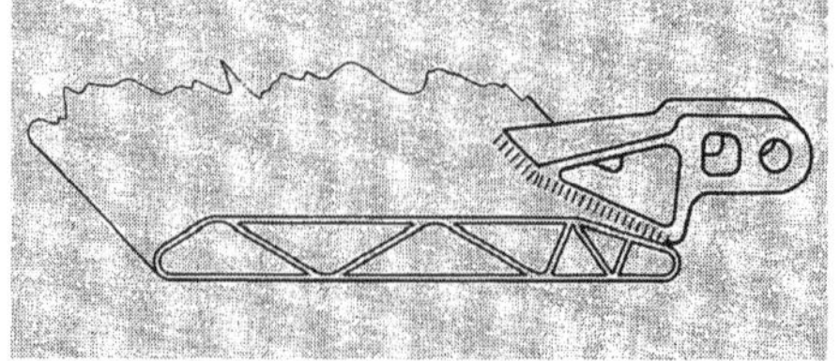

b

Fig. 15.24: Side wall designs for freight cars (a) steel and (b) aluminum.

Fig. 15.25: Aluminum side wall in a railroad coal car. (Alusuisse)

Aluminum offers a quite different solution. The load carrying element of the side wall is made from a section of internally stiffened large hollow extrusion with wall thicknesses suited to the load conditions. The hinge is made from a gravity die casting so that it can be welded onto the profile in a suitable position between the outside skin and the internal stiffening lattice. If locking and corner elements are included, this type of aluminum side wall design is made from six parts and weighs 48 kg (Fig. 15.24.a). It can be placed in the required position by one person. Because of the excellent corrosion resistance of the intermediate strength Al-Mg-Si alloys, it is not necessary to seal the ends of the extrusions, and painting is not needed. This type of aluminum side wall is virtually maintenance free.

15.6.2 Freight cars for bulk transport

An important use of aluminum occurs in countries such as the USA, Canada, and South Africa, which are rich in coal, metal ores, and other minerals but which need to transport these materials considerable distances between mine and production plant or port facility. A number of differing concepts have been tried for such trains, frequently making use of well established steel designs. Aluminum structures, however, make possible a reduction in the vehicle weight and allow an increase in the load per vehicle that can be carried. The economic reasoning behind this use of aluminum is of great significance when transporting bulk materials over large distances. A straight substitution between relatively cheap steel and more expensive aluminum for the vehicle body can be justified on

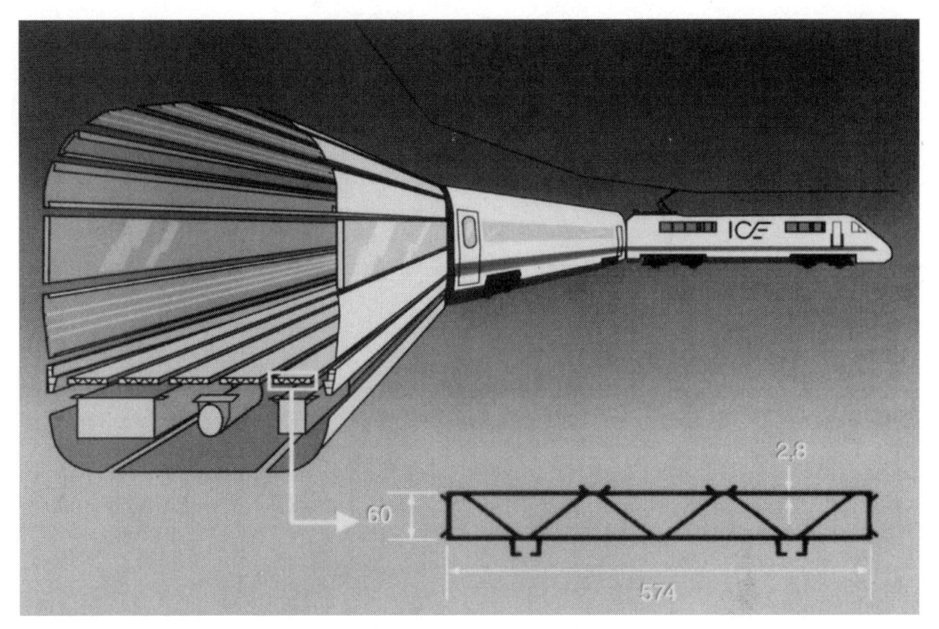

Fig. 15.26: The principle of integral construction of railway cars using wide aluminum profiles.

the basis of increased carrying capacity per vehicle and can repay the higher material costs in less than two years. One example is a freight car constructed of large extruded profiles (Fig. 15.25). An interesting aspect of this design is the positioning of the coupling on the central longitudinal support beam. Compressive and tensional stresses on the buffer system are carried by a machined multichamber hollow profile in such a way that longitudinal forces are directed down the neutral axis of the section and do not cause any secondary rotational moments.

15.6.3 Passenger vehicles

Passenger rail cars are the most important and technically demanding area of rail vehicle design. The requirements are many sided. Powered cars such as those used in underground or local trains must be able to carry all their equipment under the floor and on the roof of their structures. In addition, loads are created at varying positions in the vehicle by sitting or standing passengers. All of these loads act locally on the passenger cell. The forces from bogies and couplings during acceleration and braking have to be carried by the structure. The cars must be able to withstand the longitudinal, tensional, and compressive loads on the train set, and any tendency to develop natural vibrations must be avoided. The car must also be capable of withstanding the high aerodynamic pressure waves that occur, especially at high speeds, with a minimum of deformation. In the event of a collision, the rail car structure must protect the passengers.

In rail transportation, the development of light steel structures has been studied for approximately 60 years. Very limited opportunities remain for improvement to the current design and production costs. In contrast, the substitution of steel by aluminum provides an opportunity for major benefits. From economic considerations, it is not possible to adapt the well known steel design concepts. In order to introduce a truly successful aluminum design, ways have had to be found of making use of specific aspects of the properties of aluminum.

The use of large aluminum extrusions to form an integrated structure (Fig. 15.26) has been widely adopted as a very successful design solution. Cars are constructed from extruded profiles that run the full length of the vehicle. The extrusions are longitudinally welded together, and are supported by relatively few transverse ribs. Hollow extrusions with internal stiffening members are used to form the floor structure between the two longitudinal supporting beams. The main transverse stiffening elements are installed where the rail car rests on the bogey. In order to support equipment, pipes, and cables, the longitudinal profiles have integrated attachment channels. This reduces the amount of welding.

In a similar way, longitudinal attachment channels are integrated into the upper surface of the floor extrusions. The side walls are generally made from longitudinal profiles with integral stiffening elements. Depending upon loading, these profiles have either a single or a double skin. The stiffening elements are calculated to suit the shape and dimensions of the interior of the vehicle. Transverse profiles support the structure in the form of door and window columns. The outer surface of the roof is generally made from two longitudinal supporting members and a number of single skin profiles with integrated longitudinal stiffening elements. Transverse formed ribs provide additional support.

The end walls are made from supporting columns between the main roof members and the transverse floor elements and stiffening filler elements.

The number of individual components that make up this type of car structure is approximately one third of those required for a comparable steel vehicle. The weight is about two thirds of a steel car. Because of the many longitudinal welds, it is possible to make use of time saving automated inert gas welding systems during assembly.

The introduction of this concept has had major consequences for the railway coach construction industry. It has become essential to design the body shell and interior together in order to take full advantage of the possibilities of using large extrusions in the structure that support the interior fittings. Because of the lower number of individual component parts, the problems of logistics and manufacture are greatly simplified. The forming and working of sheet metal has almost completely disappeared. The machining of profiles has had to be approached in new ways since it makes little sense to try to machine on conventional equipment the very long profiles that are required. The introduction of similar devices to those used when working with wood directly at the point of assembly has been proven to be more economic. Adoption of the integrated design concept using extruded aluminum has achieved such major savings in assembly that the higher cost of the base material has been completely outweighed.

As far as the railway coach operator is concerned, the use of light aluminum structures provides a number of advantages:

- Energy saving in service (mainly trains with frequent stops such as metros and other urban area rail systems)
- Higher achievable acceleration from the same power unit
- Excellent corrosion resistance with the possible use of unpainted vehicles
- Low maintenance costs
- Long service life
- High scrap value when service life is ended

15.6.4 Recent substitution aluminum for steel

Whereas the first two versions of the 300 km/h TGV (the French high speed train) were built in steel, the double deck third generation, the TGV Duplex, is an aluminum design. The weight challenge that the axle load remain below 17 tonnes was the most difficult problem for the designers. Feasibility studies quickly eliminated high strength or stainless steel for the trailer bodies, as the weight saving was insufficient. The decision was made to use aluminum alloys, and a 20% weight saving compared to the previous steel trains was obtained, even though the new trains have two decks. This was achieved by the use of large extruded aluminum profiles that run the full length of each car.

15.6.5 Aluminum in magnetic levitation trains[10]

15.6.5.1 Transrapid—the German Maglev system

The Transrapid magnetic levitation system is designed for cruising speeds in the 400–500 kmph range. It uses attractive magnetic forces for suspension and guidance and a synchronous linear motor to provide propulsion and braking (Fig. 15.27). By the year 2005, the Maglev line between Berlin and Hamburg, the two largest German cities, is scheduled to be in full operation. The six section trains will cover the 292 km distance (city center to city center) in one hour.

The levitation system is based on the attracting forces of the electromagnets in the vehicle and the ferromagnetic reaction rails in the guideway. An important requirement of the train bodies is that they must be stiff and strong, but, at the same time, they must be light. Average empty weight per section in a six section passenger train set is 45 tonnes, with an average load carrying capacity of 12 tonnes per section. The vehicles are 3.7 m wide and, depending on whether it is a nose or mid section, either 27 m or 25 m long. In order to provide a light stiff structure for the body shell, a hybrid design was chosen that uses longitudinal hollow aluminum extrusions with aluminum-clad foam core panels.

An extensive test program on possible sandwich panel types showed a clear superiority for the variant with an aluminum clad foamed polyetherimide core bonded with epoxy adhesive. This type of panel exhibits high damage tolerance, excellent isotropic stiffness, and good fire characteristics.

[10] By R.J. Dean.

Fig. 15.27: The Transrapid magnetic levitation system for trains. (Alusuisse)

A Transrapid vehicle comprises a drive/levitation lower assembly that straddles and wraps around the track. This part of the vehicle contains the guidance and propulsion system of the magnetic levitation train. The passenger or cargo compartment is mounted on top of this structure and weighs, including doors and windows, approximately 8.5 tonnes. Fig. 15.28 shows a cross section of the passenger vehicle.

The main aluminum extrusions that run the full length of each section have been constructed to provide the necessary strength, but they also have a range of additional features designed into them. The profiles have been optimized for attachment of the sandwich panels, which act as the vehicle skin. The extrusions are made of an alloy similar to the 6005A type. The sandwich panels chosen for the Transrapid vehicles have a foamed polyetherimide core (thickness range 30–70 mm) clad with sheet of the 5005 type (thickness range 0.6–1.0 mm).

Fig. 15.28: A cross-section of the TRANSRAPID magnetic levitation passenger vehicle. (Alusuisse)

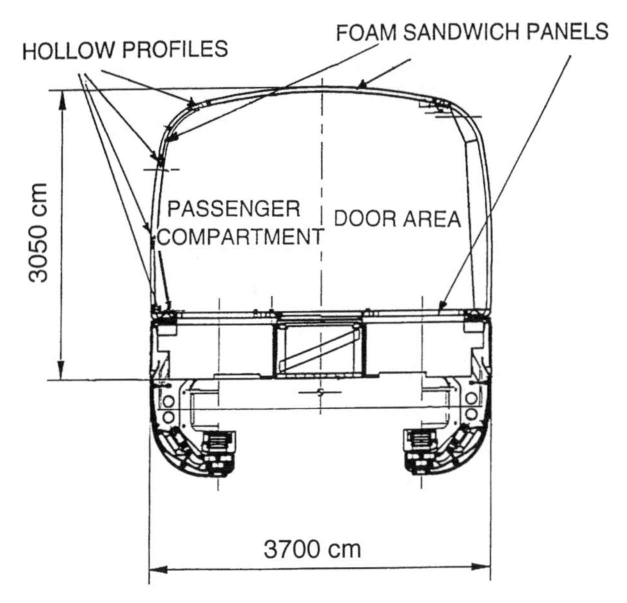

15.6.5.2 *Sandwich variants*

Extensive trials showed the advantages of using a foamed polyetherimide core material (density 80 kg/m^3). The tests showed superior static and dynamic properties and much improved damage tolerance as compared to sandwich variants with a phenolic-coated aramid paper honeycomb core (density 48 kg/m^3). Excellent fire test results and satisfactory performance after accelerated corrosion and weathering were also demonstrated.

Two methods of joining the sandwich panels to the structure were selected: laser welding and riveting. Laser welding is used to secure and seal the aluminum cladding panels onto the integrated sandwich edging profiles. The same technique is used to join the sandwich elements to the main structural extrusions. Each Transrapid vehicle cell is built from three main laser-welded sandwich structural profile elements, a complete roof, and a left and right L-shaped side wall/floor assembly. Each of these three major structural components runs the full length of the vehicle.

15.6.5.3 *MLX01—the Japanese superconducting maglev vehicle*

This project, which is currently not as advanced as the Transrapid, is intended for a new transportation artery between Tokyo and Osaka. MLX01 relies on repulsive rather than attractive forces for the levitation effect. Its superconductive coils need very low temperatures to function. Bundles of extremely fine niobium-titanium alloy wire embedded in a copper matrix are cooled with liquid helium (–269°C). The MLX01 is designed to operate at speeds up to 550 km/h.

The cross section is slightly smaller than the existing Japanese Shinkansen trains to lower aerodynamic drag and reduce weight. The train body uses aircraft technology to provide a light but very stiff structure capable of withstanding the frequent pressure changes as it runs through tunnels. The extremely high magnetic fields from the superconducting magnets makes necessary a magnetic shielding layer, which is built into the walls of the passenger cell. Alloys 5083 and 7075 are used in the body shell. Extensive use is also being made of CFRP. The cabin structure is made from four component parts, and each part comprises an outside skin with frames and stringers made from profiles. The component parts are joined together by spot welding or riveting at the frames.

15.6.6 Future prospects

In summary, the major success of aluminum in the rail sector is due to its properties as a material and the advantages it offers during vehicle construction coupled with its low maintenance requirements in service. The deciding factors in its favor are its weight saving potential and the freedom of design in two dimensions offered by the extrusion process. These factors make possible light structures with a high degree of stiffness and provide the opportunity for incorporating cost saving multifunctional design features. It is this design possibility that offers the greatest challenge for the future.

For the Transrapid vehicle body, the hybrid design, with its mainly longitudinal hollow extrusions combined with aluminum sandwich panels, has made possible the combination of high structural integrity with low cost manufacturing techniques.

Japanese Railways have operated their aluminum-bodied Shinkansen high speed trains for more than three decades. Originally the top speeds were 220 km/h. This was upgraded to 270 km/h, and double deck train sets were introduced. These vehicles are all lightweight aluminum body shell designs. With the advent of the 550 km/h superconducting Maglev, the Japanese designers are moving toward aircraft construction techniques.

15.7 Road Vehicles

15.7.1 Aluminum in the automobile industry

The motor car industry is one of the most important sectors in modern industrial society. In countries with a large car manufacturing industry it contributes a major portion of the gross national product. This, together with the fact that the mass use of this individual transport system is causing serious problems in society, has made the motor car in highly industrialized countries an economic factor of the highest importance.

The automobile industry is confronted by three environmental problems caused by their products:

- Energy consumption
- Environmental pollution, especially of the air
- End-of-life disposal of used vehicles

As will be shown, aluminum provides a key to the solution of important aspects of these problems.

On the three continents in which automobiles are manufactured, the pressures of public opinion and the resulting reaction of law makers are providing a strong impulse to change the general perception of the automobile; it is not now being viewed as a "sell and forget" product. Development efforts are directed to the concept of the private car as a system, which is seen as being the only way to overcome the three key problems mentioned earlier. This means that a very large number of new developments are necessary by both the car manufacturers and their component and sub-system suppliers. In what follows, some of these trends will be described in order to indicate their significance to suppliers of aluminum products.

Aluminum makes up approximately 7-8% of the total mass of an average car. In the USA in 1996, private cars contained, on average, 114 kg of aluminum. In Europe in 1994, the typical figure was 65 kg per vehicle.

From Table 15.5 it can be seen that cast components make up the major portion of aluminum used in automobiles. Aluminum die castings are important in both suspension and steering systems as well as in the engine and gear box units.

Aluminum casting alloys containing the elements Si, Mg, Cu, and Zn in varying proportions were introduced relatively early as engine components. A good example is aluminum's early and continuing use for pistons. The alloys that are used are described in Chapter 6. Starting in racing cars and progressing via sports cars, cast aluminum wheels can be found today in almost all vehicle models. They reduce a vehicle's unsprung weight and, hence, contribute to a smoother ride and greater driving comfort.

15.7.1.1 Internal combustion engines

There is one key reason for the use of aluminum in engines for the transport sector: their physical properties, notably light weight and high thermal conductivity. Two significant applications and the alloys utilized are outlined below.

Piston alloys

Hypereutectic alloys with up to 25% Si are used for casting pistons. The crystals of primary silicon create a wear resistant surface and also serve to reduce the thermal expan-

Table 15.5: Main uses of aluminum in automobiles in Europe in 1994

	Proportion in various groups			
Component Groups	**(%)**	**(kg)**	**Cast alloys (%)**	**Wrought alloys (%)**
Chassis and suspension	30	20	95	5
Engine/drive	46	30	90	10
Body	16	10	60	40
Accessories	8	5	60	40
Totals	100	65	85	15

a

b

Fig. 15.29: A Mercedes-Benz engine with first application of Al-Si liner technology: (a) two views of cast block with liners and (b) an assembled V6 (3.2 liter) engine. (Mercedes-Benz)

sion of the piston. The addition of up to 3% Ni and sometime up to 1% Co increases strength at elevated temperatures. Small but nonetheless important additions of several other alloying elements are also used, such as Cu, Mg, and Pb; the latter makes for easy machining. There are now about 30 frequently used piston alloys, and the subject has become a minor branch of the metallurgical sciences.

Alloys for engine blocks, cylinders, and cylinder liners

Hypereutectic alloys are also prominent in this application for the same reasons as for pistons. Vibration damping is another important consideration. The silicon content is usually in the range of 16–20%. The pioneering work on hypereutectic alloys was done by the Reynolds Metals Company, resulting in the widely used casting alloy 390.0, with 17–18% Si. Besides cylinders and cylinder liners, this alloy was also used for casting monolithic engine blocks.

The increasingly severe demands of the newer engine designs require yet higher silicon levels, and these are being attained by means of processes such as powder metallurgy (i.e., compaction of rapidly solidified powders containing higher than normal alloying constituents).

Fig. 15.29 shows a Mercedes car engine of the latest type. These four-stroke engines weigh less than 150 kg, some 50 kg less than the iron-based V6 engines that were formerly common. This weight reduction results from using aluminum and magnesium die-cast parts; the crankcase, oil sump, cylinder head, pistons, and engine control system housing are aluminum, and the air-inlet manifold, cylinder-head cover, and rocker-arm cover are magnesium. For the crankcase alone, there was a 50% reduction in weight, from 55 kg for cast iron to 26 kg for aluminum.

Cylinder liners of very high-silicon alloys offer an outstanding example of tailor-made alloy and product development. Reynolds, followed by others, have successfully extruded such alloys into seamless tubes. Cylinder liners made from these alloys reduce weight by 500 g per cylinder and provide high performance. Similar advantages can be had by depositing hard surface layers on cylinders of aluminum engine blocks; it is still uncertain which technology will provide the greatest cost effectiveness.

15.7.1.2 Aluminum wheels[11]

Automobile wheels are an important market segment for aluminum. Surprisingly, however, weight is frequently not the main factor behind the choice of aluminum; styling is often the main reason for a switch away from steel. In most cases, original equipment wheels are lighter than steel wheels, but replacement aluminum wheels do not generally save weight and some are heavier than equivalent steel wheels.

Forged wheels are not widely used because of their high cost, but because of their high strength, high ductility, and freedom from defects, they nevertheless show the best potential for weight savings. Their low porosity allows designs to be developed with very thin walls.

[11] By H.P. Erz.

Cast wheels are very common, and 90% of these are produced in the A356.0 alloy. The wheels are usually supplied in the T6 heat-treated condition. In Europe, an 11% silicon alloy has been tried for wheels. It is supplied in the non-heat treated condition but is usually annealed to improve ductility. Although the alloy has excellent castability, its mechanical properties are not as good as those obtained in A356.0 wheels. Low-pressure die casting (LPDC) is the technique most frequently chosen, and this technique is used to make 90% of total aluminum wheel production.

In Japan, thixoforming (squeeze casting) has shown promising results for low weight truck and car wheels. High capital investment is required for this production method and has proved a barrier to its wider use. Although productivity and properties are slightly improved, the favored method remains LPDC. In LPDC, solidification proceeds under an over pressure of about 0.8 bar, which is comparable to the feeding pressure in a riser about 3 m high. Many developments have improved quality and productivity of the LPDC process in recent years.

Aluminum wheels produced from rolled sheet stamped, drawn, and welded together using the same technology as used for steel wheels have been tested. The wheels were 30% lighter, but offered few styling opportunities and were more expensive than steel wheels.

Spin forging is being used as a wheel production technique. The face of the wheel is made by a forging process, and a spin forming operation is then used to shape the rim. This technique combines low weight and low cost, but offers limited styling possibilities.

In summary, styling has always been an important feature of aluminum wheel design, but the possibilities of aluminum use for weight savings have become a more important issue in recent years. Although LPDC wheels account for most of the aluminum wheels on vehicles, the need for higher ductility to improve behavior in the event of a crash has favored the use of alloys with low iron contents. This has meant increasing use of primary aluminum, which is essential if iron levels below 0.2% are to be achieved. This restriction also applies to material used for nodes made by vacuum high pressure die castings for use in space frames.

15.7.1.3 Body panels[12]

Technical and environmental criteria

Steel sheet remains the dominant material for automobile body work. The introduction of aluminum, however, offers great advantages when attempting to lower body weight and improve recyclability. Intense development effort is, therefore, being applied to further aluminum as a replacement for steel.

[12] By P. Furrer.

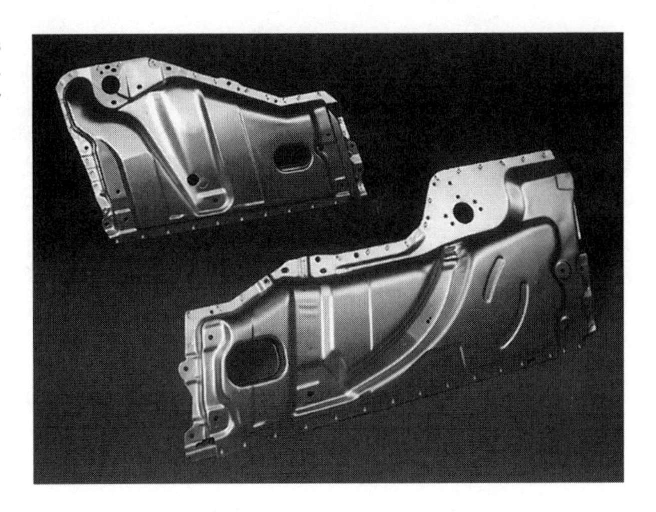

Fig. 15.30: Door components for the Audi 80 made from 5754 sheet joined together by clinching. (Alusingen)

Alloys and Properties

In recent years, aluminum's potential as a body sheet material has been the source of considerable interest. The non-heat treatable Al-Mg alloys are most frequently used for inner body parts that do not have high requirements in terms of surface quality. The main alloys are 5052, 5754, and 5182. All of these alloys offer very good formability in the soft condition (Fig. 15.30). Because of the possibility of the formation of flow lines (Lüders lines) during metal forming, these alloys are not used for exterior vehicle parts.

Alloys that form the outer skin must have five main attributes:
- High strength
- Good formability
- Suitability for spot welding
- Good corrosion resistance
- Good appearance after painting

There is no single alloy that fully meets all of these requirements, but the various markets have weighed the considerations and developed three differing alloy preferences: Al-Mg-Si-Cu in the USA, Al-Mg-Si in Europe, and Al-Mg-Cu in Japan (see Tables 15.6-8).

In the USA, the copper-containing Al-Mg-Si alloys, such as 6111, have achieved a degree of predominance. After furnace lacquering, they demonstrate higher mechanical properties than can be obtained from Al-Mg-Si alloys. They are, however, less formable than the European variants and have lower corrosion resistance.

In Europe, the principal direction of alloy development has been toward the optimization of formability. The heat treatable alloy 6016 has established itself as the most favored alloy variant for external body sheet applications. The alloy shows its best formability in the solution heat treated condition. During subsequent lacquer baking, the mechanical properties of the material increase, and, as a result, there is considerable improvement in dent resistance. This is an important property for vehicle exteriors.

The recently developed Japanese alloys, such as 5022, are based on the Al-Mg system with an addition of copper. They are generally corrosion resistant but, when viewed in terms of surface requirements, are not free of flow line defects (type B Lüders lines) after forming. Therefore, they could not be used for exterior body sheet applications in European designed vehicles. When compared with non-heat treatable conventional Al-Mg alloys, the Al-Mg-Cu materials show a mild heat treatment response.

The normal mechanical strength of aluminum body sheet in the as supplied condition is somewhat lower than conventional deep drawing steel. However, when taking into consideration the higher stiffness, which is a consequence of the greater thickness of aluminum sheet (1.2 to 1.4 times thicker than steel sheet), and the fact that the aluminum alloys age harden during stove lacquering, the final mechanical properties of the finished aluminum component are very similar to those obtained from steel.

Formability

The forming of sheet into car body components is, in practice, a relatively complex process. During the forming of a component, a single piece of sheet material may have to undergo deep drawing, stretch forming, compression forming, aperture expansion, or fold forming, or sometimes all of these together. The actual formability of sheet material is only one of a number of factors, including type of press tooling, deformation parameters, lubricants, and tool and sheet surface morphologies that influence press performance. Because of this, there are no generally applicable factors to describe formability.

Practical experience has shown that when comparing drawing behavior, the limiting elongation ratio (TYS/UTS) should, if possible, be smaller than 0.5, particularly for the more difficult products. In order to avoid necking and tearing during stretch forming, a high work hardening coefficient n (n = 0.3) is especially important. For difficult deep drawing

Table 15.6: Typical mechanical and formability values of aluminum sheet

(see Tables 15.7 for compositions and Table 15.8 for additional mechanical property information)

Alloy and Temper	$TYS_{0.2}$ (MPa)	El_5 (%)	n	r
Group 1: Al-Cu; Heat-treatable				
2008-T4	125	28	0.25	0.70
2010-T4	130	25	0.23	0.74
2036-T4	195	24	0.22	0.90
Group 2: Al-Mg-Mn; Nonheat-treatable				
5754-O	100	28	0.30	0.75
5182-O	140	30	0.31	0.75
Group 3: Al-Mg-Cu; Heat-treatable				
5022-T4	135	30	0.30	0.65
Group 4: Al-Mg-Si(CuMn); Heat-treatable				
6009-T4	125	27	0.22	0.64
6016-T4	120	28	0.27	0.6
6111-T4	150	26	0.28	0.70

components, it is important to consider the anisotropic factor r. Because of strong textural effects, it is generally necessary to use minimum and not average r values. Materials with good deep drawing characteristics have r values equal to or greater than 0.6 (Table 15.6).

Significant improvements in formability have recently been obtained with the introduction of special surface morphologies, the aim being to optimize the tribological conditions during pressing operations. Normal mill finish surfaces show anisotropic surface effects with marking from the ground work rolls producing a pattern aligned in the rolling direction. By adopting new work roll preparation methods, it is possible to produce a more or less isotropic morphology on the surface of the aluminum sheet (e.g. "Lasertex" finishes) by the use of lasers and electrical discharge topography (EDT) surfaces by spark erosion treatment of the work rolls.

"Closed lubrication pockets" are formed that ensure forced lubrication during the deformation processes and prevent the creation of localized cold welds and consequent tool pick up. High hold back pressures during forming are unnecessary. The deformed surface must, however, meet all of the requirements of the lacquering processes. The EDT surfaces have shown themselves to be particularly well suited in this respect.

Corrosion Behavior

The corrosion behavior of bare aluminum car body sheet in all types of corrosive environments can be considered as good. However, this is not generally relevant since car body components are usually painted. The application of a suitable pre-treatment before painting is a prerequisite for optimal in-service corrosion resistance. Recent developments include use in the production of pretreated and precoated aluminum sheet and coil.

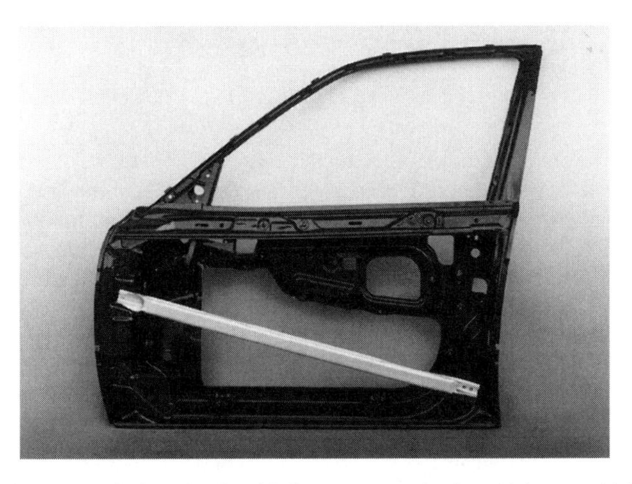

Fig. 15.31: Aluminum extruded section for side impact protection in vehicles, combining light weight with high energy absorption. (Alusuisse)

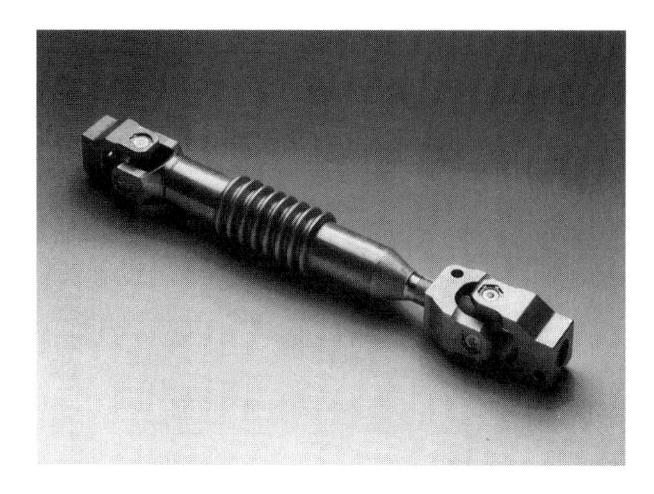

Fig. 15.32: Propeller shaft between steering column and steering gear, consisting of a forged part and three extruded sections. (Alusuisse)

Fig. 15.33: Automotive wheel produced from wrought aluminum products. (Alusingen)

Conclusions

Car bodies made from aluminum, when considered purely in terms of the cost of components plus the cost of finishing, are more expensive than the current steel car body types. Cost considerations must, however, be viewed in terms of environmental inter-relationships including the total energy balance and costs of recycling; when considered in this light, aluminum can have advantages as a body stock material over steel.

15.7.1.4 Other components made from wrought aluminum alloys

Because of the material's good thermal conductivity, aluminum alloy heat exchangers were introduced relatively early on in the development of motor vehicles. Their first serious use was as the oil cooler in the air-cooled VW Beetle. This unit was made of brazed aluminum sheet. Aluminum has steadily displaced copper for conventional radiators. Designs include both mechanically bonded and brazed fins and tubes.

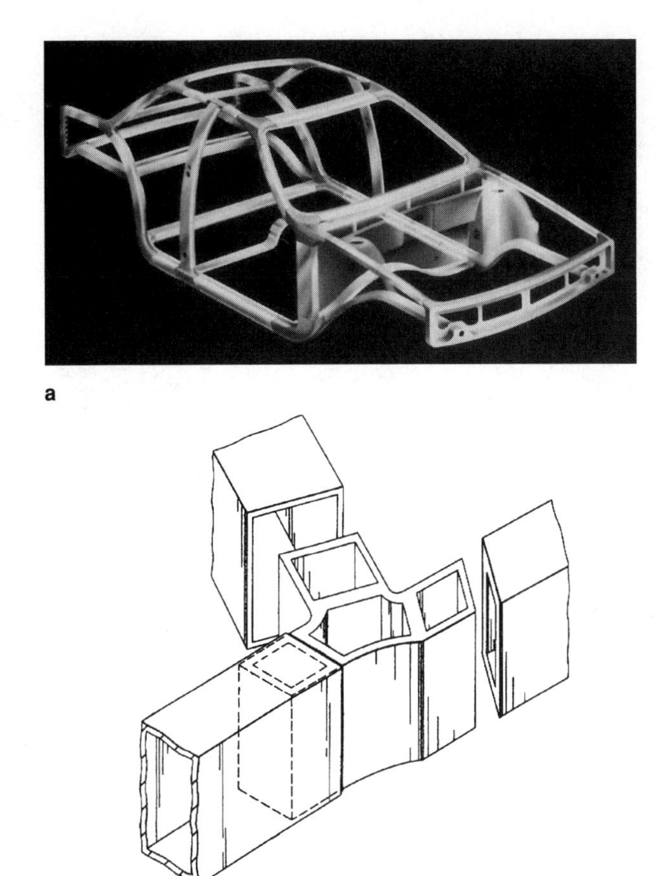

a

b

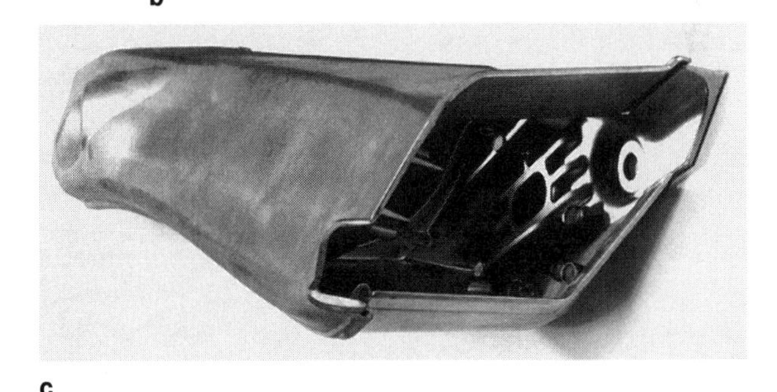

c

Fig. 15.34: An aluminum space frame made from approximately 100 parts, extruded profiles and aluminum pressure cast nodes, welded together by robots: (a) assembled space frame, (b) a schematic of an alternative node design employing extruded profiles, and (c) the cast node. (Alcoa Automotive)

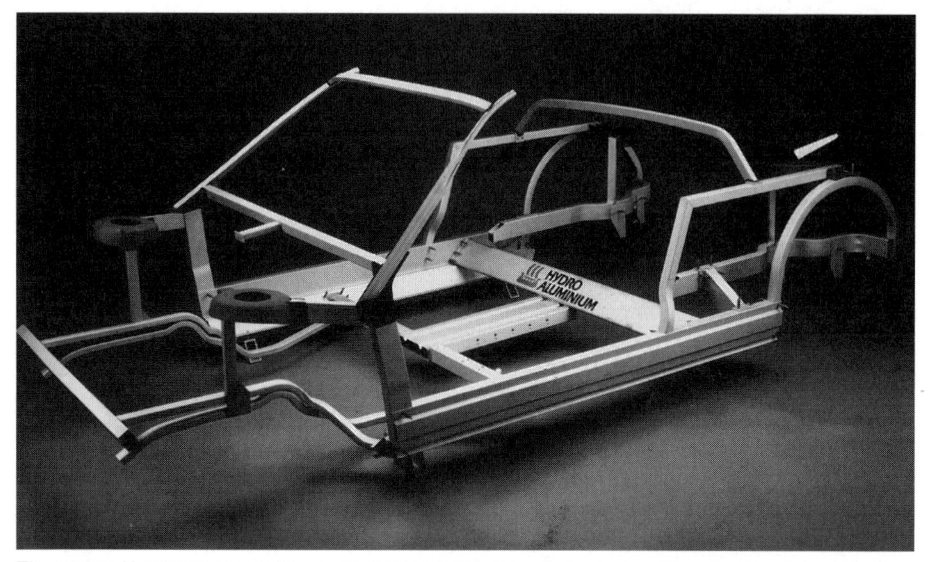

Fig. 15.35: Aluminum space frame made completely from extruded profiles. (Hydro-Automotive)

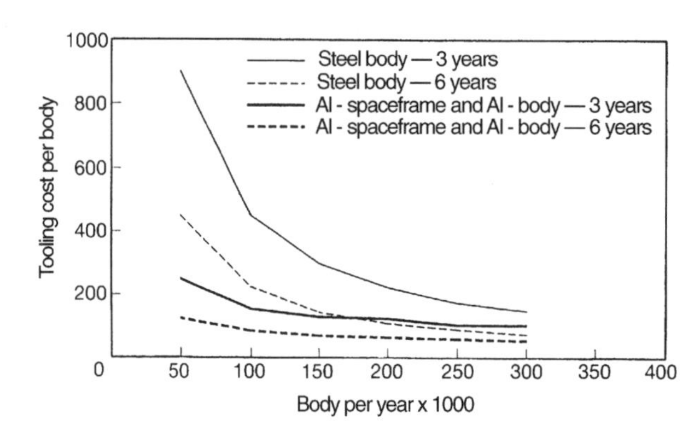

Fig. 15.36: Tooling cost of steel bodies and an aluminum space frame in relation to the number of vehicles and the rate of depreciation.

Recently, safety components for side impact, so called intrusion bars of the type illustrated in Fig. 15.31, have become standard in series production vehicles. These are manufactured from high stiffness extruded profiles. Forging use is increasing; Fig. 15.32 illustrates a drive-train shaft combining forged and extruded sections. Also, the production of wheels from wrought materials (Fig. 15.33), either rolled or extruded stock, continues to merit exploration.

Industry specialists expect a major breakthrough for extruded Al-Mg-Si profiles in the near future with the adoption of the space frame concept.

15.7.1.5 Space frame—one way to the all-aluminum car

The efforts to produce a completely aluminum car go back to the first decades of this century. The first prototype series was built in France shortly after 1950. The Pechiney Company produced the model "Dyna-Panhard," which made extensive use of aluminum instead of steel sheet. This now legendary motor car weighed only 650 kg and was extremely economical. Up to 1953, AlMg3 sheet was used. This was later followed by an Al-Cu-Mg alloy, particularly in the Citreon DS19. The alloy was a good compromise between strength and elongation with property values of approximately 200 MPa tensile yield stress and 20% elongation. In the German automotive industry, a number of short run series were also built with aluminum body work. These vehicles were predominantly hand finished and, therefore, the transition to medium and large scale production was not made.

The space frame concept involves a frame construction (Fig. 15.34.a). The load bearing components are made from extruded profiles joined by special jointing nodes. The version used as an illustration requires between 130 kg and 150 kg of aluminum, of which two thirds is in the form of cast nodes. These are manufactured by vacuum die casting in order to achieve high fracture toughness (Fig. 15.34.c). The extruded profiles are made of an Al-Mg-Si alloy with approximately 0.7% magnesium and silicon, which have been water mist quenched at the extrusion press so that the different parts of the cross section cool at different rates and avoid distortion. In another variant, the nodes are manufactured from extruded elements (Fig. 15.34.b). Yet another innovative space frame system (Fig. 15.35) is based exclusively on extruded profiles. Systems that replace the previously mentioned cast or extruded nodes with nodes made from formed sheet or forged elements are also under study.

The cladding of this stiff framework can be carried out with any of a number of materials; aluminum is one possibility because of its weight advantages and recycling characteristics. It offers possible savings of up to 200 kg in weight for the average midsize car. The space frame concept allows a greater portion of components to be obtained from subsuppliers and makes possible a higher degree of automation in production. There is a considerable reduction in the amount of spot welding. The building block principle and the much lower tooling costs make model changes easier. Adjustments in styling and the major advantage of much shorter development times for new models are very significant advantages for the space frame concept. The main type of body work that has been used by the automobile industry in recent decades is a structure of steel sheets with internal stiffening elements so that the body is able to take on a load bearing function (Monocoque Design). The tooling and machining of the steel dies mean that it is necessary to manufacture long series runs in order to recoup the high costs of tooling.

The space frame construction is very suitable for medium length series. Extrusion dies are much cheaper than the tools required for sheet forming in steel. Aluminum is, however, appreciably more expensive than steel. For the aluminum space frame concept to be viable, the advantage of lower tooling costs has to be set against higher material costs. Fig. 15.36 provides a sketch of the cost vs. quantity relationship.

Fig. 15.37: The stamped aluminum body structure of the Ford AIV. (Alcan International)

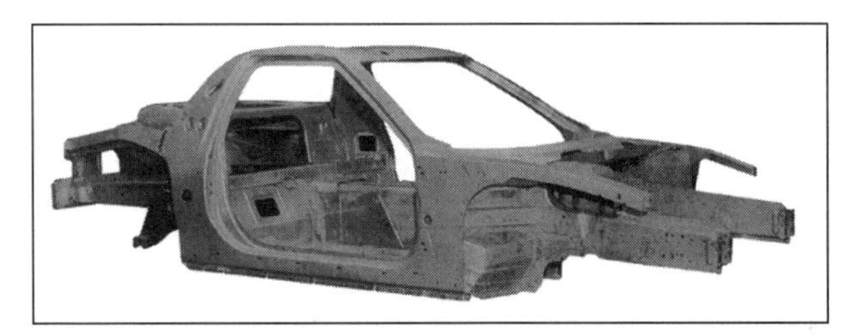

Fig. 15.38: The body structure of the GM EV1 electric vehicle. (Alcan International)

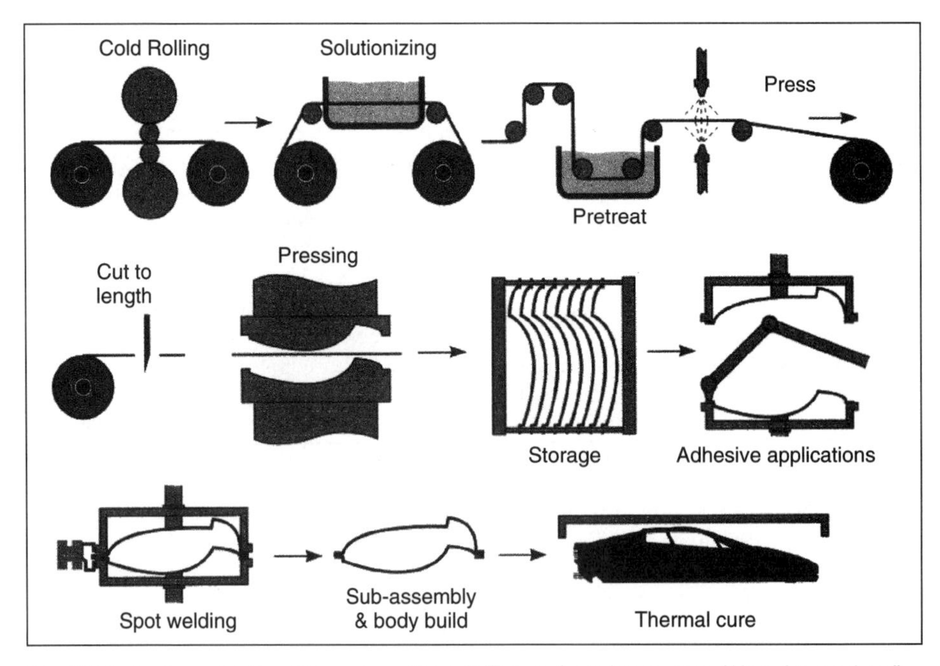

Fig. 15.39: A schematic description of the Alcan AVT manufacturing system. (Alcan International)

15.7.1.6 *Aluminum formed sheet frame*[13]

Two design approaches have emerged for aluminum vehicle structures. One, as described above, utilizes a space frame, where formed extrusions are joined together to provide the basic load bearing structure coupled with stamped sheet or polymeric panels to complete the structure. The other approach involves a stamped sheet unibody as used today with sheet steel for essentially all high volume production vehicles. This approach has been used by Honda for the aluminum-intensive Acura NSX, where the joining is through a combination of MIG spot and conventional spot welding. It has also been used by Ford in its AIV (Fig. 15.37) and Synthesis 2010 vehicles and by General Motors in its EV1 electric vehicle (Fig. 15.38), where the weld bonding system developed by Alcan (Fig. 15.39) was employed to increase the structural stiffness and fatigue life of the structures.

In both approaches, the skin or closure panels may be made of stamped aluminum sheet or polymeric material, the former being the case in the Honda and Ford examples cited.

The major advantages of stamped and weldbonded unibody structure vehicles include:

- Primary structural weight savings of up to 50% compared with today's spot welded steel structures.
- Manufacturing methods that are mostly known and now practiced by the automotive industry, including the use of existing installed stamping facilities.
- Use of a single inexpensive sheet product (e.g., 5754) for which there are existing volume production facilities.
- A cost structure that, as shown in Fig. 15.40, drops significantly at volumes above about 80,000 vehicles per year and becomes less expensive than the space frame system above about 60,000–80,000 vehicles per year.

Both aluminum-intensive approaches can provide benefits of primary and secondary weight savings, which typically could be 140 Kg (300 lbs) or more for a mid-sized vehicle,

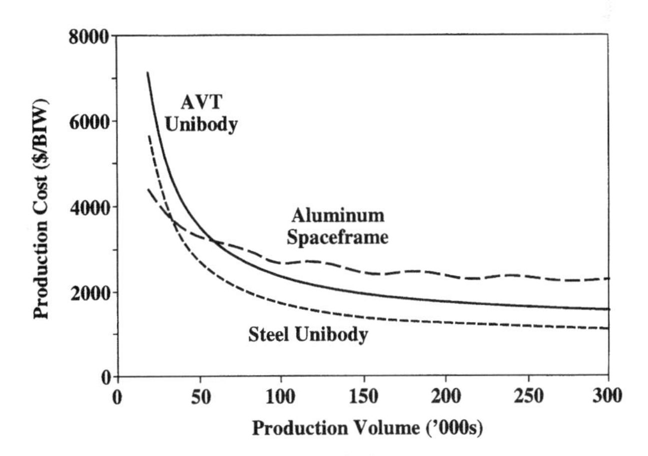

Fig. 15.40: Production costs of alternative vehicle designs.

[13] By D. Moore.

Table 15.7: Composition limits for commonly used automotive sheet alloys

(Aluminum Association Publication AT3, 1996; maximum unless a range is shown)

Alloy	Si	Fe	Cu	Mn	Mg	Cr	Zn	Ti	Other
2008	0.50–0.8	0.40	0.7–1.1	0.30	0.25–0.50	0.10	0.25	0.10	0.05V
2010	0.50	0.50	0.7–1.3	0.10–0.40	0.40–1.0	0.15	0.30	—	—
2036	0.50	0.50	2.2–3.0	0.30–0.6	0.30–0.6	0.10	0.25	0.15	—
5022	0.25	0.40	0.20–0.50	0.20	3.5–3.9	0.10	0.25	0.10	—
5182	0.20	0.35	0.15	0.20–0.50	4.0–5.0	0.10	0.25	0.10	—
5454	0.25	0.40	0.10	0.50–1.0	2.4–3.0	0.05–0.20	0.25	0.20	—
5754	0.40	0.40	0.10	0.50	2.6–	0.30 3.6	0.20	0.15	Mn+Cr 10–0.6
6009	0.6–1.0	0.50	0.15–0.6	0.20–0.8	0.40–0.8	0.10	0.25	0.10	—
6016	1.0–1.5	0.50	0.20	0.20	0.25 0.6	0.10	0.20	0.15	—
6022	0.8–1.5	0.05–0.20	0.01–0.11	0.02–0.10	0.45–0.7	0.10	0.25	0.15	—
6111	0.6–1.1	0.50	0.50–0.9	0.10–0.45	0.50–1.0	0.10	0.10	0.10	—

Table 15.8: Typical mechanical properties of the commonly used automotive aluminum sheet alloys

(Aluminum Association Publication AT3, 1996)

Alloy and temper	Ultimate tensile strength (UTS, MPa)	Tensile yield strength ($TYS_{0.2}$, MPa)	Elongation in 50 mm (El_{50}, %)	Ultimate shear strength (SS, MPa)	Modulus of elasticity (E, 10^3 MPa)
2008-T4	250	125	28	145	70
2010-T4	240	130	25	145	70
2036-T4	340	195	24	205	70
5022-T4					71
5182-O	275	130	24	165	71
5454-O	250	115	22	160	70
5754-O	220	95	26	130	71
6009-T4	220	125	25	130	69
6016-T4	240	120	28	140	69
6022-T4	255	150	26	150	69
6111-T4	290	150	26	170	69

coupled with improved torsional stiffness compared with today's spot welded steel vehicle. The weight savings lead, in turn, to reduced fuel consumption, better acceleration, reduced emissions, and better braking. The improved stiffness combined with the reduced weight and lower center of gravity lead to better handling and improved noise, vibration, and harshness (NVH) characteristics.

Table 15.7 lists the nominal compositions for the commonly used automotive sheet alloys that are both heat treatable and non-heat treatable. Generally, the 2xxx and 6xxx heat treatable alloys are used exclusively for skin and general external applications. They are supplied in the T4 temper, stamped, and then increased in strength by aging during the normal paint bake cycle that is applied to the assembled body structure. The most commonly used closure alloy sheet used in North America is 6111, while in Europe it is 6016. The non-heat treatable aluminum-magnesium alloys are supplied in the annealed (O) temper. The medium strength aluminum-magnesium alloy 5754 is preferred for structural applications to the stronger 5182 alloy, since the latter can be susceptible to stress corrosion cracking. However, 5182 sheet is commonly used in some structural applications, such as brackets, heat shields, and other locations. Here, its use will not make it sensitive to stress corrosion. It is also used in a few instances for external closure panels, but the formation of Lüders lines on forming in the O temper generally makes it unsuitable for class A surface panels.

Table 15.8 lists the typical mechanical properties of these alloys. For the heat treatable alloys, the amount of strengthening that occurs in a typical paint bake process is usually considerably less than the maximum strengthening that is available. A typical paint bake response (simulated by 2% prestrain plus 30 minutes at 177°C) for 6111 is 230 MPa; whereas, the maximum T6-T8 strength can be as high as 330–390 MPa, depending on the prestrain and aging conditions. Hence, examining just the T6 properties can be misleading, and what is required is a rapid strengthening response for the times and temperatures typically used in today's finishing lines.

15.7.1.7 Aluminum in cars as a "metal bank"

In the same way that a bank allows deposited money to circulate within the framework of the market, the value of the aluminum built into a motor car is part of the cycle that governs the use of this material. The actuators that influence value and added value operate at every level and keep the cycle in motion. The forces driving the reuse of aluminum from private cars in Europe can be summarized as follows.

1. Quantities are large. Currently, 12 million cars are scrapped in Europe each year. These cars contain, on average, 50 kg of aluminum, which means that per year 600,000 tonnes of used aluminum, partly via the dismantler's yard and partly via shredder facilities, are made available for recycling. The tendency is upward. It is estimated that by the year 2000 recycled aluminum from used cars will have reached one million tonnes.
2. The aluminum used in cars has the ecological advantage, when compared to steel, that it saves considerable quantities of fuel throughout its period of use in a car. When compared to plastics, which are lost as shredder waste and can only be recycled with high cost, up to 90% of the aluminum components can be recovered and recycled.

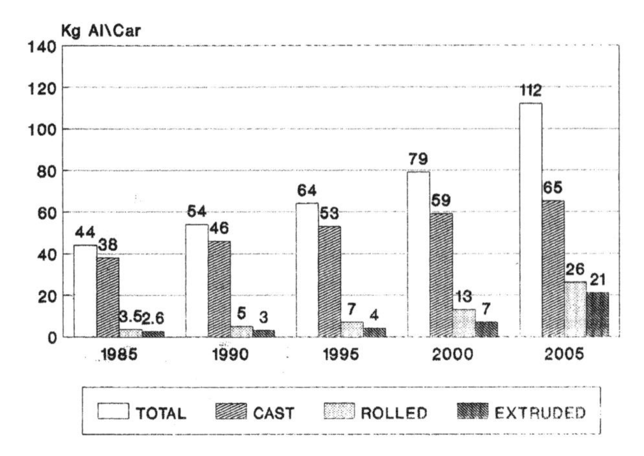

Fig. 15.41: Estimated use of aluminum in private cars up to the year 2005. (F. Ostermann, Aluminum Tech. Service)

3. Today in Europe, approximately 400,000 tons of primary aluminum and 600,000 tons of secondary aluminum are being used in new cars. From this quantity, the wrought materials and the sand- and die-cast components are principally made from primary material, whereas the vast majority of pressure die cast components are produced from secondary aluminum. New cars being built now in Europe will require at least as much secondary metal as is available from the cars that are being scrapped. The use of aluminum in vehicles is increasing the demand for recycled aluminum; this will not change in the foreseeable future. It is quite likely that the requirement for primary metal will fall as soon as the first generation of cars with higher aluminum content is recycled.
4. The technology involved in the recycling of aluminum from old motor cars is in its infancy and will develop rapidly in the next few years. This will not be driven, as is the case for plastics, by ecological legislation, but will be dominated by the market demand for this high value metal. The basic requirements for such a development are, for example, improved general and specialized separation techniques for aluminum shredder scrap. In addition, improved separation during scrapping of old cars is to be expected because the motor car manufacturers are making conscious efforts to design their vehicles to improve the opportunities to remove aluminum components at the end of the vehicle's life. In the future, upgraded remelting techniques will be available for secondary aluminum components. With the increasing volume of the material, the more specialized equipment, which is necessary to improve economics, will become available.

15.7.1.8 Future prospects

The proportion of aluminum in private cars will continue to grow in the immediate future. The reasons for this can be found in the innate characteristics of the material. The ecological pressure of the industrial society will continue. If the question "Why should the proportion of aluminum in private cars rise in this decade?" is posed, one is faced with an array of estimates and projections from a variety of sources. It is difficult to obtain reliable forecasts. Two factors are significant: the size and price class of the car and the number of vehicles to be produced in one series. The larger and higher the first factor, then the smaller is the second factor.

The use of aluminum in cars is rising. The projected average amount and distribution by product form of aluminum in cars is presented in Fig. 15.41. In this university study, it was estimated that by 2000 the quantity of aluminum in medium sized cars could reach 112 kg. Castings have a dominant share, but the forecast shows significant growth in wrought products. If the space frame concept is widely adopted, an increase of the aluminum content per vehicle will occur. On the basis of information available today, such a rise seems a distinct possibility.

Larger amounts of aluminum will definitely be used in small series of vehicles in the top price range.

15.7.2 Aluminum in commercial vehicles[14]

15.7.2.1 Advantages of aluminum

The advantages to be gained by the use of aluminum have led to its increasing use in commercial road transport vehicles.

- Low vehicle weight provides higher load carrying capacity or alternatively lower energy consumption when operating empty or only partly loaded.
- Corrosion resistant aluminum gives a longer service life and needs less corrosion protection. Maintenance and repair costs are reduced.
- The excellent formability of aluminum makes possible full use of the multifunctional design possibilities of extruded profiles.

Up to 75% of van and commercial vehicle bodies as well as tanks for bulk road transport of gas and heating oils are now made of aluminum. Bus bodies, platform bodies, tippers, and trailer chassis are manufactured in aluminum in those countries with a thriving aluminum sector. Many new developments are underway that are strongly promoting the use of aluminum in commercial vehicles.

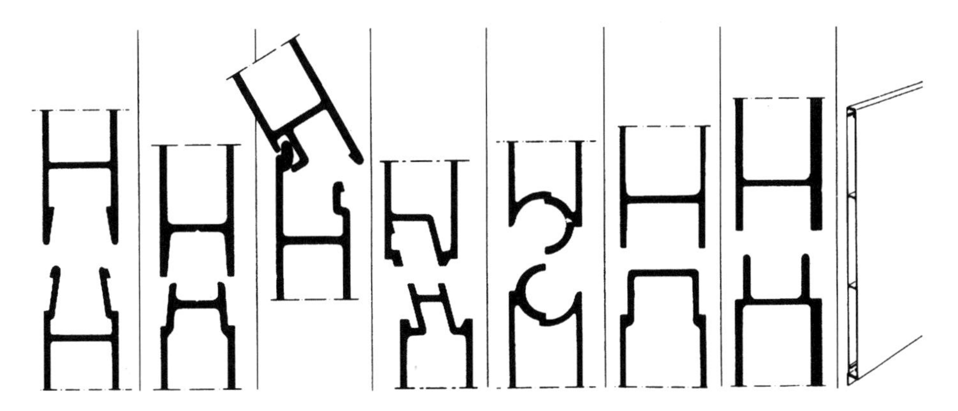

Fig. 15.42: A selection of side board systems illustrating a variety of joint types.

[14] By the late G. Angehrn.

Fig. 15.43: Aluminum design for a platform body system. (Alusuisse)

Some of the more significant advantages can be summarized as follows:

- A 6% energy savings per tonne of weight reduction, 95% energy saving on recycling, approximately two thirds of the total energy requirements provided by environmentally friendly hydroelectric power sources.
- Workplace friendly manufacture, low or limited welding (reduced ozone contamination), little or no grinding (no dust and noise in the manufacturing shops), and limited need for surface protection (less disposal problems for solvent cleaners, lacquers, and sealing agents). Use of anodizing (less environmental pollution than would be caused by zinc coating) and aluminum floor planking (no formaldehyde contamination as would be the case with wood).
- Environmentally friendly recycling. No CFCs from plastic materials, 90% of aluminum components are already recycled.

15.7.2.2 Side boards

Aluminum side boards for commercial road transport vehicles were introduced approximately 50 years ago and are usually made from two or more extruded profiles that are fastened, clipped, or otherwise fitted together (Figs. 15.42 and 15.43). For some years, there has been a trend for the smaller vehicles to move to single extrusion side walls. Modern extrusion press technology has made possible the manufacture of hollow profiles up to 800 mm wide in intermediate strength aluminum alloys. This has eliminated the need for welding. Typical applications for this type of extrusion are docking boards. These are subjected to very rough use in service and have to withstand high stresses during loading and unloading.

15.7.2.3 Platform bodied vehicles

Although the aluminum platform body (Fig. 15.43) has been available for many years, it

Fig. 15.44: Dump truck with an aluminum back and chassis.

Fig. 15.45: A joining element for extruded profiles.

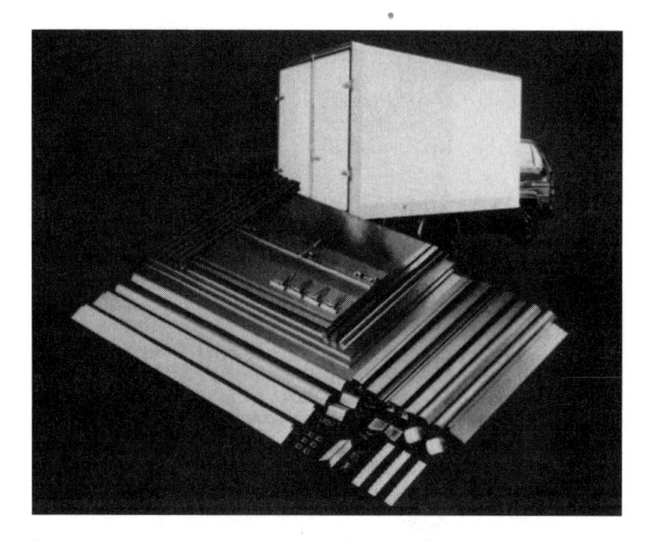

Fig. 15.46: Van body available in kit form. (Alusuisse)

has not been universally introduced. The aluminum design offers three times the service life of a conventional vehicle with a steel and wood platform. The aluminum design comprises essentially a substructure of longitudinally or transverse floor planking made from simple extruded sections. This type of vehicle, which is frequently loaded using platform trucks with small radius wheels, is very dependent on the correct choice of floor planking. Relatively small wheels create point loads that are easily capable of deforming incorrectly dimensioned planking.

15.7.2.4 Dump truck bodies

Aluminum has become established as the material for long haul tipper semi-trailers. The bodies on trailers are more frequently made in aluminum than are truck chassis. The usual tipper construction consists of sheet flooring panels stiffened with profiles. Side and front walls and the tailgate are also made from sheet extruded section combinations (Fig. 15.44). Before aluminum can achieve general acceptance for tipper bodies, the abrasion problem must be solved. Short haul vehicles that are used for frequent tipping operations show much more rapid abrasion of their flooring elements when the structure is made from aluminum as opposed to steel.

15.7.2.5 Van bodies

In this category of vehicle, there are many special types to meet specialized transport needs and the transport of specific types of goods. The support structure is made in the main from extruded sections with specially designed joining elements to simplify assembly (Fig. 15.45). Using the principle of this type of connection, it is possible to develop a wide range of nodes and corner elements.

The structures are clad with adhesively bonded sheet. The use of the concept for vehicles without or with only limited insulation is very general in Western Europe. The market is supplied with complete, ready to assemble systems in kit form; the various profile and sheet components can be joined by riveting. The systems frequently make use of steel tail gates. Body builders assemble the main elements (front wall, sides, roof, and rear frame) from a kit and build the floor and sub-structure themselves. There is a special kit (Fig. 15.46) that uses anodized profiles, sandwich elements for the sides and the roof, profiles for the sub-structure, and an aluminum tail gate. The assembly follows a simple sequence and does not require welding. It makes use of adhesive bonding and mechanical fixings, and repairs are very easy. The anodized profiles and stove-lacquered sheets eliminate the need for painting.

The new design for securing the sandwich wall elements avoids riveting and produces attractive outside surfaces that are easy to maintain. In addition, any damage during loading and off loading usually occurs on the inner surface and is not visible on the outer surface of the sandwich element. Interesting possibilities are afforded by adhesive technology, which can be used to secure panels in the extruded profile frame work. This technique makes use of relatively thick strips of adhesive that can accept a degree of elastic deformation throughout the working life of the vehicle.

15.7.2.6 Chassis

Aluminum chassis are currently almost wholly restricted to trailer and semi-trailer vehicles. The use of aluminum chassis in trucks is limited at the present time to prototype versions. Trailer designs usually involve two double-T beams with welded or riveted transverse members. In order to improve the stiffness of the longitudinal members, it is possible to weld a plate insert between the two T-profiles of the beam.

15.7.2.7 Economics of aluminum commercial vehicles[15]

A test vehicle was built by Alusuisse to study the economics of aluminum in commercial vehicles. The aims of its designers were:

- To reduce the weight of the vehicle and, hence, improve load carrying capacity and transport efficiency
- To reduce operating costs by use of materials with a long service life
- To lower maintenance and repair costs
- To lower energy consumption and so reduce costs and minimize environmental pollution

Wide use was made of aluminum; the drivers cab, previously steel, was replaced by an aluminum structure. Aluminum was also used for the bumpers, chassis, air pressure tank, fuel tanks, battery case, wheel rims, exhaust chamber, under ride guard, and platform body. The result was an aluminum portion of 1550 kg of a total weight of 16,000 kg.

After well over a decade of service the following results are worthy of mention. A relatively low proportion of savings was due to lower fuel consumption. In addition to the improved earning power because of the higher payloads, the most important saving results from lower maintenance costs. Savings on maintenance are higher than the benefits gained from lower fuel consumption.

[15] By K. Wöhrer.

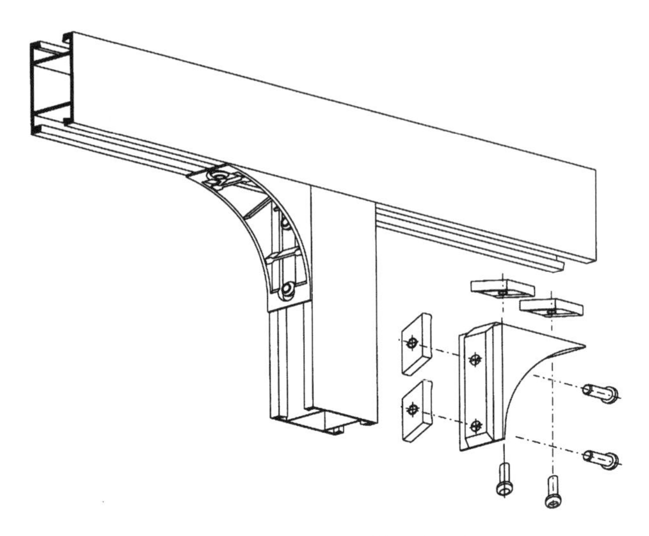

Fig. 15.47: Bolted frame system for bus construction. (Alusuisse)

Fig. 15.48: The Scania NF. (Alusuisse)

15.7.3 Buses[16]

Aluminum has been used for the construction of buses for many years. First, it was used as an outer skin on steel framed bodies. With the development of higher strength aluminum alloys, bus bodies began to be built using stressed skin designs similar to those used in aircraft construction. The framework of these light and stable vehicles was made using simple cross-section profiles (Top-hat sections, simple angles, and "U" profiles). The outer skin was fixed onto the frame by riveting. The number of worker hours and the cost of jigs for this type of structure are very high; nevertheless, the technique is still used by a few bus builders.

[16] By K.D. Waldeck.

In order to reduce production costs, riveted sheet structures were replaced by welded rectangular hollow profile frame structures onto which sheets were fixed by riveting. At first, this design was not widely accepted because of problems in making the welded frameworks, difficulties in avoiding distortion, and loss of strength in weld zones. The development of joints optimized for welding occurred at the same time as large extruded profiles became generally available. It was then possible to improve the quality of welded joints by the use of improved techniques. The large profiles enabled the number of parts to be reduced and lowered production times.

In order to repair the framework structure without having to make use of specially qualified personnel, a "repair friendly" bolted system was developed (Fig. 15.47). For well over a decade, because of the exceptional properties of these bolted joints, completely bolted skeleton structures of all types and dimensions have been used for buses. The system has been used in Europe as well as in other areas of the world.

The higher material costs of aluminum in comparison with steel can be balanced by use of the bolted aluminum system because of low costs and minimal investment. This is especially true in countries with high wage rates. The various types of bus design have undergone considerable change in recent years. With the introduction of low floor buses, the bus body has had to take on some of the loadings that were previously carried by the chassis (Fig. 15.48).

The use of aluminum in buses, particularly in combination with well designed jointing systems, offers the following advantages:

- Flexibility of design
- Cost efficiency
- Low weight
- High corrosion resistance
- Clean good-looking surfaces
- Simple and rapid repairs

Savings by aluminum buses because of lower energy consumption compared to steel are greater if the bus is operated with frequent stops and starts. The advantage is less significant if the bus is used for long distance travel.

15.7.4 Future prospects

The use of aluminum in commercial vehicles and buses will continue to rise. In recent years, truck and bus builders have been increasing the proportion of aluminum in bodies and trailers; truck and bus manufacturers are increasing the proportion of aluminum components in their designs (fuel tanks, air pressure tanks, support brackets, wheels). Their projections for the next generation of vehicles indicate greater use of aluminum. In the future, the relatively large differences that have developed between individual countries will tend to even out. The range of aluminum components in commercial vehicles is currently more varied than that found in passenger cars.

15.8 *Packaging*[17]

15.8.1 Properties and advantages of aluminum packaging systems

When compared with other materials, aluminum possesses some properties that are ideal for packaging applications. The European consumption of aluminum for packaging has, in recent years, shown an increase of approximately 4% per year. Nevertheless, the aluminum processing companies are very sensitive concerning the correct use of aluminum as a packaging material. There is no wish to push the metal into applications where technical requirements do not justify its use, especially if alternative materials would function equally as well.

Before the advent of the self-service retail outlet, packaging was only required to provide a limited protective function. Today, however, the situation is very much changed, and the package has to supply:

- Portion packaging and distribution
- Protection of the product against deterioration and mechanical damage
- Information concerning the contents and its shelf life

Under the pressure of strong competitive forces, the packaging industry has developed a wide variety of products that have served to achieve a continuous optimization in both economic and functional terms. Recently, environmental concern regarding use and disposal of packaging has become a subject of central interest.

When compared with other packaging materials, aluminum has the following combination of advantages:

- A perfect barrier against gases, moisture, contaminants, fats, oils, and light
- A neutral taste
- Low density
- Good mechanical properties at low and high temperatures
- Good thermal and electrical conductivity
- Good corrosion resistance
- Good formability at room temperature
- Good printing and embossing characteristics
- Easily combined with other packing materials such as paper and plastics
- Excellent recyclability (this aspect requires further development in the case of laminated packaging combinations)

15.8.2 Packaging systems

15.8.2.1 Beverage cans

The most remarkable development in the packaging sector is the aluminum beverage can. Its usage has seen spectacular growth over the last 30 years as evidenced by the

[17] In collaboration with H. Severus-Laubenfeld and R. Kamal.

manufacture in 1995 of more than 100 billion cans in the USA alone. Production of these cans consumes more than 2 million tonnes of aluminum annually and constitutes, in volume terms, over 50% of the aluminum sheet and plate market. The popularity of aluminum cans stems from a number of factors including ease of recycling, low cost, light weight, good thermal and mechanical properties, and proven technology. Consumption of cans in North America has reached a plateau, but demand for cans in South America, Asia, and Europe is growing. The standard 12 oz. can has been joined in recent years by larger and smaller sizes, for example the isotonic beverage and beer cans.

The development of beverage cans has been focused on continuous cost reduction with special emphasis in three areas: reduction of can body and lid stock thickness; improved alloys in terms of strength, formability, and recyclability; and the development of new can shapes (for details see Chapter 13). An important part of the optimization has been to ensure that the return and recycling of used beverage cans is economic. Currently in the USA more than 60% of used beverage cans are recycled.

15.8.2.2 Food cans

Like beverage cans, aluminum alloy food cans are made in two parts (alloys: 3004, 5182); that is to say, they have a can body and a lid. At present, this type of container is relatively flat with a depth-to-diameter ratio considerably under one. They are typically used as cans for fish and meat products. Current developments are moving toward larger and deeper aluminum food cans. In most cases, these are being deep drawn out of 0.20–0.30 mm thick sheet. Chromate-free pretreatment is followed by lacquering and decoration in the flat form so that after deep drawing the image takes on the required shape. The three-stage deep drawing operation followed by corrogation of the side wall gives a light, stiff food can. The aluminum tear-off lid makes opening simple, and both parts, can and lid, are completely recyclable.

Combi-cans, in contrast to conventional beverage and food cans, are made from three parts, that is, a closure, a body, and a base. The body is a multi-ply laminate manufactured by winding techniques. Generally, the laminates are made with five layers, the inside layer being aluminum foil 9 microns thick to provide a gas, water vapor, and taint barrier. This is hot sealed or plastic laminated to 40–60 g/m^2 paper as a stiffener. Fig. 15.49 shows the make up of such a combi-can. The end closure can be a so-called "aluminum membrane" 50-60 microns thick that has been stamped, formed, and then sealed on to the can inner wall by means of a hot sealing agent. Whereas in earlier versions of this type of container the membrane had to be cut in order to release the contents, there is now an easy-opening variant with a tear-off tab.

15.8.2.3 Aerosol containers

Aerosol containers, also known as spray cans, are generally made from pure aluminum. The one part can bodies are produced by impact extrusion of a circular slug to form a tube closed at one end. A necking operation is carried out on the open end of the tube, and the rim is roll formed. The containers are stove lacquered inside and out and are externally printed. The change from CFC gas to environmentally friendly aerosol gases

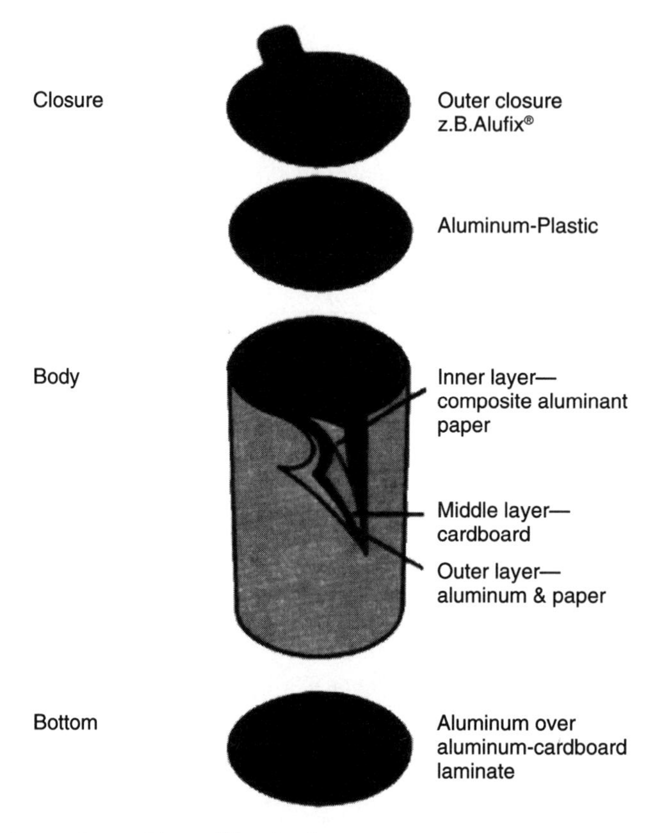

Fig. 15.49: Make-up of a combi-can. (Alusuisse)

such as propane/butane mixtures or dimethylether is already almost complete. Many packers have used new formulations to suit the contents of the aerosol. The development of double-chamber spray cans that make use of a different design principle is continuing. Plastic pistons are included in the inside of aerosol. These separate the contents from an environmentally friendly compressed gas, usually nitrogen, CO_2, or air, which is used to push the piston upward and discharge the product. So-called "roll bag containers" function in a similar way. The rim of a pouch that contains the product is sealed to the container rim.

15.8.2.4 Bottle tops

Screw bottle tops made in aluminum are showing a marked increase in growth. The well known pilfer-proof types are roll formed onto the thread of the glass bottle at the end of the filling and sealing process. At the lower end of the closure on the bottle neck a perforated ring of aluminum or plastic is separated from the rest of the cap when the bottle is opened for the first time. The undamaged ring is proof that the bottle is in its original unopened condition. The latest aluminum screw caps offer, in addition to the original safety device, a system that shows the state of the vacuum. Any gas evolution, for ex-

Fig. 15.50: Aluminum roll formed caps with a special safety system that gives a warning if an “over pressure” is developing. (Haist)

Fig. 15.51: Pharmaceutical packaging. (Alusuisse)

ample because of fermentation of fruit juices, that could lead to an over pressure in the bottle, is clearly shown up by the bottle closure. A new development is the special safety closure known as the over pressure safety cap (Fig. 15.50).

15.8.2.5 *Light containers*

Food trays made from plain aluminum thin strip formed into so-called wrinkle-walled containers are widely used for home made or purchased food for deep freezing, chilling, baking, and grilling. Smooth walled light containers are made with the lower part consisting of lacquered or plastic coated aluminum thin strip. They are sealed with a coated foil lid. This type of container is used for portion packs, especially for long life cream for coffee, individual jam packs, and honey. They are used for heat sterilized finished meals and for animal foods. An interesting development in relation to microwave cooking is a new type of lid that solves the problem of different heating rates for different parts of the meal. By means of a plastic lid with carefully ordered aluminum segments, the effect of the microwaves and the resulting heating of the food is so controlled that the entire meal is warmed uniformly and can even be selectively braised.

Heat sterilizable finished meals and partly cooked meals are packed in containers made from aluminum plastic laminates or trays made of plastic with multilayer high barrier property plastics. Because of the specified shelf-lives when storing at room temperature a large part of the aluminum and plastic trays are provided with lids that contain aluminum as a gas barrier. Containers with volumes of up to approximately 200 ml very frequently have peelable lids made from heat seal, lacquer coated aluminum thin strip. For trays with volumes greater than 200 ml, the so-called "caterers containers," the lids have to be cut away near the edge in order to gain access to the contents. In recent years, aluminum plastic laminates in various forms have become available and are used for heat sterilizable finished meals or partly cooked meals. In Europe, approximately 1000 million of such food packaging containers with contents in the 200–800 ml range are used each year.

15.8.2.6 *Collapsible tubes*

In the 1920s and 1930s, aluminum established itself as the material for impact extruded tubes. It replaced tin and lead in these applications since they were both more expensive. These two metals also did not fill the necessary physiological requirements for containers for food and pharmaceutical products.

In the early 1950s, tubes out of plastic were developed; these were usually manufactured out of low density and high density polyethylene. Plastic tubes are elastic and, to a very large extent, acid and solvent stable. Polyethylene is not a perfect gas barrier and, thus, the tubes are not suited for products that are air and oxygen sensitive. Because of this problem, multilayer or laminated tubes with much higher resistance to aggressive contents were developed. The polyethylene layer that comes into contact with the contents satisfies the absence of taint requirements for an inner layer; the aluminum layer provides gas and contamination barrier properties, In recent years, the so-called "Duplo" laminate, which contains two thin aluminum layers, has been developed. This enables the laminate tube to be deformed in practically the same manner as would be possible with a completely aluminum tube. This type of tube does not exhibit the crack sensitivity which too often occurs in aluminum tubes.

15.8.2.7 Pouches

Thin aluminum foils (6–20 microns) in combination with various plastics are used in pouch packaging systems where high barrier properties are needed against gases, moisture, contaminating substances, and temperature. In the area of food packaging, the use of part formed sterilizable pouches for finished meals is increasing. A further trend is for coffee to be packed in pouches made from an aluminum laminate and sealed under vacuum. Here, the product is first subjected to a vacuum; then, the vacuum is partially broken by the introduction of a special aroma containing gas mixtures that help to preserve the product.

15.8.2.8 Pharmaceutical packaging

Sadly, many people have to accept as part of their daily lives the single-dose, push-through pack from which they obtain the pharmaceutical products upon which they depend. These packaging systems for tablets, capsules, and suppositories usually take the form of transparent plastic blisters to contain the product and a push through aluminum foil covering. Since many new pharmaceutical products require increased protection from moisture and light, aluminum blister packaging was developed to provide a complete barrier pack (Fig. 15.51). One form of this new type of package consists of a tray, formed from an aluminum laminate, with a complete plastic blister sealed over it. A second type, which is completely water vapor resistant, consists of single aluminum laminate blister packs. The formability of the laminate and, hence, the depth of these single blisters is being steadily increased. Development efforts are being directed at improving the elongation of the aluminum plastic laminate itself. Many different shapes are already offered, and in most European countries single-dose, push-through packaging for tablets etc. has caused bottles and aluminum tubes to be moved to the side lines.

15.8.2.9 Special decorative packaging materials

The packages that have been described use aluminum in the form of circles, coil, or foil made by the rolling process. There are, however, an increasing number of packaging materials that are manufactured with a thin vacuum deposited aluminum layer on plastic or paper. Since these vacuum deposited layers are only 0.02–0.8 microns thick, they do not provide a complete light or gas barrier, especially if they have only had a single vacuum coating pass. The gas transmission rates of vacuum coated plastics are high, and these materials are unsuitable for storage of sensitive products for extended periods in humid conditions. Protection is further reduced if the coated plastic film has been stretched on a packaging machine. Plastic foils only provide a decorative effect; they do not offer the same protection as aluminum foil.

15.8.3 Use of packaging waste

Packing and packing materials with a high proportion (i.e., more than 50% by weight) of aluminum are suitable for recycling by remelting. Packaging materials with a lower proportion of aluminum, for example thin aluminum foils between plastic or paper layers, can be disposed of using various techniques. The following methods of disposal are being tested industrially or are in development: use of low temperature carbonizing or pyrolysis by which means the organic portion of the packaging material is converted into combustable gases. The energy obtained from burning these gases can be used as process energy, and the aluminum fraction can be recovered. For example, it can be used for aluminum powder. The separation of paper from plastic is also possible by grinding at low temperature or by use of liquid media that do not change during the separating process and can be recycled and re-used. After separation, the plastic, depending upon its type and purity, can be re-used. The aluminum itself as explained earlier finds use as aluminum powder that can be added as a coloring agent to plastics and paints, for welding purposes, or for the deoxidation of molten steel. Waste in very fine form produced from multilayer aluminum laminates can also be used successfully as a fuel in rotating furnaces for the manufacture of cement, especially in those cement works where there is a shortage of Al_2O_3 in the raw material. Finely divided laminates made from aluminum/paper or plastic can be used as additions for the manufacture of insulating bricks (instead of wood dust). This use has been well tested and shown good results.

Work is being carried out on the use of aluminum laminate scrap in the aluminum electrolysis process. Further development work will be required to determine whether or not the high pressure hydration process to break down the organic layers of aluminum plastic laminates to a useful hydrocarbon will ever be economic. The burning of laminates, including those containing aluminum foil, in waste incineration units is already widely practiced. In the future, however, other methods must be found to recover the metal.

The laws concerning packaging in Europe state that an increasing proportion of material used in packaging must be recycled. Aluminum is basically suited to this concept. The question is, will it be economically viable to sort the relatively small quantities of waste packaging material containing aluminum in comparison to other raw materials? The ecological balance for aluminum materials is only favorable when the material is re-used (i.e., recycled to secondary aluminum). This remains a main challenge for aluminum in packaging. A number of development projects are in progress to resolve this question.

15.8.4 Development trends

Food packaging will remain a main application for aluminum foil and thin strip, but the emphasis will be on "doing more with less." A typical example is the use of aluminum foil for products that are to be converted by printing, lacquering, or laminating. These foils used to be supplied in the thickness range of 9–10 microns. The application now uses 6 micron and 7 micron thick foils. This sets much higher requirements on the raw materials and on the foil itself. The width of foil has dramatically increased. Many modern machines are over 2 m wide. In addition, there has been an increase in rolling speeds that are now, for some products, up as high as 2500 m/minute. The development efforts of foil manufacturers are being directed toward material quality, surface finish, and dimensional accuracy to meet the ever rising requirements.

Material quality refers to a number of properties such as:

- Ultimate tensile strength (UTS) and percentage elongation
- Formability
- Flatness and porosity

Development in recent years has been aimed at raising both strength and elongation. In Table 15.9 the values of UTS and elongation for aluminum alloys together with the grain size in grains/mm^2 are listed. For foils in the 12–15 microns gage in the soft condition, strength values are between 65 MPa and 120 MPa with elongation values between 4% and 6%.

In thin strip (21–350 microns), UTS values between 90 MPa and 140 MPa with elongations between 18% and 30% can be achieved. New Al-Fe-Mn alloys such as 8101 combine high strength with higher elongation and make possible gage reductions for the same performance. The grain size (grains/mm^2) is the important factor. The porosity in thin rolled products has been much reduced because of progress in casting and rolling methods. This has made possible a reduction in foil thickness from 9 microns down to 7 microns with a lower level of porosity compared to earlier times. Today, the average porosity is 0.5 pores/dm^2 for 7 micron foil. The gage of the foil can not, however, be easily reduced further for technical and economic reasons. In order to be able to economically manufacture very thin and wide foils, improved metal quality, rolling techniques, and thermal treatments are all necessary. The higher costs involved in making possible savings in terms of lower thickness and increased area per given weight cannot otherwise be justified. The economic thickness limit for converter foil today is at about 6 microns.

Table 15.9: Important aluminum foil and thin strip alloys used as packaging materials

Foil	Aluminum Alloys	UTS (MPa)	Elongation (%)	Grain Size (grains/mm^2)
12–15	Pure Al 99.2	65–80	4	2500
microns	Pure Al 98.6	55–58	5–6	6000
soft	8014	110–120	5–6	30000
Thin strip				
70–100	Pure Al 99.2	90	30	2500
microns	Pure Al 98.3	95	30	4000
soft	3003	120	23	5000
	8014	125	30	30000
	3004	130	18	5000
	5182	140	18	6000

The quality of thin foil is very dependent on the foil stock quality. This has been the subject of intensive development efforts. There are two commonly used methods for the manufacture of foil stock:

- From hot rolled slab
- From coils produced by a roll caster

The cross section of the hot mill plate obtained from conventional hot rolling is very important when determining the quality of foil stock. There is considerable advantage to be gained by the foil roller if the foil stock has a chamber, that is, if the center gage is greater than the gage at the edges. Foil rollers are generally wary of "long edges."

Roll casting of foilstock has recently acquired new momentum. Cast gages between 3 mm and 2 mm have been shown to be very suitable for the manufacture of high quality foil (See also Chapter 8). This has considerable significance when comparing the costs of rolling down to foil gage from an approximately 500 mm thick hot mill slab and the costs of rolling to foil from roll cast plate gages.

Of great significance are tailor made structural properties in the foil stock coil. The Japanese development of so called "Bespa foil" illustrates this concept. In an Al-Fe(Ti) alloy with a low Si content, very fine AlFe and AlTi precipitates are present in the structure after homogenization of the rolling slab.[18] After rolling, this precipitate form is retained (temper annealing is used instead of recrystallization annealing). This is achieved by the phenomenon of work softening during foil rolling in order to retain the temper annealed state. Bespa foils, in contrast to conventional foils, exhibit very high elongation at low gage combined with very low porosity levels.

The manufacture of packaging laminates using aluminum foil or thin strip in combination with paper, plastics, lacquers, and inks for decoration, with the object of providing protection for a particular product, has reached a very high technical level of development. Current development is focusing on:

- Environmentally friendly techniques for pretreatment of aluminum surfaces, lacquering, laminating and extrusion coating, with the aim being to achieve a product-specific optimal concept in terms of both process and equipment
- Attempts to achieve a higher synergy and so obtain optimized properties from combinations with plastics, paper, and cardboard

Aluminum's properties make it an excellent packaging material. This, combined with its exceptional formability, has led to its use in so many applications as coated and/or laminated foil or thin strip.

15.8.5 Future prospects

In today's world, the subject "Aluminum in Packaging" has become very complex. Technical specifications that were formerly influenced mainly by the shelf life requirements of products must today meet a much wider range of standards. This situation is promoting further development.

[18] The foilstock has a silicon content of only 0.07% with an iron content of 1.2% (selected primary metal is, therefore, required and can be a certain obstacle).

In some European countries, certain packaging foils are tolerated with some reluctance (e.g., cigarette or chocolate foils). Worldwide, however, the advantages of aluminum as a packaging material and the need to distribute food and beverage mean that aluminum will have increasing importance.

15.9 Machine Tools[19]

15.9.1 Advantages of aluminum

1. Handling costs are lower because of aluminum's low weight.
2. The machinability of aluminum is very good.
3. The high thermal conductivity of aluminum rapidly distributes heat in tools and molds and, therefore, reduces those cycle times that are dependent upon heating or cooling.
4. Alloys of the Al-Zn-Mg-Cu type (e.g., 7075) in the precipitation treated and aged (T6) condition achieve strengths comparable to common structural steels. For identical dimensions their relatively low elastic modulus (1/3 that of steel) results in a loss of rigidity, but this can be compensated by use of greater wall thicknesses. In spite of this overdesign in terms of strength, the aluminum component can still provide 50% weight saving compared to steel.
5. High strength aluminum alloys are easy to machine with normal cutting tools but better results are achieved with higher cutting speeds. Tool wear is lower than for steel; this reduces time and machining costs. These gains are only possible if the cutting tools and machining conditions are optimized to suit the material.
6. Machining by use of electroerosion poses no problems. Volume removal is two to four times that of steel without any surface degradation. Aluminum is very suitable for mechanical polishing.
7. Modern fabrication methods in rolling mills allow the production of plates up to 300 mm thick in the quenched and precipitated condition. Shape stability is maintained during and after milling operations, and because the aluminum components do not require heat treatment or stress relief after machining, the risk of any distortion is practically zero (Fig. 15.52).

15.9.2 Profiles as feed stock

Aluminum profiles are finding ever more application possibilities in the machine tool industry and are being used to replace very complex welded steel assemblies. These welded assemblies have internal stresses that make precise machining very expensive. Aluminum profiles can be supplied with low levels of internal stresses and, therefore, together with the advantages that have been mentioned earlier, they can be used in many instances to provide very appreciable cost savings.

Extrusion technology makes it possible for the designer to optimize his or her profiles, not only in terms of the shape required by the application and the associated static and dynamic strength requirements, but also to design details to make entire processing steps unnecessary. This can provide considerable cost advantages.

[19] By R. Gitter and A. Stelzer.

Different surface treatments can produce a range of visual and / or functional characteristics. Profiles have almost unlimited assembly possibilities (e.g., T-grooves, channels for bolted fittings, and various profile openings). Simple changes of the profile can be used to meet static and geometric requirements. The many joining possibilities available range from simple tongue and groove designs to clipjoints. All of these possibilities ensure a high degree of functionality and optimized low weight. Many can be used to simplify assembly and installation. The fact that the advantages of extruded profiles have been so easy to exploit is due in part to the negligibly low die tooling costs that apply to the small quantities of material that can be economically supplied by the extruder.

15.9.3 Plate as feed stock

Plate as feed stock for molds, tools, and machine components is typically hot rolled and heat treated and in the gage range of 300–20 mm thick. Plates over 300 mm are usually not rolled, but are used in the "as cast" condition, or they may be forged to improve the structure if a deformation operation would be beneficial. Thinner plates are sometimes cast instead of rolled, using the lower ductility of the cast structure to improve machinability by reducing the chip length.

15.10 Electrical Conductors[20]

Energy technology worldwide uses a considerable tonnage of aluminum each year. The principal application for aluminum in the energy sector is for the transmission of electrical energy, particularly over long distances. There is a clear distinction in terms of aluminum consumption between different continents and countries:

- In Third World countries, especially those entering a phase of industrialization, aluminum consumption for the creation of an energy grid may be a major use for many years. This will continue until the energy network is virtually completed.
- In contrast, in industrialized countries, new electrical networks are not required as often, and consumption of aluminum for this application remains steady.

Nevertheless, on a global basis, aluminum's use in energy technology has a significant volume effect on both production and processing.

15.10.1 Properties and advantages of aluminum conductors

The properties and advantages of aluminum conductors are well documented. We will mention only a few key aspects, refer to the main conductor alloys, and discuss some new technologies.

Among the key characteristics of aluminum that are major advantages for electrical conductor applications are:

1. Conductivity: this is more than twice the conductivity per kilogram of copper.
2. Lightweight: this makes handling of overhead cables easier with longer spans and low insulation costs.

[20] By K. Gregor.

Fig. 15.52: (a) A range of machine tool components. Sections of large profiles and of parts machined out of low internal stress heavy gage rolled plate. The parts have many integrated functions.(Alusingen); (b) Assembled punch tool. (Alusuisse)

a

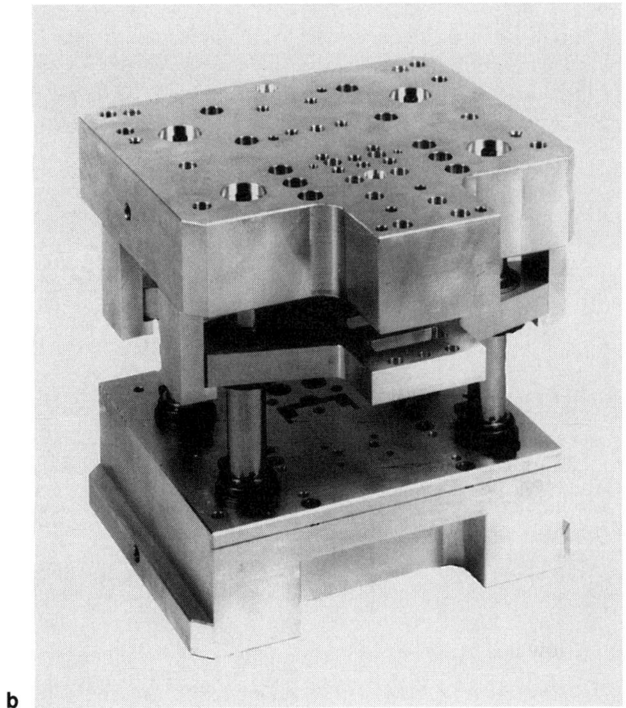

b

3. Strength: conductor alloys have a range of mechanical properties from soft to strengths comparable to mild steel.
4. Workability: this permits a wide range of processing methods for aluminum conductors from wire drawing to extrusion or rolling. The bend characteristics are excellent.
5. Creep: like all metals under sustained stress, there is a gradual deformation over years. With aluminum conductors, design factors take this into account.

Table 15.10 compares the weight and conductance of three frequently used aluminum conductor materials with copper. Looking at the data for 6101-T65 conductor alloy in more detail and comparing it with the unalloyed 1350, one can recognize an increase in minimum yield strength from 25 N/mm^2 to 175 N/mm^2. The reduction in conductivity associated with this massive increase in strength is only from 61.0% down to 56.5% IACS.

15.10.2 Processing stock for wire and field conductors

The stock for wire production is today produced almost exclusively from continuous rotary casting machines (Properzi wire, see Chapter 6). This type of equipment is normally installed close to a liquid metal source in a smelter.

Field conductors for the transmission of three phase current are a special case. This important market segment is largely supplied with wrought material processed in Conform machines. These machines appear similar in their construction to wire casting machines and are fitted with one or two wheels. The starting material, continually cast strip or granulate, is introduced by means of a rotating wheel or wheels into an extrusion chamber where frictional forces raise the temperature. This is followed by continuous extrusion through a hard metal nozzle. In a single processing step, a high precision end product can be manufactured very economically. This type of equipment is also used for the manufacture of tubes and profiles. Over 70 such machines exist worldwide.

Table 15.10: Relative weights of other conductor materials to provide equal direct current conductance (at 20°C) as compared to 1350 aluminum conductor

(1350 has an electrical conductivity of 61% IACS, the International Annealed Copper Standard)

Metal	% IACS Volume Conductivity	% IACS Mass Conductivity	Relative Weight
Aluminum			
1350	62	204	70
6201-T81	54	180	116
6101-T65	58	191	108
Copper			
Commercial HD (Hard Drawn)	96	96	209

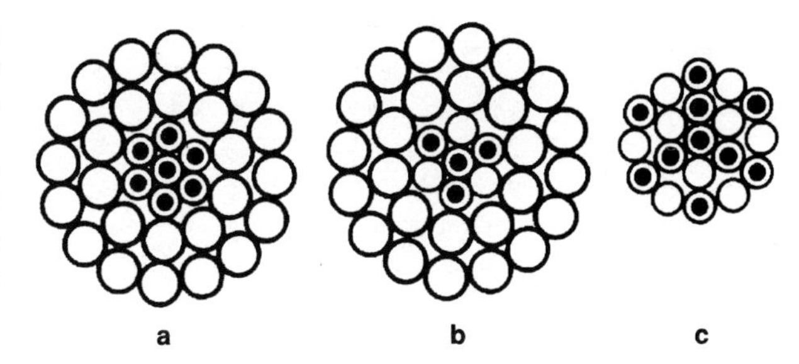

Fig. 15.53: Various TAL/Stalum overhead cables: (a) a steel aluminum cable. The steel strands (black) are coated with aluminum; (b) a cable of similar diameter with higher electrical conductivity; (c) cable with higher mechanical strength. (Berndorf)

15.10.3 Overhead cables

Aluminum alloys 1350, 6201, and 6101 are chosen for a large number of applications in this field. In recent years however, new alloys or alloy combinations have begun to play an increasing role (Fig. 15.53).

15.10.4 Aluminum cable steel reinforced (ACSR)

Although worldwide there are several manufacturing processes, extrusion techniques have become established as being the main methods for coating aluminum on to 5–8 mm diameter high strength carbon steel wire. A process very similar to the Conform system is often used. A homogeneous aluminum layer, which usually has a minimum thickness of 10% of the radius at all points, is coated on the steel wire. The resulting cross-sectional relationship is not changed even after cold deformation by hydrodynamic drawing when processing to the required finished wire diameter. The result is a wire that has nearly the same strength as conventional zinc coated steel wires; but, the aluminum coated wire has much better corrosion resistance. If zinc-coated steel is used as the cable core material in combination with aluminum, the well known problem of corrosion between exposed steel and aluminum (sacrificial corrosion of aluminum) occurs. Use of ACSR wire avoids this problem.

15.10.5 Al-Zr conductor alloy (TAL)

Aluminum-zirconium alloys were first used in Japan. The alloy has been made available in Europe since 1985 by an Austrian cable manufacturer. The fact that purity of aluminum has a major effect upon the material's electrical conductivity is well known. Elements from the first transition (Cr, Zr, Mn, V, and Ti) all cause a rapid increase in the electrical resistivity. Some elements, however, such as B, Zr, Si, Mn, and Fe, when added in increasing percentages, show a distinct change in the rate of increase in resistivity. The solubility of these individual elements in aluminum is usually only a few tenths of a percent. Higher concentrations result in precipitates of specific intermetallic phases. The shape and distribution of these precipitates is of major importance.

The change in the rate of increase in resistance caused by additions of zirconium can be explained by the solubility limit. The Al-Zr phase diagram shows solubility of zirconium at 660°C as being 0.28%. At 500°C, the solubility has fallen to 0.05%. The size and distribution of the Al3Zr phase can be influenced by use of controlled thermal treatments and can be used to hinder the movement of dislocations at higher temperatures and so pro-

Table 15.11: Data for conventional aluminum conductors and TAL

Conductivity (a—% IACS) (S.m, mm²)	UTS (MPa) (minimum–maximum)	Maximum Allowable Operating Temperature (°C)		
		Long	Short (30 min.)	Flash
TAL				
(a) 60	160–190	150	80	260
(b) 34.8				
1350-H14				
(a) 61	100–150	80	(80)	160
(b) 35.38				

vide a precipitation hardening effect. It can also inhibit movement of large angle boundaries and delay the onset of recrystallization. The alloy can be used in cables at temperatures of 150°C for extended periods. It is capable of withstanding for one second temperatures as high as 260°C with an imperceptible loss in mechanical properties. This occurs in conventional aluminum conductor alloys such as 1350 or 6101 at temperatures as low as 80°C. In Table 15.11, the relevant data for aluminum conductors are presented.

15.10.6 "Hot cables" with TAL as conductor

The higher operating temperature, up to 150°C, causes a change in the transmission resistance. It can also affect cable sag. It is important to look at the construction details in each individual case and to study the total economics of the proposed solution. This is especially important when renovating old networks or when introducing new cables to solve bottle-neck problems. There are often considerable cost advantages, and in some cases, it is only possible to provide a solution if hot cables are introduced.

A development that in part alleviates the above mentioned disadvantages of higher operating temperatures is the coating of cables. By use of a special process, the surface of the wires is thoroughly degreased. Residues from the drawing process are removed. An extremely thin layer of multicomponent polyurethane is then coated on to the coil surface, which, as a result, has much better radiation characteristics. This has the affect of lowering the cable temperature (blackbody radiation). A further process for cable coating is to apply a coloring layer for aesthetic purposes. This enables newly constructed cabling to be camouflaged to a very large degree (optical environmental protection).

Typical advantages of the use of TAL / ACSR cables can be summarized as follows:

- Higher safety standards during power surges from breakdowns or from load peaks
- Increased cable strength to raise the safety reserve
- Removal of bottlenecks in cabling systems
- Higher safe conductive capacity for standard operation of cables and switch gear
- Possibility of postponing major new investment for 10–15 years

The use of "hot cables" can enable a system to conduct up to 50% more energy than would be possible with conventional cables.

Chapter 16. Aluminum and Its Environment

16.1 Ecological Issues

16.1.1 The "issue triangle"

The issue triangle (shown in Fig. 16.1) represents the interrelationship between the aluminum industry and its environment. This concept can be applied either to the metals industry as a whole or used to illustrate individual materials. It is important to realize that every corner of the triangle offers both risks and opportunities. The opportunities exist in the "green market" (i.e., with the development and sale of environmentally friendly processes, products, and systems). The diagram also represents the risks that the material industries have faced for the last two decades for a number of reasons, such as the energy crisis, the green movement, or the general public reaction against the so-called uncontrolled growth of technology. The issue triangle can be compared to an ice flow that is subjected to the action of currents from continuously changing directions.

At the beginning of the 1970s, the raw materials corner was the subject of much serious discussion. Recently, attention has focused on energy and environmental issues. The corners of the triangle are not, however, independent of each other. It has to be stressed that in reality the triangle exists as a linked entity. To illustrate this point, we can consider environmental pollution caused by an unnecessary waste of energy or through the incorrect use of materials. It is irrelevant whether contaminants take the form of solids, dust, gas, or liquid. They are waste emissions and a burden on nature. This also applies to a product that at the end of its life ends up as unusable waste or has to be incinerated. In every case the result is the same—waste of raw materials, environmental pollution, and waste of energy.

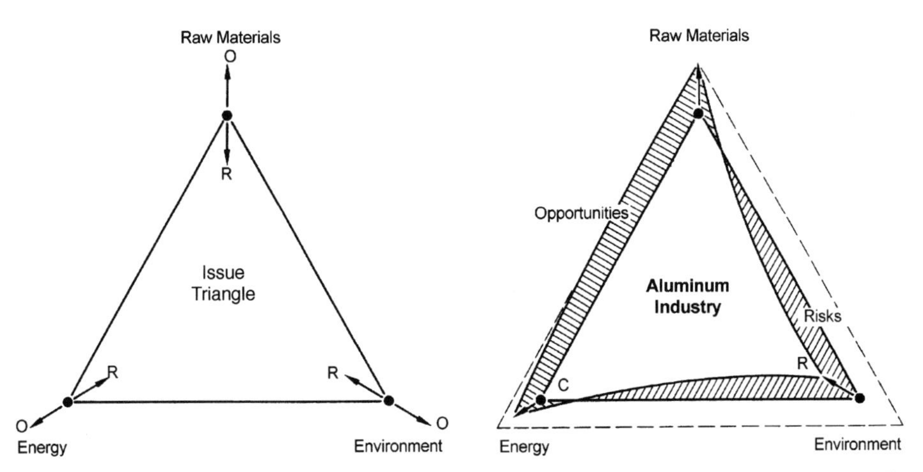

Fig. 16.1: A representation of the aluminum industry and its environment (the issue triangle). The current situation is shown on the right. O = opportunities, R = risks, C = chances.

Up to the middle of this century there was a widespread lack of awareness concerning consumption of resources. The term ecology was known only to a few specialists.

The issue triangle, in principle, was first presented to the general public in 1972 in the book *The Limits to Growth,* which was published by the Club of Rome. The quintessence of this book was the prediction that in a defined time period we would be faced with serious shortages of important raw materials. Whereas this statement caused positive results by sensitizing the public at large to ecological opportunities, the report also caused a certain degree of misunderstanding since it did not differentiate between scenarios and forecasts. In the 25 years since the publication of the report, this misunderstanding has not been completely removed from the public mind. There has been an avalanche of "concerned literature" that has affected, through educated circles and the press, the ordinary citizen and the politicians who represent him or her. At the center of this literature was the thesis of the Club of Rome that energy and raw material sources could soon be exhausted. From this idea came a whole series of alarming statements and misunderstandings. The following text will present facts on aluminum's ecological balance sheet.

16.1.2 The raw material situation of the aluminum industry

For some decades, tropical bauxite that contains the mineral bayerite, the trihydrate (Al_2O_3,$3H_2O$), has been the dominant ore for alumina production. The two main reasons for the preference for tropical bauxite are because it is easy to extract alumina from this ore using the Bayer process and there are vast reserves of this ore on four continents (Latin America, Australia, Africa, and South East Asia).

In contrast, the refining process for rock bauxite is more expensive. It is necessary to use relatively high temperatures and pressures in an autoclave in order to get the rock-like monohydrate bohmite into solution in caustic soda. Nevertheless, some Mediterranean countries, especially Greece, are using this technique. The largest activity of this type occurs in China. Bayerite does not generally occur in northern latitudes because it depends upon a weathering process that is favored by high atmospheric temperatures and humidity. All of the best reserves of trihydrate bauxite occur in the tropics. Based upon modern extraction methods, there is very little usable bauxite in North America.

In the following text, we concentrate on high grade tropical bauxite with 35–50% aluminum oxide content. In the Club of Rome report of 1972 it was stated that bauxite reserves would be exhausted in about 60 years (and that iron ore would be largely consumed in 93–173 years). The reason for these alarming projections was that only ore deposits in use at that time and constant growth in consumption were used in the computer model. Both assumptions have since been shown to be false.[1] The available bauxite reserves in 1990 rose to 140 thousand million tons, sufficient to produce more than 25 billion tonnes of primary aluminum (Fig. 16.2). Generally, the ratio 4:2:1 can be applied to high grade bauxite, aluminum oxide, and primary metal to evaluate quantities of raw materials required for production of the next stage. It can be concluded that sufficient bauxite reserves exist, without inclusion of lower grades, to supply world primary aluminum requirements.

[1] A spinoff of the Club of Rome scenario was the foundation of the IBA (International Bauxite Association) cartel in Jamaica around 1975; it practically disappeared a few years later.

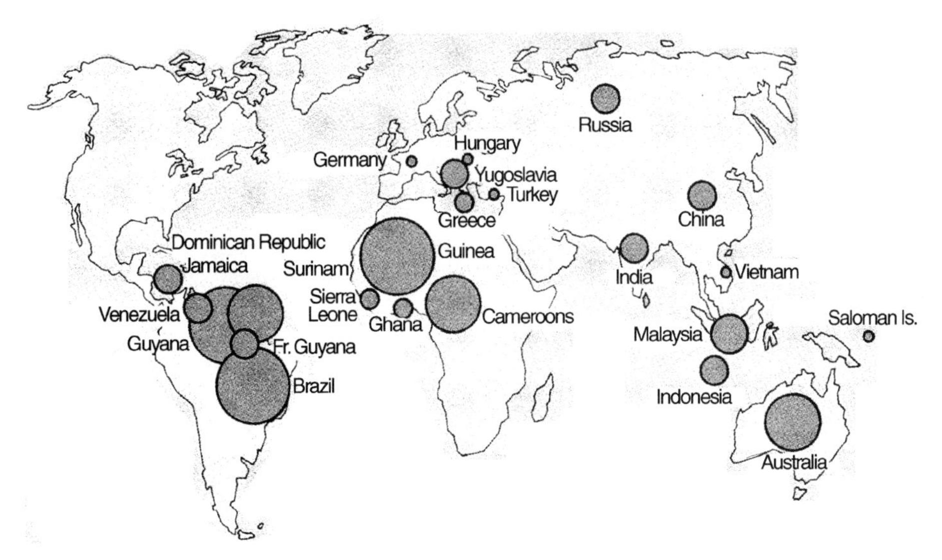

Fig. 16.2: Regional distribution of economic bauxite deposits of 140 billion tons. (1990–EAA).

In the 1972 report, the low values projected for available bauxite reserves resulted primarily from the use of a constant exponential growth factor to predict growth in consumption (at that time, aluminum consumption was rising at 8% per year). Since 1972, the growth in consumption of primary aluminum has fallen and is currently about 3.4% per annum. In addition, large new bauxite reserves have been discovered, for example in Equatorial Africa, Australia, Brazil, and Venezuela. The possibility of a bauxite shortage causing difficulties for the primary aluminum industry can be neglected.

Not many decades ago bauxite with under 40% aluminum oxide was described as being low grade. High grade bauxite with an alumina content of 50% was the ore used for alumina production. In the intervening years, however, other factors have come into play:

- The Bayer process has been modified to accept what were, under the old definition, low grades of bauxite. A large throughput of ballast materials such as quartz, in particular, can now be tolerated both technically and economically. Alumina plants are sited very close to bauxite deposits to cut the costs of transporting materials.
- The importance of the content of silica (SiO_2), which is the anhydride form of the silicic acid (SiO_2,xH_2O), in reducing the efficiency of alumina extraction has been more clearly recognized and is now an important negative aspect in assessing bauxite quality. Silica reduces the efficiency of the Bayer process by effectively lowering the alumina content of the bauxite and causing sodium to precipitate out as sodium aluminate. Today, bauxites that contain more than 5% SiO_2 are considered to be uneconomic.
- On Australia's west coast, one-fifth of the world alumina production is today based on trihydrate bauxite with only 35% alumina content, but with a very low silica/silicic acid content.
- Among well known but, to date, virtually untouched bauxite reserves there is a wide range of deposits with oxide contents that, with today's techniques, are worthy of exploitation. For example, in India there are very large bauxite deposits such as those

in Orissa with 40% alumina contents. There are massive deposits in Guinea with alumina contents of 50% or more.

It can be seen from Table 16.1 that in 1976 the proven bauxite reserves increased in one decade by a factor three. This is because of the grouping together of suspected and proven reserves. If, from Fig. 16.2, the world bauxite reserves for 1990 are taken as being 140 thousand million tons, it is clear that there is a strong trend toward exploitation of medium grade bauxite deposits. The enormous jump from 17–140 thousand million tons of reserves at the beginning of the 1990s is, to a very large extent, due to the lowering of the "cut off" grade of the bauxite.

Information concerning size of ore deposits is neither correct or incorrect; it has to be interpreted. Specialists in mineral resources have long recognized that the "cut off" grade of an ore is very frequently dependent on the price of the pure metal. In aluminum's case, however, it has been rather less dependent upon the open market price of aluminum than on the development of the processes to produce aluminum oxide from bauxite. Newer alumina refining methods have lowered the economic extraction level of the percentage of alumina in a bauxite deposit.

The study by the Club of Rome probably committed its most serious error in ignoring the trend, which had already begun in the 1970s throughout most of the Western world, of trying to achieve higher performance with less energy and material input. This concept was perfectly expressed in the phrase "doing more with less." This idea, which was first put forward by the genial architect Buckminster Fuller 25 years ago, is very relevant today and can be applied to all of the common metals and the use of energy. This is illustrated in Fig.16.3, which presents a schematic representation of a trend that has been evident since the 1970s and has affected all materials. The vector shows the extent of this trend in the last few decades.

In order to be fair to the Club of Rome report, it must be said that because it created a degree of concern it contributed appreciably to the movement dedicated to "doing more with less." In general, it is possible to say that today, 20 years after the much quoted study, all available mineral reserves and primary energy sources have **increased rather than been depleted**.

Table 16.1: Reserves and usage of four common metals 1966/1976

(In millions of tons)

	Proven reserves in 1966	In the decade 1966–1975		Proven reserves in 1976
		Consumption	New reserves	
Copper	195	–63	+324	456
Aluminum (Bauxite)	5964	–605	+11913	17272
Lead	93	–33	+115	175
Zinc	75	–54	+164	185

Source: United States Bureau of Mines, Washington, D.C., Federal Department for Geological Science and Raw Materials, Hannover.

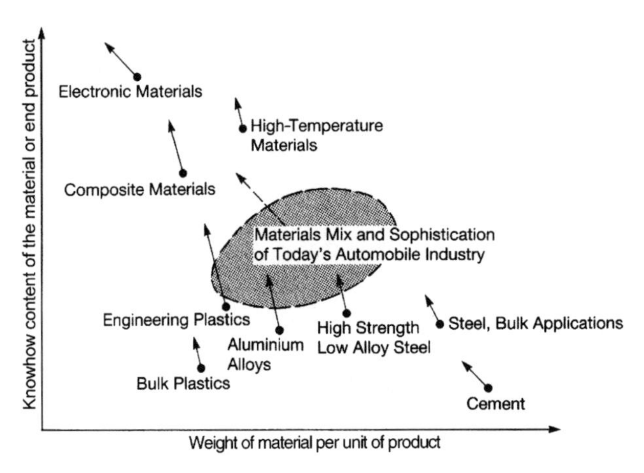

Fig. 16.3: A model for the development of material intensive industries.

16.1.3 Environmental pollution from alumina and primary aluminum production

The environmental challenge of an alumina plant is due to the fact that after processing up to 50% of the bauxite, input material has to be deposited as "red mud" waste. This red mud is similar to a low viscosity clay that owes its red color to its iron oxide content. In the last two decades, enormous efforts have been made to improve the environmental acceptability of red mud deposits. The red mud is now held in shallow lakes that prevent seepage of the fluid alkaline components into the ground water or into other fresh water sources. After some years, the red mud deposit site is closed and allowed to dry out. It is then planted with suitable vegetation with or without a topsoil layer.

The environmental problems of aluminum smelters have been a sensitive issue in Western industrialized countries for the last 20 years. Fluoride emission levels of approximately 20 kg of fluorine / tonne of primary aluminum were considered to be particularly serious. Since the 1960s, these emissions have been progressively reduced, first by use of waterspray washing systems. In the 1980s, dry absorption techniques were introduced. The method involves bringing a fresh, highly active alumina called "gamma sandy" into contact with the exhaust fumes. This technique enables fluoride emissions to be reduced by 98% by absorption of the highly reactive gas into the aluminum oxide. The oxide is then used in the electrolysis process itself. In all newly installed smelters, this dry absorption technology represents the state of the art. In the OEDC countries, all of their old smelters have had their cells encapsulated and systems for the dry absorption of gases installed. There are, however, a proportion of the smelters in use in the world that operate without efficient fluorine absorption, for example those in Russia and in China.

A developed world aluminum smelter is required to satisfy strict legal environmental standards in terms of its waste water, exhaust gases, and solid waste deposits. Every smelter should be required to satisfy this complete list of requirements irrespective of whether it is still operating or has already been shut down. The pressure for all smelters, including those in the Third World, to operate under strict environmental controls is increasing, and restrictions by such organizations as the World Bank are being used to refuse credit for unclean smelter development.

16.1.4 Energy consumption during aluminum production

In Table 16.2 a comparison is made of the energy required for bauxite mining, alumina production, and an aluminum smelter.

The alumina manufacturing process uses thermal energy obtained from fossil fuels (i.e., oil, coal, or natural gas); it uses similar quantities of energy to that required in a cement plant. In both cases, a large part of the energy is used in the so-called "calcination furnaces." In the alumina plant, there is also a large amount of thermal energy needed for the chemical reaction with the bauxite.

As can be seen in Table 16.2, the major part of the energy needed to make aluminum is consumed in the smelter itself. Old smelters needed 15–19 kWh per Kg aluminum; modern smelters need approximately 13 kWh. Following the 1973 energy crisis, the high energy consumption in smelters has been the subject of much debate, especially in western Europe and Japan. The Japanese obtain the bulk of their electrical energy from imported oil. As a result of the energy crisis, their power costs increased dramatically. Therefore, they closed down approximately 90% of their aluminum smelter capacity, which had been capable of producing more than one million tons of metal annually. In Europe, the smelter energy required for the production of primary aluminum has become increasingly expensive, albeit with a delay when compared to Japan. This has led to an increased number of smelter closures, a trend that is likely to continue.

The increase in free-market energy costs has had a major impact on smelter operations. As a result, aluminum production is increasingly being concentrated on so-called "energy islands," sites where very large quantities of hydroelectric power are freely available. Primary aluminum is the most common product from these areas but their cheap power is, on occasion, also used for electric steel making, magnesium production, or electrolysis of copper.

Table 16.2: Total process energy required for the production of one tonne of primary aluminum

(Average values for the Western World, in MJ., ref. Alusuisse)

Primary metal	Bauxite processing	2700
	Caustic soda production	9801
	Alumina production	26,796
	Anode production	3310
	Primary aluminum smelting*	117,861
Secondary metal	Scrap aluminum recycling	6913

* A value of 14.5 kWh/kg of primary metal was assumed for the electrolysis process. The total energy content of the anodes was converted into electrical units.

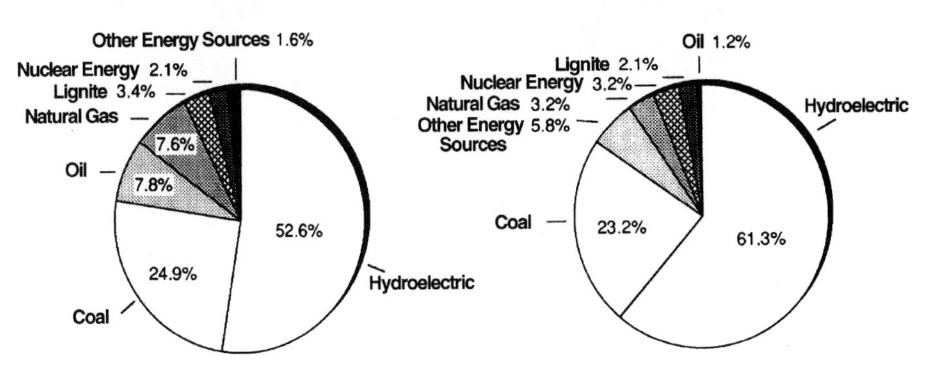

Fig. 16.4: The proportion of energy sources used in the Western world for the production of primary aluminum.

Fig. 16.4 gives an overview of the energy usage in aluminum smelters. The increasing dominance of hydroelectric power as an energy source is clearly shown. It is presently used to produce approximately 66% of all primary aluminum.

In thinly populated areas of the world there are large coal reserves that can be converted on site into energy. In order to make use of these energy sources, economic conversion into aluminum is a vital technique. Such important sites for coal-powered aluminum smelters are found in Australia, China, South Africa, the former East Bloc countries, and parts of the United States.

16.1.4.1 Rational energy use

Fig. 16.5 shows that aluminum and energy can be considered from two differing view points:

- As the energy required to produce and process the metal.
- As the energy savings that can be achieved by the use of aluminum, especially in transportation.

The expression "rational energy use" (REU) applies equally to both considerations. This aspect will be discussed further.

REU during the manufacture of aluminum

As already stated, the smelting process takes the lion's share of the energy necessary to produce primary aluminum. Every detail of the use of smelter energy is under continual scrutiny to achieve any reductions possible. In the last two decades, 5–10% of additional energy savings have been achieved by use of improved sensors linked directly to the process that control the electrolysis in detail. For example, in each electrolysis cell, the cell potential is measured to millivolt accuracy. With a small rise in measured potential, automatic feeding systems supply optimal quantities of alumina through so-called "point

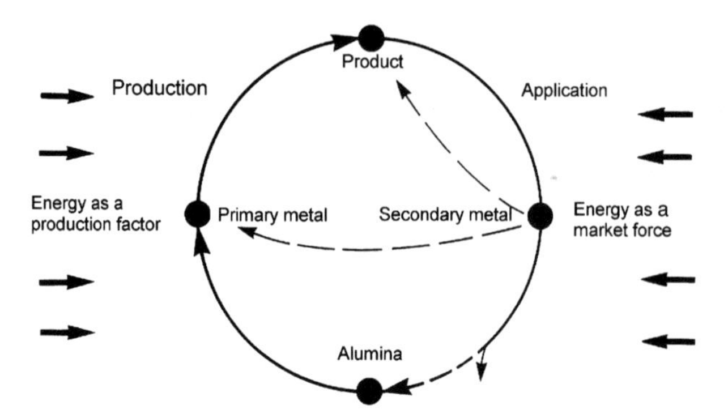

Fig.16.5: Aluminum as an energy bank through its manufacture and service life.

feeders," thereby eliminating the "anode effect" (the discontinuity in operation caused by a shortage of alumina that resulted in extra emissions). This provides a number of advantages: electricity consumption is lowered, the labor cost is lowered, and environmental pollution is reduced. If a comparison is made with the smelter practice of earlier decades:

- Manual operation of cells was practiced in earlier times and is still in use in obsolete smelters in the former East Bloc countries. In cells under manual control, large temporary increases in electrical potential could occur, which could cause a so-called "flash over." The increase in voltage was the result of a lowering of conductivity, often because of the formation of gas concentrations in the electrolysis bath. This could reduce the efficiency of a cell by up to 10%.
- Whereas sudden rises in voltage in electrolysis cells (anode effects or flash overs) were formerly an almost daily occurrence, the use of point feeders has enabled the operation of cells to be so well controlled that such events seldom occur, and the associated emission of fluorocarbon gases (CF_4 and C_2F_6) is drastically reduced.

A further important area for REU relates to the consumption of anode coke; 500 Kg of anodes are required per tonne of primary aluminum. Efficient use of energy in the plants producing the block anodes is a significant factor because the energy content of the coke anodes and the energy used in the calcination furnaces to manufacture these anodes are important parts of the smelter's total energy balance. Anodes account for 20% of the total energy use.

An interesting adjunct to the smelter process is that since the electrolysis metal is transferred into the casthouse at temperatures over 900°C, it provides a very large proportion of the total casthouse thermal energy requirements.

REU by the use of aluminum

As indicated on the right side of Fig.16.5, many aluminum products provide opportunities for the rational use of energy during the life of the product. Because of its energy-relevant characteristics, aluminum, when used in important market segments, has effects on the energy bank that can be of very great significance.

Transportation provides the most important examples of rapid repayment of the capital invested in aluminum's energy bank. The advantages have been well documented in a wide variety of publications. In rail cars, a reduction of 40% of the body weight through the use of aluminum can rapidly provide an inservice energy payback. The payback time is especially short for stop / start, short journey road vehicles. In the case of the short haul vans used for stop and go activities, energy savings can pay back in a few months the total energy used to manufacture the primary aluminum and which is stored in the product. Energy savings after this payback period are simply an energy bonus resulting from the use of aluminum.

In Fig. 16.6 it is possible to see that in private cars, within a fraction of the total time of a vehicle's life, the aluminum that has been used has paid back the original investment in energy as savings in fuel consumption. This is especially significant in commercial vehicles where lower fuel consumption has a direct effect on energy costs, especially when the vehicle is traveling unladen. This aspect has been covered in Chapter 15.

- Energy Technology: Aluminum is used in many segments of energy technology. It is often responsible for massive cost savings, for example, its use as cable contrasted with the cost and / or weight of copper cables and its important role in the exploitation of solar energy (See Chapter 15).
- Building and Architecture: In the last 20 years, aluminum has played an increasing role in the building sector because of the many new laws and standards related to energy conservation. Aluminum is now usually seen as a component part of a system rather than a material. There is increasing interest in so-called "energy facades" and winter gardens, which are useful as a passive means of storing solar energy.
- Packaging: Aluminum remains an ideal packaging material when protection of food products and extended shelf lives are important factors. Beverage cans, made largely from aluminum today, in contrast to steel cans or plastic containers are considered as an efficient part of a recycling concept (See Chapter 15).

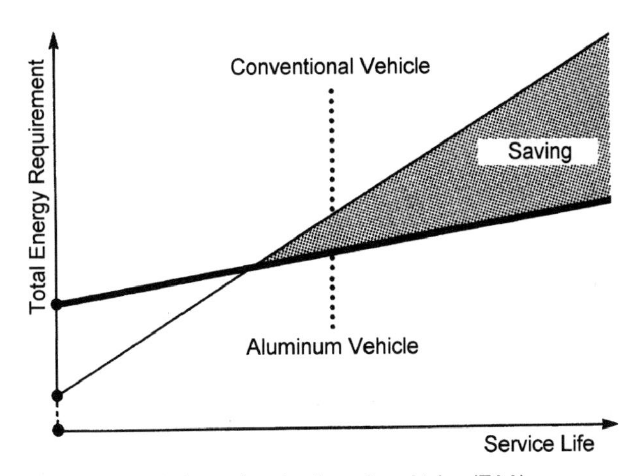

Fig. 16.6: Positive total energy balance for aluminum in vehicles (EAA).

16.1.4.2 *Aluminum: an "energy bank"*

In the caption for Fig.16.5, aluminum is described as an "energy bank;" why is this so? It refers to the energy capital that is stored in the metal itself. In the alumina processing plant and the smelters, energy is "paid" into the product and stored in the material as an energy bank. During the life of the product this "paid in" capital from the primary aluminum metal is very often repaid with interest (See Fig.16.6.). A very beneficial transaction is made with this energy bank when a product at the end of its useful life is recycled as secondary aluminum. With only 5% of the energy used in the original electrolysis, new aluminum products can be made from the secondary aluminum plant and be returned to service.

An aluminum smelter with a production capacity of 100 thousand tonnes per year requires approximately 200 MW of electrical power, 24 hours a day. This power requirement means that aluminum smelters have to run as continuously as possible. Unscheduled shutting down of electrolysis cells has to be avoided. A cell that is denied power very rapidly freezes solid, and a return to production is expensive.

This apparent disadvantage has, however, a considerable advantage in that aluminum smelters on an electricity network provide a very welcome base load, whereas most other industrial processes and domestic electricity consumption typically have periods of peak load. This requirement for a large baseload creates the tendency for aluminum smelters to be located on the so-called "energy islands" mentioned earlier.

In Fig.16.7, the world hydroelectric power potential and the present level of exploitation is shown by continent. In Western Europe, hydroelectric power sources are largely exploited, with the exception of a few smaller power sources and the still available larger reserves in Scandinavia. What is striking is the relative limited use of hydroelectric power in Africa, Asia, and Latin America. These three continents have 1.4 million MW of unused hydroelectric power reserves. By using only one quarter of the untapped hydroelectric power of these three continents, the current total world production of smelter aluminum could be increased by a factor ten. (This would require 350,000 MW of electrical power.)

For more than two decades, especially in German speaking countries, the argument has been frequently repeated that the use of aluminum, for example in the packaging sector, creates unnecessary strains on limited energy reserves. Here it must be stated clearly that by far the largest portion of primary aluminum used in Europe is manufactured with hydroelectric power from one of the many energy islands. Western European production of smelter aluminum, because of the high price of energy, will fall from year to year, and the remaining capacity will often be needed to provide an electrical base load.

16.1.4.3 *Aluminum recycling*

The terminology in this field requires some explanation. A modern phrase is "the secondary metal industry." This term covers the industry involved with the reprocessing of used aluminum products. The resulting metal is by implication considered to have un-

Figure 16.7: Hydroelectric power potential in millions of kW per continent (1990 EAA).

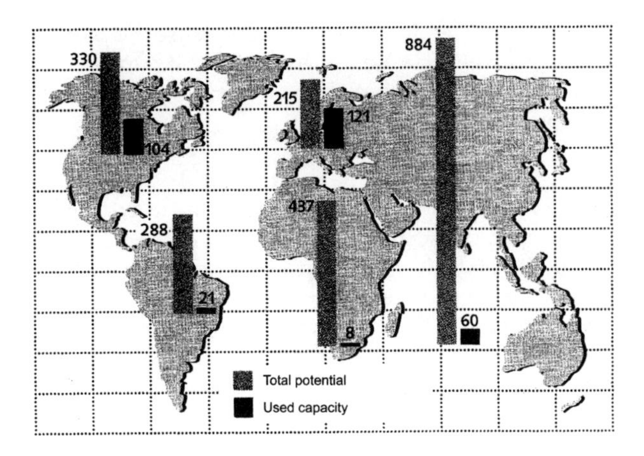

dergone a degree of down-cycling when compared to primary smelter material. The expression "down-cycling" is associated with a negative image, and the necessity of using such a term is open to question. The correct use of down-cycling is to describe the processing of the common mixed bag of some 18 different types of plastic that occur as waste from a motor car shredder. Since this type of down-cycled plastic occurs in greater volume than the availability of suitable products, such as flower boxes, most of it ends up on a dump or is incinerated. (This activity is currently questionable and may be banned in the future.) When referring to the use of aluminum following recovery by separation and remelting, it is correct to speak of "recycling."

Recently, in most of Europe, America, and Japan, legal restrictions have been introduced (or notification has been given that they will shortly be introduced) that will require recycling the materials in important consumer products back into the products themselves. The most important application field for aluminum, apart from packaging, is the automobile industry. Car manufacturers will have to ensure that approximately 90% of used materials from old cars are returned for reuse. Here, however, it is necessary to differentiate between a return of valuable components and a genuine recycling. Whereas it is well-proven practice to manufacture beverage cans from recycled beverage can scrap, in the case of motor cars, the opportunities associated with sorting the aluminum components out of an old vehicle present major technical and commercial challenges. Up to the present time, used cars have been passed through a shredder that tears the vehicle into centimeter sized fragments. The modern generation of automobile shredders are, however, steel dominant; in order to function they need a very high steel content. The sorting out of light metal (and plastics) has received low priority up to now. The nonferrous part of a used car can, however, by use of the correct flotation separation techniques be segregated so that the light metal fraction can be made available for recycling.

The aluminum scrap that comes out of a shredder has a relatively high silicon content (because of the high proportion of castings). Magnesium and other alloying elements have to be added to the levels required in casting alloys. As a result of the shredding operation on the car, there is inevitably a high iron content in this type of secondary metal. It is, therefore, necessary either to use this metal for relatively noncritical castings or, alternatively by means of an addition of primary metal, to lower the iron content to a level suitable for the broad range of casting alloys required for motor car components (Chapter 7).

In conclusion, it should be pointed out that the ease with which aluminum can be recycled plays a decisive role in determining its economic and ecological advantages. This gives aluminum scrap a high value per unit weight in contrast to scrap iron. As long ago as the end of the last century the value of used aluminum was recognized. As Fig. 16.8 shows, manufacturers at that time were concerned that the products they were selling could be recycled, a situation which still applies to the aluminum of today.

16.2 Technological Issues

16.2.1 Aluminum: A material from the first and second Industrial Revolutions

A summary presented in Table 16.3 shows a geographical breakdown of the various milestones and the consequent sequential stages of industrial development. The aluminum industry was born when the first Industrial Revolution was reaching its first prolonged high. A number of innovations took effect, often quite independently of each another. Aluminum production by electrolysis occurred at the same time as Siemens and (today's) General Electric Company developed the electric generator/motor with obvious synergistic benefits. The dynamo provided the basis for industrial electrometallurgy, and the birth of the commercial aluminum industry occurred in France and the United States. The first countries to take out licenses for the new aluminum production process were the U.K. and Switzerland. At the beginning of the second Industrial Revolution, that is to say in the first four decades after World War II, the aluminum industry experienced its golden age. For almost three decades there was a breathtaking growth in both production and consumption. Use of aluminum, the 13th element, grew by about 8% per year. This equates to a doubling in size every nine years. A central opportunity in this turbulent development phase of the second Industrial Revolution was, however, neglected: energy consumption and environmental pollution.

In the raw material corner of the issue triangle, the risk of shortages of ores for the production of aluminum is relatively small. Taken together with the trend toward recycling, it can nowadays be seen to be unimportant. The situation is different, however, in the other two corners of the triangle—energy consumption and environmental pollution. For the whole metal industry and also for the automotive sector, these two opportunity areas have taken on the deepest significance. Debates in the press have very frequently been conducted around false premises. The general public is flooded with information on the subjects of both energy consumption and environmental pollution, but there is a poor understanding of the basic relationship between these topics.

Halbheft 25. Die Gartenlaube. 1893.

Illustriertes Familienblatt. — Begründet von Ernst Keil 1853.

Jahrgang 1893. Erscheint in Halbheften à 25 Pf. alle 12—14 Tage, in Heften à 50 Pf. alle 3—4 Wochen vom 1. Januar bis 31. Dezember.

Neues vom „Metall der Zukunft".

Kochgeschirr aus Aluminium.

Bei der ökonomischen Werthschätzung des Aluminiumkochgeschirrs ist jedoch noch ein Umstand in Betracht zu ziehen. Nutzt sich das Geschirr ab, was schließlich bei jedem Geräthe einmal der Fall ist, so ist es durchaus nicht so werthlos wie altes Eisen, es wird vielmehr, wie das Kupfergeschirr vom Kupferschmied, von Aluminiumfabrikanten für den Metallwerth wieder eingelöst.

Ob das „Silber aus Lehm" den richtigen Stoff für das „Kochgeschirr der Zukunft" bildet, das läßt sich heute noch nicht sicher sagen.

Fig. 16.8: Historical evidence that the idea of the value of recycling aluminum is more than 100 years old.

16.2.2 Industries that consume energy and influence the ecology in a state of change

Certain industries have a "common denominator" in terms of their energy and environmental significance. In Fig. 16.9.a there is a simplified ABC representation that can be applied, for instance, to:

- Basic industries (such as the aluminum industry)
- Thermal power plants
- Housing
- Automobiles
- Packaging

From this ABC representation it is possible to differentiate between three distinct periods in terms of environmental pollution.

- Phase A represents the "first generation" of energy intensive processes and products. Since the mid-1970s, this technology has been the subject of much concern.
- In Phase B, there was an adjustment to energy saving or lowering of environmental pollution.
- In Phase C, "completely redesigned new systems" emerge for optimization of energy use and environmental requirements.

These three stages involve significant change in those industries that consume large amounts of energy and have a significant impact on the environment. The reader will have noted that these A, B, C time phases are not at all related to the A B C structure of the edifice.

Table 16.3: Selected milestones in the development of materials and energy

Central theme (Innovation cluster)	Invention new product	Geographical location
1760 1st Industrial Revolution		
1760–1830 (phase 1) 1800 Steam power	1779 steam engine (James Watt) 1783 steam ship 1821 glass production (semi-industrial)	Britain 1760–1830
1830–1870 (phase 2) Railways	**Aluminum production** 1854 (chemical process) St. Claire de Ville	1830–1879 Spread to USA and western Europe
Steel		1848 Industrial Revolution in France
Food production	1856 steel production Bessemer 1876 telephone bell 1879 electric light	1870 Germany begins industrial expansion
1880–1930 (phase 3) Electricity Chemical industry	Dynamo and electric motor 1880–1885 Siemens	
Gasoline engine	**Aluminum electrolysis** **1886–1888 Hall-Heroult process**	
Production line (Ford)	**1889–1893 alumina Bayer process** 1899 radio 1903 first airplane 1909 plastic bakelite	Japan, Russia Canada, Asia catching up 1920–1930 USA increase in capacity (including R&D)
Depression	1929 penicillin	
Inflation 1939 polyester	1930 petrochemicals	1930 USSR first 5 year plan
1945-1985 2nd Industrial Revolution	1940–1980 plastics industry 1943 rockets	from 1950 on
Television Atomic power Green revolution mentality by all metal producers	1947 transistors 1950–1975 "tonnage" 1957 first satellite (Sputnik)	Japan industrialized; China less efficiently

Central theme (Innovation cluster)	Invention new product	Geographical location
1957 "Doing more with less" Buckmaster Fuller		
	1966 personal computer	
Stagflation 1970 *Small is Beautiful* by E Schumacher 1972 *Limits to Growth* report by Club of Rome Autumn 1973 first oil crisis (Yom Kippur War) 1974–1982 rise in energy price (2nd & 3rd oil crises) End of tonnage mentality		From 1970 on Latin America
1985 (approximate) 3rd Industrial Revolution		
	1979 closure of Japanese smelters begins	From 1980 on Southeast Asia
Responsiveness revolution (an eco-conscious society)	1980 robots	
Crisis mood in metal industries readiness to change emphasis	1983 mobile telephones 1985 C+C (communications plus computation) 1990 ISDN informatization	South America From 1990 on
Lean Production Production of systems (C Technology) Genetechnology	genetechnology 1985-95 **cut-backs and closures of West European and US smelters**	Eastern Europe China Mexico

Phase	Distinquishing Features	Effects
A Up to the Oil Crisis (about 1975)	The Machine Age Resources wrongly allocated Technology now completely obsolete	Energy Waste and Pollution
B Since the Mid 70s	Damage Repair Using redesigns "End of Pipe" methods	Energy Saving, Reducing Environmental Damage
C The Future Determinant of Future Investment	Damage Avoidance By use of totally new systems	Absolute Minimum Energy Consumption and Pollution

a

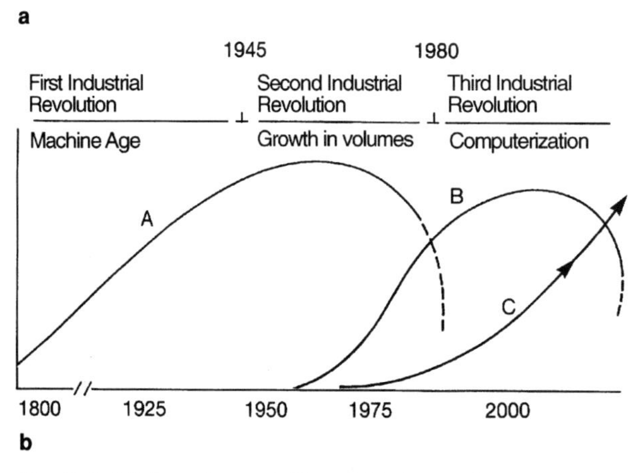

b

Fig. 16.9: (a) Time phases A, B, and C in terms of energy consumption and environmentally polluting capital equipment and production. (b) The correlation of three Industrial Revolutions with time phases A, B, and C.

16.2.3 Impact of the various time periods on the aluminum industry

The A phase according to Fig. 16.9.b began within the first Industrial Revolution of "Steam and Steel" and achieved a dominant position in the second Industrial Revolution from 1945 until the middle of the 1970s. The A period, therefore, continued for a very long time. Its related technologies were wasteful of energy and not very concerned with environmental pollution. It was the period of the throwaway product (planned obsolescence). Following the end of World War II, very large sums of money were invested for almost three decades in what would today be considered to be obsolete technology. It was a period of rapid investment in power stations, raw material industries, and energy wasting buildings.

The general opinion in the aluminum industry, however, was that the 1950s and 1960s represented a golden age. It was a period in which the prevailing attitude was to think in terms of tonnes sold. This involved the whole of the industry from the bauxite mine up to sales of semi-finished product. Energy consumption and environmental pollution were very secondary issues right up until the first United Nations Environmental Conference in Stockholm in 1972 and the first report from the Club of Rome of the same year. In the autumn of 1973 the first oil crisis occurred, and practically overnight the euphoria of the A period evaporated. Until the beginning of the 1970s, cheap and abundant supplies of energy were an engine behind industrial development for production and use of basic materials. This changed into a situation where high energy consumption was a seriously limiting factor for energy intensive processes and products.

In the B phase, which occurred after the oil crisis in the mid-1970s, it is perhaps surprising that enormous efforts were dedicated to conversions of all types. With closer scrutiny it is possible to see that at this time many temporary solutions and so-called retrofits were applied. Typical of this period are measures referred to as being "at the end of the pipe." This also applies to aluminum smelters, with the late encapsulation of electrolysis cells. It is typified in the automotive sector by the introduction of catalytic converters.

In the product area, there were also many changes in order to accommodate the new requirements for energy savings and less environmental damage. For example, there were modifications to extruded window sections that involved the inclusion of an insulating insert in order to be able to reduce the energy losses through window frames and lower the K-value of the complete window (the K-value gives the loss of thermal energy in buildings in terms of area and time). Many other developments that have been covered in this book were born in this transition phase B. Typical examples are the pressures toward recycling and the general tendency to "do more with less."

It is not fair to criticize the B phase in general as a period of piecemeal solutions. It is more important to recognize the progress that was made. It was very sensible in terms of economy, efficiency, and reduced risk to proceed with the changes involved in the switch from the A to B technology as a series of small discrete steps. The effect of changes could be precisely assessed for cost-benefit effectiveness.

How then can the change from B phase to C phase proceed? In general, the C phase requires considerable lead time for expenditure on development. A positive cash flow can often only be achieved after a lengthy period. From our diagram it is possible to understand the situation facing the aluminum industry and to appreciate how the current state of development has been reached. The aluminum industry, up to the middle of the 1970s, operated largely with production concepts of the past similar to the situation in the steel industry and, in fact, that of most commonly used materials.

During the B transition phase a broad variety of impressive results were achieved. The leaders in this period of change were definitely the Japanese. The oil crisis created overnight an opportunity that threatened the very existence of the majority of their aluminum smelters, since they used energy obtained from power stations run on imported fuel. The Japanese, however, did not allow themselves to become discouraged. Their changed attitude can be illustrated by a statement of the president of Nippon Light Met-

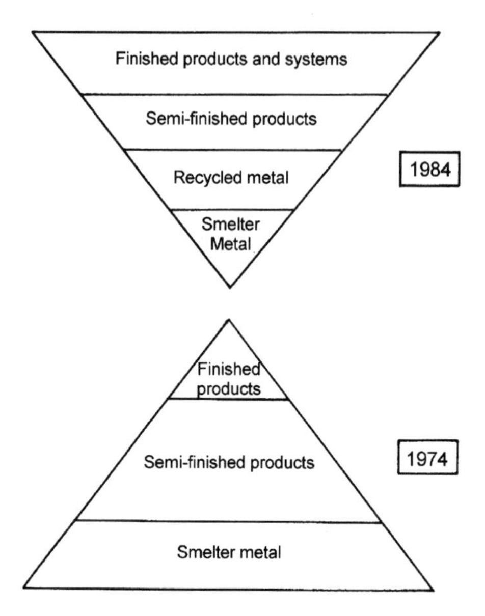

Fig. 16.10: Quantitative and qualitative change in the product mix of integrated Japanese aluminum companies caused by changes in raw material and energy cost structures between 1974 and 1984—"turning Mount Fuji upside down" (areas are proportional to turnover).

als (1970) who said "We are in the process of turning Mount Fuji upside down" (Fig. 16.10). Whereas the company in the 1970s emphasized the production of primary smelter aluminum, by the 1980s the transition was well on the way toward a focusing of effort onto material processing. Many aluminum producers in the Western world began to follow suit shortly afterwards. During this transition period in phase B, painful situations occurred, but there were many original and surprising results brought about by the change to intelligent and quality conscious processing.

We are now in the third Industrial Revolution, a period of new horizons. The opportunities of the past have to be recognized and activated. This applies to the aluminum industry. The solution for the near future of "doing more with less" is a necessity in a continuing developing industrial society. The change over to the full C phase technology will be a socio-political factor of the first order.

16.2.4 Future prospects

For the last 20 years, there has been worldwide concern over the use of energy and consumption of materials. This concern has been the driving force behind the general objective to improve performance. The "green" movement, for all its positive aspects, has left the population at large with some misleading ideas. The only way to remove these misconceptions is to clearly present the facts. Industry in general and especially the packaging, automotive, and aluminum industries are challenged to be untiring in their efforts to present the true situation in terms of energy and the environment. This is a clear international trend. In all industries that manufacture energy consuming and environmentally sensitive materials, attention has to be paid to the whole life cycle of products. A life cycle analysis (LCA) is mandatory in industrialized countries. Government agencies and industry work closely together to study the relevant state of the art and to develop scenarios in which the choices of materials and life cycles of products are improved. In such ecological assessments, the general prospects for the aluminum industry as a whole are optimistic. The aluminum industry is moving toward a new golden age.

At the end of the last century there was a cluster of synergistic innovations. This situation is again true in terms of the processing and use of aluminum stressing energy and environmental advantages. This background is providing new opportunities. The production and use of aluminum should continue to a period of sustainable development.

Chapter 17. Aluminum in Change

The aluminum industry's current state of knowledge and its positioning in terms of the environment were described in the earlier chapters. Using this as a basis, the final chapter covers possible future changes in the industry, existing as it does in a state of interaction with its technical and economic environment.

Processes and products of the aluminum industry are always in a state of change. In the two most recent decades, the main driving forces have been

- Energy saving and environmental protection
- More efficient use of resources (doing more with less)
- Moving from products to systems (higher added value)

A logical interrelationship exists between these factors and the challenges that can already be perceived for the future. In addition, the third Industrial Revolution, driven by information technology, will mean that much of what is currently accepted may be obsolescent after the turn of the century. The result of all these changes will be a move to a new paradigm. The following chapter will discuss why this is so, what the effects could be, and the way in which subsequent restructuring could proceed.

17.1 From Products to Systems

Fig. 17.1 shows the changes in development goals over time. The finished object, because of features designed into it, must satisfy the requirements of the market—the so called market pull. The driving forces are shown in the lower part of the figure. Product orientated development of aluminum alloys is shown in the upper part of Fig. 17.1 set against a time axis. Product development resulting from market requirements can be seen to be a main feature of the second Industrial Revolution. It will become clear in the course of this chapter that there is a trend to move away from the supply of semi-finished and end products toward the development and manufacture of systems.

The automobile industry provides an interesting example of this idea. It has re-engineered its previous throw away product and is changing toward energy and environmentally optimized concepts. For example, the need to reduce body shell weight in a way compatible with recyclability has set in motion both alloy and process development. This has created the need for new joining techniques and surface treatment processes, a situation that has been signaled to important suppliers such as the aluminum industry.

Quite independently, similar developments are taking place in other product segments, such as packaging. Originally, the idea was to produce one-way packaging (i.e., model A) throw-away products with no environmental pollution concerns. This concept has been modified by the introduction of transition technology type B, with the change to a form of packaging frequently described as use of containers, the starting point for the trend toward re-use. The introduction of transition technology B has provided a steady improvement in the ecological balance.

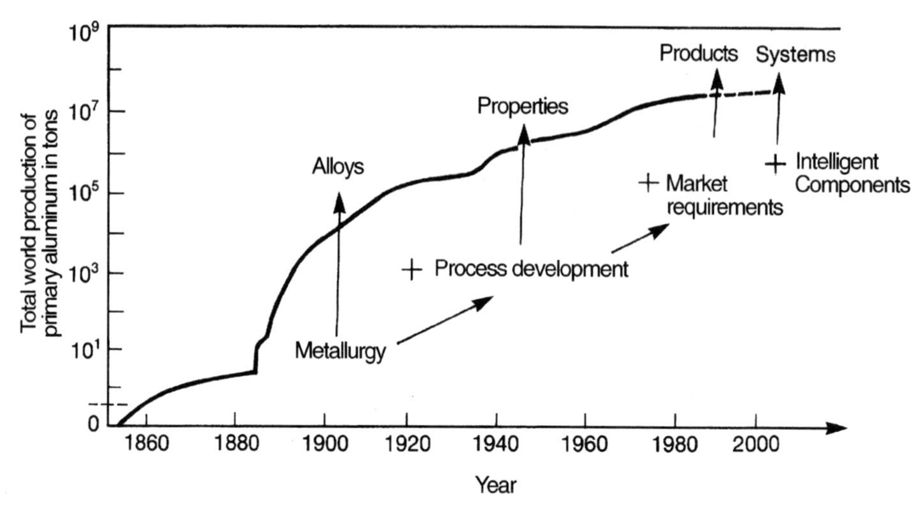

Fig. 17.1: Development of aluminum technology. (P. Furrer)

The aluminum beverage can did not emerge by steady evolution. It was developed from its conception as a product that would be recycled. This is a rare example of a consistent systems approach for the creation of a new product.

For industrialized countries, it can be generally accepted that the transition from products to systems is now underway and will be an important criterion for the development of manufacturing processes and markets.

17.2. Externally and Internally Induced Developments

Both external and internal forces can be recognized in Fig. 17.1. The key elements are market pull and properties of aluminum materials. These are seemingly independent from each other.

By looking at the "Edifice of Knowledge" (Chapter 13, Fig. 13.2) it can be seen that developments of synergistic characteristics are usually initiated by the aluminum industry, but linked with external factors such as pressure for recycling. In the following text, we have chosen a few examples to illustrate such relationships that are a vital element in the exchange between producer and user of the metal.

17.3 Components with Systemic Criteria

There are good reasons for studying the desired behavior of a component at the start of development and adjusting production and in-service use of the part to meet the requirements of the system taken as a whole. Using this technique, it is possible to use synergistic effects between the various component elements. By an intelligent application of technology, optimized use can be made of a particular component so that the design requirements are met but not exceeded by a wide margin above any desired safety factor. The result for the designer is a method of planning that will provide a narrower distribution of properties and costs and make possible a precise use of valuable component parts.

Just at the time when the car industry is making increasing use of high value cast aluminum alloy components, it is an interesting coincidence that microelectronics can be used to support intelligent processing. This makes more accurate production of component parts possible and has enabled the component designer to cooperate closely with the customer to achieve a joint objective.

17.4. The Role of "Intelligent Technologies"

The three elements—material, design, and process—traditionally important for component properties are shown in Fig. 17.2. "Intelligent technologies" are used today in all three of these segments and are applied in production processes that require rapid assimilation of complex information. The following definition applies:

"Intelligent technology provides components and solutions for applications where the processing of complex information is essential for function and/or result. Furthermore, on-line sensors are used to provide data to control industrial processes."

An important application is production control linked to on-line testing. Even changes in material structure can be detected by means of sensors. This makes possible an entirely new level of quality control and is an important distinguishing factor for the "intelligent" processing of materials. The latest methods of developing and making aluminum alloy die-cast components provide an example of the use of "intelligent" manufacturing techniques.

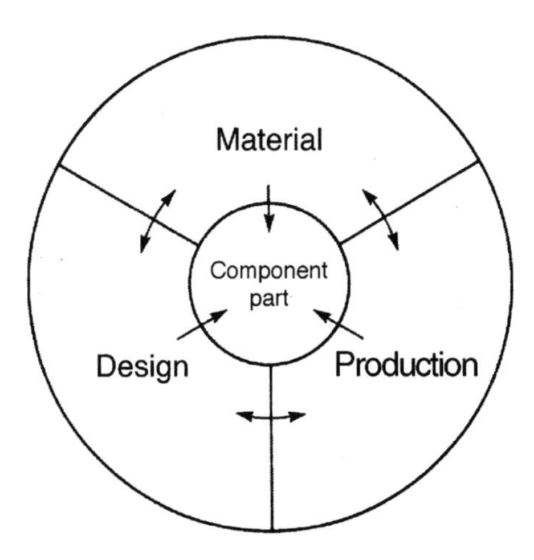

Fig.17.2: Creation of a component part. (R. Helms)

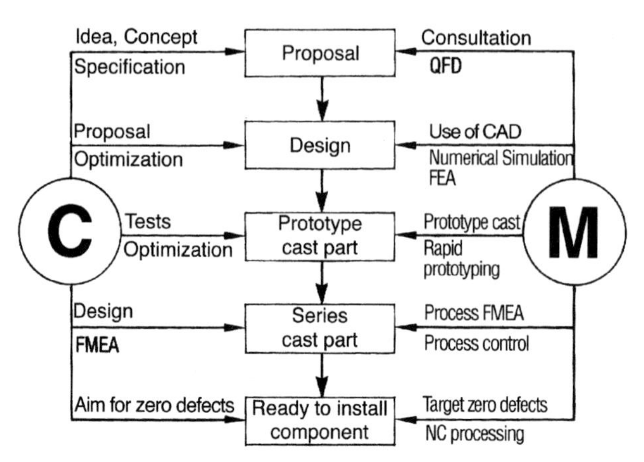

Fig. 17.3: Simultaneous engineering between customer (C) and manufacturer (M). (Honsel)

The use of computer aided engineering (CAE) in design and computer integrated manufacturing (CIM) is the current state of the art in the aluminum industry. Components and sub-systems for the car industry are produced in parallel; design, manufacturing, and quality control are linked at every stage with input from the customer (simultaneous engineering). From the first production stage, quality and function of a component or sub-system are optimized in parallel (Fig. 17.3). For the aluminum industry, this is an interesting and challenging leap forward on the way from products to systems. The customer's designers could modify, for example, their requirements in terms of cross section and strength in order to make possible an optimization of the casting process (or vice versa). In later stages of simultaneous product development, the number of opportunities for use of linked computers between supplier and customer increase.

When producing a series of cast parts, an analysis of scrap at both manufacturer and customer can be used to discover possible sources of defects and to prompt changes in process parameters with the ultimate objective of achieving zero defects. An example that is currently relevant is the production of cast nodes for the spaceframe cars of the future. From the component manufacturer's side, alloy, casting process, and thermal treatment have to be optimized in order to be able to produce a component part with both high mechanical strength and high elongation.

17.5. Challenges for the Future

The aluminum industry is in the same situation today as other major industries that are involved with energy and the environment; it is in a transition phase (B) before the effect of the third Industrial Revolution becomes general (see Chapter 16). In order to justify the changes to totally optimized systems (phase C), there is a search by decision makers for motives and reasons for adopting new technology. Typical reasons are:

- Market forces from leading companies
- Forces initiated by products that are perceived as having a positive influence on the eco-balance

- Pressure from users to have the benefit of completely redesigned new systems that have the clear hallmarks of the third Industrial Revolution

A question that it is not easy to answer is, will the rate of change to C-technologies proceed as a series of individual single steps to allow time for specific products to achieve a period of market dominance, or will the changes proceed more discontinuously? A typical example of a quantum leap forward is the aluminum spaceframe car. The question is whether such steps proceed in isolation or whether they are part of a change to a new paradigm.

17.6. The New Paradigm for the Aluminum Industry

The word paradigm was redefined in a somewhat narrower sense in 1962 by Thomas Kuhn. He stated:

"A paradigm results from a set of pre-conceived ideas . . . and is based on previous experience."

A change in paradigm is like reaching a water-shed. Studies have shown that as soon as a large number of experts and top decision makers have moved to a new paradigm, those left behind lose credibility and have less chance of future success. The change, however, can take much longer than a decade.

One of the most famous changes in paradigm occurred between the Middle Ages and modern times. The paradigm was concerned with whether the Earth or the sun was the center of the planetary system. The earlier, earth centered paradigm was strongly supported at that time by the Catholic Church. Many of its educated clergy at first opposed the new ideas of Copernicus. The acceptance of the sun as the center of the planetary system was a "revolutionary" move to the new paradigm. In the years since there have been many such changes of paradigm in the world of science and technology.

In terms of the aluminum industry, the transition from phase B to phase C development can clearly be discussed in terms of a move to a new paradigm. The energy crisis (1973-1974) was, in fact, a "watershed" event for most of the materials industries. Before this,

Table 17.1: Change of paradigm for the aluminum industry

Old (Until approximately 1975)	New (Emerging this decade)
1. Manufacturing technology often from first Industrial Revolution ("Heat and beat")	1. Intelligent processing "Simultaneous engineering" with customer
2. Low emphasis on energy consumption and environment	2. "Doing more with less" (Materials and energy)
3. Ton pushing	3. Specialties for highly demanding market niches
4. Dominance of specialists	4. Integralistic concepts "From products to systems"
5. Importance of large suppliers	5. Flexible, innovative, medium and small suppliers

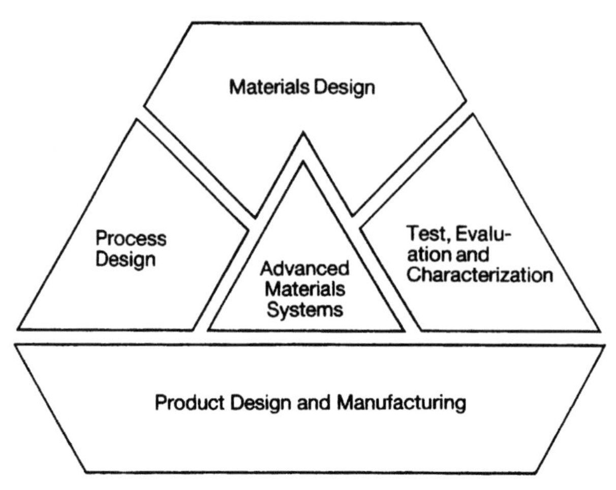

Fig. 17.4: New paradigm of the competitive situation in the materials industries showing the integration of various activities. (P.R. Bridenbaugh)

cheap abundant energy was a main industrial driving force. After the energy crisis, energy-intensive industries suffered because of their dependence. This set in motion a whole new set of attitudes and resulted in a change of paradigm. In Table 17.1, we can see how earlier attitudes were matched to the new situation.

It is well worth fully understanding what happens when a paradigm change, such as that defined by Thomas Kuhn, takes place. A number of clearly recognizable steps occur:

- A new paradigm exists and acts as an invitation to potential participants.
- Those involved in the old paradigm group divide themselves into two camps: those who want to persist with the old successful principles and those who want to join the new paradigm group.
- There is an atmosphere of crisis in the previously dominant group.

At the beginning, there is a shortage of experience in the new area, but the motto "nothing succeeds like success" can be applied to the pull in the new direction.

In Fig. 17.4, an example of a new paradigm for the new competitive materials industries is presented. In this paradigm two words are particularly noticeable: the word "design," which has also been illustrated in the sketch of simultaneous engineering (Fig. 17.3) and, in the center of the new paradigm, the expression "advanced material systems," which summarizes today's developments within different segments of the "Edifice of Knowledge."

17.7. Tailor-made Products for Demanding Market Niches in Relation to Economies of Scale

Looking at Fig. 17.4, one perceives that integrating three different levels of design with large amounts of evaluation data into an advanced materials system remains a domain of a rather large firm. But in Table 17.1 the increasing role of flexible small suppliers is mentioned within the new paradigm, which requires some interpretation.

The basic difference between the old and the new paradigm is that the old paradigm (B technology) is typified by large semi-fabricating manufacturing plants and a mental attitude dominated by tons of metal. The change that can be seen with the advent of the new paradigm is the recognition of the importance of tailor-made products for specific market niches. This situation can act in favor of so called mini-plants. The advantage of mini-plants for material processing lies in their flexibility when faced with the need to manufacture high value specialties that meet narrow market requirements. Peter Drucker made a comparison in this context between the earlier large production units, which he called "big battleships," and the needs of the market of tomorrow, which were better met by a "flotilla" of smaller, more maneuverable vessels that can react more rapidly to changing market situations.

This concept can apply to companies in the aluminum sector that wish to stay competitive or become market leaders. One point, however, has to be stressed; a maneuverable middle-sized company needs a data base and a system to precisely supply even small product volumes in order to satisfy its market and product segment.

To predict the changes that will occur as we move into the new millennium, we can make use of Peter Drucker's flotilla metaphor and refer to his middle-sized ships that have on board high-tech hard- and software. These can operate under two different types of flag:

- Middle-sized companies that sail under their own flag
- Flotillas that operate under the control of a flagship or out of a particular safe harbor

The two types of flags can be recognized today and their suitability for entering narrow channels in the coastline of the market makes the advantages of different ship types and command structures very significant.

Mini-plants have always existed in the aluminum industry, but their number and importance has increased in the last three decades. There is no strict definition of a mini-plant or a mini-mill. The simplest way would be to use equipment cost for a unit of production. Although mini-plants very frequently begin with relatively simple products, they acquire very high levels of specialized know-how. Thus, mini-mills may have advantages for specific market niches.

The car industry, with its increasing requirements for aluminum parts, is a test case. A specific aspect is the outsourcing trend to the supply of complete sub-systems. Here, two or three mini-plants can join together very successfully to manufacture, sell, and lower costs in a small efficient organization unit.

The economies of scale, however, still apply to bauxite, alumina, and smelter plants. Here, the very large financial outlay involved in building a smelter with an annual capacity of several 100,000 tonnes per year usually involves the creation of a consortium to spread the risk.

These factors taken together create the relatively complex technical and ecological environment for the change of paradigm in the aluminum industry.

17.8. Aluminum on the Way to Sustainable Development

In terms of energy considerations, the use of energy islands to provide the energy for the production of primary aluminum will reduce criticism based on the material's consumption of energy in industrialized societies.

The aluminum industry of the future is not threatened by limits to growth. Much more, the challenges are provided by the change to the new paradigm in which the central requirement is sustainable development. As discussed in Chapter 16, raw material supply to the aluminum industry is being very much affected by the increasing availability of valuable recycled aluminum scrap. "Aluminum is the recyclable material *par excellence.*" (J.Schirner).

There is, however, a component in the triangle of environmental issues that will continue to challenge the aluminum industry; this involves the use to be made of aluminum from scrap from various sectors, but mainly from packaging and transportation. There is no doubt that the innovative aluminum industry will deal with these challenges in an effective manner. The way is already clear and will involve a switch from products to systems within these industrial sectors. This change will involve a close working relationship with customers in order to provide sustainable development.

With this background, the challenges and opportunities for the aluminum industry are quite clear. The conclusions can be drawn from the new paradigm and relate to three general areas:

- Further integration of the technology in society, particularly in terms of environmental questions. Creativity and inventiveness will be needed to meet the challenges of the future. Aluminum's ecological advantages will be particularly significant.
- Improved competitiveness during processing. There will be a change in attitude away from the idea that capital should mainly be used to create production capacity in terms of tonnes per year.
- New products, markets, and systems. There will be the steady development in aluminum applications in terms of the third Industrial Revolution by means of a steady introduction of information technology.

From these three main points, it can be seen that future development needs a cluster of innovations to create a synergy of existing knowledge with new ideas.

In the near future, there are hundreds of different products that need to be developed into systems. The opportunities are there for a large number of aluminum producers and users to participate in the general innovative advance. There are, of course, risks, but equally so, there are risks in not recognizing that changes will occur.

Conventional resources such as raw materials and energy are no longer the central issue. In the future the use of information combined with the use of the resource of motivated and well informed personnel will be needed to achieve sustainable development. The aims of this type of development are many-sided and demanding, but they are also rewarding. This has been very well summarized in a statement by P.R. Bridenbaugh:

The only way to create wealth in this, or any society, is through the manufacture of technically-rich, socially-useful, competitively-priced, environmentally-sound, and energy-efficient products: The only way to satisfy these product requirements is through the integration of very sophisticated materials, manufacturing and product design processes.

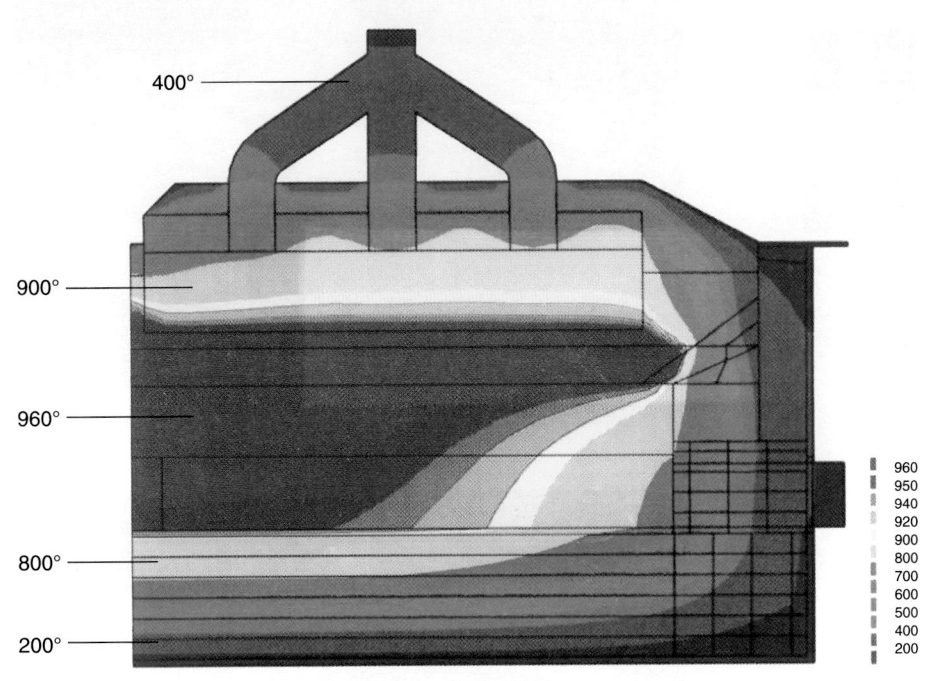

Fig. 2.5: Temperature distribution in a 180 kA electrolysis cell, drawn using a computerized mathematical model. (VAW)

Temperature
in °C

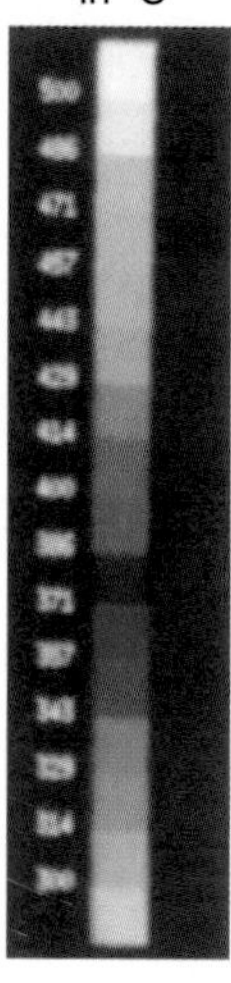

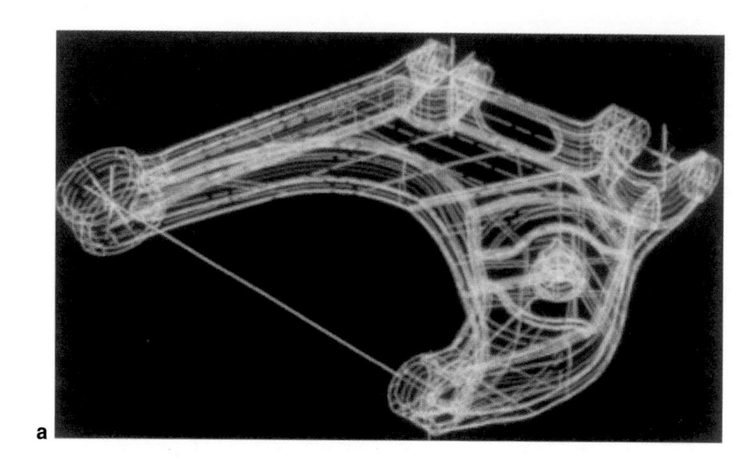

a

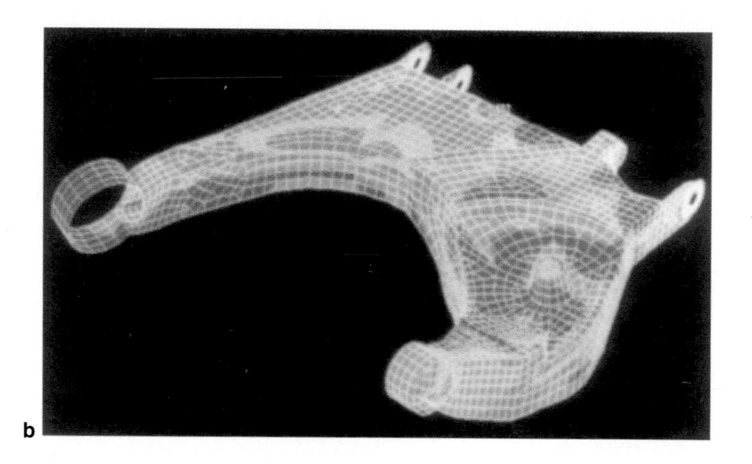

b

Fig. 6.35: (a) CAD model of the surface shape of the steering arm. (b) stress distribution in the steering arm by FEM analysis. (c) Temperature distribution in a solidification simulation as represented by different shades.

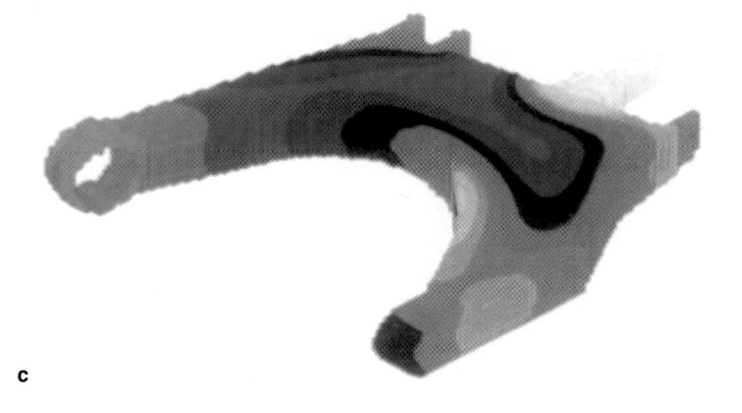

c

Appendix A. Explanation of Important Terms

Chapter 2. Production and Processing of Aluminum

Bauxite: Principal ore from which aluminum is extracted, named after the town of "Les Baux" in southern France. Bauxite contains about 40–55% aluminum oxide together with impurities, mainly iron ore with impurities, mainly iron oxide.

Alumina Plant: A large chemical plant in which bauxite is converted into aluminum oxide ("alumina"). Almost all alumina plants use the Bayer Process. In this process, the bauxite is dissolved in hot caustic soda, and the impurities are separated as "red mud" or "bauxite residue." The alumina is precipitated and filtered, then calcined.

Smelter or Primary Reduction Plant: A plant producing primary aluminum from alumina generally using the Hall-Heroult process and consisting of rows of cells called potlines.

Commercial Purity Aluminum: Commercial purity or primary aluminum, the product of aluminum electrolysis, contains 99.0–99.8% aluminum (i.e., with up to 1% impurities, mainly iron and silicon).

Hall-Héroult Process: In 1886, an American, Charles Martin Hall, and a Frenchman, Paul L.T. Heroult, independently, almost at the same time, discovered the technological process by which aluminum is still produced today. In the process, aluminum oxide, dissolved in a molten salt bath, is electrolyzed at about 950°C. A historical curiosity: Both men were born in the same year (1863) and died in the same year (1914).

Electrolysis: Electrochemical process in which a conducting electrolyte is decomposed by direct electric current. Positively charged ions (cations) migrate with the electric current to the cathode, and negatively charged ions (anions) migrate to the anode.

Cryolite: Cryolite is a mineral containing sodium aluminum fluoride (Na_3AlF_6). It occurs as natural mineral in Greenland, but today it is mostly produced synthetically. Cryolite is the main component of the electrolytic bath in the Hall-Heroult process.

Anodes: Anodes serve to introduce the direct current into the electrolyte. The anodes are produced from a mixture of petroleum coke and coal-tar pitch. Prebaked anode blocks with a volume of approximately 1 m^3 are predominant today. They are produced in a separate anode plant at a temperature of about 1300°C. Söderberg continuous anodes are also used. They consist of a single carbon mass contained in a steel shell. The anodes are formed from a carbon paste (petroleum coke and pitch), which moves downward by gravity within the steel shell and is baked in-situ by the heat of the cells. Söderberg cells are now used less and less in countries with strict regulations for workers health and pollution control because of the emission of coal-tar pitch volatiles.

Petroleum Coke: The main material used for producing the anodes utilized in the Hall-Heroult process. Petroleum coke is produced from oil refinery residue through a coking process. It must have a sufficiently high purity to produce commercially pure aluminum of 99.0–99.8%. From the environmental standpoint, petroleum coke must have a low sulfur content.

Aluminum Fluoride (AlF_3): Aluminum fluoride is an important component of the electrolyte in the Hall-Heroult process. Aluminum fluoride is either produced from the mineral fluorspar or as a by-product of the phosphate fertilizer industry.

Fluoride Emissions: Aluminum reduction plants emit small amounts of fluoride in the form of gases or solid particles. In all highly developed countries where aluminum smelters now operate, strict standards have been imposed to limit these emissions.

Hooded Cells: Individual electrolysis cells, in modern aluminum smelters, are encapsulated by "hoods" that capture all effluents, either gases or particles. These effluents are treated in abatement equipment.

Scrubbing Systems: An abatement system to capture fluoride emitted from electrolysis cells. Hooded cells are used in conjunction with a dry scrubber in which alumina is used as an adsorption material. The fluoride-laden alumina is later used as a feed stock in the cells. Some smelters use wet scrubbers where water or an aqueous solution is used to treat the potgases. In many cases this system is used in conjunction with cryolite recovery.

Aluminum Chloride Electrolysis: A new process announced by Alcoa in 1974 in which aluminum chloride is electrolyzed instead of the classical Hall-Heroult electrolyte containing alumina and fluorides. The electrolysis of aluminum chloride saves up to 30% of the electrical energy in the reduction step versus the Hall-Heroult process. Alcoa's pilot operation started in 1976 but was later abandoned.

Super Purity Aluminum: Super purity aluminum contains at least 99.99% aluminum. In the past it was produced by a second electrolysis (Hoopes cell). Today, zone refining and recantation after slow solidification are used.

Secondary Aluminum: Aluminum metal produced from scrap. The feedstock is either new scrap, run-around scrap, or "old scrap," which was previously a product in the market.

Recycling: The systematic collection of used aluminum products or scrap for reprocessing into useful products. Using recycled metal saves up to 95% of the energy required to produce primary aluminum.

Energy Effectiveness: A comparison of energy requirements for a material's full economic cycle, that is, from the ore to the final product as well as reuse or recycling of a material. Only in this way can comparisons be made between materials for a given application.

Ingot Casting: Pouring an aluminum melt to produce rolling ingot, extrusion ingot (billets), or wire bar ingot almost exclusively by continuous casting (DC casting).

Alloying: Addition of other elements to molten aluminum.

Castings: Objects at or near finished shape obtained by solidification of a melt in a mold by one of three processes: sand casting, permanent-mold casting, or pressure die casting.

Semis-plant: A fabricating plant receiving ingots either from a smelter or from its own casthouse and using either rolling mills, extrusion presses, or forging presses to produce semi-finished products.

Aluminum Semi-finished Products: These are the products of a fabricating plant (sheet, coil, foil, extrusions, or forgings) that are used by other plants to produce finished products.

Shaping or Fashioning with Cutting Tools, "Machining": The desired shape of the work piece is obtained by turning, planing, milling, drilling, boring, sawing, filing, etc.

Forging: Plastically deforming metal, usually hot, into desired shapes with compressive force, in one or more steps, with or without dies. The starting material may be cast or wrought (forging stock).

Impact Extrusion: A part formed in a confining die from a metal blank (slug), usually cold, by rapid single stroke application of force through a punch, causing the metal to flow around the punch and / or through an opening in the punch or die. This method is used to produce cans, collapsible tubes, etc., for the packaging industry as well as pressurized cylinders and condenser housings.

Spinning: A chipless forming technique used to produce symmetrical round pieces. Spinning is labor intensive and requires a rotating device in which a circle can be mounted and formed by the proper application of pressure during rotation.

Etching: Chemical surface treatment technique. For aluminum, there are a number of etching processes that serve different purposes, for example, the removal of the as-cast or as-rolled surface or the pretreatment of the surface prior to the application of a coating.

Brightening: A chemical or electrolytic surface treatment that is used, with or without mechanical polishing, to obtain highly brightened surfaces on suitable material. Special alloys or high purity aluminum of more than 99.8% Al with or without a magnesium addition are generally required. Brightened aluminum finds application in reflectors, car bumpers, automotive trim, exterior and interior architecture, etc. In order to preserve the finish, a protective oxide layer is usually applied after brightening, normally by anodic oxidation in sulfuric acid.

Anodic Oxidation ("Anodizing"): A process for artificially increasing the thickness of the oxide layer on aluminum by applying a direct current while the article being treated is suspended in an electrolyte. The resulting layer is 0.005–0.05 mm thick. This layer increases the corrosion resistance of aluminum significantly. Anodic oxidation serves to retain the reflectivity and protects the brightened or polished surface (e.g., against weathering).

Aluminum Alloy: The term aluminum alloy is used to describe super purity or commercial purity aluminum to which metallic additions were made. The additions, which are called alloying elements, are made to impart certain properties such as an increase in mechanical properties. The most important alloying elements for aluminum include: magnesium, copper, silicon, manganese, and zinc.

Wrought Alloys: Alloys for mechanical working are called wrought alloys. They are usually cast by the DC method into ingots, billets, continuous cast strip, or wire bar and subsequently fabricated by hot deformation processes such as extrusion, forging, rolling, etc. Wrought alloys have good formability combined with medium to high strength.

Casting Alloys: Casting alloys are used for cast parts. The casting alloys are characterized by their ability to fill a mold cavity and their low cracking tendency just after solidification.

Chapter 3. The Internal Structure of Aluminum

Crystallographic Structure: This term is used to define the system in which the atoms in solid metal are arranged in a three dimensional lattice.

Crystalline: Solid bodies in which the atoms are arranged in a definite crystal lattice are called crystalline. All solid metals are crystalline. A typical characteristic of the crystal is its "anisotropy" (i.e., the directionality of its properties, such as strength or formability).

Amorphous: Substances in which the atoms have no definite arrangement are amorphous, for example, liquids, molten metals, glass, rubber.

Chapter 4. Formation of the Cast Structure

Grain Structure: Depending on previous treatment, the grain structure may be referred to as a cast structure, cold-worked structure, or annealed ("recrystallized") structure. Every structure is composed of individual grains or crystals. All the atoms within a grain are arranged in a continuous uniform "lattice." various types of grain include:
1. As-cast grains: They originate during crystallization from the melt. See "cast grain."
2. Deformed grains: These form during hot or cold working and are usually elongated.
3. Recrystallized grains: They form during annealing. The size of these grains plays an important role in the subsequent fabrication of the metal. The grain size for most products is between 0.01–1 mm, but may be much larger under certain circumstances. Coarse grain is basically undesirable. The grains may be revealed by etching the surface with appropriate reagents. See also "subgrains" and "recrystallization."

Specific Heat: Specific heat is the number of joules required to raise the temperature of one gram of a body one degree Kelvin. Aluminum requires approximately 1.05 joules/g K in the range between room temperature and the melting point.

Heat of Fusion: The amount of heat required to convert one gram of a substance from solid to liquid after it has reached the melting point with no temperature change. For aluminum, the latent heat of fusion is 396 J/g. The heat of fusion supplies the aluminum atoms enough kinetic energy to release them from the metal lattice (i.e., to transfer them to the liquid state). See also "heat of solidification."

Heat of Solidification: The amount of heat liberated when a gram of metal solidifies after having been cooled to the melting point in the transformation from liquid to solid. The absolute amount is the same as the heat of fusion, or 396 J/g for aluminum.

Cast Grain: During solidification, numerous crystallization nuclei form at different sites in the melt. Each nucleus grows rapidly into a grain. The individual grains eventually touch one another, which hinders further growth. The size and shape of the cast grains depend on the solidification conditions, as well as the alloying elements present. All of the cast grains together make up the cast structure.

Unit Cell: The unit cell of aluminum is face-centered cubic and consists of one atom at each corner of a cube and an additional atom in the center of each cube face. There are eight corner atoms in each unit cube, but since each of these is shared by eight adjacent cubes, only one-eighth of each atom belongs to any one unit cube. The six atoms in the face are each shared by two adjacent cubes, so only one-half of each atom belongs to a single cube. Therefore, the total number of atoms per unit cube is $8 \times 1/8 + 6 \times 1/2 = 4$. A "grain" or crystal consists of numerous unit cells arranged together.

Lattice Plane: Any plane cutting through the center of at least three atoms is called a lattice plane. The most important crystallographic planes are those containing the highest density of atoms.

Residual Melt: This is the last liquid metal to solidify. It does so at the grain boundaries. This melt residual is enriched in alloying elements and is why grain boundaries are usually the weakest points in the structure.

Grain Etching or Macroetching: The structure is revealed by etching the surface with definite reagents. Some lattice planes are exposed preferentially by the etching, so that individual grains reflect light to different degrees, according to their orientation.

Columnar Crystals: These are elongated cast grains that form near the surface of the mold due to rapid directional heat removal.

Dendrite: A crystal that has a tree-like branching pattern, being most evident in cast metals rather slowly cooled through the solidification range.

Homogeneous Alloys: Alloys in which the alloying elements are below the maximum solubility so that the structure has a uniform composition after a solution-heat-treatment near the solidus temperature (homogeneous structure). Below the solidus temperature, the solubility decreases. Therefore, homogeneous alloys may show a "dual-phase" structure (e.g., through grain segregation or after a heterogenization anneal).

Solid Solution: If a metal (B) is soluble in aluminum (A) in the solid state, the atoms of A and B form a solid solution. The foreign atoms (B atoms) distribute themselves randomly within the aluminum lattice. In aluminum alloys, the alloying elements in solution basically form a "substitutional solid solution" (i.e., some of the atomic sites normally occupied by aluminum atoms are occupied by foreign atoms). On the other hand, hydrogen dissolved in aluminum forms an interstitial solid solution. The extremely small hydrogen atoms can arrange themselves in the interstices in the lattice without disturbing the arrangement of the aluminum atoms.

Heterogeneous Alloys: Alloys in which the content of the alloying element is greater than its limit of solubility in solid aluminum are called heterogeneous alloys. In the structure, there are crystals enriched in the alloying element (heterogeneities) between and within the aluminum grains. In heterogeneous alloys, two or more different crystal types exist in the solid metal.

Foreign Atoms: The atoms of the alloying elements as well as the natural impurities are called foreign atoms.

Macrostructure: Structure that is visible without magnification after grain etching.

Microstructure: In order to investigate the microstructure, a polished metal section is etched with an appropriate reagent, and examined at a magnification of 50–1000× under a light microscope or with special preparation under an electron microscope with enlargements up to 150,000×.

Chapter 5. Melt Quality and Treatment

Microsection: A sample is taken from the material and the cross-section to be studied is ground and polished until it has a flat mirror-like surface. After this preparation, the microstructure can be studied.

Heterogeneity or Inclusion: A foreign crystal embedded in the aluminum "matrix" is called a heterogeneity. Heterogeneities can arise during solidification or during an anneal. In the latter case, the particle may also be called a precipitate and is usually fine as compared to the former, which is usually coarse. Nonmetallic particles such as oxides and nitrides in a solid metallic matrix are generally called inclusions.

Cooling Curve: A cooling curve may be established as follows: A metal is heated above the melting point and allowed to cool slowly. A thermocouple, inserted in the melt, is used to monitor the temperature during cooling. When solidification begins, there is an interruption in the cooling cycle while the heat of solidification is liberated. A plot of the temperature as a function of time is called a cooling curve. This procedure may also be referred to as "thermal analysis."

Freezing or Melting Range: The temperature range between the liquidus and solidus temperatures. All alloys except the eutectics have a discernible solidification range that may be over 100°C. The same range must be passed through upon melting.

Liquidus Temperature: That temperature at which solidification begins during the cooling cycle or the temperature at which the alloy becomes completely molten when heated.

Solidus Temperature: That temperature at which solidification is complete during the cooling from the molten state or, upon heating, the temperature at which the alloy begins to melt.

Equilibrium Diagram: Equilibrium diagrams are constructed from a large number of individual cooling curves. Equilibrium diagrams show the solidus and liquidus temperatures for a given alloy system as well as phase transformations in the solid state. The results show the phases as a function of temperature and alloy composition.

Grain Segregation (Microsegregation): Grain segregation designates an uneven distribution of alloying elements within a grain after solidification. For the most commonly used hypoeutectic alloys, the content of the alloying element is usually less in the center of the grain and higher near the grain boundary than the average for the whole grain.

Solubility Limits: The maximum addition of an alloying element that is soluble in aluminum at equilibrium at a given temperature, according to the phase or equilibrium diagram is called the solubility limit. In solid aluminum, some of the solubility limits include magnesium, 17.4% maximum; copper, 5.7% maximum; silicon, 1.65%. Maximum solubility decreases with decreasing temperature.

Coring: A solidified crystal segregated or "cored" in such a way that the concentration of the foreign atoms varies in concentric rings. The core often takes the form of a dendrite. It results from cooling an alloy rather rapidly through a two-phase field.

Eutectic: A fine-celled structure, usually banded or lamellar, resulting from the simultaneous solidification of two or more phases. In the eutectic structure there are no primary crystals. A eutectic alloy solidifies at the eutectic temperature.

Eutectic Alloy: An alloy that has the exact composition corresponding to the eutectic point on an equilibrium diagram. Example: A casting alloy of aluminum with approximately 12–13% silicon. Typical characteristics: absence of a freezing range during solidification or melting, very fine cellular structure.

Primary Crystal: During solidification of a heterogeneous alloy, the first crystal type to solidify is called the primary crystal.

Divorced Eutectic: If one crystal type dominates during eutectic crystallization, a divorced eutectic results in which the desirable fine-celled structure is partly replaced by larger crystals.

Hypoeutectic Alloy: Any binary alloy in which the content of the alloying element is less than the eutectic composition. In a hypoeutectic, heterogeneous alloy, the primary crystals that solidify first are poorer in the foreign atoms than the average composition of the alloy. The residual melt, therefore, is enriched in the alloying element. All wrought alloys and a number of casting alloys are hypoeutectic.

Hypereutectic Alloy: Binary alloys whose content of the alloying element is greater than the eutectic composition. The primary crystals contain more of the added element than the overall composition would indicate. Example: High silicon (>13% Si) casting alloys, like piston alloys.

Cells: In as-cast structures, the grains are often divided into cells. At a cell boundary, the concentration of the alloying elements shows a discontinuous change (usually a maximum), whereas the orientation of the aluminum lattice does not change at the cell boundaries.

Chapter 6. Industrial Casting Processes

Mold Casting: Molten metal is poured into a form that gives the casting its desired shape. Normally, finishing is some type of machining, but in a few cases, there may be a small amount of plastic deformation. Specific casting alloys are used for mold casting.

Mold: A rigid form in which molten metal is allowed to solidify.

Permanent Mold Casting: Molten metal is poured into heated metal (permanent, i.e., reusable) molds, usually made from steel.

Sand Casting: The melt is poured into a sand mold. Sand castings solidify slower than permanent mold castings. Each cast requires a new mold.

Pressure Die Casting: In pressure die casting, molten metal is injected into a steel mold under elevated pressures of up to 100 atmospheres.

Risers: Risers serve as reservoirs for molten metal and are connected to the casting. They provide additional metal to prevent the cavity that tends to form, due to shrinkage, during solidification of the casting.

Cavity, Porosity: The volume shrinkage of 4–7% which an aluminum alloy undergoes during solidification can lead to voids in the cast structure. If these voids can be easily seen with the eye, they are called cavities. Interior voids (voids in a casting or billet) should be differentiated from exterior voids. If the internal voids are very small, they are referred to as porosity or microporosity.

Ingots: An interim or temporary cast form suitable for remelting or for further fabricating by hot deformation (e.g., rolling, extrusion, forging). Wrought alloys are used in the production of such ingot. Direct chill casting is used predominately in the production of ingots.

Segregation: The non-uniform distribution of alloying elements, impurities or microphases within a metal structure is called segregation.

Macrosegregation: Macrosegregation refers to an enrichment of the alloying elements in the segment of the casting which solidifies last. "Inverse segregation" may occur in ingot production, which is an enrichment of the alloying elements in the outer zones of the ingot, caused by rapid solidification. The variations in analytical composition across the ingot cross-section caused by macrosegregation cannot be removed by subsequent heat treatment as is feasible within one grain during grain segregation.

Direct Chill Casting (DC Casting): A semi-continuous process for casting, principally of rolling ingots and extrusion billets. One or more strands may be cast in lengths of several meters and, in the case of horizontal casting, the process may be fully continuous. In vertical casting, the aluminum melt is poured into a water-cooled mold about 10 cm high, open at the bottom. At the beginning of the cast, the bottom of the mold is sealed with a block attached to a vertical hydraulic piston. As the molten metal solidifies, the piston descends slowly and additional molten metal is added until the ingot attains the desired length. In DC casting, solidification and heat removal proceeds very rapidly, which favors a desirable structure (fine grain, macrosegregation only close to surfaces).

Sump: The liquid metal pool within the mold cavity bounded by the solidification front from beneath and towards the mold wall. The sump depth and shape is dependent on the feeding system and heat removal.

High Temperature Anneal of Ingots or Billets (Homogenization): In order to equalize grain segregation, which arises due to rejection of foreign atoms from the dendrite during solidification, an anneal is carried out at the highest possible temperature (near the solidus temperature) to produce an equilibrium condition. This high temperature anneal between 500–600°C is therefore called homogenization. Precipitation often occurs during this treatment, in which case it would be more correct to call it heterogenization. The expression homogenization is also used for solution-heat-treatment.

Diffusion: The exchange of position of foreign atoms or aluminum atoms within the aluminum matrix is called diffusion. (In the latter case, movement of the aluminum atoms within the aluminum structure is called self-diffusion.). Diffusion can occur at room temperature such as in natural aging, but is of special significance at temperatures above 100°C. Depending on alloy composition and treatment temperature, either precipitation may occur or precipitates may dissolve.

Strip Casting: In strip casting, endless strip or plate, 2–25 mm thick with a width of up to 2.0 m, can be continuously cast. Preferably, molds are utilized which can move for some distance together with the newly solidified surface.

Chapter 7. Properties of Aluminum under Mechanical Stress and during Deformation

Deformation: A change in shape caused by applying sufficient stress.

Elastic Deformation: In elastic deformation, the deformed body resumes its original shape when the applied stress is removed. Rubber is a typical elastic material

Plastic or Permanent Deformation: The change in shape is permanent after the applied stress is removed. Plasticine is a material that only undergoes plastic deformation.

Hard Condition: Semi-finished material (for example, sheet) that has a high degree of cold work is designated "hard."

Soft Condition: Aluminum semi-finished product, for example, sheet, which has been annealed at 300–500°C is designated "soft." Such an anneal causes a complete removal of the work hardening through recrystallization and partially removes the alloy hardening through coarse precipitation of the alloying metals (heterogenization).

Tensile Test: Under precisely defined conditions, a sample of the material to be examined is tested for its behavior under tensile loading. The mechanical properties, as well as the formability of the material, can be determined by the tensile test, for example, the elastic limit, the ultimate tensile strength, and total elongation after fracture. (Reference ASTM Standards Test Methods B557 and E8)

Stress: Stress is defined as the force per unit area. The stress may be either in tension, compression, or shear depending on the direction and orientation of the loading. The results of the tensile test are normally reported in the international system of units as megapascals (MPa) or, in the English system, in pounds per square inch (psi) or kilopounds per square inch (ksi).

Elastic Limit, Proportional Limit: In the tensile test, the elongation of the sample is determined under steadily increasing load. At the beginning, with a limited load, the elongation increases in proportion to that applied load (i.e., if the load doubles, the elongation doubles, etc). Within this elastic limit there is no permanent strain, which means that if the load is removed, the sample will return to its original length. The proportional limit is the maximum stress at which strain remains directly proportional to stress.

Modulus of Elasticity: The ratio of stress to corresponding strain below the proportional limit (i.e., within the limits of Hooke's Law). (Reference ASTM Standard Method E111) As there are three kinds of idealized stress states, so there are three kinds of moduli of elasticity for any material: modulus in tension (also known as Young's Modulus; abbreviated E), in compression (abbreviated E_C), and in shear (abbreviated E_S) . All are expressed in MPa.

Tensile Yield Strength: In the tensile test, the engineering stress at which a material exhibits a specified limiting deviation from the proportionality of stress and strain or permanent set, usually 0.2%. In some countries, the tensile yield strength is referred to as the "proof stress," and an offset of 0.1% or 0.01% is sometimes used. Yield strength is measured in MPa and usually abbreviated TYS, sometimes with a subscript indicating the offset for measurement.

Cold Deformation (Cold Working): All forming techniques that are carried out at ambient temperature yield a cold-worked structure whether the method is cold rolling, drawing, deep drawing, etc. In addition to imparting shape, the purpose of cold deformation is to add strength to the structure. The demarcation between cold and hot working is not sharp. One can still speak of cold working at temperatures up to 250°C because a certain hardening will still take place.

Hardening: Increase of the mechanical properties of a metal through cold work (work hardening) or by the addition of alloying elements (alloying hardening).

Deformation Hardening or Work Hardening: During plastic deformation, dislocations move preferably on a fixed slip path. Anchoring and interaction between an increasing number of dislocations makes further slip along these paths within the metal structure more difficult. The formability of the metal is reduced by previous cold work.

Ultimate Tensile Strength: In the tensile test, the maximum tensile stress that a material is capable of sustaining, calculated from the maximum load during a test carried to breaking divided by the original cross-sectional area of the specimen. It is measured in MPa or N/mm^2 and is usually abbreviated UTS or simply TS.

Degree of Cold Work: The degree of cold work, for sheet, is calculated as follows:

[(Starting Thickness–Final Thickness)/Starting Thickness] × 100 (%)

Aluminum sheet with 60–90% cold work is called hard (H18), 20–50% as half-hard (H14), and 0% is defined as soft (annealed).

Hardness: The resistance of a material to deformation, particularly permanent deformation, indentation, or scratching. Different methods of hardness measurement (for example, Brinell, Rockwell, or Vickers methods) give different values and may give different ratings for different materials. There is no absolute scale for hardness, and such measurements give only a rough estimate of the strength or formability. (Reference ASTM Standards Test Methods E10, E18, and E92)

Elongation: In the tensile test, the total increase in gage length of a specimen referenced to an original gage length prescribed on the specimen, usually expressed as a percentage of the original gage length. Total elongation is dependent upon the initial gage length. For cylindrical samples, the test length is usually 4, 5, or 10 times the sample diameter; for sheet-type specimens, it is usually 50 mm or 2 inches. Elongation is usually abbreviated El, often with a subscript indicating the gage length.

Ductility: The ability of a material to deform plastically rather than crack or fracture under stress, especially in the presence of notches or other local stress raisers. It is sometimes used to reflect elongation and/or reduction of areas, but may also be used to encompass notch toughness and fracture toughness. (Reference ASTM Standards Terminology E8, E616 and E1150)

Toughness: A broad term for the ability of a material to deform plastically rather than crack or fracture under stress, especially in the presence of notches or other local stress raisers. It usually encompasses notch toughness and fracture toughness as well as ductility.

Notch Toughness: One measure of the ability of a material to deform plastically under stress in the presence of stress raisers, measured by the ratio of the tensile strength of a notches specimen divided by the tensile yield strength (NTS/TYS, or Notch Yield Ratio). (Reference ASTM Standards Test Methods E338 and E602)

Creep: The time-dependent increase in strain in a solid resulting from an applied stress. The slope of a creep-time curve at any point is called the creep rate. The combination of applied stress and time to rupture at that stress are called the rupture stress and rupture life, respectively. (Reference ASTM Standard Methods E139 and E1457.

Fracture Toughness: A generic term for various measures of resistance to the extension of a crack to fracture under loading. The most commonly used measure of fracture toughness is the plane strain fracture toughness, K_{Ic}, the crack extension resistance under plane-strain crack-tip conditions. K_{Ic} is usually determined by ASTM Standard Method E 399, and has the units of $MPa \cdot m^{1/2}$.

Crash Testing: A type of testing to determine relative energy absorption in automotive structures or components in order to obtain a relative measure of crashworthiness and to enable the design of crash energy management into a vehicle.

Fatigue: A generic term for the repeated periodic or non-periodic loading applied to a test specimen or to a structure, usually utilized in conjunction with measurements to define the number of cycles of loading at a given stress or stress ratio that will cause the specimen or structure to fail. (Refer to ASTM Standard Definitions of Terms Relating To Fatigue E1150)

Moment of Inertia: The moment of inertia of a body with respect to an axis is the sum of the products obtained by multiplying the mass of each elementary particle by the square of its distance from the axis. Hence, the moment of inertia of the same body varies according to the position of the axis. It has its minimum value when the axis passes through the center of gravity. The moments of inertia of surfaces are especially useful in calculating the strength of beams. Moment of inertia is designated by I (capital i). Together with the modulus of elasticity of the material, the moment of inertia determines the degree of deflection of a beam, for example, under a given load. In order to keep the elastic deflection low, the moment of inertia must be high. This is accomplished by keeping as much of the mass as possible on the periphery of the profile cross-section in the direction of the main loading component.

Chapter 8. Fundamentals of Cold Working and Forming

Slip: The fundamental process of plastic deformation of a metal. Individual regions of the crystal slip, relative to each other, along crystallographic glide planes. The slip lines are often visible on the surface of the deformed metal, this is referred to as "macroscopic slip."

Dislocation: A linear defect in the uniform lattice structure. On an atomic scale, strengthening can be explained by the anchoring of dislocations.

Vacancy: Absence of an atom in the crystal lattice. Vacancies mainly arise from cold deformation or heat treatment and are prerequisite to atom movement in the solid state.

Orange Peel: A pebble-grained surface that develops in forming coarse-grained aluminum sheet. Origin of the orange peel roughening is linked with the less constrained deformation of individual grains, with different orientations, at a free surface. Orange peel is related to grain size and is not noticeable on fine-grained material, since the individual crystals produce only extremely fine slip lines.

Single Crystal: A metal specimen in which all parts of the crystal lattice have exactly the same orientation, thus consisting of one single grain.

Slip Planes: The location of slip planes is precisely defined in the lattice structure. Dislocations moving along the slip planes cause entire groups of atoms to be relocated. This forms the basis for plastic deformation of metals.

Texture: Texture is defined as the same or similar orientation of numerous crystals within a structure. Texture is synonymous with preferred orientation. Crystals tend to arrange themselves in "preferred" positions during solidification (cast texture) or deformation (deformation texture), which may change direction during annealing (recrystallization texture).

Lüder Lines: On the surface of deep drawn parts, occasional lines or bands may be observed that originate in the structure of the sheet being formed. This is observed mainly after forming annealed Al-Mg alloy sheet.

Looper Lines: A surface roughening that takes the shape of loops on deep drawn aluminum sheet and is due to non-uniform deformation resulting from structural irregularities in the metal.

Earring: During deep-drawing of sheet, uneven flow of the metal is often observed, which leads to four or eight "ears" on the upper edges of the deep-drawn part. These ears may be located at 45° or 90° to the rolling direction and cause additional scrap in the production of deep-drawn parts. Origin of the ears is related to texture. Therefore, a texture-free sheet has no earring.

Isotropic or Anisotropic State: Amorphous substances are isotropic (i.e., their properties are independent of direction). For example, the physical properties or the deformability are the same in all directions. On the other hand, a crystal is anisotropic (i.e., its properties have a significant directionality). Deformation takes place only along the slip planes, the positions of which are precisely located within a crystal.

Quasi-isotropic State: This is the state of a polycrystal having a fine-grained texture-free structure in which the small crystals are randomly oriented. The properties of the crystals average out so that a structure composed of numerous crystals does not exhibit any directionality during deformation.

Chapter 9. Industrial Forming Processes

Shaping Processes: In the production of wrought semi-finished or finished products, the metal is plastically deformed, for example by rolling, extruding, drawing, pressing, forging, etc. No chips are created with these techniques.

Plastic Deformation: During plastic deformation, the cross-section of the ingot or semi-finished product is greatly reduced (for example, during rolling, extruding, drawing, impact extrusion).

Sheet Forming: During sheet forming, the thickness of the sheet is not changed significantly (bending, deep drawing, corrugating), except during stretch-forming.

Extrusion: Before extrusion, the billet is heated to approx. 400–550°C and then, by application of pressure, forced to flow through a die with the opening having the desired shape of a rod, bar, profile, or tube. Hydraulic presses of a few thousand tons pressure are normally used.

Fiber Structure: The pattern of long, thin grains after a given deformation process, for example, extrusion or forging. The original cast crystals are greatly elongated during the deformation so that numerous parallel "fibers" may be seen. Heterogeneities, existing at the grain boundaries of the original cast structure, can be found between the fibers.

Deep Drawing: A chinless, shaping process in which a blank or workplace (usually controlled by a pressure plate) is forced into and / or through a die by means of a punch to form a hollow body or component whose wall thickness is substantially the same as that of the original material.

Cast Skin: After casting, an irregular skin is found on the surface of castings or ingots. Liquation lying directly under the oxide skin gives a rough appearance to casting skin also. Since this surface layer is enriched in alloying elements, it is often removed by scalping.

Rolling Mill: The term rolling mill refers to an installation consisting of a roll stand containing the rolls, usually either two or four rolls stacked vertically.

Two-High Mill: A rolling mill that contains only two work rolls.

Four-High Mill: In a four-high mill, four rolls are stacked vertically. Both of the inner rolls (work rolls) have a considerably smaller diameter than the two outer rolls (back-up rolls).

Tandem Mill: A tandem mill consists of multiple rolling stands arranged in sequence, which are used to give successive reductions to aluminum strip, for example.

Hot Rolling: Hot rolling takes place between 350–550°C. In the first stage, the ingots are hot-rolled to plate (thickness = 5–15 mm) or to a hot band 2–5 mm thick, which subsequently is usually cold rolled.

Cold Rolling: Cold rolling takes place at room temperature. Through rolling, the temperature in the rolled product may increase up to 150°C.

Chapter 10. Removal of Work Hardening through Heat Treatment

Hot Deformation: Deformation at 350–550°C. Hot rolling, forging, or extrusion are typical techniques. The working serves to shape the material. The coarse heterogeneities of the cast structure are broken up. The resulting wrought structure is either recrystallized and / or has a certain degree of cold deformation.

Recrystallization: Recrystallization is the transformation of a cold-worked grain structure, at sufficiently high temperatures, into a soft metal structure with new grains. While the coldworked grains are usually elongated, the new recrystallized grains are approximately equiaxed. Recrystallization takes place in the structure through nucleation and growth, the latter taking place through migration of grain boundaries.

Full Anneal: A full anneal takes place above the recrystallization temperature. During annealing, the strengthening which took place during cold working is removed. In heat-treatable alloys an annealing temperature is selected such that in addition to the recrystallization, a complete-as-possible precipitation of the dissolved alloying components takes place. Slow cooling enhances this effect.

Recrystallization Temperature: A temperature above which the formation of new grains within the deformed structure takes place. The recrystallization temperature is dependent on several factors. It is lowered with increasing cold work and increasing anneal time.

Critical Reduction: The critical reduction corresponds to a cold deformation of approx. 3–10%. A structure deformed within this range yields an especially coarse grain upon subsequent annealing, since few new grains are nucleated.

Secondary Recrystallization: If an anneal at a high temperature is carried out for extended times after severe deformation, the grain growth takes place in two steps. During secondary recrystallization, individual grains that were completely recrystallized during the first step continue to grow at the expense of neighboring grains until finally there are a few individual giant grains.

Stress Relief (Recovery): Stress relief refers to the partial removal of cold work through a thermal treatment below the recrystallization temperature in the range of 150–250°C. During stress relieving, strengthening from previous cold work is removed by a maximum of 50–60% and there is no visible texture change as is the case in recrystallization.

Chapter 11. Alloy Hardening

Alloy Hardening: By the addition of alloying elements to commercially pure or super purity aluminum, a strengthening of the metal is achieved. The movement of dislocations in the aluminum lattice is more difficult due to the foreign atoms. There are two different mechanisms of alloy hardening. Certain foreign atoms in solution, like Cu, Mg, Si, and Zn, have a substantial strengthening effect. The strengthening is greatly increased, through fine precipitates or "zones" of the foreign atoms. The foreign atoms that precipitate in coarse heterogeneities offer very little strengthening in hypoeutectic wrought alloys.

Non-Heat-Treatable Alloys: This group of alloys cannot be strengthened through heat treatment. The alloy hardening of these alloys comes from the atoms in solution. Typical examples: Al-Mn or Al-Mg alloys.

Heat-Treatable Alloys: The mechanical properties of this group can be greatly improved through appropriate heat-treating processes which include three basic steps:
1. Solution-heat-treatment
2. Quenching
3. Aging

During aging, the strength increases through the precipitation of constituents, which had been in supersaturated solution, into very fine particles. Aging may be carried out at room temperature (natural aging) or at 120–180°C (artificial aging).

Intermetallic Compound: A compound of two or more metals in which the atoms are grouped in a special lattice arrangement. Example: the compound $FeAl_3$ in which case the lattice contains three aluminum atoms for each iron atom. Other intermetallic compounds contain no aluminum atoms like Mg_2Si, $MgZn_2$ and many others.

Ternary Alloy: An alloy that contains three principal elements.

Binary Alloy: An alloy containing two principal elements.

Zones: Collection of foreign atoms in very small regions, which do not disturb the continuity of the aluminum lattice (coherent precipitation).

Heterogenization Anneal: During heterogenization annealing, the foreign atoms in solution precipitate into crystals that may be seen microscopically. These foreign crystals are considerably smaller than those found in the cast structure. If an anneal at a given temperature has a heterogenizing effect, an increase in electrical conductivity may be seen.

Solution-Heat Treatment: Heating to take into solution the alloying additions in heat-treatable alloys. The proper temperature selection may be determined from the phase diagram. Depending on the alloy, that temperature may be 450–565°C.

Chapter 12. Corrosion Resistance and Protection

Corrosion: The deterioration of a metal through environmental influences.

Pitting: A type of localized corrosion in which the attack on the metal results in small cavities.

Galvanic Corrosion: Galvanic corrosion results when two metals of differing electrochemical potential are brought into contact in the presence of an electrolyte. In the case of aluminum, galvanic corrosion can occur: (a) through conducting metallic contact with a nobler metal (for example, copper); (b) through heterogeneities present in the metal structure that are nobler than aluminum; or (c) by the action of heavy metal ions (for example, copper ions) in an aggressive liquid (electrolyte).

Galvanic Cell: Two metals with differing potential in conductive contact and immersed in an electrolyte form a galvanic cell, which creates a current. For example, the voltaic cell made with zinc and copper electrodes.

Electrochemical Series of the Metals (Galvanic Series): The metals can be arranged in an activity series, which corresponds to how noble they are. The electrochemical potential compared to a standard reference electrode is given in volts. The reference generally used is the standard hydrogen electrode. Metals may have a negative or positive potential depending on their reaction in dilute acids.

Potential Difference: The greater the distance (potential difference) between two metals in the activity series, the greater the attack on the less noble metal if the two form a galvanic cell.

Local Cell: A galvanic cell of very small dimensions. Such a localized cell may arise on a metal surface for a variety of reasons, for example, if a small foreign sliver or an inclusion is present in the structure or if there is a weak point in the oxide layer.

Local Current: The electrical current flowing in a local cell is called local current, which corresponds to the galvanic cell current and indicates the amount of localized corrosion.

Intergranular Corrosion: Intergranular corrosion occurs along the grain boundaries. This type of corrosion takes place preferentially in high strength copper or magnesium-containing alloys if, through unsuitable thermal treatment, continuous precipitation is present at or near the grain boundaries that is less noble or significantly more noble than the aluminum matrix. In the latter case, the attack will take place in the immediate vicinity of the grain boundary.

Stress Corrosion Cracking: This is a special case of corrosion that occurs in certain alloys in certain structures if significant sustained tensile stress is present at the surface combined with chemical or electrochemical effects of the service environment, resulting in intercrystalline cracks.

Exfoliation: A type of corrosion that progresses parallel to the outer surface of the metal, causing layers of material to be elevated by the formation of corrosion product.

Oxide Layer: The natural oxide layer protects aluminum against corrosion. In dry air it is around 0.005 Am thick, but can grow to 1 μm through reaction with moisture, in which case this layer may be considered a useful corrosion product. An artificial oxide layer may be achieved chemically or electrolytically. In the latter case (anodizing), the thickness is normally around 5–50 Am and increases corrosion resistance considerably.

Cladding: For corrosion-sensitized alloys, especially those containing copper, cladding is applied to the base material to offer protection. The cladding is often commercial purity aluminum, Al-Mn, or an alloy with approximately 1% Zn, which is especially effective for cathodic protection because of its less noble potential.

Chapter 13. Aluminum's Edifice of Knowledge, Its Main Impacts

Edifice: The vast body of knowledge about aluminum for developing new or improved processes and new end uses, represented in a structure relating process, structure and properties to the steps in manufacturing and using materials.

Inert Gas Welding: In inert gas welding, an electric arc between a tungsten electrode (GTAW or TIG welding), a welding rod of the filler alloy (GMAC or MIG welding), or a continuously fed filler alloy wire (also GMAW or MIG welding) and the material being welded serves as a heat source The arc and the weld zone are protected against the intrusion of oxygen by an inert gas stream as a shield (e.g., argon, helium, or a mixture of the two). This technique makes it possible to weld without a flux.

Chapter 14. Aluminum in the Materials Competition

Composite Material: A combination of a metal and non-metal or of two or more metals or polymers produced in such a way that they perform as a unit for specific applications. Examples are metal matrix composites employing graphite or alumina fibers.

Chapter 16. Aluminum and Its Environment

Issue Triangle: A representation of the opportunities and limits for a material in terms of its environment, source, and energy requirements.

Sustainable Development: This term originates among forestry management experts concerned with assuring that new forests are planted to compensate for the harvesting of trees for commercial operations. Sustainable development was the key issue in the Brundtlandt Report to the United Nations in 1987 and emerged as the guiding principle for future worldwide economic development at the Rio conference on ecology in 1992. The concept suffers from the large number of different interpretations. In Chapters 16 and 17, sustainable development is outlined for the aluminum industry in a specific non-political copy summarizing the healthy long-term growth anticipated for the industry under the new paradigm.

Life Cycle Analysis (LCA): LCA encompasses an in-depth study of energy utilization; solid, liquid, and gaseous emissions; material balance; and cost involved in the production and use of specific product. Its principal value is in assisting process and application improvements to enhance ecological balance.

Intelligent Technology: Intelligent technology provides components and solutions for applications where the processing of complex information is essential for function and/or result. Furthermore, on-line sensors are used to provide data to control industrial processes. Examples are computer-aided engineering, design or manufacturing (CAE, CAD, CAM) and networking like simultaneous engineering between supplier and user of aluminum components for automotive applications.

Chapter 17. Aluminum in Change

Paradigm: A pattern or model, in this case a representation of the situation in a metals industry considering all aspects of its process technology, application, maturity, and ecological impact.

Metallurgical Processes in the Production of Aluminum Semi-finished Products

In order to help the reader become oriented in regard to structural changes due to specific measures, a summary is added in Appendix C that refers to rolled products.

With the technological processes used in the fabrication of aluminum semis, one should differentiate between the primary purpose of the procedure and those results or effects that occur as a matter of course. But, too often the individual results cannot be distinguished. For example, rolling, forging, or extrusion may be utilized for shaping, working, or strengthening depending upon the application. For this reason, the purpose and the result of a definite process are presented together. In addition, information is presented concerning changes in structure from an atomic standpoint as well as macroscopically and microscopically.

Some of the more important processes referred to in Appendix C should be recapitulated.

Starting Material for Production of Semi-Finished Products: Billets, ingots, and coils that are almost exclusively produced by DC casting. Characteristic of the structure: the individual grains are surrounded by heterogeneities.

Preheating Prior to Hot Deformation: Depending on the temperature selected, the preheating may bring the work piece to the proper temperature for better formability or may be used to uniformly distribute the alloying elements in solution within the cast structure at temperatures near the solidus temperature (high temperature anneal or homogenizing). Incidentally, the preheating serves a third purpose, which is the removal of high stresses in the casting.

Hot Working: In hot working, it is possible to work and shape the material with relatively modest energy consumption. The structure subsequent to hot working may vary widely depending on alloy, hot-working temperature, and forming technique. After hot working, the structure is either recrystallized or stress-relieved to a great extent. If semi-finished products with exactly defined properties are desired, the preheat treatment before hot working, as well as the hot working itself, must be exactly controlled in terms of temperature and time since the thermal history has a decisive effect on the structure of the semi-finished products. After hot deformation, a fine-grained, segregation-free and uniform structure is ideal for cold working. However, hot deformation is often the final process in semi-finished products production (extruded products, forgings, hot-rolled construction sheets).

Cold Working: For cold working, a previously wrought structure (i.e., hot worked) is usually required. In cold working ingot or billet, the cast structure would break up with relatively small degrees of cold work since the large, brittle heterogeneities act as stress risers during cold working. During hot working, the heterogeneities in the cast structure fragment without material fracture. Subsequent to hot working, cold working is carried out until the energy required for further work becomes too great. If further reduction in thickness of the material is required, the material is either annealed or stress relieved. Either of these heat treatments may represent the final step in the production of semi-finished goods if a soft or stress-relieved material is required for further shaping (for example, for deep drawing).

Full Annealing: During annealing, several processes may take place at the same time. The simplest case is the annealing of commercial purity or super purity aluminum subsequent to cold working. In this case, the main process involved is the transformation of the grain structure (i.e., recrystallization). In the case of alloys, there is also a rearrangement of the foreign atoms added as alloying elements in addition to the recrystallization. Since a recrystallization temperature of 300–450°C is usually selected, which is lower than the previous hot working, a precipitation of the alloying elements takes place during annealing.

Stress Relieving (Recovery): The same may be said for stress relieving as was said above for annealing. The classical case of a pure stress relief may take place for commercial or super purity aluminum. With alloys, precipitation processes may take place. Moreover, stress relief processes take place automatically during any hot deformation.

Solution Heat Treating and Quenching: The purpose of solution heat treatment is to take into solid solution the maximum practical concentration of the hardening elements such as copper, magnesium, silicon, and zinc. If the material being treated was previously cold worked, recrystallization can take place simultaneously. Recrystallization can be suppressed through a short-time solution heat treatment and / or the addition of elements that inhibit recrystallization in order to retain a fibrous structure, which is often desirable ("press effect") to reduce the susceptibility to stress corrosion. The objective of quenching is to preserve as nearly as possible the solid solution formed at the solution heat treating temperature and to maintain a certain minimum number of vacant lattice sites. Other phenomena are described in the appendix.

Aging: The starting point for aging is the presence of hardening elements in a supersaturated solid solution and quenched in vacancies in the aluminum lattice. During the course of aging, precipitation processes take place that bring about an increase in strength. Depending on the alloy type, precipitation may be carried out at room temperature (natural aging, e.g., Al-Cu-Mg) or at an elevated temperature about 150°C (artificial aging, e.g., Al-Mg-Si).

Appendix B. Alloy and Temper Designation Systems for Aluminum

B1. Scope

This is a summary description of the alloy and temper designation systems for aluminum alloys, based upon ANSI H35.1, which provides systems and detailed rules for designating wrought aluminum and wrought aluminum alloys, aluminum and aluminum alloys in the form of castings and foundry ingot, and the tempers in which aluminum and aluminum alloy wrought products and aluminum alloy castings are produced. For more detail on the designation systems and applicable rules, refer to either *Aluminum Standards and Data* or ANSI H35.1. Specific limits for chemical compositions and for mechanical and physical properties to which conformance is required are provided by applicable product standards.

B2. Wrought Aluminum and Aluminum Alloy Designation System[1]

A system of four-digit numerical designations is used to identify wrought aluminum and wrought aluminum alloys. The first digit indicates the alloy group as follows:

- Aluminum (99.00 percent and greater) 1xxx

---

[1] Chemical composition limits and designations conforming to this standard for wrought aluminum and wrought aluminum alloys and aluminum and aluminum alloy castings and foundry ingot may be registered with the Aluminum Association provided: (1) the aluminum or aluminum alloy is offered for sale, (2) the complete chemical composition limits are registered, and (3) the composition is significantly different from that of any aluminum or aluminum alloy for which a numerical designation already has been assigned.

[2] For codification purposes, an alloying element is any element that is intentionally added for any purpose other than grain refinement and for which minimum and maximum limits are specified.

[3] Standard limits for alloying elements and impurities are expressed to the following places:
Less than .001 percent 0.000X
.001 but less than .01 percent 0.00X
.01 but less than 0.10 percent
Unalloyed aluminum made by a refining process 0.0XX
Alloys and unalloyed aluminum not made by a refining process 0.0X
0.10 through 0.55 percent 0.XX
(It is customary to express limits of 0.30–0.55 percent as 0.X0 or 0.X5)
Over 0.55 percent ....0.X, X.X, etc.
(except that combined Si + Fe limits for 1xxx designations must be expressed as 0.XX or 1.XX)

[4] Standard limits for alloying elements and impurities are expressed in the following sequence: Silicon; Iron; Copper; Manganese; Magnesium; Chromium; Nickel; Zinc; Titanium (Additional specified elements having limits are inserted in alphabetical order according to their chemical symbols between Titanium and Other Elements, Each, or are listed in footnotes.); Other ("Other" includes listed elements for which no specific limit is shown as well as unlisted metallic elements. The producer may analyze samples for trace elements not specified in the registration or specification. However, such analysis is not required and may not cover all metallic "other" elements. Should any analysis by the producer or the purchaser establish that an "other" element exceeds the limit of "Each" or that the aggregate of several "other" elements exceeds the limit of "Total," the material shall be considered non-conforming.) Elements, Each; Other Elements; Total; and Aluminum (Aluminum is specified as minimum for unalloyed aluminum and as a remainder for aluminum alloys.).

Aluminum alloys grouped by major alloying elements:[2,3,4]

- Copper 2xxx
- Manganese 3xxx
- Silicon 4xxx
- Magnesium 5xxx
- Magnesium and silicon 6xxx
- Zinc 7xxx
- Other element 8xxx
- Unused series 9xxx

The assigned designation is in the 1xxx group whenever the minimum aluminum content is specified as 99.00 percent or higher. The alloy designation in the 2xxx through 8xxx groups is determined by the alloying element (Mg_2Si for 6xxx alloys) present in the greatest mean percentage, except in cases in which the alloy being registered qualifies as a modification or national variation of a previously registered alloy. If the greatest mean percentage is common to more than one alloying element, the choice of group will be in order of group sequence Cu, Mn, Si, Mg, Mg_2Si, Zn, or others. The last two digits identify the aluminum alloy or indicate the aluminum purity. The second digit indicates modifications of the original alloy or impurity limits.

B2.1. Aluminum

In the 1xxx group for minimum aluminum purities of 99.00 percent and greater, the last two of the four digits in the designation indicate the minimum aluminum percentage.[5] These digits are the same as the two digits to the right of the decimal point in the minimum aluminum percentage when it is expressed to the nearest 0.01 percent. The second digit in the designation indicates modifications in impurity limits or alloying elements. If the second digit in the designation is zero, it indicates unalloyed aluminum having natural impurity limits; integers 1–9, which are assigned consecutively as needed, indicate special control of one or more individual impurities or alloying elements.

B2.2 Aluminum alloys

In the 2xxx through 8xxx alloy groups, the last two of the four digits in the designation have no special significance but serve only to identify the different aluminum alloys in the group. The second digit in the alloy designation indicates alloy modifications. If the second digit in the designation is zero, it indicates the original alloy; integers 1–9, which are assigned consecutively, indicate alloy modifications.

---

[5] The aluminum content for unalloyed aluminum made by a refining process is the difference between 100.00 percent and the sum of all other metallic elements plus silicon present in amounts of 0.0010 percent or more, each expressed to the third decimal before determining the sum, which is rounded to the second decimal before subtracting; for unalloyed aluminum not made by a refining process, it is the difference between 100.00 percent and the sum of all other metallic elements plus silicon present in amounts of 0.010 percent or more, each expressed to the second decimal before determining the sum. For unalloyed aluminum made by a refining process, when the specified maximum limit is 0.0XX, an observed value or a calculated value greater than 0.0005 but less than 0.0010% is rounded off and shown as "less than 0.001;" for alloys and unalloyed aluminum not made by a refining process, when the specified maximum limit is 0.XX, an observed value or a calculated value greater than 0.005 but less than 0.010% is rounded off and shown as "less than 0.01."

B2.3 Experimental alloys

Experimental alloys are also designated in accordance with this system, but they are indicated by the prefix X. The prefix is dropped when the alloy is no longer experimental. During development and before they are designated as experimental, new alloys are identified by serial numbers assigned by their originators. Use of the serial number is discontinued when the X number is assigned.

B2.4 National variations

National variations of wrought aluminum and wrought aluminum alloys registered by another country in accordance with this system are identified by a serial letter following the numerical designation. The serial letters are assigned internationally in alphabetical sequence, starting with A but omitting I, O, and Q.

B3 Cast Aluminum and Aluminum Alloy Designation System[1]

A system of four-digit numerical designations is used to identify aluminum and aluminum alloys in the form of castings and foundry ingot. The first digit indicates the alloy group as follows:

- Aluminum (99.00 percent minimum and greater) 1xx.x

Aluminum alloys grouped by major alloying elements:[2,3,4]

- Copper 2xx.x
- Silicon, with added copper and / or magnesium 3xx.x
- Silicon 4xx.x
- Magnesium 5xx.x
- Zinc 7xx.x
- Tin 8xx.x
- Other element 9xx.x
- Unused series .... 6xx. x

The alloy group in 2xx.x–9xx.x, excluding 6xx.x alloys, is determined by the alloying element present in the greatest mean percentage, except in cases in which the alloy being registered qualified as a modification of a previously registered alloy. If the greatest mean percentage is common to more than one alloying element, the alloy group will be determined by the sequence shown above.

The second two digits identify the aluminum alloy or indicate the aluminum purity. The last digit, which is separated from the others by a decimal point, indicates the product form (i.e., castings or ingot). A modification of the original alloy or impurity limits is indicated by a serial letter before the numerical designation. The serial letters are assigned in alphabetical sequence starting with A but omitting I, O, Q, and X, the X being reserved for experimental alloys.

B3.1. Aluminum castings and ingot

In the 1xx.x group for minimum aluminum purities of 99.00 percent and greater, the second two of the four digits in the designation indicate the minimum aluminum percentage.[5] These digits are the same as the two digits to the right of the decimal point in the minimum aluminum percentage when it is expressed to the nearest 0.01 percent. The last digit, which is to the right of the decimal point, indicates the product form: 1xx.0 indicates castings, and 1xx.1 indicates ingot.

B3.2. Aluminum alloy castings and ingot

In the 2xx.x–9xx.x alloy groups, the second two of the four digits in the designation have no special significance but serve only to identify the different aluminum alloys in the group. The last digit, which is to the right of the decimal point, indicates the product form: xxx.0 indicates castings, xxx.1 indicates ingot that has chemical composition limits conforming to 3.2.1, and xxx.2 indicates ingot that has chemical composition limits that differ but fall within the limits of xxx.1 ingot.

B3.3. Experimental alloys

Experimental alloys are also designated in accordance with this system, but they are indicated by the prefix X. The prefix is dropped when the alloy is no longer experimental. During development and before they are designated as experimental, new alloys are identified by serial numbers assigned by their originators. Use of the serial number is discontinued when the X number is assigned.

B4. Temper Designation System[6]

The temper designation system is used for all forms of wrought and cast aluminum and aluminum alloys except ingot. It is based on the sequences of basic treatments used to produce the various tempers. The temper designation follows the alloy designation, the two being separated by a hyphen. Basic temper designations consist of letters. Subdivisions of the basic tempers, where required, are indicated by one or more digits following the letter. These designate specific sequences of basic treatments, but only operations recognized as significantly influencing the characteristics of the product are indicated. Should some other variation of the same sequence of basic operations be applied to the same alloy, resulting in different characteristics, then additional digits are added to the designation.

[6] Temper designations conforming to this standard for wrought aluminum and wrought aluminum alloys, and aluminum alloy castings may be registered with the Aluminum Association provided: (1) the temper is used or is available for use by more than one user, (2) mechanical property limits are registered, (3) the characteristics of the temper are significantly different from those of all other tempers that have the same sequence of basic treatments and for which designations already have been assigned for the same alloy and product, and (4) the following are also registered if characteristics other than mechanical properties are considered significant: (a) test methods and limits for the characteristics or (b) the specific practices used to produce the temper.

B4.1. Basic temper designations

F—as fabricated applies to the products of shaping processes in which no special control over thermal conditions or strain hardening is employed. For wrought products, there are no mechanical property limits.

O—annealed applies to wrought products that are annealed to obtain the lowest strength temper and to cast products that are annealed to improve ductility and dimensional stability. The O may be followed by a digit other than zero.

H—strain-hardened (wrought products only) applies to products that have their strength increased by strain-hardening, with or without supplementary thermal treatments to produce some reduction in strength. The H is always followed by two or more digits.

W—solution heat-treated applies to an unstable temper applicable only to alloys that spontaneously age at room temperature after solution heat-treatment. This designation is specific only when the period of natural aging is indicated; for example, W2hr.

T—thermally treated to produce stable tempers other than F, O, or H applies to products that are thermally treated, with or without supplementary strain-hardening, to produce stable tempers. The T is always followed by one or more digits.

B4.2. Subdivisions of basic tempers

B4.2.1 Subdivision of H temper: Strain-hardened

The first digit following the H indicates the specific combination of basic operations as follows:

H1—strain-hardened only applies to products that are strain-hardened to obtain the desired strength without supplementary thermal treatment. The number following this designation indicates the degree of strain-hardening.

H2—strain-hardened and partially annealed applies to products that are strain-hardened more than the desired final amount and then reduced in strength to the desired level by partial annealing. For alloys that age-soften at room temperature, the H2 tempers have the same minimum ultimate tensile strength as the corresponding H3 tempers. For other alloys, the H2 tempers have the same minimum ultimate tensile strength as the corresponding H1 tempers and slightly higher elongation. The number following this designation indicates the degree of strain-hardening remaining after the product has been partially annealed.

H3—strain-hardened and stabilized applies to products that are strain-hardened and whose mechanical properties are stabilized either by a low temperature thermal treatment or as a result of heat introduced during fabrication. Stabilization usually improves ductility. This designation is applicable only to those alloys that, unless stabilized, gradually age-soften at room temperature. The number following this designation indicates the degree of strain-hardening remaining after the stabilization treatment.

H4—strain-hardened and lacquered or painted applies to products that are strain-hardened and subjected to some thermal operation during the subsequent painting or lacquering operation. The number following this designation indicates the degree of strain-hardening remaining after the product has been thermally treated as part of the painting/lacquering cure operation. The corresponding H2X or H3X mechanical property limits apply.

The digit following the designation H1, H2, H3, and H4 indicates the degree of strain-hardening as identified by the minimum value of the ultimate tensile strength. Numeral 8 has been assigned to the hardest tempers normally produced.

Tempers between O (annealed) and HX8 are designated by numerals 1–7.

- Numeral 4 designates tempers whose ultimate tensile strength is approximately midway between that of the O temper and that of the HX8 tempers.
- Numeral 2 designates tempers whose ultimate tensile strength is approximately midway between that of the O temper and that of the HX4 tempers.
- Numeral 6 designates tempers whose ultimate tensile strength is approximately midway between that of the HX4 tempers and that of the HX8 tempers.
- Numerals 1, 3, 5, and 7 designate, similarly, tempers intermediate between those defined above.
- Numeral 9 designates tempers whose minimum ultimate tensile strength exceeds that of the HX8 tempers by 2 ksi or more.

The ultimate tensile strength of intermediate tempers, determined as described above, when not ending in 0 or 5, shall be rounded to the next higher 0 or 5.

The third digit,[7] when used, indicates a variation of a two-digit temper. It is used when the degree of control of temper or the mechanical properties or both differ from, but are close to, that (or those) for the two-digit H temper designation to which it is added or when some other characteristic is significantly affected.

B4.2.2 Subdivision of T temper: Thermally treated

Numerals 1–10 following the T indicate specific sequences of basic treatments, as follows:[8]

T1—cooled from an elevated temperature shaping process and naturally aged to a substantially stable condition applies to products that are not cold worked after cooling from an elevated temperature shaping process or in which the effect of cold work in flattening or straightening may not be recognized in mechanical property limits.

---

[7] Numerals 1–9 may be arbitrarily assigned as the third digit and registered with the Aluminum Association for an alloy and product to indicate a variation of a two-digit H temper (see note 6).

[8] A period of natural aging at room temperature may occur between or after the operations listed for the T tempers. Control of this period is exercised when it is metallurgically important.

T2—cooled from an elevated temperature shaping process, cold worked, and naturally aged to a substantially stable condition applies to products that are cold worked to improve strength after cooling from an elevated temperature shaping process or in which the effect of cold work in flattening or straightening is recognized in mechanical property limits.

T3—solution heat-treated,[9] *cold worked, and naturally aged to a substantially stable condition* applies to products that are cold worked to improve strength after solution heat-treatment or in which the effect of cold work in flattening or straightening is recognized in mechanical property limits.

T4—solution heat-treated[9] *and naturally aged to a substantially stable condition* applies to products that are not cold worked after solution heat-treatment or in which the effect of cold work in flattening or straightening may not be recognized in mechanical property limits.

T5—cooled from an elevated temperature shaping process and then artificially aged applies to products that are not cold worked after cooling from an elevated temperature shaping process or in which the effect of cold work in flattening or straightening may not be recognized in mechanical property limits.

T6—solution heat-treated[9] *and then artificially aged* applies to products that are not cold worked after solution heat-treatment or in which the effect of cold work in flattening or straightening may not be recognized in mechanical property limits.

T7—solution heat-treated[9] *and overaged/stabilized* applies to wrought products that are artificially aged after solution heat-treatment to carry them beyond a point of maximum strength to provide control of some significant characteristic.[10] It also applies to cast products that are artificially aged after solution heat-treatment to provide dimensional and strength stability.

T8—solution heat-treated,[9] *cold worked, and then artificially aged* applies to products that are cold worked to improve strength or in which the effect of cold work in flattening or straightening is recognized in mechanical property limits.

T9—solution heat-treated[9] *artificially aged, and then cold worked* applies to products that are cold worked to improve strength.

T10—cooled from an elevated temperature shaping process, cold worked, and then artificially aged applies to products that are cold worked to improve strength or in which the effect of cold work in flattening or straightening is recognized in mechanical property limits.

---

[9] Solution heat-treatment is achieved by heating cast or wrought products to a suitable temperature, holding at that temperature long enough to allow constituents to enter into solid solution, and cooling rapidly enough to hold the constituents in solution. Some 6xxx series alloys attain the same specified mechanical properties whether furnace solution heat-treated or cooled from an elevated temperature shaping process at a rate rapid enough to hold constituents in solution. In such cases, the temper designations T3, T4, T6, T7, T8, and T9 are used to apply to either process and are appropriate designations.

[10] For this purpose, characteristic is something other than mechanical properties. The test method and limit used to evaluate material for this characteristic are specified at the time of the temper registration.

Additional digits,[11] the first of which shall not be zero, may be added to designations T1–T10 to indicate a variation in treatment that significantly alters the product characteristics that are or would be obtained using the basic treatment.

B4.2.3 Variations of O Temper: Annealed

A digit following the O, when used, indicates a product in the annealed condition having special characteristics. As the O temper is not part of the strain-hardened (H) series, variations of O temper shall not apply to products that are strain-hardened after annealing and in which the effect of strain-hardening is recognized in the mechanical properties or other characteristics.

---

[11] Additional digits may be arbitrarily assigned and registered with the Aluminum Association for an alloy and product to indicate a variation of tempers T1–T10, even though the temper representing the basic treatment has not been registered (see note 6). Variations in treatment that do not alter the characteristics of the product are considered alternate treatments for which additional digits are not assigned.

Appendix C. Synopsis of Modifications to the Structure and Atomic Arrangement for the Principle Fabrication Stages of Rolled Products

The two lower rows apply only to the age-hardening alloys.

<table>
<tr><th>Operation</th><th>Starting material</th><th>Product obtained</th><th>Purpose of operation</th></tr>
<tr><td rowspan="2">1. Preheat before hot deformation
2. High temperature thermal treatment

Temperature:
300–640°C</td><td>cold</td><td>about 300–640°C hot</td><td rowspan="2">1. Improves formability
2. Eliminates grain segregation and supersaturation in the cast structure (through long heating times, e.g., 3–10 hours as close as possible to the solidus)</td></tr>
<tr><td colspan="2">rolling ingot</td></tr>
<tr><td>Hot deformation
e.g., hot rolling
Temperature:
300–550°C</td><td>preheated ingot</td><td>hot-rolled plate
hot-rolled coil</td><td>1. Working
2. Shaping</td></tr>
<tr><td>Cold deformation
e.g., cold rolling
Temperature
20–150°C</td><td>hot-rolled plate (5–15 mm thick)
hot-rolled coil (2–5 mm thick)
as well as annealed sheet or coil (about 0.5–5 mm thick)</td><td>Cold-rolled sheet, coil, and foil
High degree of cold work gives hard material
less cold work gives material with intermediate hardness</td><td>1. Shaping
2. Strengthening</td></tr>
</table>

All other data are valid for every wrought alloys and commercial purity aluminum

	Principal Effects of the Operation		Micrograph after the operation (schematic) Enlargement: several hundred times
	On the atomic arrangement	On the visible structure	
	Loosens lattice, thus 1. Eases movement of atoms on slip planes 2. Facilitates diffusion, establishes equilibrium	1. *Macrostructure* none 2. *Microstructure* Slowly dissolves heterogeneities, removes grain segregation, spheroidization of second phase particles	**Cast structure** grain boundaries 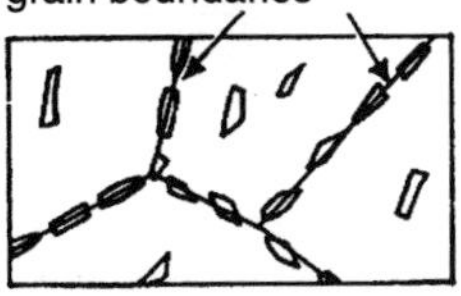 The typical characteristics of the cast structure remain after a high temperature thermal treatment i.e., relatively coarse grain, heterogeneities mainly in grain boundaries
	1. Creates lattice defects through plastic forming, later partial removal through recovery 2. Hot deformation above the recrystallization point: eliminates lattice defects through recrystallization	*Macro and Microstructure* Breaks up the coarse heterogeneities of the cast structure. Also, elongates cast grains. Recrystallized structure if degree of deformation and temperature is high enough	**Wrought structure** 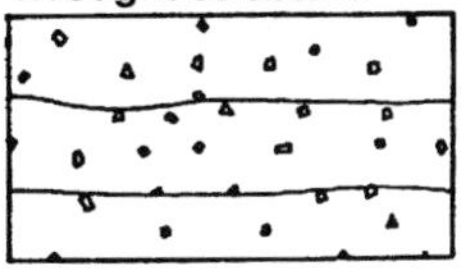 stress relief (recovery) or recrystallized 
	Produces heavy dislocation pile-ups through strain hardening	*Macro and Microstructure* Elongates the recrystallized or stress relieved grains in the longitudinal direction	**Cold worked structure** 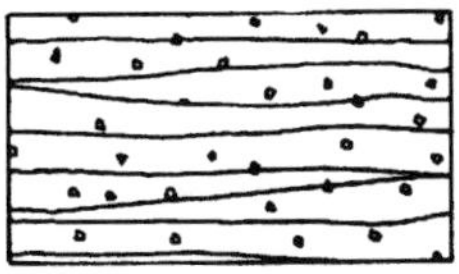

Operation	Starting material	Product obtained	Purpose of operation
Partial annealing not with age-hardening alloys or with Mg contents over 4% Temperature: about 200–300°C	cold-rolled sheet mainly more than 40% cold work	Partially annealed sheet (half-hard, for instance)	Produces medium formability by reducing strain-hardening up to 50% of initial values
Annealing Temperature: about 250–550°C	cold-rolled sheet minimum 25% usually with more than 40% cold work	Annealed sheet	Restores good formability by 1. eliminating strain hardening through recrystallization 2. reducing alloy strengthening through precipitation of the the alloying elements
Solution heat treatment then quenching Only with age-hardening alloys Temperature: 450–550°C	hot rolled, cold rolled, and annealed material	Solution heat-treated material (quenched to room temperature)	To obtain the condition for age-hardening
Aging Only with age-hardening alloys Temperature: room temperature to 185°C	solution heat-treated and quenched material	Age-hardened material	To achieve maximum strength values

Principal Effects of the Operation		Micrograph after the operation (schematic) Enlargement: several hundred times
On the atomic arrangement	**On the visible structure**	
1. Removes part of the dislocation pile-ups 2. Allows diffusion of foreign atoms	1. *Macrostructure* No modification 2. *Microstructure* Fine precipitation on grain boundaries (when foreign atoms are present in supersaturated solution)	**Partially annealed structure** 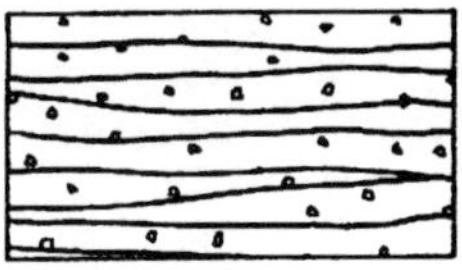
1. Almost completely eliminates dislocations 2. Rearranges foreign atoms according to the equilibrium diagram	1. *Macrostructure* Forms a recrystallized structure 2. *Microstructure* Numerous, fine precipitates occur when alloying elements precipitate at the annealing temperature	**Recrystallized structure** 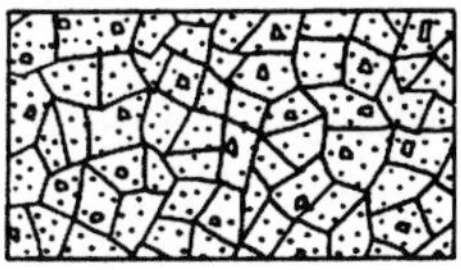 Medium-sized heterogeneities come from the cast structure, fine heterogeneities are precipitated during annealing
1. Age-hardening constituents go into solution (form a solid solution). Quenching retains them in solution 2. Eliminates dislocation pile-ups formed during strain hardening	1. *Macrostructure* Recrystallizes (under proper conditions) 2. *Microstructure* Dissolves those heterogeneities that contain age-hardening atoms	**"Homogeneous structure"**  The homogeneous matrix contains heterogeneities coming mainly from the cast structure. In these are precipitated undissolved Fe, Mn, and Si, for instance.
Atoms in supersaturated solid solution diffuse (e.g., Cu atoms with AlCuMg) whereby they form very fine clusters or "zones" greatly strengthening the lattice	1. *Macrostructure* No visible structural change 2. *Microstructure* Very fine agglomerates of foreign atoms can be seen through an electron microscope	**"Homogeneous structure"**  For the heterogeneities see remarks directly above

Appendix D. Chemical Compositions and Properties

Table D.1: The Physical Properties of Pure Aluminum 99.5%

(From *Aluminium Taschenbuch,* 13th Edition and *Aluminium und Aluminiumlegierungen* by Altenpohl)

Atomic weight	26.98
Atomic number	13
Crystal structure, face-centered cubic, atomic spacing (length of unit cube)	$4.0496 \cdot 10^{-8}$ cm
Density at 20°C	2.71 g/cm^3
Thermal conductivity	2.1–2.3 W/cm · K
Coefficient of thermal expansion 60–100°C (293–373 K)	23.5 1/(K · 10^6)
Increase in volume on changing from solid to liquid	6.5%
Melting point	658°C (931 K)
Specific heat at 20°C (293 K)	0.9 J/g K
Specific heat at 658°C (931 K) solid	1.13 J/g K
Specific heat at 700°C (973 K)	1.045 J/g K
Average specific heat from 0–658°C (273–931 K) solid	1.1×10^4 J/g K
Heat of fusion	396 J/g
Heat of vaporation at 1.01325 bar (= 1 atm)	396 J/g
Boiling point	2270°C (2543 K)
Electrical conductivity	34–36 m/Ohm · mm^2
Resistivity at 20°C	$2.65 \cdot 10^{-6}$ Ohm · cm
Temperature coefficient of resistance	1.15 10^{-8} Ohm · cm/K
Electrochemical equivalent	$9.3167 \cdot 10^{-5}$ g/A · s
Modulus of elasticity E	$7.2 \cdot 10^4$ N/mm^2
Shear modulus G	$2.7 \cdot 10^4$ N/mm^2
Poisson's ratio	0.34

Table D.2: Chemical Composition Limits for Representative Aluminum Alloys

(The Aluminum Association)

Designation	Si	Fe	Cu	Mn
Pure Aluminum				
1100	0.95 Fe + Si		0.05–0.20	0.05
1350	0.10	0.40	0.05	0.01
Wrought Aluminum Alloys[1]				
2014	0.50–1.2	0.7	3.9–5.0	0.40–1.2
2024	0.50	0.50	3.8–4.9	0.30–0.9
2219	0.20	0.30	5.8–6.8	0.20–0.40
3003	0.6	0.7	0.05–0.20	1.0–1.5
3004	0.30	0.7	0.25	1.0–1.15
4032	11.0–13.5	1.0	0.50–1.3	—
5052	0.25	0.40	0.10	0.10
5083	0.40	0.40	0.10	0.40–0.10
5454	0.25	0.40	0.10	0.5–1.0
6061	0.40–0.8	0.7	0.15–0.40	0.15
6063	0.20–0.6	0.35	0.10	0.10
7005	0.35	0.40	0.10	0.20–0.7
7050	0.12	0.15	2.0–2.6	0.10
7075	0.40	0.50	1.2–2.0	0.30
Cast Aluminum Alloys[1]				
201.0	0.10	0.15	4.0–5.2	0.20–0.50
295.0	0.7–1.5	1.0	4.0–5.0	0.35
319.0	5.5–6.5	1.0	3.0–4.0	0.50
332.0	8.5–10.5	1.2	2.0–4.0	0.5
356.0	6.5–7.5	0.6[6]	0.25	0.35
A356.0	6.5–7.5	0.20	0.20	0.10
347.0	6.5–7.5	0.15	0.05	0.03
443.0	4.5–6.0	0.8	0.6	0.50
512.0	1.4–2.2	0.6	0.35	0.8
535.0	0.15	0.15	0.05	0.10–0.25
705.0	0.20	0.8	0.20	0.40–0.6
850.0	0.7	0.7	0.7–1.3	0.10

[1] Remainder is aluminum.
[2] Also contains Vn 0.05–0.15, Zr 0.10–0.25
[3] Also contains Zr 0.08–0.20
[4] Also contains Zr 0.08–0.15
[5] Also contains Ag 0.240–1.0
[6] If Fe exceeds 0.45 percent, Mn shall not be less than one-half the Fe content
[7] Also contains Be 0.003–0.007, 0.005 percent maximum
[8] Also contains Sn 5.5–7.0

Mg	Cr	Ni	Zn	Others Ti	Others Each	Total	Al
—	—	—	0.10	—	0.05	0.15	99.00
—	0.01	—	0.05	—	0.03	0.10	99.50
0.20–0.8	0.10	—	0.25	0.15	0.05	0.15	
1.2–1.8	0.10	—	0.25	0.15	0.05	0.15	
0.02	—	—	0.10	0.02–0.10	0.05[2]	0.15	
—	—	—	0.10	—	0.05	0.15	
0.8–1.3	—	—	0.25	—	0.05	0.15	
0.8–1.3	0.10	0.50–1.3	0.25	—	0.05	0.15	
2.2–2.8	0.15–0.35	—	0.10	—	0.05	0.15	
4.0–4.9	0.05–0.25	—	.25	0.15	0.05	0.15	
2.4–3.0	0.05–0.20	—	0.25	0.20	0.05	0.15	
0.8–1.2	0.04–0.35	—	0.25	0.15	0.05	0.15	
0.45–0.9	0.10	—	0.10	0.10	0.05	0.15	
1.0–1.8	0.06–1.20	—	4.0–5.0	0.01–0.06	0.05[3]	0.15	
1.9–2.6	0.04	—	5.7–6.7	0.06	0.05[4]	0.15	
2.1–2.9	0.18–0.28	—	5.1–6.1	0.20	0.05	0.15	
0.15–0.55	—	—	—	0.15–0.35	0.05[5]	0.10	
0.03	—	—	0.35	0.25	0.05	0.15	
0.10	—	0.35	1.0	0.25	—	0.50	
0.50–1.5	—	0.50	1.0	0.25	—	0.50	
0.20–0.45	—	—	0.35	0.25	0.05	0.15	
0.25–0.45	—	—	0.10	0.20	0.05	0.15	
0.45–0.6	—	—	0.05	0.20	0.05	0.15	
0.05	0.25	—	0.50	0.25	—	0.35	
3.5–4.5	0.25	—	0.35	0.25	0.05	0.15	
6.2–7.5	—	—	—	0.10–0.25	0.05[7]	0.15	
1.4–1.8	0.20–0.40	—	2.7–3.3	0.25	0.05	0.15	
0.10	—	0.7–1.3	—	0.20	—[8]	0.30	

Appendix E. Subject Index*

*Glossary page numbers are given in parentheses.

Appendix F. A List of Key References

1. *Aluminum Standards & Data* (Washington, D.C.: Aluminum Association, published periodically).
2. *Standards for Aluminum Sand and Permanent Mold Castings* (Washington, D.C.: Aluminum Association, published periodically).
3. *The Aluminum Design Manual* (Washington, D.C.: Aluminum Association, published periodically).

About the Author:

Prof. Altenpohl was Vice President Technology at Swiss Aluminium Ltd. (Alusuisse), headquartered in Zurich, Switzerland. His career began in the 1950s with scientific investigations related to the physical metallurgy of aluminum. From 1957–1959, he was head of operational research at Aluminium Foils, Inc. (then Conalco) at Jackson, TN. Thereafter, he built up and headed a department for process and product development for Alusuisse at Zurich. Since the 1970s, he has been involved with the problems and opportunities surrounding the aluminum industry. Energy, ecology, and new technologies are subjects that he covers in his books and numerous publications. Dr. Altenpohl tackles these key areas pertaining to the aluminum industry with an integralistic approach. Utilizing his knowledge from industry, he has developed into a "specialist in cohesion."

He has held industry leadership positions as chair of the International Primary Aluminium Institute (IPAI) Environmental Committee and was a longtime member of their Energy Committee. In 1967, Dr. Altenpohl was awarded the Sainte-Claire Deville medal by the Societé Francaise de Metallurgie in recognition of his scientific and academic work to broaden the knowledge of aluminum. In 1977, he served as an advisor to the Office of Technology Assessment on their programs on Materials Conservation and Future Availability of Material Imported by the United States.

The University of Virginia elected Dr. Altenpohl a Visiting Professor of Materials Science in 1978. In 1979, Massachusetts Institute of Technology appointed him a Visiting Senior Fellow. In 1981, Dr. Altenpohl was awarded the K-J, Bayer-Medal in recognition of his work in the research of light metals by the Austrian Verband der Metallindustrie. In 1983, Business School St. Gall elected him as Professor for Technology.

Other books written by D.G. Altenpohl

1956—*Aluminum Viewed from Within* (Dusseldorf: Aluminum Verlag), 200 pages.

1965—*Aluminum and Aluminum Alloys* (New York: Springer Verlag), 900 pages (over 2000 literature references).

1976—*Technology Planning; Formula for the Future* (German) (Frankfurt, Umschau Verlag), 170 pages.

1980—*Materials in World Perspective* (New York: Springer Verlag), 220 pages.

1985—*Informatization: The Growth of Limits* (Dusseldorf: Aluminium Verlag), 180 pages.

About The Aluminum Association

Founded in 1933, the Aluminum Association is the trade association for the aluminum industry in the United States. Its members are producers of primary and secondary aluminum, aluminum alloys, semifabricated wrought and cast aluminum, and related products. The primary missions of the Association are to provide effective leadership to the aluminum industry; enhance the position of aluminum as "the material of choice;" remove impediments to the fullest use of aluminum; and help the aluminum industry achieve its environmental, societal, and economic objectives. In support of these aims, the Association addresses emerging public issues of concern to the industry; develops appropriate industry policy positions; and represents the industry's views on relevant legislative and regulatory matters to appropriate government bodies. In addition, the Aluminum Association publishes statistics on the aluminum industry, its products, and its markets while informing users and specifiers about the advantages of aluminum and disseminating information on the properties, characteristics, fabrication techniques, and applications of aluminum and its alloys. For more information, contact:

The Aluminum Association
900 19th Street, N.W., Suite 300
Washington, D.C. 20006
(202) 862-5100; fax (202) 862-5164
http://www.aluminum.org

About The Minerals, Metals & Materials Society

Headquartered in the United States but international in both its membership and activities, The Minerals, Metals & Materials Society (TMS) is a member-run professional organization that encompasses the entire range of materials and engineering, from minerals processing and primary metals production to basic research and the advanced applications of materials. Included among its nearly 13,000 professional and student members are metallurgical and materials engineers, scientists, researchers, educators, and administrators from more than 70 countries on six continents. In support of its mission, TMS sponsors meetings and publishes books and journals; promotes technology transfer; promotes the education and development of current and future professionals; represents the profession in the accreditation of educational programs and in the registration of professional engineers (a U.S.-grounded activity); encourages professionalism, ethical behavior, and concern for the environment; and stimulates a worldwide sense of unity in the profession. For more information, contact:

The Minerals, Metals & Materials Society
420 Commonwealth Drive, Warrendale, Pennsylvania 15086 U.S.A.
Toll Free: 1-800-966-4867; Fax: (412) 776-3770
E-Mail: csc@tms.org; http://www.tms.org.